A Robot Engineering Textbook

A Robot Engineering Textbook

Mohsen Shahinpoor

UNIVERSITY OF NEW MEXICO

HARPER & ROW, PUBLISHERS, New York
Cambridge, Philadelphia, San Francisco, Washington,
London, Mexico City, São Paulo, Singapore, Sydney

1817

Sponsoring Editor: Cliff Robichaud
Project Editor: Thomas R. Farrell
Cover Design: 20/20 SERVICES INC.
Cover Illustration: Massoud Ahghar
Text Art: Fineline Illustrations, Inc.
Production Manager: Jeanie Berke
Production Assistant: Brenda DeMartini
Compositor: Composition House Limited
Printer and Binder: The Maple Press Company
Cover Printer: New England Book Components

Library of Congress Cataloging-in-Publication Data

Shahinpoor, Mohsen.
A robot engineering textbook.

Includes bibliographies and indexes.
1. Robotics. I. Title.
TJ211.S45 1987 629.8′92 86-18319
ISBN 0-06-045931-X

86 87 88 89 9 8 7 6 5 4 3 2 1

To the memory
of my parents

Contents

CHAPTER 8 Robotic Sensing and Vision 385

CHAPTER 9 Robotic Motion Programming 430

Preface

This book has been written to provide a modern text for engineering students who are taking an initial course in robot engineering. *Robot engineering* is defined as the analyses and functional design of robots, robotic manipulators, and flexible manufacturing systems. This book has emerged from a compilation of my class notes and lectures, at both the senior and graduate levels, on the fundamentals of robotics and the design and control of industrial robots. My intense interest in the subject was one reason for writing this book; however, the prime motivation was the growing acceptance by engineering professors of the importance of robot engineering education in universities and the growing demand for increasing productivity across the nation. Robot engineering is a very complex subject and requires a diversified technical background. The engineering students and staff who will be using this book as a first college text in robot engineering should have a solid knowledge of engineering statics, Lagrange–Euler dynamics, the theory of machines and mechanisms, machine design, control, differential calculus, matrix analysis, computer programming, digital electronics, microprocessor engineering, and mathematics of coordinate transformations. An added feature of the book is the use of bond graph dynamic modeling of robotic systems for those students who have some familiarity with bond graph modeling of complex systems. Even more esoteric engineering subjects, such as modern transducer engineering, computer vision, and sensing, are involved in robot engineering. These subjects constitute the foundation for the practice of robot engineering. Most robotics experts agree that robot engineering integrates and uses a greater number of mathematical tools and the sciences than any other professional engineering study. Probably the only engineering subject that comes close to it in this regard is "engineering design."

Each chapter is complemented by the material covered in previous chapters, so it is best to cover the material in the order given.

Chapter 1 deals with a general introduction to computer-based robotic manipulators and their definitions and areas of application, a history on robotics evolution and revolution, robot geometrical configurations and workspace, the wrist and gripper subassemblies, robot drive systems, and the importance of resolution, accuracy, and repeatability for robotic manipulators as well as some examples and problem assignments. It is essential for students and staff to obtain an introductory knowledge of robotics evolution and fundamentals through the material covered in Chapter 1. The remaining chapters have been designed in accordance with ASEE's stated goal:

> The technical goal of engineering education is preparation for performance of the functions of analysis and design or of the functions of construction, production, or operation with full knowledge of analysis and design of the structure, machine, or process involved.

Chapter 2 discusses the kinematics of robotic manipulators and the significance of robotic coordinate transformations. The notion of homogeneous transformations of manipulator joint coordinates and the derivation of Denavit–Hartenberg representations are discussed fully. Direct and inverse kinematics problems are covered in depth to familiarize students with the significance of robotics kinematics. A large number of illustrative examples pertaining to various types of robot are presented. A number of inverse kinematics problems, which are at the core of automated manufacturing, are solved in this chapter.

Chapter 3 is intended to familiarize the students with the differential transformations encountered in dealing with moving objects and moving robotic manipulators. This is an important chapter that emphasizes the problems encounterd in robotic differential chase in the presence of digital cameras that are either attached to the manipulator or mounted elsewhere. Again a number of examples are presented and discussed thoroughly.

Chapter 4 deals with the mathematics of robotic workspaces, work envelopes, constraints imposed on robot extremities, and the numerical construction of robotic workspaces. The problem of robotic motion trajectory planning is discussed fully, and a universal algorithm for the design of 4–3–4 and 3–5–3 robotic trajectories is presented along with a set of numerical data pertaining to such trajectories for six-axis robots. It is essential that students obtain a minimal knowledge of the design of robotic workspaces, trajectories, and task descriptions for object manipulation, so a number of examples pertaining to these topics have been worked out and discussed.

Chapter 5 is concerned with the dynamics of robotic manipulators. This is an extremely important and involved area of robot engineering. For an undergraduate course on the fundamentals of robotics, a good portion of this chapter can be skipped. However, in a graduate course on the design and control of industrial robots, this chapter should be covered in full. The reason is that undergraduate students find this chapter a very difficult one. To compensate for the lack of complete coverage of this chapter, undergraduate students should be given more hands-on experience in the robotics laboratory. Chapter 5 also discusses three different methods of obtaining the equations for the dynamics of robotic manipulations. The first approach and probably the easiest is the bond graph modeling of robotic dynamics. This needs some prior knowledge on bond graph modeling of complex systems (which some undergraduate students may not have). The technique is then used to derive the governing dynamic equations of a number of robotic manipulators. Then, the Lagrangian dynamics of robot manipulators is elaborated upon with a number of illustrative examples. Another approach in formulating the robotics dynamics, namely, the Newton–Euler formulation, is fully discussed and the significance of recursive formulations are highlighted with worked out examples. The problem of inverse dynamics and its connection with the inverse kinematics is also discussed.

Chapter 6 concentrates on the problems associated with the control of robot motions, velocities, and accelerations. Obviously, robot control is at the heart of flexible automation and deserves detailed consideration. Essentially, the robot control problem is inherently nonlinear and thus rather hard to digest for undergraduate students. However, the subject of robot motion control is presented here in a manner comprehensible for both undergraduate and graduate students. First, the robotic open- and closed-loop control systems are discussed and then the significance of error-controlled robotic dynamics is considered. Second, the chapter further discusses the control structure of the amplifier as well as a generalized algorithm for the control of slow-moving robotic manipulators and the important computed-torque technique. Third, because of their importance, the problems of coordinated movement for both variable- and constant-speed controllers are discussed. Finally, a discussion is presented on typical joint torque control for robotic manipulators. On the whole, this chapter is an indispensible part of any text on modern robot engineering and the treatment given is adequate for a first course on robotics.

Chapter 7 discusses the significance of robotic forces, moments, and torques, inverse statics, and compliant robotic motions. These topics are felt to be extremely important in the design of robotic manipulators for higher load carrying capacities and more compliance with disturbing forces in assembly operations. Again, this is a chapter that must be included in any first text on robot engineering. Specifically, the importance of compliant motion cannot be overlooked and must be given top priority. This is achieved in this chapter by a number of detailed examples. The chapter covers the equilibrium of robotic forces and moments, the inverse statics problem, compliance analyses, force and kinematically constrained motions, and force sensing systems. Finally, problems associated with robotic jamming are covered and the Simunovic jamming criterion is discussed.

Chapter 8 concentrates on sensors for robotic manipulators, an important addition to any first course in robotics. Students should gain a basic knowledge on how to make robots more intelligent by equipping them with various sensing devices. These devices are force sensors, digital TV cameras, and computer vision. Some currently used tactile, force, and vision systems also are discussed to familiarize students with current technology on robotic sensing and vision. Obviously, for a broader knowledge on these topics the students should take a course on robotic sensing and vision.

Chapter 9 deals with robot motion programming. It first discusses the significance of "manual teaching," or "show-and-teach," and "lead-and-teach," or "guiding." It then discusses the programming languages and examines their historical evolution. For a number of popular robot programming languages, such as the MINI, WAVE, AL, VAL, and AML, the fundamental programs of the insertion of a peg in a hole are presented. Finally, a complete set of robotic teach-assembly programs are presented. These programs include the robotic "RESET," the "TEST," the "Create a Move File," the "Play Back a Move File," the "Edit a Move File," and the "Inverse Kinematics," all applied to an educational robotic manipulator.

I would like to thank my colleages and students, both at Clarkson University Robotic Center and the University of New Mexico's Robotic Laboratories, without whose help this textbook could not have been written. Among them are Professor R. A. Schilling of Clarkson University, Professor M. Jamshidi of the University of New Mexico, and Dr. H. Seraji of the Jet Propulsion Laboratory who helped me with some of the material covered in Chapter 6; Professor G. Starr of the University of New Mexico, and graduate students, Ali Meghdari, Ali Akbarzadeh, Ali Tabesh, Mike Caporali, David Campbell, Robert Cook, Young Tao Kim, and Bing Ching Chiou. I also thank my secretaries, Gayla Angel, Terri Grandon, and Rhonda Price, for the preparation of the manuscript; and Mr. Massoud Ahghar, who created the original drawings on the CADAM system. Last, but not least, I thank Mr. Cliff Robichaud, for his encouragement and support; Thomas R. Farrell, for his superb technical supervision of the project; and the staff of Harper & Row, Publishers, who did an excellent job of preparing the final text.

Mo Shahinpoor

Chapter 1

General Considerations of Robotic Manipulators

1.1 INTRODUCTION

Humankind appears to be on the brink of another industrial revolution, one that will involve the use and implementation of increasingly flexible, intelligent, sensitive, and efficient robots in the factory, field, office, and every other aspect of our everyday lives. This fact combined with a more acute awareness of human safety has been instrumental in creating the demand for a larger number of robotic manipulators. As a result, in recent years, the use of robotic manipulators has increased rapidly, particularly in industrial settings where the quest for greater productivity and tighter quality control has been paramount because of intense international competition.

The word *robot* seems to have first become popular when the Czech playwright Karel Čapek's play entitled *R.U.R.* (or *Rossum's Universal Robots*) was first performed in Paris, France, in the 1920s. In that play, small, artificial, and anthropomorphic creatures strictly obeyed their master's orders. In Czech and Russian these creatures were called *robotnic*, from *robota*, which is the Czech and Russian word for "drudgery" and "hard work." In the play *R.U.R.*, the robots eventually turned against their creators, but in doing so they set the stage for their own destruction with the appearance of two robots named Primus and Helina who fell in love with each other. The reader with an interest in the folkloric history of robotics is referred to Cohen (1966), Kato (1983), and "Surveys Reveal Robot Populations and Trends" (1982).

The first robots were thought to be evil human-looking machines. Isaac Asimov in the 1970s introduced the idea of good robots, whom he called droids, and popularized the word *robotics*. Some droids such as R2-D2 and C3-PO became famous in recent years thanks to the motion picture industry. Recently, real droids with high intelligence have been manufactured in the United States (Heath's Hero) and Japan (Sumitomo's Android; see Cohen, 1985).

Webster's dictionary defines a *robot* as "an automatic apparatus or device that performs functions ordinarily ascribed to human beings." However, we all know that this is not a very accurate description. In fact, many robots can perform certain tasks in a far superior and more sophisticated manner than human beings can. It is also true that by the above definition even washing machines can be considered as programmable robots. Thus, it is necessary to devise a universal definition for robotic manipulators. In this book we adopt the following definition:

> A robot is a reprogrammable multifunctional manipulator designed to move materials, parts, tools, or specialized devices, through variable programmed motions for the performance of a variety of tasks.

This definition covers a broad spectrum of robot manipulators, and within this definition there are different classes of robots. These classes include the following:

1. *Automated or Flexible Manufacturing.* These "industrial" robots are employed in a wide range of manufacturing processes, including parts assembly and inspection, materials handling, welding, and materials painting.
2. *Remote Exploration.* This class of robotic manipulators is designed to survive in environments that humans cannot tolerate. They can be used to explore the unknown from the edge of our solar system to the depths of earth's oceans.
3. *Prosthetic and Biomedical.* Robotic technology and its accompanying sensor technology can be teamed to provide prosthetic limbs with touch sensation. Someday miniature robots under a surgeon's control may probe the human body to operate on patients from within their bodies.
4. *Hazardous Material Handling.* Robots in this class have been employed to remove bombs and handle hazardous materials.
5. *Service.* Some present uses for service robots include security, janitorial, mail delivery, fire sentry, fire fighter, and voice-commanded errand robots. The use of service robots will increase as technology progresses and manufacturing costs fall.

1.2 A BRIEF HISTORY OF TELEOPERATORS AND MANIPULATORS

The robot manipulator made its debut with the onset of the nuclear age. The first manipulators were known as teleoperators. These devices allowed an operator to perform tasks from a safe distance, such as handling radioactive materials (Goetz, 1963). The first automatic electric-powered teleoperator was developed in 1947. The teleoperator was constructed so that the operator had control of the manipulator by means of a slave unit; that is, the manipulator repeated the motions made by the operator. This first manipulator did not give any form of force feedback, and therefore tasks such as turning a wrench or placing objects on work surfaces were difficult to perform. In 1948, General Electric developed a teleoperator (Handy Man) that provided force feedback by backdriving the slave with forces felt at the manipulator (Goetz, 1963). Later, the basic concept of this teleoperator was reintroduced by General Electric (Krsnak and Howe, 1975) in the form of their Man-Mate manipulator. The Man-Mate allowed an operator to move objects that were heavy or dangerous by moving a small lever. The Man-Mate translated the information from the lever into signals to the manipulator. The operator also had a feel for the operation being performed by means of the force feedback to the lever.

Whereas the teleoperator was controlled by a human, the computer-controlled systems for present day robots had their advent in the numerical-controlled machines. M.I.T. demonstrated the first servo-controlled numerical milling machine in 1953, with support from the U.S. Air Force (Rosenberg, 1972). The dimensional information of the piece to be milled was stored in digital form on paper tape, and an optical reader converted the holes in the tape to signals that ultimately controlled the mill. The most important aspect of this new technology was that tooling up to produce another piece involved little more than putting a new tape on the mill. This was the dawn of flexible tooling, in which a machine was simply reprogrammed to perform multiple tasks (Engleberger, 1980). Ernst (1961) incorporated the two technologies in a thesis entitled, "A Computer Operated Mechanical Hand." A teleoperator equipped with touch sensors was coupled with a computer to perform tasks by processing the tactile information received from the touch sensors.

The robot developed by Ernst led to the development of robots that were able to repeat a finite number of predefined moves indefinitely. These limited-sequence robots could be used to perform pick-and-place operations as described by Potter (1975). The pick-and-place tasks must be well defined. Incoming parts must be placed by the robot in a uniform manner, parts must be preinspected to ensure uniformity, and the placing operation should not allow parts to get jammed. Although the constraints are many, the limited-sequence robot is used for tasks such as material transfer, packaging, and simple assembly.

The use of the simple-sequence robot is restricted in the number of tasks it can perform. It assumes that the working conditions will be almost perfect. Unfortunately, conditions seldom are. In 1962, Tomoric and Boni (1962) developed a robotic hand with pressure sensors that could determine the size and weight of objects, and Reswick and Meyler [12] developed a device for quadriplegics (paralysis of four extremities) that enabled them to handle objects. Roberts (1963, 1965) demonstrated in 1963 the feasibility of processing digital halftone images by means of homogeneous transformations. In 1966, McCarthy, at Stanford's Artificial intelligence laboratory (SAIL), developed a manipulator robot with ears, eyes, and hands (microphones, TV cameras, and manipulators) that could recognize spoken commands and respond accordingly (McCarthy, 1967). Further advances in computer vision were made by Wichman (1967). Pieper (1968) developed a technique of obtaining mathematical solutions to the inverse kinematics problem by using the theories of closed-link chains and homogeneous transformations.

One of the early applications of manipulators was in the installation of underwater power cables using mechanical arms attached to a submersible vehicle (see "Cable Connectors Pave Way for Underwater Electric Power System," 1970).

In the 1970s, the instant insanity puzzle (four cubes with different colored faces must be stacked so that no identical colors appear on any side of the stack!) was solved at Stanford University (see Feldman et al., 1971) by means of a robot equipped with a digital TV camera and a microcomputer. In Japan Inoue (1971) was the first to create a constant-torque manipulator joint by turning a constant-torque crank using a 10-bit force feedback and control signal. In 1972, Bolles and Paul demonstrated the use of a robot in assembling an automotive water pump using visual and force feedback (see Bolles and Paul, 1973). In the same year, Will at IBM developed an arm with touch and force sensors to assemble a 20-part typewriter unit (see Will, 1973). Another breakthrough occurred in Japan when Ejiri et al. (1972) developed a robot that could automatically assemble pieces of machinery from plane drawings. Following the work of Inoue (1971), Paul (1972) developed a system in which manipulator joints were automatically chosen by the computer and force-activated rather than position-activated, in response to direction of compliance.

As technology improved Flatau (1973) argued that manipulators are a viable means of extending our dextrous capabilities, whether it be by using a 40 ft arm or an 8 in. arm.

Konstantinov and Zankov (1975) recorded the gradual change from conventional forging processes to total automation. Their paper concentrated on the use of manipulators in automated mass production, as well as the use of industrial robots in small-scale productions.

Drake et al. (1977) describe a method which allows machines to perform many assembly operations quickly, simply, and economically. A robot arm performs the motions of fetching parts and tools and removing completed assemblies under computer control. The process was demonstrated in the assembly of a 17-part automobile alternator by using a robotic arm with four degrees of freedom to fetch parts and tools.

Favareto (1978) describes the design, operation, and features of an industrial robot designated "POLAR 6000" which was developed to incorporate a welding transformer in the robot arm and is adapted specifically for spot-welding applications.

The use of robots in inspection is discussed by Kirsch (1979). Noncontact optical sensors combined with computer control and continuous-path robots allow inspection and decision-making capabilities with greater reliability than that of humans. A robot arm can position a camera to various inspection points on a part and perform video gauging functions at each preprogrammed position. Dimensional checks are possible owing to the mechanical positions repeatability of the robot. In the system discussed, positional accuracy is within $\pm$ 0.001 in.

In a paper by Boykin and Warren (1979), it is mentioned that, as a result of stricter radiation dosage regulations for maintenance workers, future maintenance work could be done by remote manipulators and robotic devices.

Of course one of the most well-known manipulators in use today is the Remote Manipulator System (RMS) on board the NASA space shuttle. Described by Borrowman (1979), the RMS is a mechanical duplicate of the human arm. When fully utilized, the RMS can perform many different tasks, such as transferring astronauts from place to place and repairing, recalibrating, and even retrieving malfunctioning satellites. The manipulator itself is 15.2 m long and able to handle cargos up to 19.3 m long and 4.6 m in diameter. The RMS is capable of positioning a 25,000 kg payload 9 m above the orbiter in less than 9 min.

As technology improved further, it became evident that robots could do even more. Camera and Migliardi (1980) discuss the assembly and manipulation operations performed by robot arms (DEA PRAGMA A 3000) and the mandatory quality control operations that are integrated with the assembly operations. The robot is capable of performing multiple functions and can be assigned jobs such as selective picking of preordered parts, dimensional control, testing of assembled parts, and deposition into different quality classes. The obvious advantages of combining assembly and inspection into the same robot arm and computer control are product quality and ease of budgeting time and money.

With increased demands being placed on robots and automated equipment, the technology associated with robotics must expand in its scope. Initially, robots were used in simple pick-and-place operations, however, as engineering improved, the demands also increased. This allowed new designs of robot arms and hands to become a reality.

In the meantime, Japan made stunning progress in the field of robotic structure development and applications. The first robot used in Japan was a Versatran which was acquired by Toyota Automobile Company in 1967. Later, Kawasaki Heavy Industries in collaboration with Unimate of the United States started to manufacture robots in 1968. It was the great support of the Japanese government and other industries and universities that created a robotic boom in Japan and eventually made Japan number 1 in robot applications and sales in the world; the Japanese FANUC Company has more than 55% of the world market.

The following highlighted history of industrial robotic applications has been excerpted with permission from Zeldman (1984):

1. Czech author Karel Čapek's dramatization of *R.U.R.* (or *Rossum's Universal Robots*) coined the word robot as early as 1920.

2. Isaac Asimov wrote a series of stories about robots. These culminated in a book, *I, Robot*, of nine short stories published in 1950. Asimov's "Three Laws of Robotics" were developed in 1942:

   (a) A robot may not injure a human being, or, through inaction, allow a human being to come to harm.

   (b) A Robot must obey the orders given to it by human beings except where such orders would conflict with the first law.

   (c) A robot must protect its own existence as long as such protection does not conflict with the first or second laws.
3. George Devol Jr. applied for a patent for a programmable transfer device—or robot hand—in 1954.
4. Devol licensed Manning, Maxwell, and others. A preliminary product specification is outlined for the Unimate robot of 1958.
5. AMF initiated an R&D project for a Versatran (versatile transfer) robot in 1958.
6. AMF introduced Model 102, a continuous-path transfer device, and Model 212, a point-to-point transfer device, in 1962, while Unimate prepared the Mark I prototype for field testing in 1962.
7. AMF licensed two European manufacturers in 1965 and started development of a low-cost Versatran solid-state robot system.
8. Versatran introduced Model 301, a point-to-point solid-state robot in 1966.
9. Unimate introduced Mark II, 2000. Their cumulative experience with Mark II, 1900, is based on 70 industrial installations. In 1967 the first robot was imported to Japan and an AMF Versatran was used on a paint-spraying application.
10. In 1968, AMF Versatran introduced Model 302—a point-to-point robot, Japan developed two servo robots with Versatran motions and the first Unimate was imported by Japan's Kawasaki Heavy Industries.
11. The first Unimate was installed in Japan's Nissan Automobile Factory in 1969.
12. In 1970, Unimate introduced the first six-degree-of-freedom industrial robot and AMF demonstrated its model 401 improved continuous-path industrial robot.
13. AMF Japan Ltd. was formed in 1970. By this time there were 26 Versatrans installed and operating in Japanese factories.
14. The Japan Industrial Robot Association (JIRA) was formed in 1971.
15. IBM developed a rectangular Cartesian robotic manipulator (7565) in 1972.
16. The Canadian Government and Hayes-Dana built three prototype industrial robots in 1973.
17. ASEA robots developed an all-electric industrial robot in 1974.
18. Cincinnati Milacron entered the industrial robot market in 1976.
19. General Motors Corp. gave a contract to Unimate to build the PUMA (programmable universal machine for assembly) in 1978.

20. Automatrix Inc. and Cypotech, which was a joint venture between Ransburg and Renault, were formed in 1980.
21. In 1982, General Electric signed an agreement with Volkswagen and Hitachi for sale of their robots in the United States.
22. Cincinnati Milacron introduced the famous T series robots in the industrial market in 1982.
23. In 1982, General Motors and FANUC of Fujitsu signed a joint agreement to manufacture robots in the United States.
24. Westinghouse acquired Unimation in 1982.

1.3 A BRIEF HISTORY OF ROBOTIC GRIPPERS

In a comprehensive survey of gripper designs, Lundstrom (1973) notes a multitude of different gripper designs, including grippers with stiff fingers and/or flexible fingers, vacuum grippers, magnetic grippers, and specialized grippers used for handling different objects such as large boxes or sheets of paper. Obviously, most of these grippers are specialized devices suited for specific tasks. Skinner (1974) noted that revolute fingers can be rotated into prehensile modes, which offers simple, reliable solutions to the problems of manipulator design. Skinner discusses two examples of this type of finger design that have been used successfully: one installed on an industrial robot and the other used by NASA for its Skylab manned space station in the mid-1970s.

Another gripper design was proposed by French and Yraets (1974) in 1974, in which the gripper–arm combination was capable of handling parts weighing up to 5 lb (within a 1 ft^3 volume) and had a work envelope of approximately 10 ft by 10 ft by 3 ft. Parallel to the design of this gripper, work was being done by Murate (1976) on an articulated robotic arm that was able to lift relatively light payloads. In his design, Murata optimized the length of the arm based on requirements of workspace and manual dexterity. Actuators were used to counterbalance the mass of the arm and to reduce the inertial moment generated by the arm as it moved.

One major stumbling block in the implementation of industrial robots has been the fact that, for some industrial applications, a variety of grippers is required if the robot is to be flexible enough to accommodate changes in production or to account for the variety of styles of parts that may be produced. Konstantinov (1975) suggests the implementation of a gripper classification system according to the grasping motion of the jaws. This system, along with other such classification systems of grippers and workpieces, should simplify the task of selecting appropriate grippers for designated tasks.

In later works, a great emphasis was placed on the design of finger and gripper systems. Rovetta (1977) describes several different gripper designs for industrial robots. One of the more interesting designs that he mentions is a gripper modeled after the human hand, complete with mechanical phalanges and return springs to simulate the muscle–tendon interactions. In this same vein, a design has been put forth by Caporali (1982) and Caporali and Shahinpoor

(1984) in which the human hand has been modeled as a gripper, complete with five fingers and opposing thumb. Unfortunately, this design is rather large in its current incarnation, but the device does show the versatility and creativity that is often present in gripper design. Other sophisticated multifingered hands, such as the Salisbury hand (Mason and Salisbury, 1985) and the Hitachi hand (Nakano et al., 1984) have also been designed.

In Japan, much work has been done on various gripper designs in which the shape, structure, and motion of the human hand is simulated. Okada and Tsuchiya (1977) designed a gripper that used a finger subsystem very similar to the human hand. They classified the ranges of movements of their creation in terms of shoulder, elbow, wrist, and finger motions and discovered that the elbow and shoulder tended to effect relatively large ranges of motion in the position of the hand, whereas the wrist and individual fingers produced a higher degree of accuracy in a more limited range of motion. For this hand, only three digits (two fingers and one thumb) were used, since it was felt that three digits were the minimum number satisfactory to accurately model the motions of the human hand. While the control mechanism of this design proved to be more than slightly cumbersome, this hand showed a remarkable flexibility and adaptability in performing various tasks.

Hirose and Umetani (1977), also in Japan, developed a "soft" gripper that was able to conform to the outer contour of objects of irregular shapes and hold them with a uniform pressure distribution along the length of the fingers. Each finger, which resembled lengths of bicycle chain, was actuated by cables that ran internally along the length of the finger and produced a sinuous curling motion of the finger.

While worthwhile experience was gained by modeling grippers after the human hand and arm, work also continued on methods of setting parameters for gripper designs based on their intended tasks. Engelberger (1978), who has been called the father of modern robotics, put fourth the idea that gripper design would benefit from careful study and analysis of the activities that it will be called upon to perform. Kolpashnikov and Chelpanov (1981) carried this theme one step further by presenting a classification of grippers based on given quantitative factors such as size, position and orientation, and gripping force. From there, further restrictions and parameters are determined according to the final installation and activities that the robot would be performing.

With the variety of gripper designs, there is an even greater variety of ways of tackling design problems. Demaurex and Gerelle (1979) presented one method of designing grippers and arms based on certain specifications, namely, cadence, precession, range, and load. Through the use of formulas presented in their work, all these factors can be summarized into a Q factor that is basically a function of the lowest natural frequency and of the range that the arm is required to work within, but, remarkably enough, this factor is independent of the size of the arm.

Another method, formulated by Burckhardt and Gerelle (1980), based the design criteria on the torque and stiffness densities of the arm articulations. This method, as well as being used for the design of new robotic arms, can also be used to compare existing robotic arms. As proof of this, Burckhardt and Gerelle compared the design of the Trallfa TR 300 W industrial robot with the human arm using their parameters.

This does not mean to say that most design and research work has been done using the same basic ideas. There is still much work being done in robotics, as well as orthopedic design, for new gripper designs. One such orthopedic device, the Belgrade hand prosthesis by Rakic (1969), offers a great deal of additional insight into the design of both the gripper and the control systems used to operate the gripper. While it is not directly applicable to the area of robotics applications, the Belgrade hand is a five-fingered device that tries to mimic the human hand in both function and appearance, for aesthetic and practical considerations. In another paper, Rovetta (1980) discusses a mechanical hand that can grip objects of almost any shape. This hand, which consists of two fingers and a sliding plate with a return spring to provide a third point of contact, is activated by cables working against return springs, and objects are grasped by the action of the sliding plate pressing against the fingers. In his paper, Rovetta mentioned that this hand could have industrial and prosthetic applications, especially because of the relative simplicity of the design.

Parallel to the research in grippers, much work has been done in the area of robotic arm design. Fichter and McDowell (1980) devised an arm that was basically a closed kinematic chain rather than an open kinematic chain in an effort to provide an alternative to the anthropometric school of thought in arm design. Frank (1980) discusses his design of a three roll wrist that offers the advantage of a very flexible wrist that can perform more intricate and complex tasks than might usually be expected. Tabesh (1984) designed and constructed a gripper with simulated yaw for GE P50 robot.

One important aspect of gripper and manipulator design is the area of control, which can be divided into two categories: the first deals with the actual physical control (electronic, pneumatic, hydraulic, and mechanical aspects) and the second deals with the mathematical connotation of the theory of control, namely, the design and programming of robotic manipulators.

The earliest and simplest form of physical control used pneumatics as the major method of powering manipulators and grippers. In addition there are many examples of industrial robots that use a combination of pneumatic and hydraulic, pneumatic and electronic, or other hybrids as the motive force behind governing and controlling robotic actions. One example is a Canadian industrial robot (Fluidics Control Canadian Robot, 1971) that uses hydraulics to direct the arm motions and pneumatics to control the actions of the gripper. In another adaptation of pneumatic control, Hanafusa and Asada (1976) devised a pneumatic positioning system that was comprised of a three-fingered gripper that had pneumatic sensors located at the fingertips, and it was the feedback from these sensors that governed the motion of the entire manipulator arm.

However, as advances were made in microprocessor technology, electronic control became more feasible, and more to the point, more affordable. Darazan and Jeffery (1976) developed methods of using microprocessors for both point-to-point and continuous-path control and programming. By using the logic algorithms presented in their work, pneumatically driven manipulators could be directed along predetermined paths in either of the two modes mentioned earlier. Later, Batur and Kaya (1981) used microprocessors to control a pneumatic manipulator, which sensed its own position from potentiometers mounted on the manipulator that sent analog signals to the microprocessor, which in turn converted these signals to a digital form and sent the new signals to the controlling air relays to move the arm.

At the same time, mathematical algorithms were developed for predicting and generating the motion of manipulators. One, by Loeff and Soni (1974) generated incremental motions for each position of the manipulator. A recursive routine using this algorithm would automatically direct the manipulator to its specified destination while avoiding collisions with obstacles in its path.

Mathematical models also have been developed to represent the motion of the robot. Birk (1976) developed a set of empirical equations that enabled a robot having six degrees of freedom to orient and position workpieces. The following constraints apply: first, the robot can select a single part from a random collection of similar parts without a vision system, and second, there are three points from which the spatial locations of both the part and the fixture can be determined. A manipulator working under Birk's algorithm would learn to place the part in its most optimal position. Taking a different point of view, Murata (1977) developed an algorithm for the real-time control of manipulators based on the kinematics of the arm, rather than the dynamics, as was usually the case. Since this algorithm was based on the kinematics, it was no longer necessary to solve iterative equations using differential and algebraic methods that are required for the dynamics of a manipulator. This in turn allowed for faster and easier programming and a shorter run time for the processor that was controlling the robot.

Taking yet a different approach, Luh and Walker (1977) used linear programming methods to solve minimum travel time problems for manipulators. They based their method on the assumptions that the acceleration and velocity constraints were known and that the minimum positional error constraints can be determined or assumed based on the configuration of the manipulator.

As another approach to mechanical arm control, Raibert (1977) used a form of state space memory in which this system is able to control the manipulator, to adapt to mechanical changes, and to perform coordinate transformations of the manipulator implicitly. This in turn generates a table of equations of motion that are governed by state variables, and these equations are implemented using data derived from "exploratory" moves that the robot has made previously.

In addition to the determination of the robot's position in space and time, another important consideration is the ability of the gripper to grasp parts in a stable manner, to offer some feedback as to how much force is being applied, and, in some cases, to determine its proximity to the part in question. In a study of stable prehension of a robotic hand, Hanafusa and Asada (1977) found that stable prehension is generated through the use of a stability condition that is returned to the robot from the gripper itself. Along with this concept, Gerelle (1978) proposed a force feedback method of control to provide some kind of tactile sense, which he felt was essential in the implementation of control for industrial robots. He suggested that a force sensor be used to provide information for force feedback that, in turn, is used to correct the robot's movements.

1.4 A BRIEF HISTORY OF ROBOTIC SENSORS AND CONTROL

Since the early 1970s robotic research has been focused on developing sensors to provide information feedback to the robot control system about the task being performed. These sensors can be grouped into two categories:

Internal Sensors

Potentiometers
Tachometers
Strain gages
Microswitches

External Sensors

Tactile
Vision
Sonar
Speech recognition

There has been extensive research devoted to the external robotic sensors. This research has concentrated on both hardware and software. Kato et al. (1975) produced a robotic eye that was capable of distinguishing between different colors. Agin (1975) describes an experimental vision system, with industrial applications, that fuses hardware and software into a total robotic vision system. The system may be used for part recognition and placement, inspection, and feature extraction. One important goal of robotic vision researchers is to incorporate the types of tasks that humans perform in assembling—namely, part recognition and orientation—and in process inspection. Without these, the robot does not have much of an advantage over the human worker. A human can very easily detect faulty parts and scrap them in an assembly process. Present day robots do not have the capability of doing in-process inspections, and the parts that the robots are assembling must be presented to the robot in precise locations and orientations.

There is extensive work being done on other forms of sensors other that vision systems. Tactile sensors are being manufactured with resolution rivaling that of a human (Lieberman and Lavin, 1982). As the tactile sensor grids become finer, there is an increasing need to preprocess sensor information before the information reaches the controller. This will reduce the burden of the controller circuitry. Coupling a microprocessor with a sensing device creates a smart sensor that can be programmed to look for various conditions as dictated by the task being performed and to react accordingly. The development of increasingly complex and numerous smart sensors will force the robotic control architecture to become more complex to support this multiprocessing. Cain and Bolles (1982) have discussed the visual recognition of partially visible objects by the local-feature-focus methods.

Another developing sensor technology is SONAR. The greatest advantage in the use of SONAR systems is their ability to provide a measure of distance. A vision system does not have this advantage. Proximity information is very useful in noncontact tasks, such as spray painting and other coating operations. Another advantage of SONAR is that it does not require the even lighting and sharp contrasts of present vision systems. Cook (1984) demonstrated a microcomputer-based robotic SONAR imaging system that could perform contour plotting, part location and sizing, and simple part recognition. The SONAR transducer was placed at the end of a three-axis manipulator that was controlled by the microcomputer. A simple data collection circuit transmitter binary-coded distance measurements that were processed by the computer. The microcomputer also controlled the manipulator kinematics. In this manner, the small microcomputer-based manipulator could perform dynamic tracking and obstacle avoidance.

In the area of tactile sensor development, Sato et al. (1977) developed a method in which they used a three-dimensional tactile head to identify parts and their contours by binary image processing. The sensor that they developed consisted of an 8×8 array of spring-loaded probes, which when actuated by contact with an object produced an output voltage. This output was then processed into a binary data form, which in turn was fed into the controlling computer. Through algorithms that were developed to accommodate the sensor head, the robot was able to recognize and perform dimensional checking of the part.

Briot (1980) used a different approach in his method of developing tactile sensors. He developed an "artificial skin" that consisted of a printed circuit that had a matrix of isolated pressure cells inlaid on it. When pressure was exerted on these cells, the transverse impedance would change in a manner that was directly proportional to the amount of applied pressure. From this information, an outline of the part could be generated from the surfaces of contact between the grippers and the part. After being processed by the controlling computer, objects could be identified by comparisons with stored images from previous applications.

At the same time, Bejczy (1979) discussed the use of what he called "smart" sensors for the determination of proximity, force and torque, tactile, and slip parameters for the control of automated robots. Taking this concept one step further, Inoue (1979) used force feedback measurements to correct relatively minor inaccuracies in the positioning of the parts to be assembled and to compensate for inaccuracies in the positioning of the robotic arm itself. For an extensive review on robotic sensing and intelligence see IEEE (1983).

There have been other modes of control that have been put forth. In researching industrial manipulators, Vukobratovic and Stokic (1981) found that the major driving force is the coupling between the degrees of freedom and that this force could be measured directly. They stated that, once some sort of feedback could be generated with respect to this coupling, system stability could be enhanced and greater control could be achieved using this "global force" feedback. Takegaki and Arimoto (1981) also used a global control for large-scale motion to achieve higher levels of control than would otherwise be possible. By using linear feedback of generalized velocities and coordinates for these large-scale motions, they found that this process could be followed by task-oriented control from microprocessors to fine-tune the smaller-scale motions that result, especially at the manipulator tip for the final orientation of the gripper versus the part. One major drawback to this system is the extreme complexity inherent with the system, which requires either some form of coprocessing or extremely fast microprocessor to cope with the vast amount of information being generated by the system.

Raibert and Craig (1981) developed a hybrid technique of feedback control using position orientation in conjunction with force–torque control. Using a force sensor at the wrist to measure force–torque with position sensors located at the manipulator joints, control is achieved through the feedback that the task and its parameters impose on the manipulator.

Hewitt (1979) proposed a method for decoupling nonlinear dynamic systems, such as those that are often present in robotic manipulators. Using his method one obtains decoupled controls for the equations of motion for industrial robotic arms. In a later work Hewitt (1980) used a form of active force control to obtain decoupled motion in an arm. Thus, he was able to derive a control algorithm that required only partial system modeling and could be run in real time.

In another area much work has been done in the determination of the workspace in which manipulators are able to function. Murata and Hashimoto (1979) discuss a computer-oriented algorithm that computes the workspace of a robotic arm when the initial position arm length and range of motion of the joints have been fed into the computer.

In the same light, Sugimoto and Duffy (1980a,b) have presented two papers on the subject of measuring the extreme distance of a robotic arm. In their first paper (Sugimoto and Duffy, 1980a), they present an algorithm in which the extreme extensions of manipulators with arbitrary numbers of joints can be developed. However, they point out that owing to assumptions inherent in their algorithm, it can fail when applied to manipulators with unusual configurations, that is, robots with intersecting or parallel axes. Their second paper (Sugimoto and Duffy, 1980b), deals with this problem and develops a procedure to determine the extreme extensions of manipulators with parallel or intersecting configurations. This is important since there are specialized industrial robots that have parallel or intersecting axes, and the method expedites the calculation of the work envelopes of the manipulators.

Tsai and Soni (1980) developed a set of design charts that can be used to synthesize two- and three-link manipulators. They based their charts on the loci curves of two-link manipulators, and these charts can be used to determine the workspaces of the two- and three-link designs.

Once the work enveloped has been determined, a problem still exists in how to optimize the motions of the manipulator within the workspace. Luh and Lin (1981) discuss a method, based on system constraints, for obtaining minimum travel time for a manipulator. They base their system on the fact that manipulator end effectors are required to traverse a preplanned path in order to attain task completion without collisions with obstacles in the workspace. This is usually achieved when the motion of the manipulator is programmed initially in the manipulator's teach mode (Fig. 1.4.1). Vukobratovic and Kircanski (1982) have developed a method for optimal synthesis of manipulator trajectories. For further details on some recent developments, see Paul (1983), Brady et al. (1983), Shahinpoor (1984), Campbell (1984), Shahinpoor et al. (1986), and Seraji et al. (1985a,b,c). It must be emphasized that the productivity of a manufacturing system depends in large measure on the design of the product and on the way the resources of robots, machines, labor, and raw materials are brought together. There is no guarantee that a total replacement of workers by robots would have much effect on the output of any manufacturing firm or on the cost of its products if careful planning of resource utilization is not made in advance. For these reasons, the productivity can increase only by flexible automation and the introduction of data-processing technology if the capacity to link design management and manufacturing into a unified network is increased substantially. Flexible manufacturing can include a great variety of activities and products. It may comprise a continuous stream of manufacturing processes such as in the manufacture of chemicals. These processes can be controlled readily by closed-loop feedback type control systems. Other manufacturing systems deal with batches of materials. However, the emphasis in the implementation of flexible manufacturing systems is on industries that design and manufacture discrete products rather than materials that are processed either in batches or continuously. These types of industries encompass the fabrication and assembly of automobiles (Fig. 1.4.2), computers, microelectronic components of computers such as semiconductor chips, airplanes, appliances,

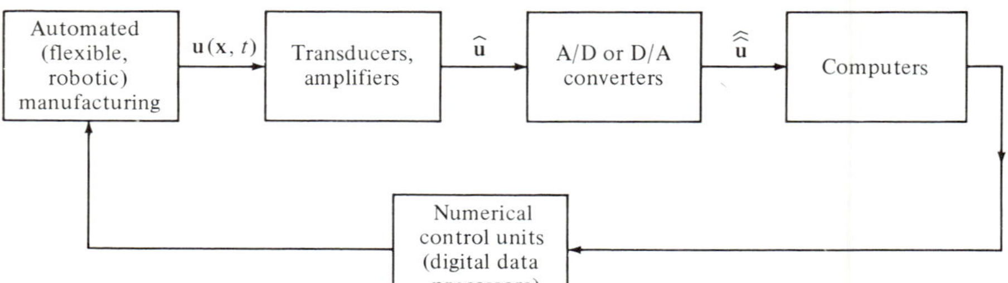

Figure 1.4.1 Numerically controlled automated (flexible, robotic) manufacturing.

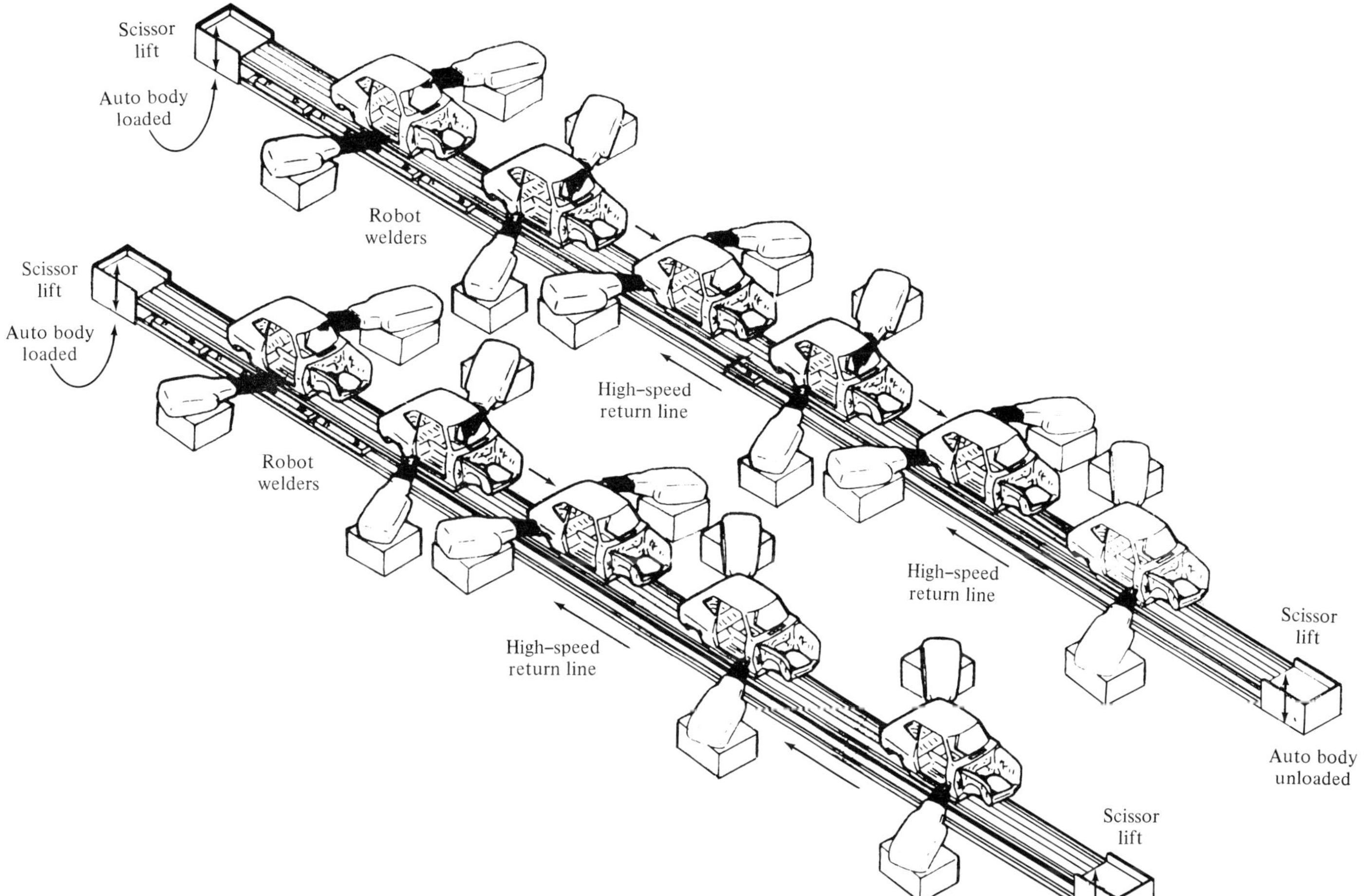

Figure 1.4.2 Typical multistation, fully automated auto body framing operations.

clothing, packaging, building materials, production machines, machine tools, and even robots. The impact of flexible manufacturing and robotics can be felt most effectively by such discrete manufacturing industries. Early impact of such flexible type design and automation has already been felt by the advent of CAD/CAM type manufacturing systems. CAD/CAM programs can carry out accurate geometric transformations so fast that the designers are no longer bound by a few sections of the product and can deal with infinitely many sections. They can observe the rotation of the parts about any axis on the screen and they can zoom in and out to see details and to visualize the object as a whole. The application of CAD drastically improves the productivity. For example, a redesign of an automobile that used to take up to 2 years now can be done in less than 8 months. Application of CAD in plastic parts manufacturing industries has increased productivity by a factor of 4 on the average.

Early application of CAM was based on the automatic storage and retrieval systems and the availability of numerically controlled machines. The earliest numerically controlled machine tools were programmed by means of punched paper tapes. However, small digital computers mounted on the machines have replaced the old punched paper tape technique. In a direct numerically controlled system, all connections between the tools are controlled electronically and the transfer of the workpiece from one machine to another may be done either manually or robotically. For products that are maufactured in large numbers, the process can be automated by means of a numerically controlled single-purpose machine. Such machines can be costly and may lack the flexibility to be used for the manufacture of other products. The use of robots as preprogrammable machines capable of moving material and performing repetitive tasks then becomes the logical alternative for flexible manufacturing systems. At first, it appears that the robots offer the following advantages:

Increased productivity
Effective equipment utilization
Reduced labor costs
Improved quality and consistency of parts
Improved flexibility
Shorter lead times
Flexibility and ease of programming
Ability to work in hazardous environments
Improved quality of work areas
Improved competitive positions
Improved quality of production areas
Minimal diagnostic and debugging costs
Low payback periods
Good return on investments

However, they also suffer from a number of disadvantages. One of the main difficulties in the use of robots for assembly is that the robot is not yet able to pick a randomly oriented part out of a bin without using a special vision system. The robot can compete economically with other machines if the orientation of the parts to be assembled is fixed or preserved at stages of the assembly process.

Current industrial use of robots includes the following tasks (Engelberger, 1980):

1. *Die Casting.* The use of robots in this field was facilitated by inherent constant orientation of parts in their dies and necessitated by the often unpleasant working environment associated with die casting.
2. *Spot Welding.* This chore is required primarily in the automotive industry and currently comprises the majority of manipulators in use.
3. *Arc Welding.* The use of robots for arc welding applications has been popularized because much of the equipment used for manual arc welding can be adapted easily to a robot.
4. *Investment Casting.* In this application, robots are used mainly for the transfer of parts between workstations.
5. *Forging.* Basically, in this area, robots are used only to handle hot metal parts or to work in areas of high temperature. Their application is limited because of the relatively low levels of batch production and the complexity of some of the parts.
6. *Press Work.* Again, robots in this area are used primarily in the automotive industry for the stamping of body panels. One main consideration in their implementation has been increased worker safety in stamping and punch press operations
7. *Spray Painting and Surface Treatment.* In this area, robots are becoming popular in industry because of the hazardous working environment for the human employees and because of the robots' greater uniformity and consistency in applying paint, sealer, and so on.
8. *Plastic Molding.* Robots are used in this field to load and unload injection molding dies and for trimming or deburring some moldings. These installations are justified economically by the faster cycle times and higher productivities over their human counterparts.
9. *Foundry Applications.* Here, robots are used frequently for the ladling of the molten metal into the molds, and in other areas characterized by hostile working conditions for the human workers.
10. *Machine-Tool Loading.* The use of robots in this area has been encouraged by the prior installation on numerically controlled (NC) machinery and the relatively easy task of interfacing the robots to the NC tools.
11. *Palletization.* Here, robots have been used in parts and production palletizing and in the resulting inventory control.

From this list of current industrial applications, it is obvious that the development and implementation of robotic manipulators has made great strides from its humble beginnings as a master–slave teleoperator developed at the Argonne National Laboratory in 1947 for the handling of radioactive materials.

1.5 COMPUTER-BASED ROBOTIC MANIPULATORS

The computer-based robotic manipulators of interest in this book consist of the following main components (see Figs. 1.5.1 and 1.5.2):

1. A mini or microcomputer with adequate memory CPU, secondary storage devices, and a terminal.
2. An n-jointed electromechanical or hydraulic manipulator arm with rotary, revolute, prismatic, and telescopic joints.
3. External sensors such as digital vision (digitizing camera and a TV), force sensors and transducers, and tactile sensors and end effectors.
4. An I/O logic controller to process information in case continuous motors with optical choppers (encoders) are to be used.

A general robotic laboratory setup, on the other hand, should consist of the following items (see Fig. 1.5.3):

1. A number of multiple-axes, that is, six-jointed, mechanical arms such as the PUMA series Rhino XR-2, GE P50, IBM 7565, or Cincinnati Milacron's T3.
2. A number of micro or minicomputers such as the PDP-11 series or the IBM PC-AT series.
3. A number of interface devices such as A/D or D/A converters and digital I/O logic controllers.
4. A number of electronic control devices such as optical encoders, choppers, or current-driven amplifiers.
5. A number of digital TV cameras or computer vision systems.
6. A number of digital microphones or computer hearing systems.
7. A number of ultrasonic image analyzers and sonar devices.
8. A large quantity of stepper motors, continuous motors with optical encoders, conveyor type modules, and carousel type modules.
9. A large quantity of machine parts such as gears, chains, pulleys, sprockets, splines, belts, sheaves, fasteners, flexible shafts, cables, and couplers.
10. A good stock of software for certain basic spatial manipulation and control.

This laboratory set provides the means to implement both robotics education and research. Of course, suitable robot engineering reference books and laboratory manuals should accompany the above mentioned tools.

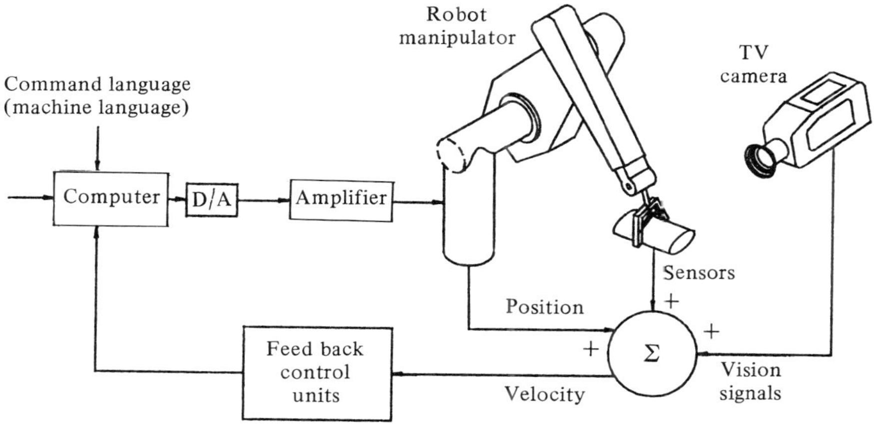

Figure 1.5.1 Organization of a computer-based robotic manipulator.

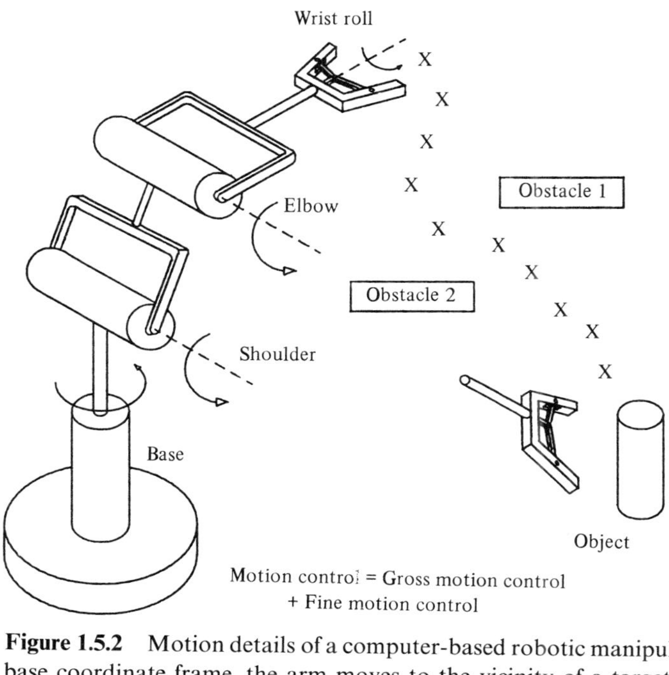

Figure 1.5.2 Motion details of a computer-based robotic manipulator. With respect to a base coordinate frame, the arm moves to the vicinity of a target object by means of a planned trajectory to perform an assembly task.

1. Rhino® XR Mark II Robotic Arm
2. Mark II Controller
3. Linear Slide Base
4. Software Manuals Rasp 1.04
5. Rasp 3.04
6. IBM Software
7. TRS-80
8. Language Card Manual
9. Teach Pendant Manual
10. Experimental Motor Kit
11. Input Output Card
12. RHINO-Com Language Card
13. CCS 7710A Card
14. "Hands-On-Introduction to ROBOTICS"
15. Teach Pendant
16. Moto-Dremel® Holder
17. Vacuum Fingers
18. Aluminum Base
19. Rotary Indexing Carousel
20. Magnetic Pickup
21. Triple Finger Attachment
22. Rhino-Rite Hand
23. X-Y Table
24. Extra Long Finger Attachments
25. Narrow Fingers
26. Clamshell Attachment
27. Shovel Attachment
28. Chain Belt Conveyor

Figure 1.5.3 A laboratory educational setup for robotics. (Courtesy of Rhino Robots.)

1.6 ROBOT GEOMETRICAL CONFIGURATIONS

The workspace or volume of a robot is its sphere of influence: the arm and wrist subassembly unit can travel to any point within those boundaries. Mechanically, most robots fall into one of five categories:

1. Cartesian or rectangular robots.
2. Cylindrical robots.
3. Spherical or polar robots.
4. Revolute or articulated robots.
5. Snakelike or tensor-arm robots.

1.6.1 Cartesian or Rectangular Robots

Examples of these types of robot are the IBM RS-1 and 7565 White Cloud or the SIGMA robot of Olivetti. They normally consist of three linear orthogonal motions along their joint axes as shown in Figs. 1.6.1 and 1.6.2. The best resolution and accuracy are achieved within the robot's workspace. Some of the advantages and disadvantages are listed below.

Advantages

1. High resolution and great accuracy.
2. Good obstacle avoidance and collision prevention.
3. Independency of gravity loading, that is, ease of joint motion control.

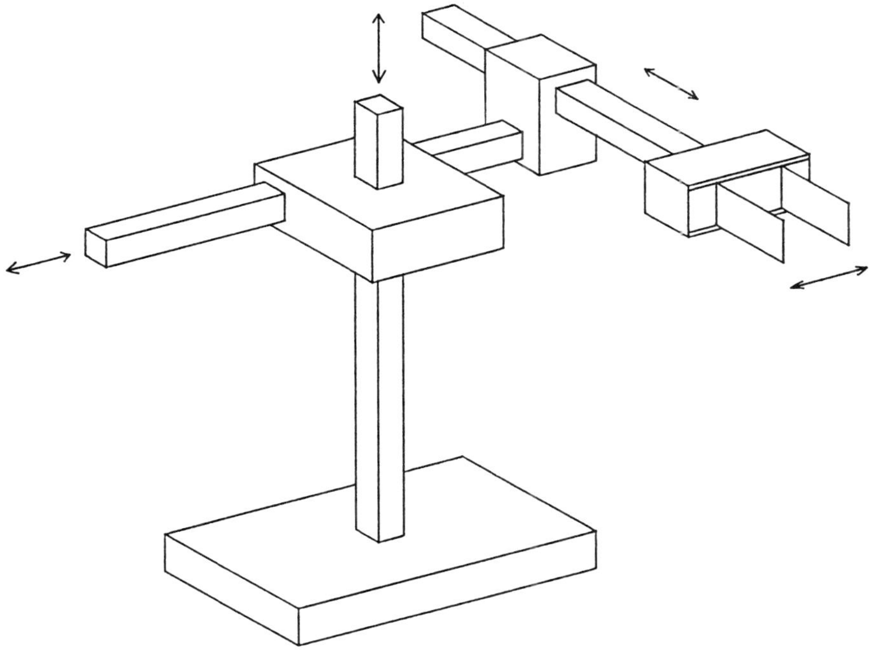

Figure 1.6.1 Cartesian or rectangular robotic manipulators.

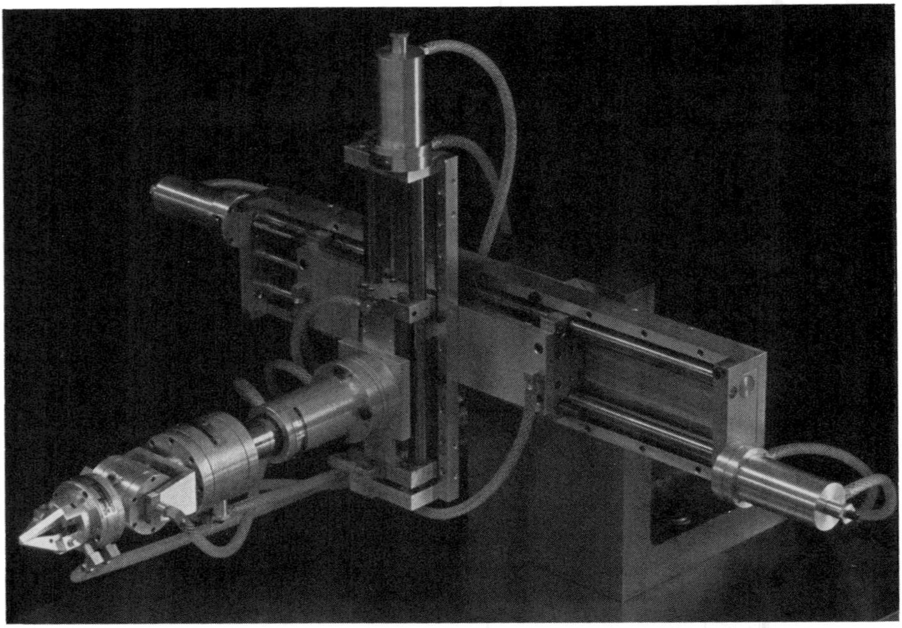

Figure 1.6.2 Rectangular Cartesian robot. (Courtesy of Mack Corp.)

Disadvantages

1. Large structural framework.
2. Confinement of the workspace.
3. Restriction on being compatible with other robotic arms in a common workspace.
4. More complex mechanical design for linear sliding motions.
5. Requirement of a larger floor space.

1.6.2 Cylindrical Robots

Examples of these types of robots are Versatran Model F600 of the Prab Company or the Stanford arm (see Paul, 1983). These robots normally consist of two linear orthogonal motions and a rotary type motion as shown in Figs. 1.6.3 and 1.6.4. The advantages and disadvantages of this type of robot are listed below.

Advantages

1. Almost independent of gravity loadings.
2. Collision-free movement.
3. Two linear axes make the mechanical design less complex than the Cartesian robots.

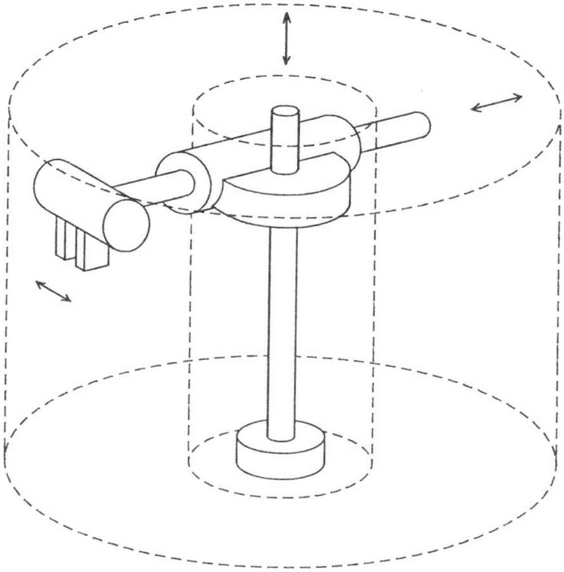

Figure 1.6.3 Cylindrical robotic manipulators.

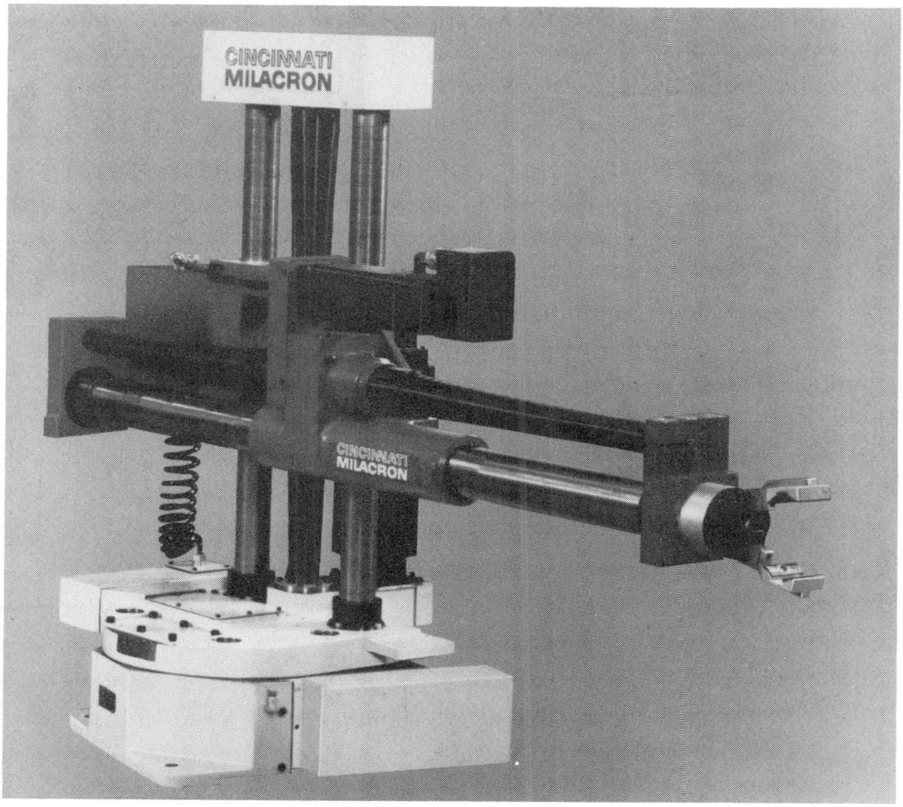

Figure 1.6.4 Cylindrical polar robotic manipulator. (Courtesy of Cincinnati Milacron.)

Disadvantages

1. Large amount of structures.
2. Restriction on compatibility with other manipulators in a common workspace.
3. Less accuracy and resolution compared with rectangular robots.

1.6.3 Spherical or Polar Robots

An example of these robots are the Unimate 2000B manufactured by Unimation Inc. These robots essentially consist of two rotary motions and a single sliding (telescopic) motion (see Figs. 1.6.5 and 1.6.6). The advantages and disadvantages of this type of robot are listed below.

Advantages

1. Lowest weight and minimal structural complexity.
2. Short joint travel for many motions.
3. Compatibility with other robots and machinery in the common workspace.
4. Good resolution because errors in positions are perpendicular.

Figure 1.6.5 Spherical robot manipulator. (Courtesy of Unimation, Inc.)

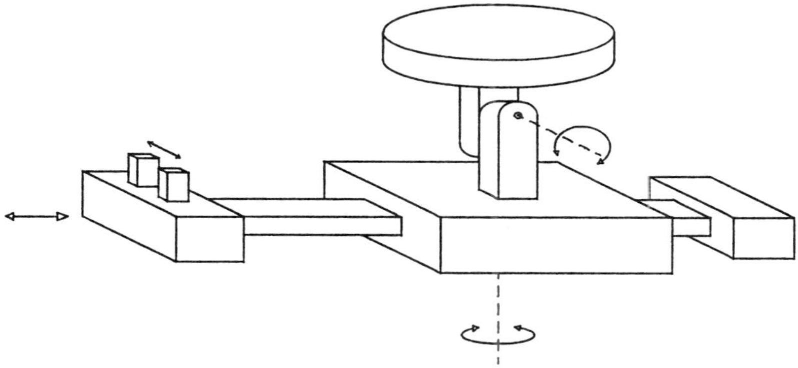

Figure 1.6.6 Spherical or polar robotic manipulators.

Disadvantages

1. Large and variable torques on second and third joint of the arm, creating a counterbalance problem.
2. Limited ability to avoid collisions with obstacles.
3. Position error is large due to rotary motion and is proportional to the radius.

1.6.4 Revolute or Articulated Robots

These are the humanoid type robot with redundant degrees of freedom. Examples of these are the PUMA 500/600/250 series or the T^3 robot of Cincinnati Milacron (see Figs. 1.6.7 and 1.6.8). The advantages and disadvantages are listed below.

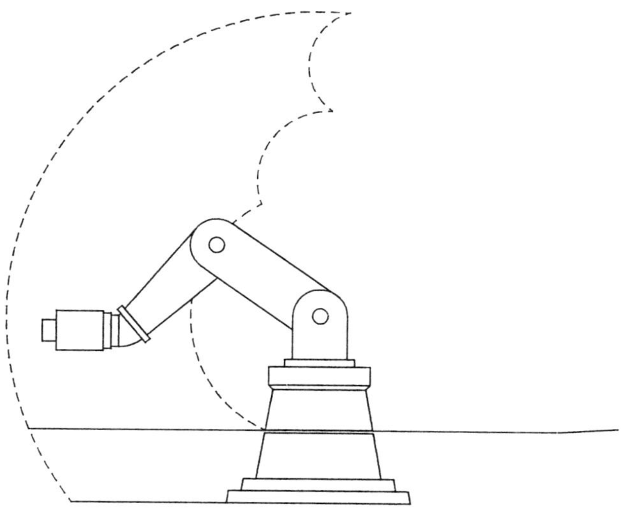

Figure 1.6.7 Revolute or articulated robotic manipulator.

Advantages

1. Flexibility to reach over or under an object.
2. Compatibility with other robots working in the same common workspace.

Disadvantages

1. Poor resolution and accuracy, that is, maximum positional error.
2. Large and variable torques on joints, creating a counterbalance problem.
3. Limited ability to avoid obstacles.
4. High moment of inertia, gravity effects, and dynamic instability (e.g., vibrations).

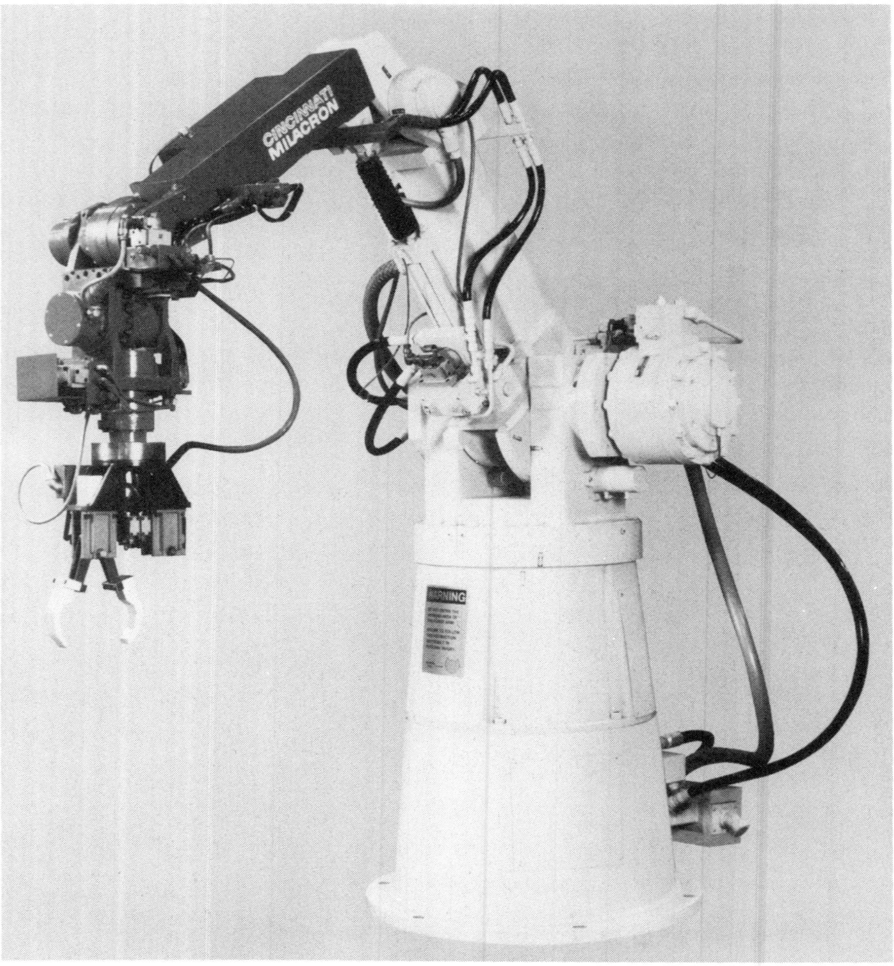

Figure 1.6.8 The Cincinnati Milacron $T^3$566 robot manipulator. (Courtesy of Cincinnati Milacron.)

1.6.5 Snakelike or Tensor-Arm Robots

Ideally, these robotic arms can take any shape in three-dimensional space. Thus, they necessarily consist of many elements that can move with respect to their adjacent elements (see Figs. 1.6.9 and 1.6.10).

As early as 1970, a number of patents on such mechanical manipulators had been awarded: for example, Fig. 1.6.9 is a tensor-arm robotic manipulator patented in 1970 (filed in 1968) by Victor C. Anderson and Ronald C. Horn (U.S. Patent No. 3,497,083). With the advent of direct-drive mechanical arms, such multielement robotic arms may gain popularity in the future. There are a number of applications that cannot be performed with conventional arms that could be performed easily with a tensor arm, for example, passing through a restricted passage such as a manhole.

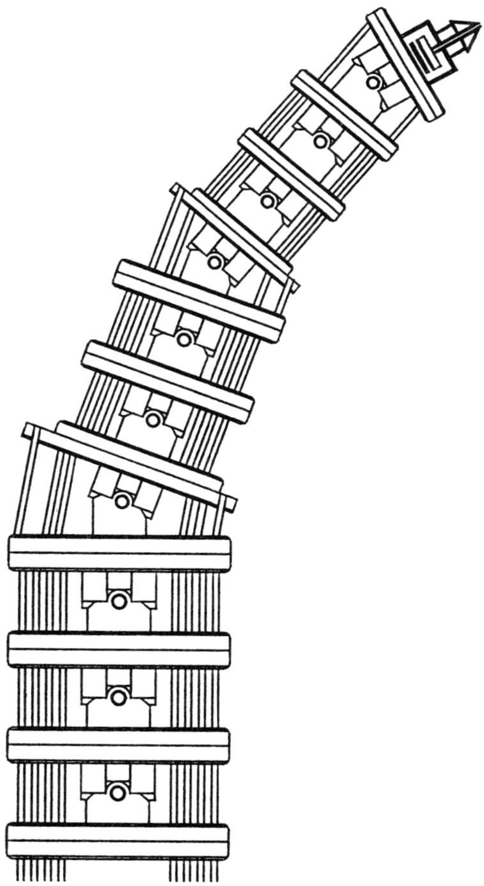

Figure 1.6.9 Tensor-arm robotic manipulator.

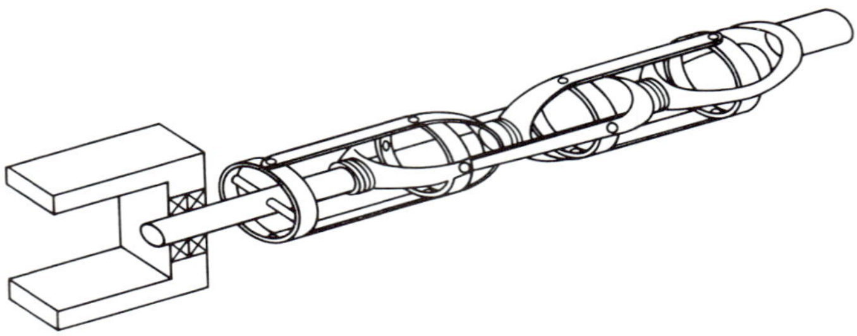

Figure 1.6.10 Flexible robotic arm.

1.7 THE WRIST AND GRIPPER SUBASSEMBLIES

All robotic link articulations are destined to position the wrist and gripper (end effectors) in a suitable or appropriate location for material or task handling. The wrist articulation is basically a roll–pitch–yaw type sequence of motion (Figs. 1.7.1 and 1.7.2).

For the gripper assembly (Fig. 1.7.3), many designs have been tried. However, it is generally believed that a three-finger gripper is sufficient for most practical applications in robotic manipulation. A five-fingered robotic hand (Caporali–Shahinpoor hand) that resembles a human hand has been designed and tried for practical applications (see Caporali, 1982). However, it does not seem to offer much more flexibility or articulation than the three-fingered gripper. For certain special applications, such as playing a musical instrument, a five-fingered hand may offer superior advantages. Figures 1.7.3–1.7.8 display various types of mechanical gripper configurations that are currently in common use. Figure 1.7.9 show a sophisticated three-fingered gripper manufactured by Hitachi of Japan, and Figure 1.7.10 shows a five-fingered gripper constructed by Caporali and Shahinpoor (Caporali, 1982; Caporali and Shahinpoor, 1984). In addition to mechanical grippers, there also exist a great variety of other types of grippers, such as (1) adhesive grippers, (2) magnetic grippers, (3) suction grippers, and (4) needlelike (poking) grippers.

Figure 1.7.3 shows a hand with a single movable finger. This is a single-acting jaw gripper and has applications where there is access underneath a part as it lies on a rack. This hand can be used to scoop up a piece quickly. Its simple design makes it one of the most economical grippers around.

Figure 1.7.4. displays a special hand for handling cartons. The dual-jaw gripper can open and grasp inaccurately located objects of light weight. The lifting, placing, and handling of cardboard cartons is a common application for this type of gripper, since the actuators and jaws can be adjusted for any size carton.

Figure 1.7.5 shows a special robotic hand for manipulating slender or abnormal-sized light objects. It has a pair of pneumatic actuators that can be adjusted to handle and grip slender bodies.

Figure 1.7.6 shows a cam-operated hand with inside and outside jaws. If a part can be reoriented between the time when the part is placed in a machine

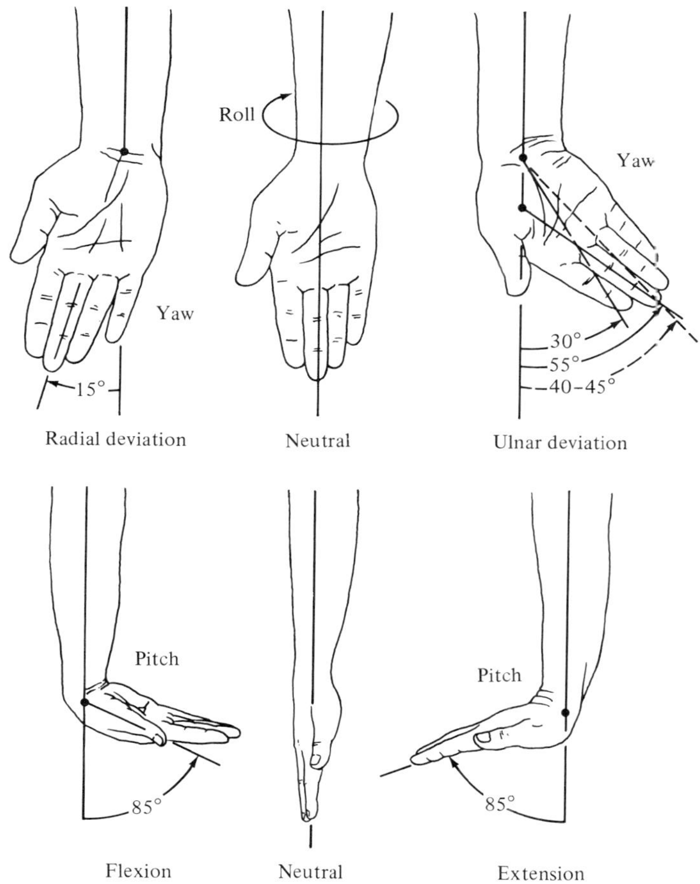

Figure 1.7.1 Human-wrist articulation.

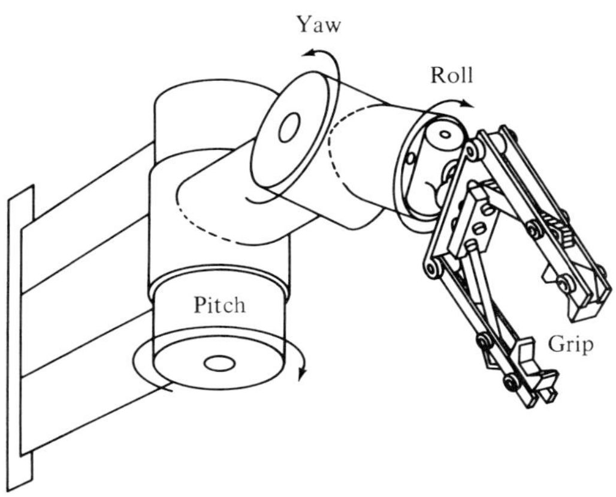

Figure 1.7.2 Details of the robotic wrist articulation.

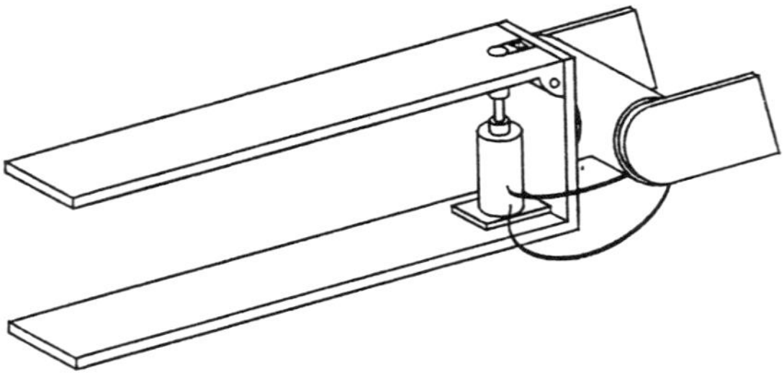

Figure 1.7.3 Single-jawed mechanical gripper.

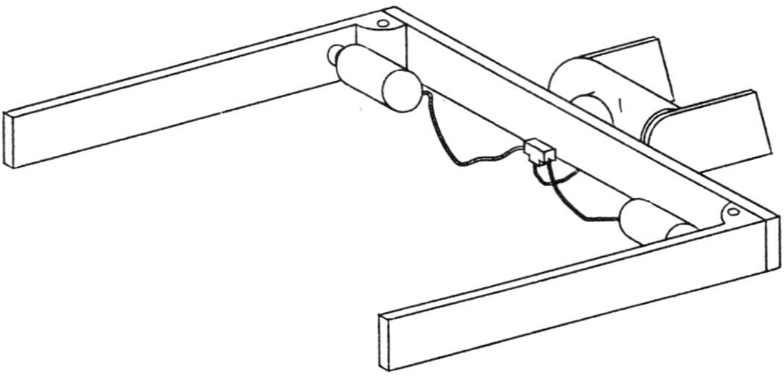

Figure 1.7.4 Dual-jawed gripper for handling cardboard cartons.

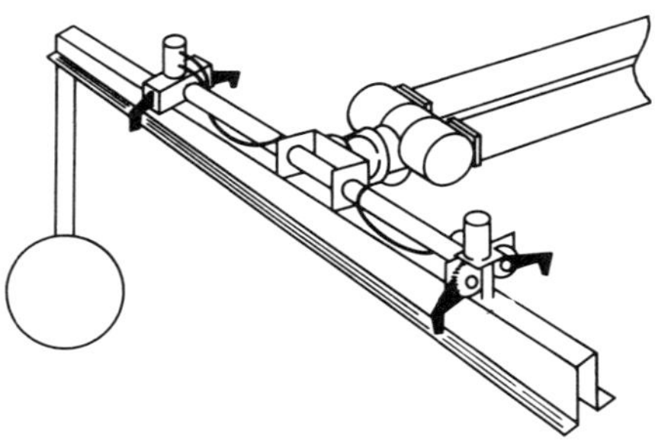

Figure 1.7.5 Special gripper with modular actuators.

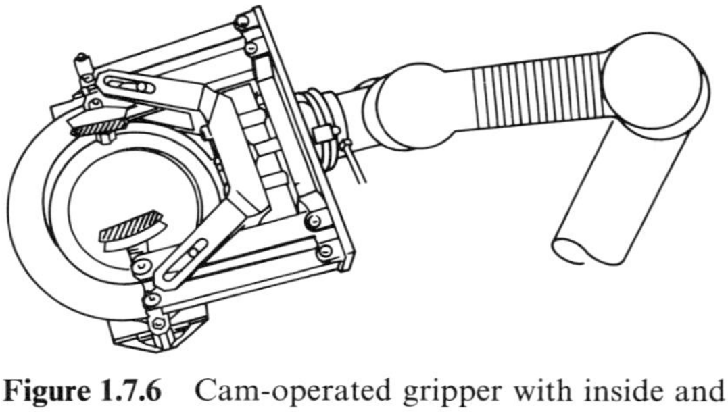

Figure 1.7.6 Cam-operated gripper with inside and outside jaws.

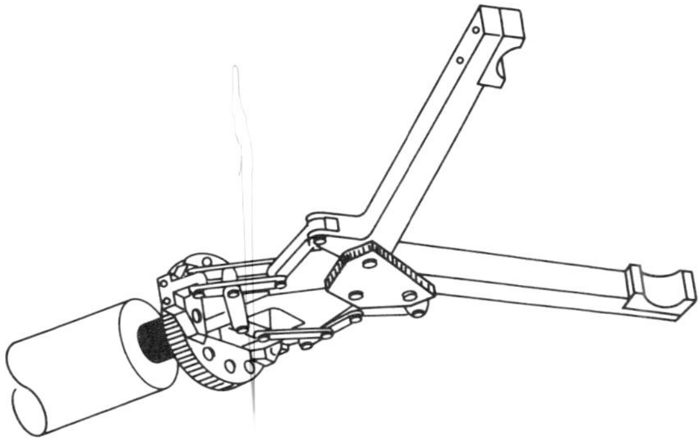

Figure 1.7.7 Gripper with wide-opening jaws.

and when it is removed, this gripper is the one to apply. When the part is oriented as shown, the hand can grasp it on the outside by applying the outer self-aligning pads. If the part is turned over, the inner pads will grasp the inside. Similarly, if the surface to be grasped is reoriented between the time when it is placed in a machine and when it is removed again, this type of gripper can handle the situation accordingly.

Figure 1.7.7 shows a wide-opening hand that is especially suitable for situations in which the part to be picked up is not always to be found in a constant orientation or at the same location. As the jaws swing closed, they sweep the inaccurately oriented and placed part into its grasp.

Figure 1.7.8 shows another version of the cam-operated hand designed to handle heavy parts of bulky objects. This gripper has the ability to keep the center of gravity of the part close to the wrist of the robot. Unfortunately, the applicability of this hand is limited to a very narrow range of object sizes.

Figure 1.7.9 shows a three-fingered gripper made by Hitachi of Japan. It is quite a versatile gripper made from lightweight alloys and exhibiting humanlike flexibility. Shape memory alloy (SMA) wires are used for finger and wrist actuation. A remarkable feature is that the SMA wires have the ability to expand and contract if a current is passed through them. In fact, this phenomenon is used to actuate the finger and wrist movements.

Figure 1.7.10 shows a five-fingered gripper designed and constructed by Caporali and Shahinpoor (Caporali, 1982; Caporali and Shahinpoor, 1984).

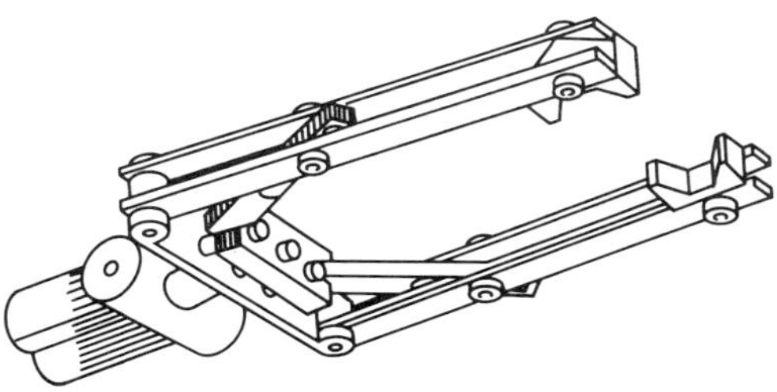

Figure 1.7.8 Dual-jawed cam-operated gripper.

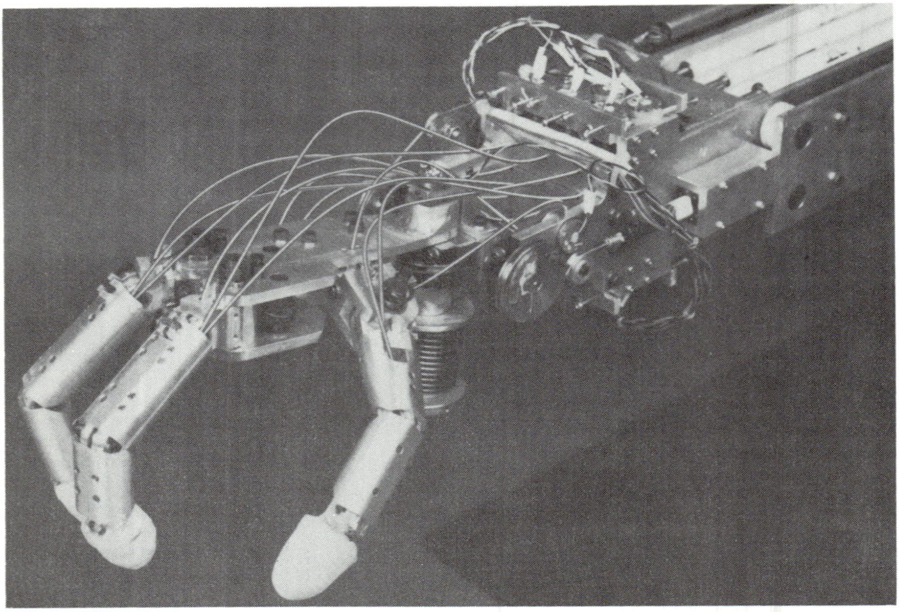

Figure 1.7.9 Three-fingered robotic gripper. (Courtesy of Hitachi, Japan.)

Figure 1.7.10 Five-fingered robotic hand. (Caporali–Shahinpoor hand, 1983).

The design was intended for modeling of sophisticated robotic manipulators. The knuckle motion of the specially designed fingers is created by cables and pulleys with interdendritic hinge joints. The cable motions are then numerically controlled by either stepper motors or controlled linear actuators. However, as mentioned before, three-fingered grippers have been determined to be sufficient for performing nearly all functions attributable to human hands while a five-fingered gripper may be suitable for simulation of playing musical instruments and similar tasks. They also may find applications as prosthetic hands for bioengineering applications. For an excellent reference on robotic hands see Mason and Salisbury (1985).

For further examples of multifingered robotic hands, see the Salisbury hand (Fig. 1.7.11) and the Utah–M.I.T. hand (Fig. 1.7.12).

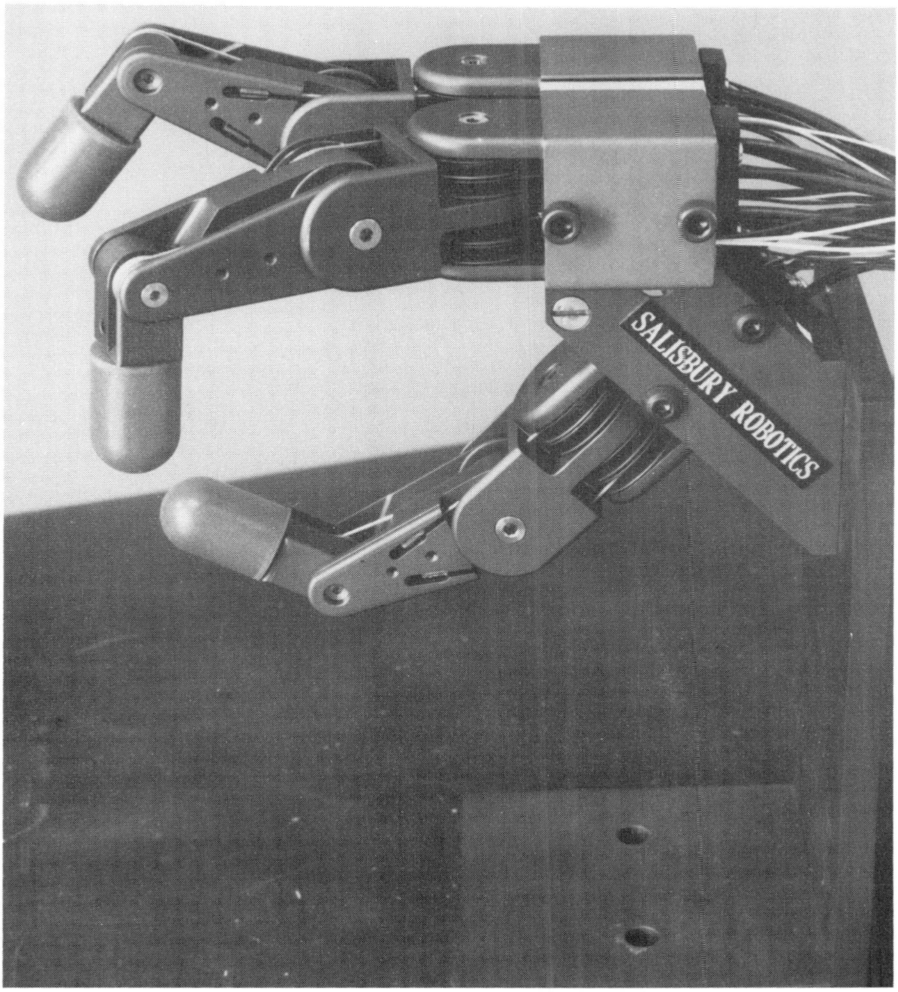

Figure 1.7.11 Salisbury hand. (Courtesy of Ken Salisbury, 1984.)

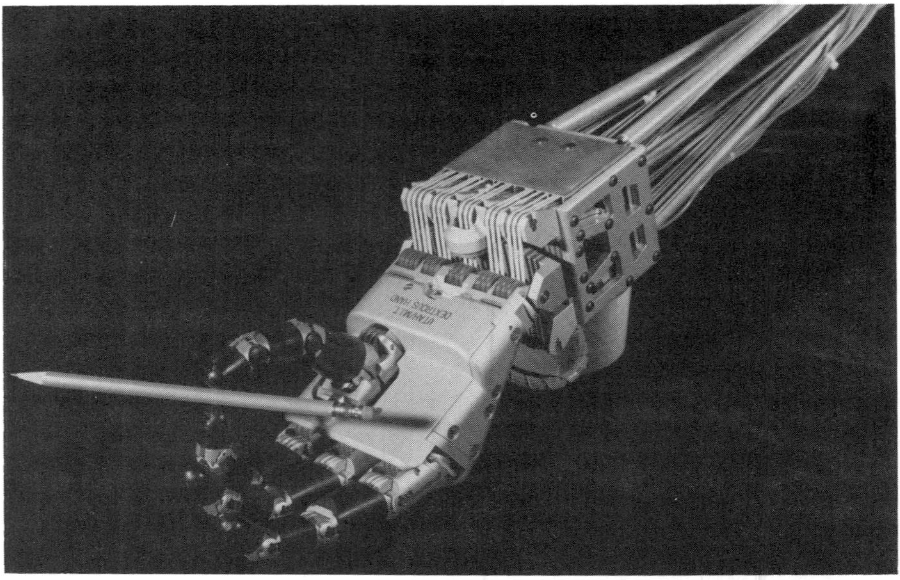

Figure 1.7.12 Utah–MIT hand. (Courtesy of the University of Utah.)

1.8 ROBOT DRIVE SYSTEMS

1.8.1 Introduction

To control and guide the joint motions in a robot, a drive system is necessary. These drive systems could be electrical, hydraulic, pneumatic, or a combination of these three drive systems.

Hydraulic drive systems for robots are simpler than other drive systems because they have so few moving parts, while at the same time they have physical strength (high payload) and high speed. Hydraulic drive systems generally use hydraulic motors with electrohydraulic servo valves and analog resolver units for control and feedback. A good example of a robot with hydraulic drive is the IBM 7565 (RS1) White Cloud. The disadvantages of hydraulic robots is that they oscillate when decelerating to a point, they require extra space to house the hydraulic pumps and accessories, they are noisy owing to the hydraulic pumps, and they suffer from a high maintenance cost for such things as changing oil filters and oil, repairing hydraulic line ruptures and oil leaks, and trapping and exhausting fumes.

Pneumatic drive systems are used mainly for limited-sequence (bang-bang motion) robots. These robots are used principally in pick-and-place type applications or quick assembling type operations. The advantage of these systems is their low cost of operation and maintenance. One main disadvantage is that the control of position and speed is very difficult due to the compressibility of air. Thus, they operate on open-loop control mode.

TABLE 1.8.1 GENERAL SPECIFICATIONS OF ROBOTIC DRIVE SYSTEMS

Hydraulic drive systems	
Use:	Hydraulic servo valves analog resolver units
Advantages:	High speed, mechanical simplicity, physical strength, higher power-to-weight ratio
Disadvantages:	May suffer from oscillations and hydraulic hammer, noisy, additional maintenance and floor space
Electric drive systems	
Use:	Current/voltage amplifiers, stepper motors, dc torque motors, optical encoders, and linear actuators
Advantages:	Easy to maintain, less expensive, easy to design, floor space and good accuracy and repeatibility
Disadvantages:	Low speed, low payload, need to convert for linear axis motion, lower power to weight ratio
Pneumatic drive systems	
Use:	Limited-sequence control, bang-bang control
Advantages:	Quick assembly, low operating and maintenance costs
Disadvantages:	Difficult position and speed control, low accuracy and repeatability

Electrically driven robots are mainly actuated dc torque motors or stepping motors. Since these motors are highly efficient with a good torque–speed relationship, they are in wide use in robotics. On the other hand, since they deliver peak power at high speeds, they must be combined with gear reduction units (speed reducers). Usually, the electrically actuated manipulators enjoy the best accuracy and repeatability compared to either hydraulic or pneumatic ones. On continuous dc motors, optical choppers or encoders are also used for position feedback and control.

Table 1.8.1 summarizes the general specifications of robotic drive systems.

1.8.2 Resolution, Spatial Resolution, Accuracy, and Repeatability

Resolution or control resolution is the smallest increment that the robot controller can control by means of D/A or A/D commands (DACS) for position feedback and control. For example, if the tip of a robot arm is to move a distance d by means of n bit DACS through the controller and subsequently the joint motor, then there exists a small increment of $d/2^n$ that the controller cannot control. This is because the distance d is digitized by 2^n and thus a control resolution is inevitable. For rotary motions, one has to divide the angular range θ by 2^n, to obtain the control angular resolution, and then multiply this by the arm length to obtain the tip control resolution.

EXAMPLE 1.8.1 ■

Determine the control resolution of a sliding arm tip if it is to traverse a distance of 40 in. while receiving 10 bit DACS for position feedback and control (see Fig. 1.8.1).

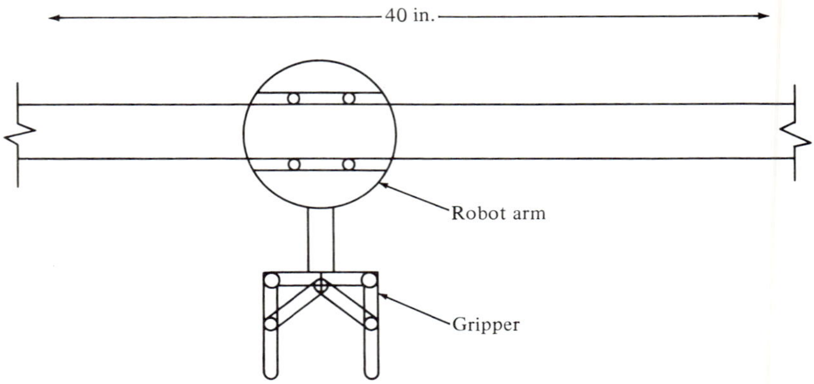

Figure 1.8.1 Prismatic robotic manipulator.

Solution. There are $2^{10} = 1024$ positions in a 40 in. length. Thus, the control resolution is

$$\text{CR} = \frac{40}{1024} = 0.03906 \text{ in.}$$ ■ ■

EXAMPLE 1.8.2 ■

Determine the control resolution of a polar robot with an arm radius of 24 in. and an angular rotation range of 270° if the joint motor is receiving 8 bit DACS for position feedback and control (see Fig. 1.8.2).

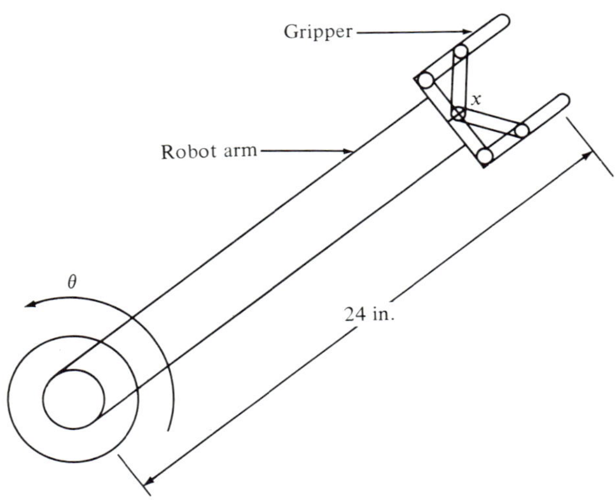

Figure 1.8.2 Polar cylindrical robot with a gripper.

Solution. Note that the angular control resolution is given by

$$\delta\theta = \frac{270/180}{256} = 0.01841 \text{ radians.}$$

Thus, the control resolution of the tip is

$$R = 24 \times 0.01841 = 0.44179 \text{ in.}$$

Note that if we had used a sliding joint, we would have a control resolution of

$$\text{CR} = \frac{24}{256} = 0.0944488 \text{ in.,}$$

which shows that sliding resolution is superior to angular resolution in this case. ■ ■

Spatial resolution is defined as the smallest increment of motion achieved by a robotic manipulator at its tool or end effector. It is the combined resolution including the control resolution and the motion inaccuracies that originate from mechanical or dynamic interaction of the robot with its environment. The sources of mechanical inaccuracies are gravitational, vibrational, accelerational (coriolis, centripetal, etc.), and deformational (elastic, plastic, viscoelastic, viscoplastic, etc.). These topics are discussed briefly in later chapters. Another source of inaccuracy must be considered when one discusses position trajectories, which are calculated by a computer from the kinematics equations of a robot manipulator. In this manner, the accuracy of the robot may be further affected by the numerical round-off errors, the truncation error, and the error resulting from trigonometric approximations. Figure 1.8.3 shows the difference between control and the spatial resolution. It also defines the accuracy as half the control or spatial resolution of a robot.

In a "taught-and-replay" mode, which is quite frequently used in industrial applications, the robot is led through a sequence of positions (teach mode) by a manual control device and the computer continuously records the joint positions during the entire teaching operation. Later, the computer commands the robot to repeat the motion (playback mode). In this case, *repeatability* is defined as the ability of the robot to return to a taught position. This can then be influenced only by the control resolutions and the mechanical inaccuracy (see Fig. 1.8.3).

EXAMPLE 1.8.3 ■

Prove that *repeatability* is generally less than the *accuracy* in a robot.

Solution. Repeatability is affected only by control resolution and mechanical inaccuracy; however it is not relevant to an arbitrary "untaught" position. The tool tip is to return to a previously taught position. Thus, the computer can sense the proximity of adjacent positions and return the robot tip to a position as close as possible to the previously "taught" position. Therefore, repeatability, at most, is equal to the accuracy. ■ ■

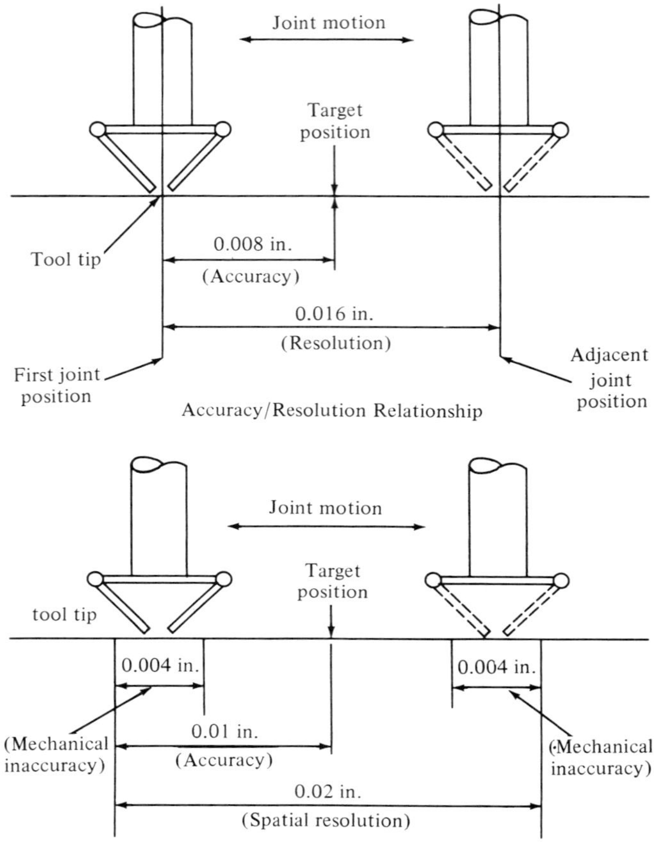

Figure 1.8.3 Control and spatial resolutions and accuracies relationships.

EXAMPLE 1.8.4 ■

Determine the control resolution and inaccuracy of the PUMA robot shown in Fig. 1.8.4. If the mechanical inaccuracy is 0.001 in., also determine the spatial resolution and inaccuracy. (Assume 10 bit DACS for position feedback and control.) Consider only the planar problem.

Solution. For joint A

$$\text{Angular resolution} = \frac{200 \times 2\pi}{360 \times 1024} = 0.0034 \text{ radians.}$$

Control resolution at B due to joint A $= 0.0034 \times 17 \approx 0.0579$ in. For joint B;

$$\text{Angular resolution} = \frac{270 \times \pi}{180 \times 1024} = 0.0046 \text{ radians.}$$

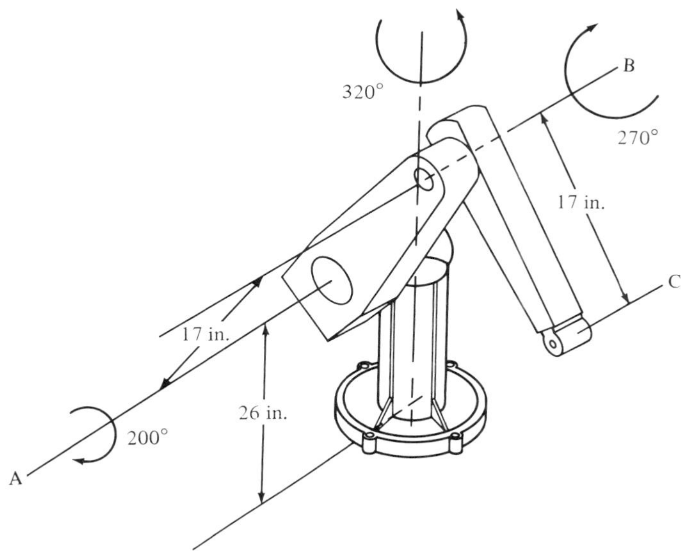

Figure 1.8.4 A configuration of PUMA 560 (not to scale).

Control resolution at C due to joint B = 0.0046 × 17 = 0.0782 in. The corresponding inaccuracies are 0.0289 and 0.0391 in., respectively. Finally, the corresponding spatial inaccuracies are 0.02945 and 0.0396 in., respectively ■ ■

1.9 STATISTICAL ANALYSIS OF ROBOTIC INACCURACIES AND DIMENSIONAL INACCURACIES

Note that robotic spatial and control resolutions and inaccuracies do indeed have a statistical distribution. We assume these distributions to be Gaussian (normal), such that there exist mean and standard deviations for inaccuracies. For a given target position, the spatial inaccuracies can be considered dimensional tolerances on the parts to be mated. A natural standard deviation can be defined as 1/3 of such tolerances, and thus one can find the percentage of parts that do interfere by the following procedure:

Find the mean position μ_r of the robot that represents the target position. The spatial inaccuracies are now interpreted as tolerances t_r, such that $\mu_r \pm t_r$ is the expected position of the robot tip. The mean dimension of the mating parts is now denoted by μ_p with a bilateral tolerance of t_p. The standard deviations, if not given, can be calculated as σ_r and σ_p, respectively, for the robot tip and the part, such that

$$\sigma_r = \frac{t_r}{3} \quad \text{and} \quad \sigma_p = \frac{t_p}{3} \tag{1.9.1}$$

TABLE 1.9.1 AREAS UNDER THE STANDARD NORMAL DISTRIBUTION CURVE

$$A = \int_0^z \frac{1}{\sqrt{2\pi}} e^{-z^2/2}\, dz$$

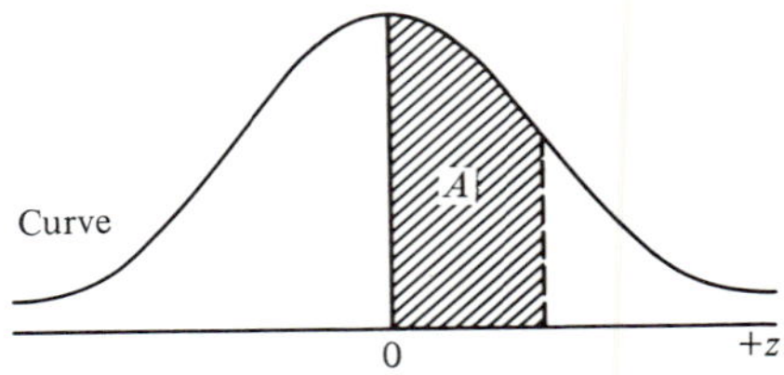

z	0	1	2	3	4	5	6	7	8	9
0.0	0.0000	0.0040	0.0080	0.0120	0.0160	0.0199	0.0239	0.0279	0.0319	0.0359
0.1	0.0398	0.0438	0.0478	0.0517	0.0557	0.0596	0.0636	0.0675	0.0714	0.0754
0.2	0.0793	0.0832	0.0871	0.0910	0.0948	0.0987	0.1026	0.1064	0.1103	0.1141
0.3	0.1179	0.1217	0.1255	0.1293	0.1331	0.1368	0.1406	0.1443	0.1480	0.1517
0.4	0.1554	0.1591	0.1628	0.1664	0.1700	0.1736	0.1772	0.1808	0.1844	0.1879
0.5	0.1915	0.1950	0.1985	0.2019	0.2054	0.2088	0.2123	0.2157	0.2190	0.2224
0.6	0.2258	0.2291	0.2324	0.2357	0.2389	0.2422	0.2454	0.2486	0.2518	0.2549
0.7	0.2580	0.2612	0.2642	0.2673	0.2704	0.2734	0.2764	0.2794	0.2823	0.2852
0.8	0.2881	0.2910	0.2939	0.2967	0.2996	0.3023	0.3051	0.3078	0.3106	0.3133
0.9	0.3159	0.3186	0.3212	0.3238	0.3264	0.3289	0.3315	0.3340	0.3365	0.3389
1.0	0.3413	0.3438	0.3461	0.3485	0.3508	0.3531	0.3554	0.3577	0.3599	0.3621
1.1	0.3643	0.3665	0.3686	0.3708	0.3729	0.3749	0.3770	0.3790	0.3810	0.3830
1.2	0.3849	0.3869	0.3888	0.3907	0.3925	0.3944	0.3962	0.3980	0.3997	0.4015
1.3	0.4032	0.4049	0.4066	0.4082	0.4099	0.4115	0.4131	0.4147	0.4162	0.4177
1.4	0.4192	0.4207	0.4222	0.4236	0.4251	0.4265	0.4279	0.4292	0.4306	0.4319
1.5	0.4332	0.4345	0.4357	0.4370	0.4382	0.4394	0.4406	0.4418	0.4429	0.4441
1.6	0.4452	0.4463	0.4474	0.4484	0.4495	0.4506	0.4515	0.4525	0.4535	0.4545
1.7	0.4554	0.4564	0.4573	0.4582	0.4591	0.4599	0.4608	0.4616	0.4625	0.4633
1.8	0.4641	0.4649	0.4656	0.4664	0.4671	0.4678	0.4686	0.4693	0.4699	0.4706
1.9	0.4713	0.4719	0.4726	0.4732	0.4738	0.4744	0.4750	0.4756	0.4761	0.4767
2.0	0.4772	0.4778	0.4783	0.4788	0.4793	0.4798	0.4803	0.4808	0.4812	0.4817
2.1	0.4821	0.4826	0.4830	0.4834	0.4838	0.4842	0.4846	0.4850	0.4854	0.4857
2.2	0.4861	0.4864	0.4868	0.4871	0.4875	0.4878	0.4881	0.4884	0.4887	0.4890
2.3	0.4893	0.4896	0.4898	0.4901	0.4904	0.4906	0.4909	0.4911	0.4913	0.4916
2.4	0.4918	0.4920	0.4922	0.4925	0.4927	0.4929	0.4931	0.4932	0.4934	0.4936
2.5	0.4938	0.4940	0.4941	0.4943	0.4945	0.4946	0.4948	0.4949	0.4951	0.4952
2.6	0.4953	0.4955	0.4956	0.4957	0.4959	0.4960	0.4961	0.4962	0.4963	0.4964
2.7	0.4965	0.4966	0.4967	0.4968	0.4969	0.4970	0.4971	0.4972	0.4973	0.4974
2.8	0.4974	0.4975	0.4976	0.4977	0.4977	0.4978	0.4979	0.4979	0.4980	0.4981
2.9	0.4981	0.4982	0.4982	0.4983	0.4984	0.4984	0.4985	0.4985	0.4986	0.4986
3.0	0.4987	0.4987	0.4987	0.4988	0.4988	0.4989	0.4989	0.4989	0.4990	0.4990
3.1	0.4990	0.4991	0.4991	0.4991	0.4992	0.4992	0.4992	0.4992	0.4993	0.4993
3.2	0.4993	0.4993	0.4994	0.4994	0.4994	0.4994	0.4994	0.4995	0.4995	0.4995
3.3	0.4995	0.4995	0.4995	0.4996	0.4996	0.4996	0.4996	0.4996	0.4996	0.4997
3.4	0.4997	0.4997	0.4997	0.4997	0.4997	0.4997	0.4997	0.4997	0.4997	0.4998
3.5	0.4998	0.4998	0.4998	0.4998	0.4998	0.4998	0.4998	0.4998	0.4998	0.4998
3.6	0.4998	0.4998	0.4999	0.4999	0.4999	0.4999	0.4999	0.4999	0.4999	0.4999
3.7	0.4999	0.4999	0.4999	0.4999	0.4999	0.4999	0.4999	0.4999	0.4999	0.4999
3.8	0.4999	0.4999	0.4999	0.4999	0.4999	0.4999	0.4999	0.4999	0.4999	0.4999
3.9	0.5000	0.5000	0.5000	0.5000	0.5000	0.5000	0.5000	0.5000	0.5000	0.5000

From Eq. (1.9.2), we can find the Z value corresponding to the desired area under the normal distribution curve; that is,

$$Z = \frac{\mu_r - \mu_p}{(\sigma_r^2 + \sigma_p^2)^{1/2}}. \tag{1.9.2}$$

Match this value of Z with the area A_z in Table 1.9.1. The probability of interference or percent rejection can be found by using $P = 0.5 - A_z$. If more than two parts are involved in assembling an item, follow the same procedure in obtaining the average dimensions of all parts that are to be fitted inside other parts. Find the corresponding standard deviations and proceed accordingly to find Z and consequently the percent rejection due to interference. The following examples clarify the situation.

EXAMPLE 1.9.1 ■

A robotic manipulator is inserting a round shaft inside a hole such that $d_s = 0.999 \pm 0.0005$ in. and $d_h = 1 \pm 0.001$ in. The total spatial resolution of the robot tip (end effector) is 0.0005 in. Determine the percent rejection due to interference if natural standard deviations are assumed.

Solution. The effective shaft diameter due to robotic inaccuracies is now

$$d_s^* = 0.99925 \pm 0.00075 \text{ in.}$$

The standard deviations of the shaft and robot and the hole are, respectively,

$$\sigma_s^* = \frac{0.00075}{3} = 0.00025 \quad \text{and} \quad \sigma_h = \frac{0.001}{3} = 0.00033$$

such that $(\sigma_s^{*2} + \sigma_h^2)^{1/2} = 0.00416$. Finally,

$$Z = \frac{1 - 0.99925}{0.000416} = 1.80$$

This corresponds (see Table 1.9.1) to $A_z = 0.4641$. Thus, $P = 0.5 - 0.4641 = 0.0359$ or the percent rejection is 3.6%. ■ ■

1.10 A BRIEF HISTORY OF ROBOTIC SOFTWARE

Early robotic manipulators or teleoperators were programmed either mechanically or electrically to repeat a task once they were taught how to perform it. The first robotic software for computer implementation was developed around 1973 at the Stanford Artificial Intelligence Laboratory (SAIL) and was an experimental research language (see Paul, 1977). Later, a more advanced robot programming language based on Algol software was developed at SAIL and was named AL (see Mujtaba and Goldman, 1981). This new software had the capability of programming assembly operations using force-sensing devices. IBM, in the meantime, was developing much more advanced robotic software among which were Emily, ML, and AML (a manufacturing language) (IBM, 1981, 1982).

Among the existing robotic software one may mention Help [97] by the General Electric Company (1981, 1982), which was developed to control the multiaxis Cartesian robots; JARS by the Robotics and Teleoperators Group of NASA's Jet Propulsion Laboratory, which was developed as high-level programming software for robotics control, MCL by McDonnell Douglas Corp. (Wright Patterson Air Force Base, 1981), which is an extension of the APT programming language for numerically controlled machines; RAIL by Automatix, Inc. (1982), which is used to control both the vision and motion of robot manipulators; PRL, by SRI International (Park, 1979), which is used to control assembly operations using multiple robots; and VAL by Unimation, Inc. (1981, 1983) which provides the ability to perform robotic tasks easily. For an excellent article on industrial robot programming languages and a comparative evaluation see Gruver et al. (1984). Accordingly, Basic is gaining popularity as a base robotic software and may emerge as the most popular robotic software in the future. We fully discuss the programming issues in Chapter 9.

BIBLIOGRAPHY

Agin, Gerald J. (1975). An Experimental Vision System for Industrial Applications, in *Proceedings of the 5th International Symposium on Industrial Robots*, pp. 135–148.

Automatic, Inc. (1982). *RAIL Software Reference Manual (ROBOVISION and CYBERVISION)*, Automatix Inc., Billerica, MA, Rev. 3.0, MN-RB-07 (Jan.).

Batur, Celal, and Kaya, Azmi (1981). Microprocessor Controlled Robot Arm, in *Proceedings of the 8th ASEE Conference, Education and Industry: A Joint Endeavor*, Los Angeles, Calif., June 21–25, Vol. 2, pp. 582–588.

Bejczy, Antal K. (1979). Manipulator Control Automation Using Smart Sensors, in *Electro 79 Conference Proceedings*, New York, New York, Apr. 24–26, pp. 32–38.

Birk, John R. (1976). Computation for Robots to Orient and Position Handheld Workpieces, *Trans. IEEE Systems, Man and Cybernetics*, **SMC-6**(10), 665–671.

Bolles, R. and Paul, R. P. (1973). "The Use of Sensory Feedback in a Programmable Assembly System," The Stanford Artificial Intelligence Laboratory, Stanford University, AIM-220 (Oct.).

Borrowman, Gerald L. (1974). Remote Manipulator Systems, *Spaceflight*, **21**(12), 495–496.

Boykin, W. H., and Warren, M. E. (1979). Application of Pattern Recognition to a Flying Robotic Arm, in *Proceedings of the 18th IEEE Conference of Decision and Control Including the Symposium on Adaptive Processes*, Fort Lauderdale, FL, Dec. 12–14, pp. 40–42, 1979.

Brady, M., Hollerbach, J. M., Johnson, T. L., Lozano-Perez, T., and Mason, M. T. (1983). *Robot Motion—Planning and Control*, M.I.T. Press, Cambridge, MA.

Briot, M. (1980). The Utilization of an "Artificial Skin" Sensor for the Identification of Solid Objects, in *Proceedings of the 9th International Symposium on Industrial Robots*, Washington, DC, Mar. 5–7, pp. 529–547, 1980.

Burckhardt, C. W., and Gerelle, E. G. R. (1980). Dynamic Design Parameters for Robot Arms: Experimental Results, in *Proceedings of the 10th International Symposium on Industrial Robots and 5th International Conference on Industrial Robot Technology*, Milan, Italy, March 5–7, pp. 321–329.

Cable Connectors Pave Way for Underwater Electric Power System. (1970). *Oil Gas J.*, **68**(39), 93–95.

Cain, R. A., and Bolles, R. C. (1982). Recognizing and Locating Partially Visible Objects: The Local-Feature-Focus Method, *Int. J. Robotics Res.*, **1**(3), 57–82 (Fall).

Camera, A., and Migliardi, G. F. (1980). Automated Inspection of Parts and Final Functional Control Performed During Assembly Process, in *Proceedings of the 5th International Conference on Automatic Inspection and Production Control and 12th IPA Arbeitstag*, Stuttgart, Germany, June 25–26, 1980, pp. 175–186.

Campbell, D. (1984). "An Iterative Algorithm for the Inverse Kinematics Solution of a General *N*-Axis Robot Manipulator Using Powell's Optimization Technique," M.Sc. thesis, Mechanical and Industrial Engineering Department, Clarkson University, Potsdam, NY, Aug. 1984; also MIE-Res. Rept. No. MIE-105 (1984).

Caporali, M. (1982). "Design and Bond Graph Dynamic Modeling of a Multi-Fingered Robot Hand," M.Sc. thesis, Mechanical Engineering Department, Clarkson College of Technology, Potsdam, NY, Aug. 1982; also MIE-Res. Rept. No. 089 (Aug.).

Caporali, M., and Shahinpoor, M. (1984). Design and Construction of a Five-Fingered Robotic Hand, *Robotics Age*, **6**(2), 14–20.

Cohen, J. (1966). *Human Robots in Myth and Sciences*, Allen and Unwin, London, England, 1966.

Cohen, C. L. (1985). Robots Get Smart in Japan, *Electronics Week*, Jan., pp. 15–16.

Cook, R. (1984). "Design and Modeling of a Three-Axis Robot Arm With an Ultrasonic Distance Sensor," M.Sc. thesis, Mechanical Engineering Department, Clarkson University, Potsdam, NY. May.

Craig, John J. (1980). JARS: JPL Autonomous Robot System, Robotics and Teleoperators Group, Jet Propulsion Laboratory, Pasadena, CA.

Darazan, P. J., and Jeffery, M. F. (1976). Microprocessor Control and Pneumatic Drive of a Manipulating Arm, in *Proceedings of the 3rd Conference on Industrial Robot Technology and 6th International Symposium on Industrial Robots*, University of Nottingham, England, March 24–26, pp. 86–96.

Demaurex, M. O., and Gerelle, E. G. R. (1980). Can I Build This Robot?, in *Proceedings of the 9th International Symposium on Industrial Robots*, Washington, DC, March 13–15, pp. 621–639.

Drake, Samuel H., Watson, Paul C., and Simunovic, Sergio N. (1977). "High Speed Robot Assembly of Precision Parts Using Compliance Instead of Sensory Feedback," in *Proceedings of the 7th International Symposium on Industrial Robots*, Tokyo, Japan, Oct. 19–21, pp. 87–98.

Ejiri, M., Uno, T., and Da, H. Y. (1972). A Prototype Intelligent Robot That Assembles Objects From Plane Drawings, *IEEE Trans. Comp.*, **C-21**(2), 199–207 (Feb.).

Engelberger, Joseph F. (1978). "Robot Arms for Assembly," ASME Paper, No. 78-WA/DSC-37.

Engelberger, J. F. (1980). *Robotics in Practice*, IFS Publications Ltd., Kempston, England.

Ernst, H. A. (1961). "A Computer Operated Mechanical Hand," Ph.D. thesis, M.I.T., Cambridge, MA.

Favareto, Marcello M. (1978). Polar 6000—A New General Purpose Robot Particularly Suited for Spot-Welding Applications, in *Proceedings of the 4th International Conference on Industrial Robot Technology*, Stuttgart, Germany, May 30–June 1, pp. 67–77.

Feldman, J., et al. (1971). The Use of Vision and Manipulation to Solve the Instant Insanity Puzzle, in *Proceedings of the 2nd International Joint Conference on Artificial Intelligence*, London, England, pp. 359–364.

Fichter, E. F., and McDowell, E. D. (1980). Novel Design for a Robot Arm, *Advances in Computer Technology*, presented at the *International Computer Technology Conference ASME Century 2—Emerging Technologies Conference*, Vol. 4, San Francisco, CA, August 12–15, pp. 250–256.

Flatau, Carl R. (1973). "Manipulator as a Means of Extending Our Dexterous Capabilities to Larger and Smaller Scales," in *Proceedings of the 21st Conference on Remote Systems and Technology*, San Francisco, CA, Nov. 12–13, pp. 47–50.

Fluidics Control Canadian Robot (1971). *Design Eng.*, **17**(11), 34–35.

Frank, Moshe (1980). "Three Wrist Roll Robot, SME Paper Ser. MS-80-699.

French, Douglas, and Yraets, William (1974). Actuating and Drive Mechanism Design for Automatic Assembly, in *Proceedings of the 11th Numerical Control Society, Annual Meeting and Technical Conference*, Toronto, Ontario, March 31–April 3, pp. 383–398.

General Electric Company (1981). *A12 Assembly Robot Maintenance and Installation Manual*, General Electric Co., Orlando, FL, P50VE005 (Dec.).

General Electric Company (1982). *Allegro Operator's Manual* (*A12 Assembly Robot*), General Electric Co., Orlando, FL, #P50VE025 (Feb.).

Gerelle, Eric G. R. (1978). Force Feedback Control, in *Proceedings of the 4th International Conference on Industrial Robot Technology*, Stuttgart, Germany, May 30–June 1, pp. 194–205.

Goertz, R. C. (1963). Manipulators Used for Handling Radioactive Materials, in E. M. Bennett (Ed.), *Human Factors in Technology*, Chap. 27, McGraw-Hill, New York.

Gruver, W. A., Soroka, B. I., Craig, J. J., and Turner, T. L. (1984). Industrial Robot Programming Languages: A Comparative Evaluation, *IEEE Trans. Systems, Man, and Cybernetics*, **SMC-14**(4), 565–570.

Hanafusa, Hideo, and Asada, Huruhiko (1977). Stable Prehension by a Robot Hand With Elastic Fingers, in *Proceedings of the 7th International Symposium on Industrial Robots*, Tokyo, Japan, Oct. 19–21, pp. 361–377.

Hanafusa, H., and Asada, H. (1976). Adaptive Control of a Robot Hand Equipped with Pneumatic Proximity Sensors, in *Proceedings of the 3rd Conference on Industrial Robot Technology and 6th International Symposium on Industrial Robots*, University of Nottingham, England, March 24–26, pp. D4,12P.

Hewitt, J. R. (1979). Decoupled Control of Robot Movement, *Electronics Lett.*, **15**(21), 670–671.

Hewitt, J. R. (1980). Dynamic Decoupled Control of Robot Arms, in *Proceedings of the IEEE Colloquium on Control of Manipulators and Robotic Devices*, London, April, England, p. 4.

Hirose, Shigeo, and Umetani, Yoji (1977). Development of Soft Gripper for the Versatile Robot Hand, in *Proceedings of the 7th International Symposium of Industrial Robots*, Tokyo, Japan, Oct. 19–21, pp. 353–360.

IBM (1981). *IBM Robot System/1: AML Reference Manual*, 2nd ed., IBM Corporation, Boca Raton, FL, SC34-0410-1 (Sept.).

IBM (1982). *IBM Robot System/1: General Information Manual and User's Guide*, 3rd ed., IBM Corporation, Boca Raton, FL, GA34-0180-2 (Mar.).

IEEE (1983). 3rd Short Course Via Satellite Transmission, "Robot Sensing and Intelligence," Nov. 7.

Inoue, H. (1971). Computer Controlled Bilateral Manipulator, *Bull. Jpn. Soc. Mech. Eng.*, **14**(69), 199–207.

Inove, Hirochika (1979). Force Feedback in Precise Assembly Tasks, in *Artificial Intelligence: An MIT Perspective*, Vol. 2, pp. 219–241, M.I.T. Press, Cambridge, MA.

Kato, H., Morinaga, S., Kato, T., and Inagoki, S. (1975). A New Integrated Robot Eye for Color Discrimination, in *Proceedings of 5th International Symposium on Industrial Robots*, Chicago, IL, pp. 135–148.

Kato, Ichiro (1983). Coming Robot New Era (A Viewpoint), *Robotica*, **1**, 9–13.

Kirsch, Jerry (1979). Robots and Their Advantage in Inspection, in *Proceedings of the SPIE Seminar, Vol. 170: Optimization in Quality Assurance 2*, Los Angeles, CA, Jan., pp. 40–42.

Kolpashnikov, Sergei, and Chelpanov, Igor (1981). Defining of the Scheme and the Parameters of an Industrial Robot Gripper by a Group of Specifications, in *Proceedings of the 11th International Symposium on Industrial Robots*, Tokyo, Japan, October 7–9, pp. 301–308.

Konstantinov, Michael S. (1975). Jaw-Type Gripper Mechanisms, SME Tech. Paper Ser. MS 1975, in *Proceedings of the 5th International Symposium on Industrial Robots*, Chicago, IL, Sept. 22–24, Book 2, Paper MS75-264.

Konstantinov, M. S., and Zankov, Z. I. (1975). Multi-grippers Hot Forge Manipulators, *Ind. Robots*, **2**, 47–55.

Krsnak, H., and Howe, M. (1975). Applications of High Performance Heavy Duty Industrial Manipulators, in *Proceedings of the 5th International Symposium on Industrial Robots*, Chicago, IL, Sept. 22–24, pp. 285–296.

Lieberman, L. I., and Lavin, M. S. (1982). AML/V: An Industrial Machine Vision Programming System, *Int. J. Robotics Res.*, **1**(3), 42–56 (Fall).

Loeff, Luis A., and Soni, A. H. (1974). "Algorithm for Computer Guidance of a Manipulator Between Obstacles," ASME Paper No. 74-DET-89.

Luh, J. Y. S., and Lin, C. S. (1981). Optimum Path Planning for Mechanical Manipulators, *Trans. ASME: J. Dynamic Systems, Measurement and Control*, **103**(2), 142–151.

Luh, J. Y. S., and Walker, M. W. (1977). Minimum Time Along the Path for a Mechanical Arm, in *Proceedings of the 16th IEEE Conference on Decision and Control, Including the Symposium on Adaptive Processes, and a Special Symposium on Fuzzy Set Theory and Applications*, New Orleans, LA, Dec. 7–9, 1977, pp. 755–759.

Lundstrom, Goran (1973). Industrial Robot Grippers, *Ind. Robot*, **1**(2), 72–82.

Mason, M. T., and Salisbury, J. K., Jr. (1985). *Robot Hands and the Mechanics of Manipulation*, M.I.T. Press, Cambridge, MA.

McCarthy, A. (1967). "Robotics Hearing with Spoken Commands," Stanford Artificial Intelligence Laboratory, Stanford University, AIM 56.

Mujtaba, Shahid, and Goldman, Ron (1981). *AL User's Manual*, 3rd ed., Computer Science Department, Stanford University, Palo Alto, CA. Rept. No. STAN-CS-81-889 (Dec.).

Murata, Hideto (1976). "Kinematic Design for Articulated Robot Arm," Engineering Faculty, Robe University, Japan, No. 22, pp. 83–92 (Mar.).

Murata, Hideto (1977). "Algorithm for Kinematic Control of Articulated Robot Arm," Engineering Faculty, Kobe University, Japan, Rept. No. 23, pp. 95–106.

Murata, Hideto, and Hashimoto, Masaharu (1979). Algorithm for Computing the Work Area of the Planar Articulated Robot Arm, Engineering Faculty, Kobe University, Japan, Rept. No. 25, pp. 43–51.

Nakano, Yoshiyuki, Fujie, Masakatsu, and Hosada, Yuji (1984). Hitachi's Robot Hand, *Robotics Age J. Intell. Mach.*, **5**(7), 18–20.

Okada, Tokuji, and Tsuchiya, Seiji (1977). On a Versatile Finger System, in *Proceedings of the 7th International Symposium on Industrial Robots*, Tokyo, Japan, Oct. 19–21, pp. 345–352.

Park, William T. (1981). The SRI Robot Programming System (RPS)—An Executive Summary, SRI International, Menlo Park, CA, Mar. 29, 1981 [see also How to Generate a New Version of the SRI Robot Programming System (RPS), SRI International, Menlo Park, CA, Aug. 5, 1979].

Paul, R. P. (1972). "Modeling Trajectory Calculation and Servoing of a Computer Controlled Arm," Stanford Artificial Intelligence Laboratory, Stanford University, AIM 177.

Paul, R. P. (1977). Wave: A Model-Based Language for Manipulator Control, *Ind. Robot*, **4**(1), 10–17 (1977).

Paul, R. P. (1983). *Robot Manipulators—Mathematical Programming and Control*, M.I.T. Press, Cambridge, MA.

Pieper, D. L. (1968). "The Kinematics of Manipulators Under Computer Control," Stanford Artificial Intelligence Laboratory, Stanford University, AIM 72.

Potter, Ronald (1975). Practical Application of a Limited Sequence Robot, in *Proceedings of the 5th International Symposium on Industrial Robots*, Chicago, IL, Sept. 22–24, pp. 55–64.

Raibert, Marc H. (1977). "Mechanical Arm Control Using a State Space Memory," SME Tech. Paper, Ser. Manuscript for Robots 2 Conference, Detroit, MI, Oct. 31–Nov. 3, 1977, Paper No. MS77-750.

Raibert, M. H., and Craig, J. J. (1981). Hybrid Position/Force Control of Manipulators, *Trans. ASME: J. Dynamic Systems, Measurement and Control*, **103**(2), 126–133.

Rakic, M. (1968). The Belgrade Hand Prosthesis, in *Proceedings of the Institution of Mechanical Engineers 1968–1969*, London, England, Vol. 183, Part 3J, pp. 60–67.

Reswick, A., and Meyler, A. (1962). "A Device for Quadriplegics to Handle Objects," M.I.T. Lincoln Laboratory, Res. Rept. No. 195.

Roberts, L. G. (1963). "Machine perception of Three-Dimensional Solids," M.I.T. Lincoln Laboratory, Res. Rept. No. 315.

Roberts, L. G. (1965). "Homogeneous Matrix Representation and Manipulation of *N*-Dimensional Constructs," M.I.T. Lincoln Laboratory, Document No. MS 1045.

Rosenburg, J. (1972). "A History of Numerical Control 1949–1972: The Technical Development, Transfer to Industry, and Assimilation," U.S.C. Information Sciences Institute, Marina Del Ray, CA, Report No. ISI-RR-72-3.

Rovetta, Alberto (1977). On Specific Problems of Design of Multipurpose Mechanical Hands in Industrial Robots, in *Proceedings of the 7th International Symposium on Industrial Robots*, Tokyo, Japan, Oct. 19–21, pp. 337–343.

Rovetta, A. (1980). "On Functionality of a New Mechanical Hand," ASME Paper, No. 80-C2-DET-124 (1980).

Sato, N., Heginbothan, W. B., and Pugh, A. (1977). A Method for Three-Dimensional Part Identification by a Tactile Transducer, in *Proceedings of the 7th International Symposium on Industrial Robots*, Tokyo, Japan, Oct. 19–21, pp. 577–585.

Shahinpoor, M. (1984). *A Robot Engineering Textbook*, University of New Mexico Press, Albuquerque.

Shahinpoor, M., Jamshidi, M., and Kim, Y. T. (1987). Exact Solutions for the Inverse Kinematics Problem of a Standard 6-Axis Robot Manipulator, *J. Robotic Systems*, **3**(1), 12–21.

Skinner, Frank (1974). "Design of a Multiple Prehension Manipulator System," ASME Paper No. 74-DET-25.

Sugimoto, K., and Duffy, J. (1980a). "Determination of Extreme Distances of a Robot Hand—1. A General Theory," ASME Paper No. 80-DET-57 (1980).

Sugimoto, K., and Duffy, J. (1980b). "Determination of Extreme Distances of a Robot Hand—2. Robot Arms with Special Geometry," ASME Paper No. 80-DET-59 (1980).

Surveys Reveal Robot Populations and Trends (1982). *Robotics Today*, Feb., pp. 79–80.

Tabesh, A. (1984). "The Design, Modeling, and Construction of a Gripping Device for the GE Model P50 Robot," M.Sc. thesis, Mechanical and Industrial Engineering Department, Clarkson University, Potsdam, NY (Dec.).

Takegaki, Morikazu, and Arimoto, Suguru (1981). A New Feedback Method for Dynamic Control of Manipulators, *Trans. ASME: J. Dynamic Systems, Measurement and Control*, **103**(2), 126–133.

Tomoric, A., and Boni, A. (1962). "Development of a Hand With Pressure Sensors," M.I.T. Lincoln Laboratory, Res. Rept. No. 192.

Tsai, Y. C., and Soni, A. H. (1980). "Accessible Regions and Synthesis of Robot Arms," ASME Paper No. 80-Det-101.

Unimation, Inc. (1980). *User's Guide to VAL*, version 12, Unimation, Inc., Danbury, CT, 398-H2A (June).

Unimation, Inc. (1981). *VAL Univision Supplement*, version 13, VSNO, 2nd ed., Unimation, Inc., Danbury, CT (July).

Unimation, Inc. (1983). *User's Guide to VAL-II, Part 1: Control from the System Terminal*, version X2 (April); *Part 2: Communications with a Supervisory System*, version X2, Apr.; *Part 3: Real-Time Path Control*, version X2 (April) Unimation, Inc., Danbury, CT.

Vukobratovic, M., and Kircanski, M. (1982). A Method for Optimal Synthesis of Manipulation Robot Trajectories, *Trans. ASME: J. Dynamic Systems, Measurement and Control*, **104**(2), 188–193.

Vukobratovic, M., and Stokic, D. (1981). One Engineering Concept of Control of Manipulators, *Trans. ASME: J. Dynamic Systems, Measurement and Control*, **103**(2), 119–125.

Wichman, M. W. (1967). "The Use of Optical Feedback in Computer Control of an Arm," Stanford Artificial Intelligence Laboratory, Stanford University, AIM 56.

Will, A. (1973). "Robot Assembler For Type Writers," IBM Thomas J. Watson Research Center, Rept. No. 37.

Wright Patterson Air Force Base (1981). "Robotic System for Aerospace Batch Manufacturing," *Task B—High Level Language User's Manual*, Wright Patterson Air Force Base, OH, Tech. Rept. AFML-JR-79-4202 (Oct.).

Zeldman, M. I. (1984). *What Every Engineer Should Know About Robots*, Dekker, New York.

HOMEWORK PROBLEMS

1. A two-axis Cartesian robot has a reach of 25 in. in the z direction and a reach of 30 in. in the y direction. The robot controller is sending 8 bit D/A and A/D signals (DACS) for control and feedback. Determine the overall resolution and accuracy of this robot (see Fig. P1.1).

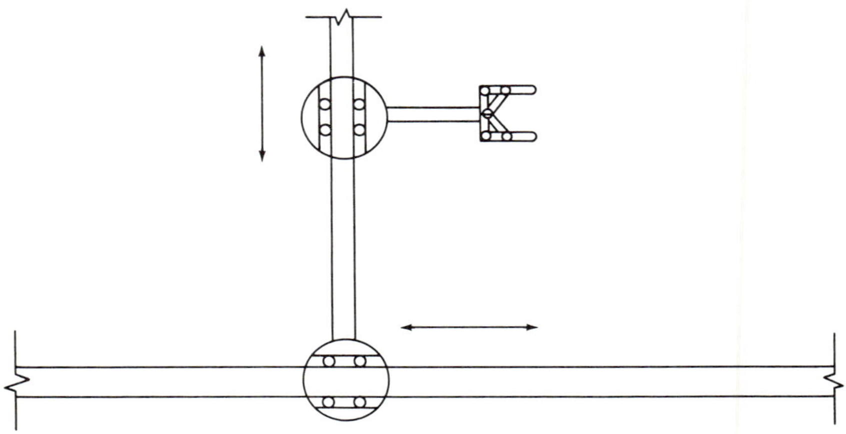

Figure P1.1 Two-axis Cartesian robotic manipulator.

2. Using the details of Problem 1, assume that the gripper is capable of finger motion parallel to the y direction with a range of 3 in. If the gripper motor is controlled by the same controller, determine the worst accuracy in a pick-and-place operation.
3. In Problems 1 and 2, assume that the robot gripper finger motion is perpendicular to both the x and y directions, that is, motion is in the z direction, with xyz as a rectangular Cartesian coordinate system. Determine the worst accuracy that may result in a pick-and-place assembly process using this rectangular robot.
4. Describe at least five different types of work environment where a tensor-arm robot manipulator may be used.
5. Discuss the possibility of moving and controlling the robotic joint by a direct-drive mechanism (motor) involving the rotor, stator, and ball bearings.
6. Discuss the possibility of moving and controlling a large number of joints in a snakelike (tensor-arm) robotic manipulator by means of direct-drive magnetic actuation.
7. An xyz rectangular Cartesian robotic manipulator has reach spans of 40, 38, and 42 in. in the x, y, and z directions, respectively. The robot controller is sending 10 bit

D/A (DACS) command signals for control and feedback. Determine the worst resolution and the accuracy of this robot in assembly processes (see Fig. P1.7).

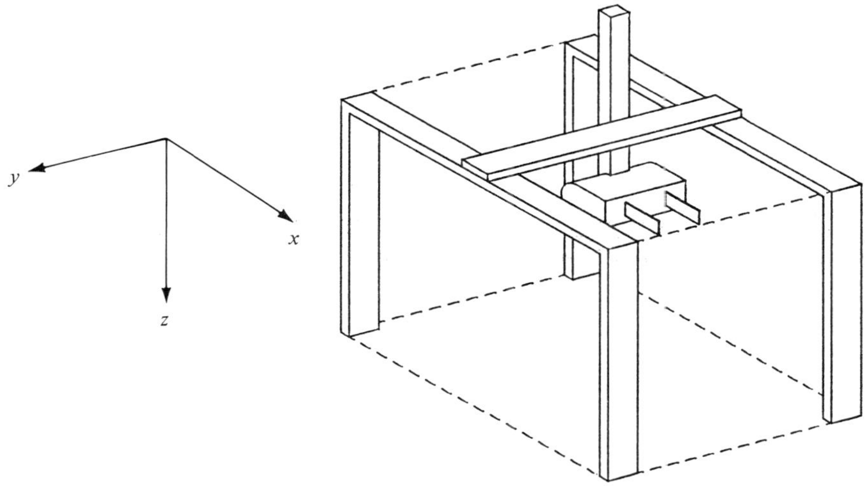

Figure P1.7 Three-axis rectangular Cartesian robotic manipulator.

8. In Problem 7, consider joint A to be a cylindrical prismatic joint such that the x arm can rotate around and slide along the y axis. Thus, one has a partly rectangular, partly cylindrical polar robot. If the same controller command DACS (10 bit) are at work, determine the overall resolution and accuracy of this robot in a pick-and-place operation.

9. Show by means of a worked out example of a single-axis rotary joint robot that prismatic robots have better resolutions and accuracies at the tool tip in assembly processes than revolute robots when other influences such as mechanical imperfections and vibrations are ignored (see Fig. P1.9).

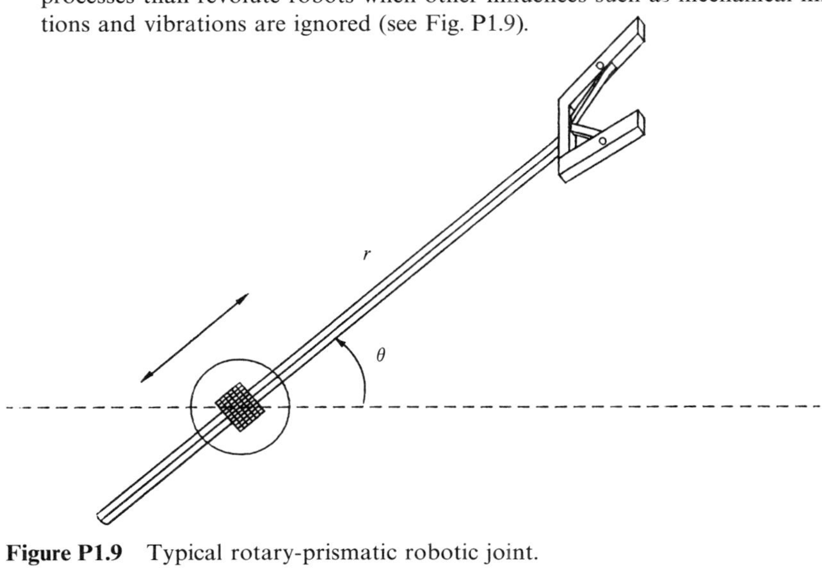

Figure P1.9 Typical rotary-prismatic robotic joint.

10. The accuracy of a tool tip approach for a welder robot is determined to be 0.0586 in. If the robot's end effector is 30 in. away from the welding spot at the end of each cycle of welding, determine how many bits of DACS for positional control and feedback at the tip are desired. How can one improve repeatability to less than 0.015 in.? (See Fig. P1.10.)

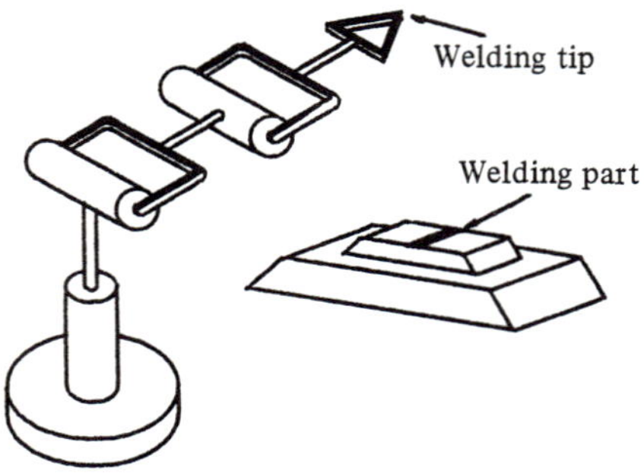

Figure P1.10 Typical robot welding operation.

11. A spherical robot arm has a prismatic arm that can expand from a minimum radius of 3 in. to a maximum radius of 4 ft. The spherical angles θ and ϕ are constrained to lie within $(0, 3\pi/4)$ and $(0, 2\pi)$, respectively. If 12 bit DACS are used to send position commands, determine

(a) the worst resolution and

(b) the worst position accuracy. (See Fig. P1.11.)

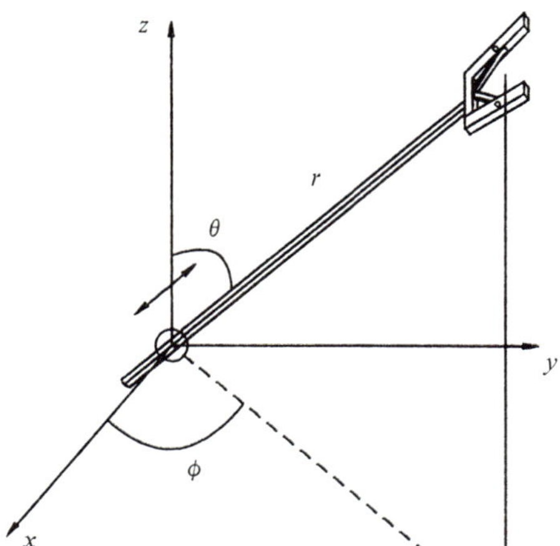

Figure P1.11 Spherical robotic arm.

12. A robot is programmed by means of control 12 bit DACS to insert shafts of $d_s = 10 \pm 0.008$ in. inside holes of $d_h = 10^{+0.008}_{-0.000}$ in. Assuming natural standard deviations, that is, 1/3 of tolerances, determine the percentage of parts that are to be rejected due to interferences in assembly processes. The robot is programmed to reject all such parts that do interfere. The span of robot gripper motion is 15 in. (see Fig. P1.12).

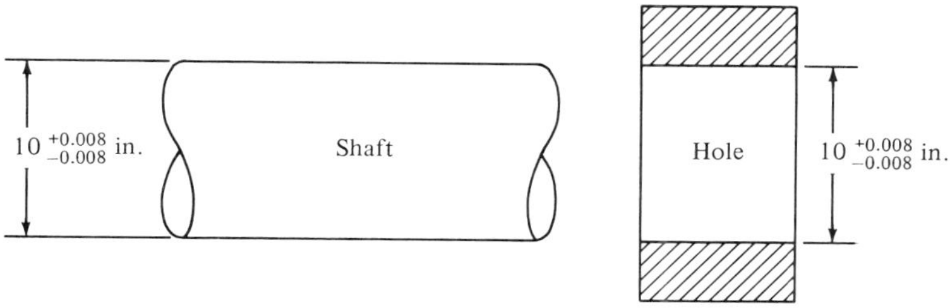

Figure P1.12

13. A robotic manipulator is mating parts A into parts B. It picks up parts A from a stack and moves them horizontally and then vertically down to fit into parts B which have come to a stop momentarily (Fig. P1.13). The arm is receiving 12 bit DACS for position feedback and control. Assuming natural standard deviations, that is, 1/3 of tolerances, determine the percentage of parts that are to be rejected due to interference. The robot is instructed to reject all parts that do interfere.

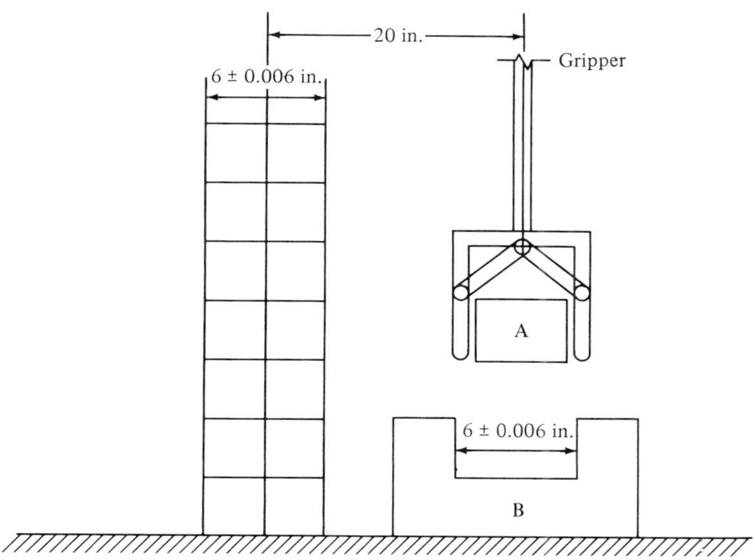

Figure P1.13 Robotic operation for mating parts.

14. Two robotic manipulators work together for a pick-and-place manufacturing task in conjunction with a single conveyer belt. The conveyer motor receives 10 bit DACS for speed control while traversing a speed range of 0–40 cm/s. Referring to Figure P1.14, note that the first robot (A) places parts on the belt while the second robot (B) picks them up and carries them to another location. The robot arms are receiving 12 bit DACS for position and 10 bit DACS for speed control. The ranges of movement

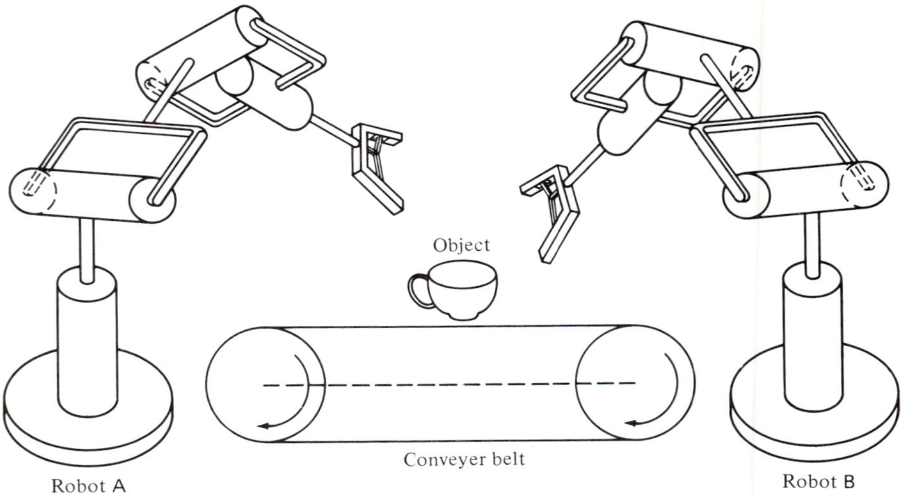

Figure P1.14 Robotic workstation with a conveyor belt.

and speed for robot A are 23 cm and 40 cm/s and for robot B are 28 cm and 5 cm/s, respectively. Determine the overall resolution of this automation process in both speed and positional manipulation.

15. In Figure P1.15, the cup is located in a frame with a transform **C** from the reference frame *xyz*. The robot base is located at transform **Z** from the *xyz* frame. The tip of the

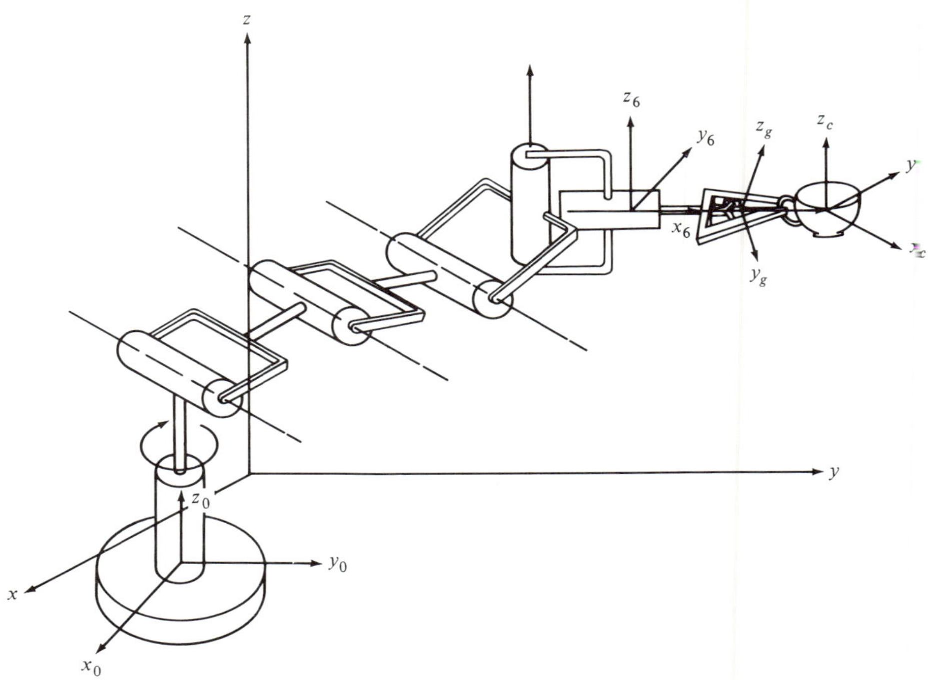

Figure P1.15

robot, where it is attached to the gripper, is at a transform $^{T_6}\mathbf{G}$ with respect to the T_6 frame. Determine the kinematic attitude of the cup with respect to the gripper, that is, $^{G}\mathbf{C}$ if

$$\mathbf{Z} = \begin{bmatrix} 0 & -1 & 0 & 12 \\ 1 & 0 & 0 & 14 \\ 0 & 0 & 1 & 1 \\ 0 & 0 & 0 & 1 \end{bmatrix}; \quad {}^{Z}\mathbf{T}_6 = \begin{bmatrix} 0 & 1 & 0 & 5 \\ -1 & 0 & 0 & 9 \\ 0 & 0 & 1 & 9 \\ 0 & 0 & 0 & 1 \end{bmatrix}$$

$$^{T_6}\mathbf{G} = \begin{bmatrix} 0 & -1 & 0 & 0 \\ 1 & 0 & 0 & 4 \\ 0 & 0 & 1 & 0 \\ 0 & 0 & 0 & 1 \end{bmatrix}; \quad \mathbf{C} = \begin{bmatrix} 1 & 0 & 0 & 11 \\ 0 & 1 & 0 & 25 \\ 0 & 0 & 1 & 32 \\ 0 & 0 & 0 & 1 \end{bmatrix}$$

16. Referring to Figure P1.16, determine the worst positional accuracy of the pick-and-place operation if all motors receive 12 bit DACS for position control and the camera receives 8 bit DACS for image recognition. Take all angular ranges of rotation as 2π radians.

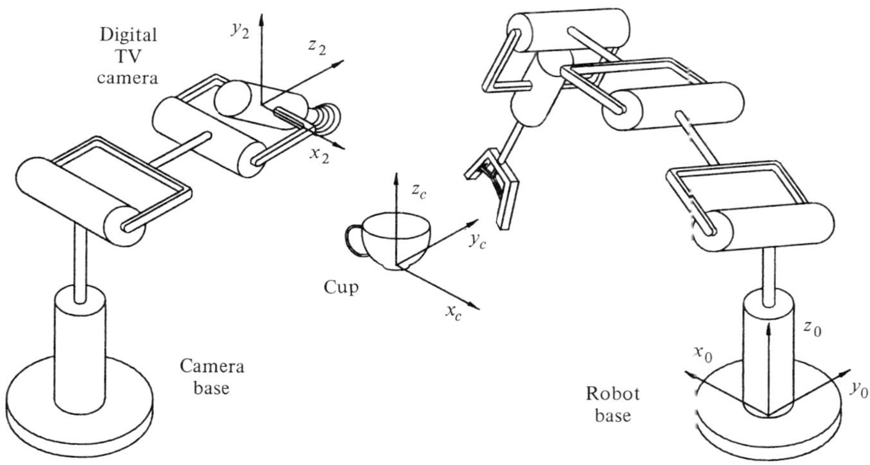

Figure P1.16 Camera-based robotic operation.

Chapter 2

Kinematics of Robotic Manipulators

2.1 INTRODUCTION

In this chapter we review the science of motion of objects regardless of the forces that generate the motion. We then apply our knowledge to robotic manipulators and the objects to be manipulated. By definition this treatment is called the kinematics of robotic manipulation, which includes the kinematics of robots and the associated objects, tools, and frames of references. Kinematics is probably the most important area of robot engineering because robotic manipulation can be achieved only by the motion of the robotic hand and associated parts, tools, and object in a Euclidean three-dimensional space. Such operations involve positioning (translation) and orientation (general rotation) of the robotic hand (gripper) and the manipulation of objects, tools, and parts. Traditionally, kinematics can be described completely by the translation and the rotation of coordinate frames attached to the tools, parts, objects, robotic hand, and other parts of the body of a robot.

We first describe the structure of a typical robotic manipulator and then introduce the mathematical notations used throughout this chapter and in later chapters. We then describe some mathematical preliminaries on vectors and matrices and discuss the applications of 4×4 and $n \times 4$ homogeneous transformations in robotic kinematics and dynamics. Subsequently, these transformations are used to describe the general position and orientation, that is, kinematic attitude, of one robotic link with respect to an adjacent robotic link. Thus, the notion of Denavit–Hartenberg homogeneous transformations are introduced and the pertinent equations are derived. Then the two problems of *direct* and *inverse* kinematics in robotics are discussed fully and many relevant problems are solved for a great variety of robotic manipulators.

2.2 GENERAL DESCRIPTION OF ROBOTIC MANIPULATORS

Kinematically, a robotic manipulator can be considered to be an open-loop articulative chain of links connected in series by either prismatic (sliding) or revolute (rotary) joints. Generally, one end is relatively free and commonly attached to a specialized tool such as a multifingered hand or gripper. The other end of the chain is attached to a supporting base that is generally stationary. It is clear that the location and orientation of the robotic hand is the result of the collective effect of translation and rotation of each joint of a robotic manipulator chain of links. Such joint motions are generated by actuators that are either electrical, hydraulic, or pneumatic. The joint actuators either are locally active at the joints or produce action at a distance by belts, pulleys, sheaves, chains, sprockets, or hydraulic fluid lines.

The central issue in industrial robotic manipulation is the ability to position the robotic hand at a specified location with a specified orientation at a given time. This issue lies at the heart of flexible and automated manufacturing. Figure 2.2.1 depicts a standard six-axis robotic manipulator possessing a base (waist), shoulder, elbow, and wrist capable of pitch, roll, and yaw motions as well as a gripping motion.

The position and orientation of the robotic end effector can be described in terms of the position and orientation of a coordinate frame attached to the end effector with respect to an inertial reference coordinate frame. Generally, this reference frame is fixed at the base of the robotic manipulator as shown in Figure 2.2.2.

The combination of the spatial location and orientation of the end effector (kinematic attitude) can be described mathematically by means of 4×4 homogeneous transformations. These homogeneous transformations are used to solve both the direct and inverse kinematics problems of robotic manipulation. The *direct kinematics* problem in robotics is to find the kinematic attitude given the vector of the joint displacements, $\boldsymbol{\theta}_i = [\theta_1\ \theta_2\ \theta_3 \cdots \theta_n]^T$, for a n-axis robotic manipulator. On the other hand, the *inverse kinematics* problem for robotic manipulators involves finding the vector of the joint displacements θ_i, $i = 1, 2, \ldots, N$, for n-axis robotic manipulators given the kinematic attitude of the gripper with respect to the base coodinates. Normally, the latter is the most desirable problem to solve because it is at the heart of flexible automated manufacturing operations and assembly processes. On the other hand, the direct kinematics solutions are used for special applications such as direct trajectory planning or obstacle collision prevention algorithms. The *kinematics attitude* of the end effector generally is given by a 4×4 matrix $\mathbf{T}$ that comprises a position vector $\mathbf{p}$ of the origin of the end effector coordinate frame as well as a 3×3 rotation matrix $\mathbf{R}$ that comprises three column vectors, $\mathbf{n}$, $\mathbf{o}$, and $\mathbf{a}$, which are, respectively, the normal, sliding, and approach vectors of the end effector (see Figs. 2.2.3 and 2.2.4).

Since the position and orientation of the end effector can be described by means of three coordinates x, y, and z and three Euler angles θ, ϕ, and ψ, a world

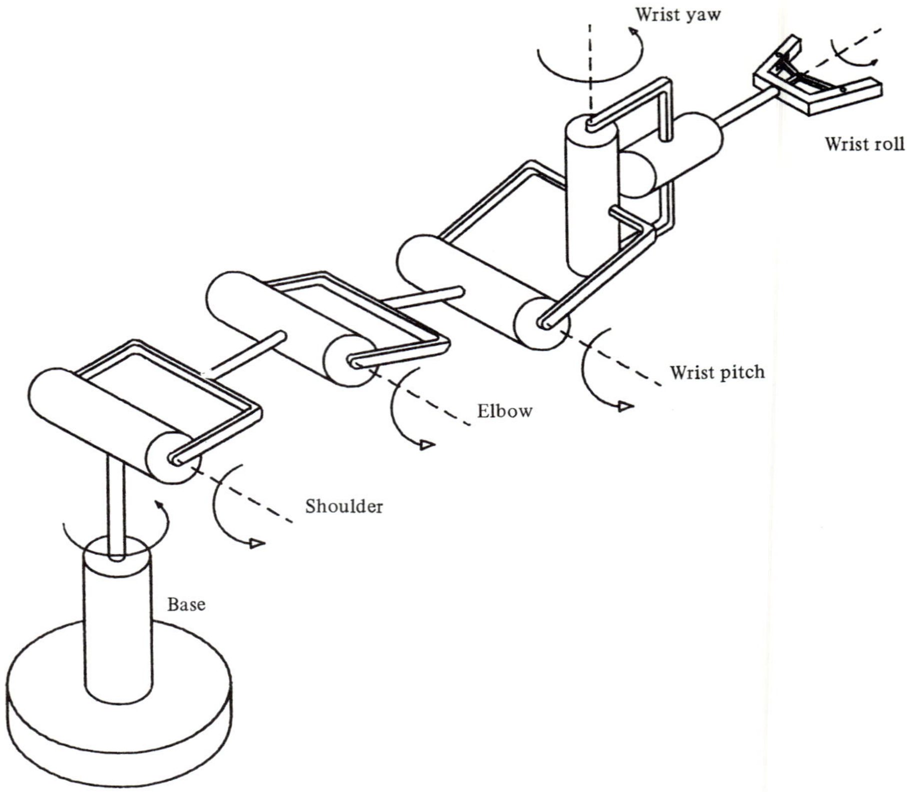

Figure 2.2.1 Typical six-axis revolute robotic manipulator.

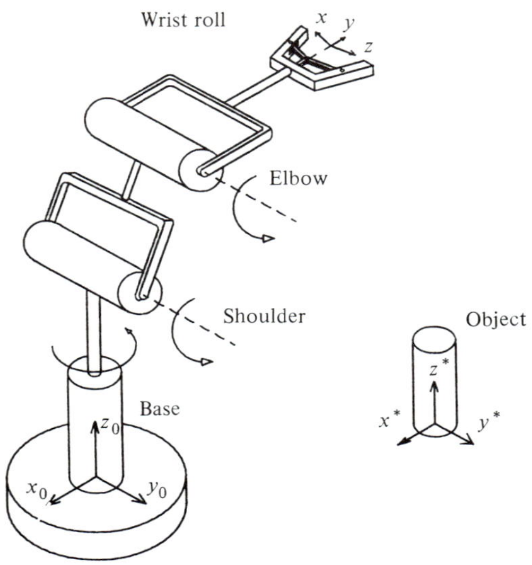

Figure 2.2.2 Relative configuration of the gripper and base coordinate frame in a robotic manipulator.

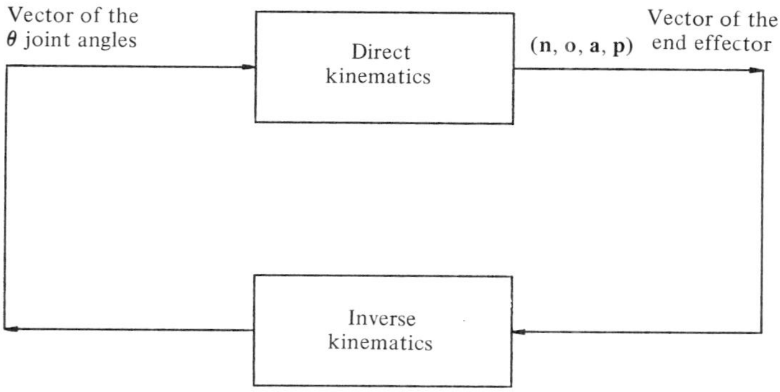

Figure 2.2.3 Direct and inverse kinematics relationship in robotic manipulators.

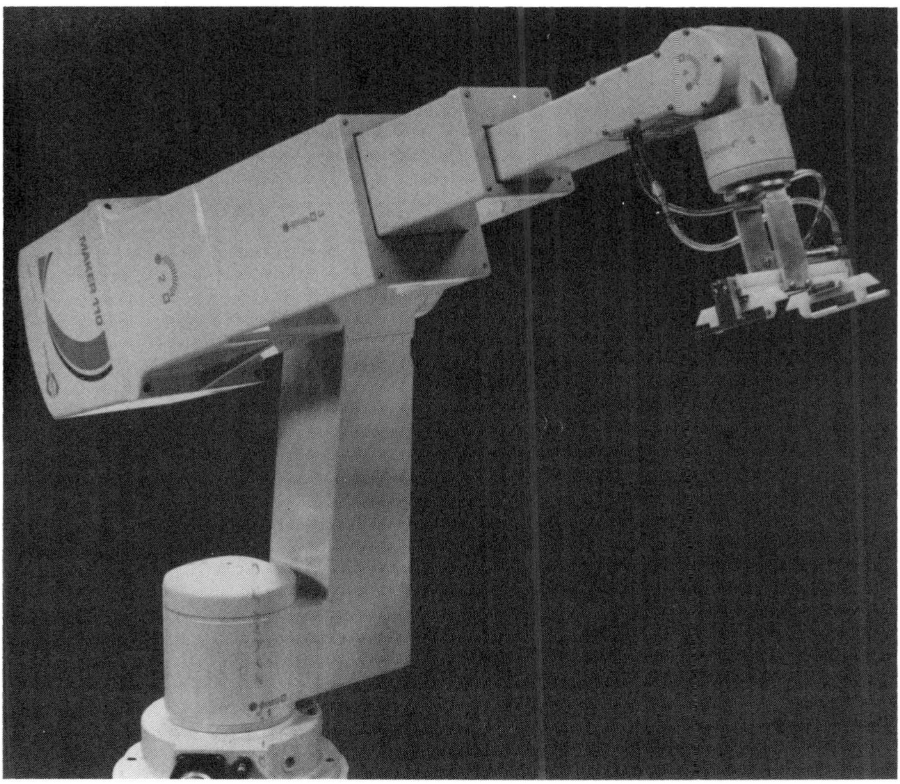

Figure 2.2.4 US Robots Maker 110. (Courtesy of U.S. Robots.)

coordinate can be defined by means of a fixed and a rotational set of coordinate systems.

To solve the direct and inverse kinematics problems for a robotic manipulator, we first review vectors and matrix algebra. We start with a brief description of the mathematical notation used in this book.

2.3 A BRIEF DESCRIPTION OF MATHEMATICAL NOTATIONS

Conventional notation for the representations of vectors, scalars, and matrices are used throughout the text. Presuperscripts denote descriptive information regarding coordinate frames. For example, ${}^{A}\mathbf{r}$ is a position vector described with respect to frame **A**. Frame **A** itself is represented by a 4×4 homogeneous transformation that describes the location of its origin and relative rotation with respect to a universal inertial reference frame. Sometimes an ith frame is denoted by $\mathbf{A}_i$ and sometimes a frame i described with respect to frame j is denoted by $\mathbf{A}_j^i$. Compact notation is also used for trigonometric functions, such as

$$C_{ijk} = \cos(\theta_i + \theta_j + \theta_k) \quad \text{and} \quad S_{ijk} = \sin(\theta_i + \theta_j + \theta_k). \tag{2.3.1}$$

Transformations generally are denoted by **T** while homogeneous transformations are given by **H**. Rotational transformations are represented by a 3×3 matrix **R** and translational transformations are denoted by **Trans**. The Euler and roll–pitch–yaw transformations are denoted by **Euler**(ϕ, ψ, θ) and **RPY**(ϕ, ψ, θ), respectively. Generally, the objects are described by their first letter denoting a $4 \times n$ homogeneous transformation; that is, a camera is denoted by **C** as a $4 \times n$ transformation.

2.4 MATHEMATICAL PRELIMINARIES ON VECTORS AND MATRICES

We consider $\mathbf{x}$, $\mathbf{v}$, and $\mathbf{u}$, as vectors, ρ, Q, and L as planes, and **X**, **A**, and **CONV** as coordinate frames. Two vectors, $\mathbf{v}$ with respect to the base coordinate and $\mathbf{u}$ with respect to a hand coordinate, can describe the same point p in three-dimensional space but basically are different vectors as shown in Fig. 2.4.1. In fact, point p could belong to an object being held by the robotic hand while its location with respect to both the hand coordinate (moving) and the base coordinate (fixed) are important for computational purposes and control. Note that

$$ {}^{\text{Base}}\mathbf{v} \equiv {}^{B}\mathbf{v} \neq {}^{H}\mathbf{W} = {}^{\text{Hand}}\mathbf{W}. \tag{2.4.1}$$

${}^{B}\mathbf{v}$ and ${}^{H}\mathbf{W}$ are different vectors describing the same point.

Right-hand rules and right-handed coordinate frames are used throughout the text.

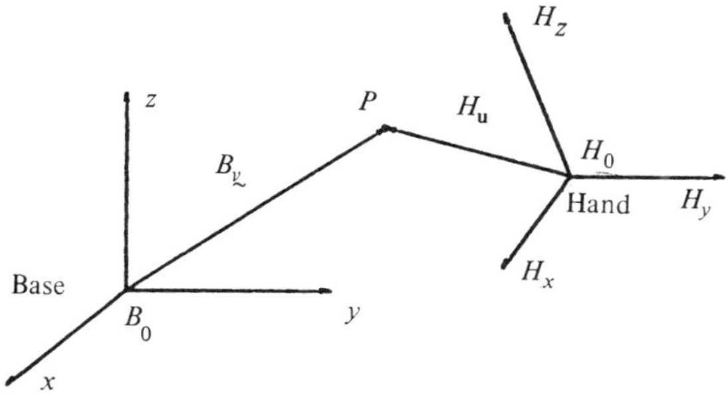

Figure 2.4.1 Description of a point p with respect to two coordinate systems—Hand and Base.

For vectors we use standard notation; for example,

$$\mathbf{a} \equiv a_i \equiv [a_1 \quad a_2 \quad a_3]^T \tag{2.4.2}$$

with respect to a rectangular Cartesian coordinate system $x_i \equiv [x_1\ x_2\ x_3] \equiv [x\ y\ z]$. If the unit vectors along the x, y, z axes are, respectively, $(\mathbf{e}_1, \mathbf{e}_2, \mathbf{e}_3) \equiv (\mathbf{i}, \mathbf{j}, \mathbf{k})$, then

$$\mathbf{a} = \sum_{i=1}^{3} a_i \mathbf{e}_i = a_i \mathbf{e}_i \tag{2.4.3}$$

where the Einstein summation notation is applied to the repeated index i and is used throughout the text. The vector dot product is defined such that

$$\mathbf{a} \cdot \mathbf{b} = \sum_{i=1}^{3} a_i b_i \equiv a_i b_i \equiv a_1 b_1 + a_2 b_2 + a_3 b_3 \tag{2.4.4}$$

The vector cross product is defined such that

$$\mathbf{a} \times \mathbf{b} = \varepsilon_{ijk} a_j b_k \mathbf{e}_i \tag{2.4.5}$$

where ε_{ijk} is the standard permutation symbol defined such that

$$\varepsilon_{ijk} \equiv \begin{cases} 1 & \text{if } i, j, k \text{ form an even permutation of 1, 2, 3,} \\ -1 & \text{if } i, j, k \text{ form an odd permutation of 1, 2, 3,} \\ 0 & \text{if any index } i, j, \text{ or } k \text{ is repeated.} \end{cases} \tag{2.4.6}$$

The determinant is defined such that for a 3×3 matrix $\mathbf{A} = a_{ij}$, $i, j = 1, 2, 3$,

$$\det \mathbf{A} = \varepsilon_{ijk} a_{i1} a_{j2} a_{k3}. \tag{2.4.7}$$

Furthermore, note that

$$\mathbf{a} \times \mathbf{b} = \begin{vmatrix} \mathbf{e}_1 & \mathbf{e}_2 & \mathbf{e}_3 \\ a_1 & a_2 & a_3 \\ b_1 & b_2 & b_3 \end{vmatrix} \tag{2.4.8}$$

The volume formed by a parallelepiped whose three principal sides are given by **a**, **b**, and **c** is given by

$$V = |\mathbf{a} \cdot (\mathbf{b} \times \mathbf{c})| \tag{2.4.9}$$

where | | stands for the absolute value. The following vector identities are used throughout the text:

$$\mathbf{a} \times (\mathbf{b} \times \mathbf{c}) = \mathbf{b}(\mathbf{a} \cdot \mathbf{c}) - \mathbf{c}(\mathbf{a} \cdot \mathbf{b}), \tag{2.4.10}$$

$$(\mathbf{a} \times \mathbf{b}) \times \mathbf{c} = \mathbf{b}(\mathbf{a} \cdot \mathbf{c}) - \mathbf{a}(\mathbf{b} \cdot \mathbf{c}), \tag{2.4.11}$$

$$(\mathbf{a} \times \mathbf{b}) \cdot (\mathbf{c} \times \mathbf{d}) = (\mathbf{a} \cdot \mathbf{c})(\mathbf{b} \cdot \mathbf{d}) - (\mathbf{a} \cdot \mathbf{d})(\mathbf{b} \cdot \mathbf{c}), \tag{2.4.12}$$

$$(\mathbf{a} \times \mathbf{b}) \times (\mathbf{c} \times \mathbf{d}) = \mathbf{b}[\mathbf{a} \cdot (\mathbf{c} \times \mathbf{d})] - \mathbf{a}[\mathbf{b} \cdot (\mathbf{c} \times \mathbf{d})], \tag{2.4.13}$$

$$(\mathbf{a} \times \mathbf{b}) \times (\mathbf{c} \times \mathbf{d}) = \mathbf{b}[\mathbf{a} \cdot (\mathbf{c} \times \mathbf{d})] - \mathbf{a}[\mathbf{b} \cdot (\mathbf{c} \times \mathbf{d})]. \tag{2.4.14}$$

Two-dimensional vectors or matrices can be treated in a similar manner. For example, if **A** is a matrix and **a** is a vector, then the product is a vector **b** such that

$$\mathbf{b} = \mathbf{A}\mathbf{a} = \begin{bmatrix} A_{11} & A_{12} & A_{13} \\ A_{21} & A_{22} & A_{23} \\ A_{31} & A_{32} & A_{33} \end{bmatrix} \begin{bmatrix} a_1 \\ a_2 \\ a_3 \end{bmatrix} = \begin{bmatrix} b_1 \\ b_2 \\ b_3 \end{bmatrix}. \tag{2.4.15}$$

EXAMPLE 2.4.1. ■

If $\mathbf{a} \equiv [-1\ 0\ 2]^T$ and

$$\mathbf{A} = \begin{bmatrix} 1 & 3 & 0 \\ 2 & 1 & -1 \\ 4 & 1 & 0 \end{bmatrix},$$

find the product **b**.

Solution.

$$\mathbf{b} = \mathbf{A}\mathbf{a} = \begin{bmatrix} -1 \\ -4 \\ -4 \end{bmatrix} = \begin{bmatrix} 1 & 3 & 0 \\ 2 & 1 & -1 \\ 4 & 1 & 0 \end{bmatrix} \begin{bmatrix} -1 \\ 0 \\ 2 \end{bmatrix}.$$ ■■

Note that $\mathbf{b} \equiv [b_1\ b_2\ b_3]$, where

$$\begin{aligned} b_1 &= A_{11}a_1 + A_{12}a_2 + A_{13}a_3, \\ b_2 &= A_{21}a_1 + A_{22}a_2 + A_{23}a_3, \\ b_3 &= A_{31}a_1 + A_{32}a_2 + A_{33}a_3. \end{aligned} \tag{2.4.16}$$

Using the Einstein summation convention, we may write

$$b_i = A_{ij}a_j, \qquad i, j = 1, 2, 3. \tag{2.4.17}$$

If **A** is denoted by A_{ij}, where $i, j = 1, 2, \ldots, n$, then A_{ij} is an $n \times n$ matrix of rank 2. Similarly, $A_{i_1, i_2, i_3, \ldots, i_m}$, where $i_1, i_2, \ldots, i_m = 1, 2, \ldots, n$ is an $n \times n$ dimensional matrix of rank m. The transpose of a matrix A_{ij} is equal to A_{ji} such that the rows and columns change their roles. Therefore,

$$A_{ij}^T = A_{ji} = \begin{bmatrix} A_{11} & A_{21} & A_{31} \\ A_{12} & A_{22} & A_{32} \\ A_{13} & A_{23} & A_{33} \end{bmatrix}. \tag{2.4.18}$$

EXAMPLE 2.4.2 ■

Find the determinant of the transpose of a matrix A_{ij},

$$A_{ij} = \begin{bmatrix} 1 & 2 & 3 \\ 4 & 5 & 6 \\ 7 & 8 & 9 \end{bmatrix}. \tag{2.4.19}$$

Solution. Note that

$$A_{ij}^T = A_{ji} = \begin{bmatrix} 1 & 4 & 7 \\ 2 & 5 & 8 \\ 3 & 6 & 9 \end{bmatrix} \tag{2.4.20}$$

and thus

$$\det A_{ji} = (5 \times 9 - 8 \times 6) - 4(2 \times 9 - 8 \times 3) + 7(2 \times 6 - 5 \times 3)$$

or

$$\det A_{ji} = -3 + 24 - 21 = 0. \quad ■■ \tag{2.4.21}$$

A matrix whose determinant is equal to zero is called a singular matrix. The determinant of $n \times n$ matrix A_{ij} is given by

$$\det A_{ij} = \varepsilon_{i_1 i_2 i_3 \cdots i_n} A_{i_1 1} A_{i_2 2} A_{i_3 3} \cdots A_{i_n n}, \tag{2.4.22}$$

where $\varepsilon_{i_1 i_2 i_3 \cdots i_n}$ is the generalized permutation symbol defined similar to expression (2.4.6). Note that the inverse of a matrix A_{ij} is defined as A_{ij}^{-1} such that

$$(A_{ij})(A_{ij})^{-1} = \mathbf{AA}^{-1} = \mathbf{I} = \delta_{ij} \equiv \text{unitary matrix}, \tag{2.4.23}$$

where δ_{ij} is the *Kronecker delta*—a matrix whose off-diagonal elements are identically zero and whose diagonal elements are unity, such that if $i, j = 1, 2, 3, 4$, then

$$\delta_{ij} = \begin{bmatrix} 1 & 0 & 0 & 0 \\ 0 & 1 & 0 & 0 \\ 0 & 0 & 1 & 0 \\ 0 & 0 & 0 & 1 \end{bmatrix}. \tag{2.4.24}$$

There exists a special relation called the $\varepsilon - \delta$ identity between ε_{ijk} and δ_{ij} such that

$$\varepsilon_{ijk}\varepsilon_{ist} = \delta_{js}\delta_{kt} - \delta_{jt}\delta_{ks}. \tag{2.4.25}$$

The inverse of a matrix A_{ij} is given by

$$A_{ij}^{-1} = \frac{(\text{Cofactor } A_{ij})^T}{\det A_{ij}}, \tag{2.4.26}$$

where T denotes the transpose and the cofactor of a matrix A_{ij} is defined such that, for $i, j = 1, 2, 3$, one has

$$\text{Cofactor } A_{11} = A_{22}A_{33} - A_{32}A_{23}, \tag{2.4.27}$$

$$\text{Cofactor } A_{12} = -(A_{21}A_{33} - A_{31}A_{23}), \tag{2.4.28}$$

$$\text{Cofactor } A_{13} = A_{21}A_{32} - A_{31}A_{22}, \tag{2.4.29}$$

$$\text{Cofactor } A_{21} = -(A_{12}A_{33} - A_{32}A_{13}), \tag{2.4.30}$$

$$\text{Cofactor } A_{22} = A_{11}A_{33} - A_{31}A_{13}, \tag{2.4.31}$$

$$\text{Cofactor } A_{23} = -(A_{11}A_{32} - A_{31}A_{12}), \tag{2.4.32}$$

$$\text{Cofactor } A_{31} = A_{12}A_{23} - A_{22}A_{13}, \tag{2.4.33}$$

$$\text{Cofactor } A_{32} = A_{11}A_{23} - A_{21}A_{13}, \tag{2.4.34}$$

$$\text{Cofactor } A_{33} = A_{11}A_{22} - A_{21}A_{12}, \tag{2.4.35}$$

Thus, for $i, j = 1, 2, 3$, one has

$$(\text{Cofactor } A_{ij})^T = \begin{bmatrix} \text{Cofactor } A_{11} & \text{Cofactor } A_{21} & \text{Cofactor } A_{31} \\ \text{Cofactor } A_{12} & \text{Cofactor } A_{22} & \text{Cofactor } A_{32} \\ \text{Cofactor } A_{13} & \text{Cofactor } A_{23} & \text{Cofactor } A_{33} \end{bmatrix}, \tag{2.4.36}$$

One can find the inverse of A_{ij} by Eq. (2.4.26) by first finding $\det A_{ij}$.

EXAMPLE 2.4.3 ■

Find the inverse of the matrix A_{ij} given in Example 2.4.1.

Solution. note that

$$\text{Cofactor } A_{ij} = \begin{bmatrix} 1 & -4 & -2 \\ 0 & 0 & 11 \\ -3 & 1 & -5 \end{bmatrix} \tag{2.4.37}$$

and

$$(\text{Cofactor } A_{ij})^T = \begin{bmatrix} 1 & 0 & -3 \\ -4 & 0 & 1 \\ -2 & 11 & -5 \end{bmatrix} \tag{2.4.38}$$

The determinant of A_{ij} is found to be -11. Thus,

$$A_{ij}^{-1} = \frac{(\text{Cofactor } A_{ij})^T}{\det A_{ij}} = \frac{-1}{11}\begin{bmatrix} 1 & 0 & -3 \\ -4 & 0 & 1 \\ -2 & 11 & -5 \end{bmatrix}. \tag{2.4.39}$$

To check we note that $A_{ij}A_{jk}^{-1} = \delta_{ik}$:

$$-\begin{bmatrix} 1 & 3 & 0 \\ 2 & 1 & -1 \\ 4 & 1 & 0 \end{bmatrix} \times \frac{1}{11}\begin{bmatrix} 1 & 0 & -3 \\ -4 & 0 & 1 \\ -2 & 11 & -5 \end{bmatrix} = \begin{bmatrix} 1 & 0 & 0 \\ 0 & 1 & 0 \\ 0 & 0 & 1 \end{bmatrix}. \quad \blacksquare\blacksquare \tag{2.4.40}$$

2.5. HOMOGENEOUS REPRESENTATION OF OBJECTS

To describe the homogeneous representation of objects we first introduce the concept of a homogeneous vector. Homogeneous vectors have the same direction but different magnitudes and components (see Fig. 2.5.1). Note that $N = a\mathbf{i} + b\mathbf{j}$, where $\mathbf{i}$ and $\mathbf{j}$ are unit vectors along the x and y axes, respectively. Note further that all $\mathbf{u}_i$ vectors belong to a set of homogeneous vectors $[\mathbf{u}_i]$ such that $\mathbf{u} \equiv [x\ y\ 0\ w]^T$ and

$$x = aw, \qquad y = bw, \qquad z = 0. \tag{2.5.1}$$

Vectors $\mathbf{u}_1$, $\mathbf{u}_2$, $\mathbf{u}_3$, and $\mathbf{u}_4$ in Fig. 2.5.1 belong to the same set. In particular, when $w = 1$, $x = a$, $y = b$, and $z = 0$, the proper definition of a vector (physical vector) with respect to the x, y, z coordinate results. The quantity w is called the scale factor. Note that $[0\ 0\ 0\ n]^T$ is a null vector at the origin, the vector $[0\ 0\ 0\ 0]^T$ is undefined for any nonzero scale factor n, and $[a\ b\ c\ 0]^T$ is a vector at infinity, generally showing a particular direction.

2.5.1 Planes

Since planes can be represented by their outward normal vector, a similar definition for homogeneous planes can be stated such that a homogeneous plane is represented by $\mathbf{p}$. Since $\mathbf{p} \equiv [a\ b\ c\ d]$ is a row vector rather than a column

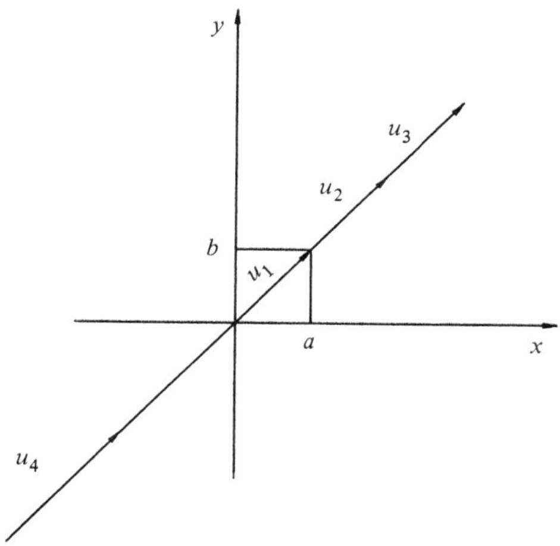

Figure 2.5.1 Representation of homogeneous vectors.

vector, if a vector $\mathbf{v} \equiv [x\ y\ z\ w]$ lies in the plane $\mathbf{p}$, then the inner product of $\mathbf{p}$ and $\mathbf{v}$ is zero; that is,

$$\mathbf{p}\cdot\mathbf{v} = [a \quad b \quad c \quad d]\begin{bmatrix} x \\ y \\ z \\ w \end{bmatrix} = ax + by + cz + dw = 0. \tag{2.5.2}$$

2.5.2 Points and Objects

Points can be represented as homogeneous position vectors such that a point p having coordinates a, b, c can be represented as a vector $\mathbf{u} \equiv \mathbf{a}i + \mathbf{b}j + \mathbf{c}k$ and generally can be represented by $\mathbf{u} \equiv [x\ y\ z\ w]^T$, where $x = aw$, $y = bw$, and $z = cw$ with w being a scaling factor. Similarly, objects can be expressed as a set of points, that is, homogeneous column vectors that normally are packed into a $4 \times n$ homogeneous matrix with n representing the number of characteristic points peculiar to the object in a geometrical sense. For example an inverted pyramid (Fig. 2.5.2) can be described easily by a 4×5 homogeneous transformation.

Then the inverted pyramid can be represented by a homogeneous transformation **A** such that

$$\mathbf{A} = \begin{bmatrix} 0 & 1 & -1 & -1 & 1 \\ 4 & 5 & 5 & 3 & 3 \\ 0 & 4 & 4 & 4 & 4 \\ 1 & 1 & 1 & 1 & 1 \end{bmatrix}. \tag{2.5.3}$$

The order of appearance of columns in **A** is arbitrary. However, once set, it must remain the same. For example, the first column is the top vertex of the

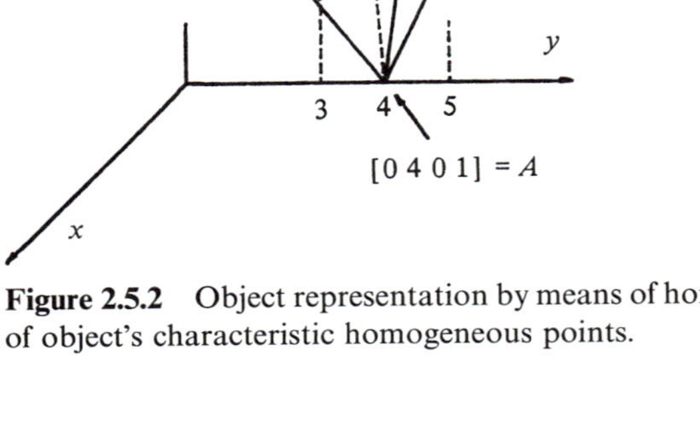

Figure 2.5.2 Object representation by means of homogeneous transformations made up of object's characteristic homogeneous points.

pyramid and no matter what transformations the pyramid has undergone the first column will remain as the top vertex. Therefore, under a translation of one unit along the x direction the proper transformation is

$$\mathbf{T}(a, x) = \begin{bmatrix} 1 & 0 & 0 & a \\ 0 & 1 & 0 & 0 \\ 0 & 0 & 1 & 0 \\ 0 & 0 & 0 & 1 \end{bmatrix} \tag{2.5.4}$$

Once the transformation is applied to the inverted pyramid, it translates the pyramid along the direction by one unit and the new object is given by **A** such that

$$\mathbf{A}_{\text{new}} = \mathbf{T}(a, x)\mathbf{A} = \begin{bmatrix} a & 1+a & -1+a & -1+a & 1+a \\ 4 & 5 & 5 & 3 & 3 \\ 0 & 4 & 4 & 4 & 4 \\ 1 & 1 & 1 & 1 & 1 \end{bmatrix}, \tag{2.5.5}$$

Now the pyramid top vertex or the first column is the point $[a\ 4\ 0\ 1]$. Similar transformations may be applied to the object to rotate it around and translate it. However, all columns preserve their original identity as if we had labeled them with certain names such as A, B, C, D, and E.

2.5.3 Perspective Projection

To an observer, different points of a three-dimensional object appear to be at different distances. Thus, the concept of perspective transformation can find applications in robotics, especially when robots are equipped with vision systems. We will discuss these kinds of transformation fully in Chapter 8.

2.5.4 Homogeneous Transformations

A homogeneous transformation is defined in this text as a 4×4 matrix **H** that generally represents translation, rotation, stretching (scaling), and perspective projection of vectors, points, objects (matrices), and other types of geometrical entity. It generally consists of a 3×3 rotation matrix **R**, a 3×1 translation vector **p**, a 1×3 perspective projection vector **f**, and a 1×1 scaling (stretching) matrix **w**. Thus,

$$\mathbf{H} = \begin{bmatrix} \mathbf{R} & \mathbf{p} \\ 3 \times 3 & 3 \times 1 \\ \mathbf{f},\ 1 \times 3 & w \\ & 1 \times 1 \end{bmatrix}, \tag{2.5.6}$$

where **R** stands for a 3×3 pure rotation, **p** stands for the vector of translation, and **f** stands for a 1×3 row vector of perspective projection.

EXAMPLE 2.5.1 ■

In Figure 2.5.3 determine the proper homogeneous transformation **H** that produces an object (pyramid) manipulation from the initial position and orientation given by Figure 2.5.2.

Solution. As can be seen from the figure, the object first has translated in the negative z direction by -4 units, then rotated about the z axis by $-90°$, then rotated about the x axis by 180°, and finally translated along the x axis by one (1) unit. Thus, the overall transformation **H** is given by

$$\mathbf{H} = T(x, 1)R(x, 180°)R(z, -90°)T(z, -4); \tag{2.5.7}$$

that is,

$$\mathbf{H} = \begin{bmatrix} 1 & 0 & 0 & 1 \\ 0 & 1 & 0 & 0 \\ 0 & 0 & 1 & 0 \\ 0 & 0 & 0 & 1 \end{bmatrix} \begin{bmatrix} 1 & 0 & 0 & 0 \\ 0 & -1 & 0 & 0 \\ 0 & 0 & -1 & 0 \\ 0 & 0 & 0 & -1 \end{bmatrix} \begin{bmatrix} 0 & 1 & 0 & 0 \\ -1 & 0 & 0 & 0 \\ 0 & 0 & 1 & 0 \\ 0 & 0 & 0 & 1 \end{bmatrix} \times \begin{bmatrix} 1 & 0 & 0 & 0 \\ 0 & 1 & 0 & 0 \\ 0 & 0 & 1 & -4 \\ 0 & 0 & 0 & 1 \end{bmatrix}, \tag{2.5.8}$$

or

$$\mathbf{H} = \begin{bmatrix} 0 & 1 & 0 & 1 \\ 1 & 0 & 0 & 0 \\ 0 & 0 & -1 & 4 \\ 0 & 0 & 0 & 1 \end{bmatrix}. \quad ■■ \tag{2.5.9}$$

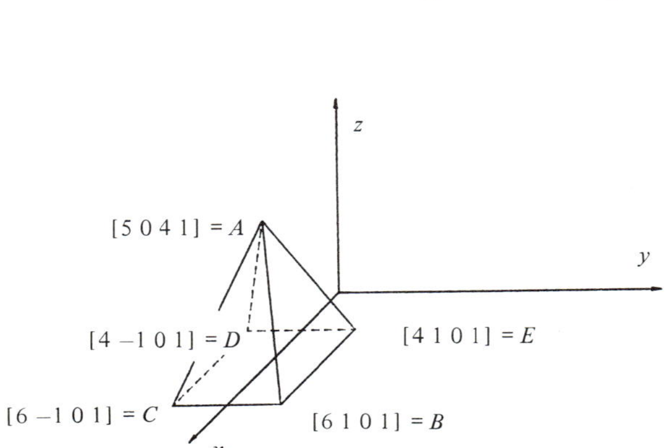

Figure 2.5.3 Object manipulation.

Note that the object in its new location and configuration is represented by a matrix $\mathbf{A}_{\text{new}}$,

$$\mathbf{A}_{\text{new}} = \mathbf{HA} = \begin{bmatrix} 0 & 1 & 0 & 1 \\ 1 & 0 & 0 & 0 \\ 0 & 0 & -1 & 4 \\ 0 & 0 & 0 & 1 \end{bmatrix} \begin{bmatrix} 0 & 1 & -1 & -1 & 1 \\ 4 & 5 & 5 & 3 & 3 \\ 0 & 4 & 4 & 4 & 4 \\ 1 & 1 & 1 & 1 & 1 \end{bmatrix}, \tag{2.5.10}$$

such that

$$\mathbf{A}_{\text{new}} = \begin{bmatrix} 5 & 6 & 6 & 4 & 4 \\ 0 & 1 & -1 & -1 & 1 \\ 4 & 0 & 0 & 0 & 0 \\ 1 & 1 & 1 & 1 & 1 \end{bmatrix}. \tag{2.5.11}$$

Clearly, this checks with the arrangement of points A, B, C, D, and E in Figure 2.5.3. Note that any rotation by an angle θ about the y coordinate can be expressed by a transform $\mathbf{Rot}(y, \theta)$ such that

$$\mathbf{Rot}(y, \theta) = \begin{bmatrix} \cos\theta & 0 & \sin\theta & 0 \\ 0 & 1 & 0 & 0 \\ -\sin\theta & 0 & \cos\theta & 0 \\ 0 & 0 & 0 & 1 \end{bmatrix} = \begin{bmatrix} C\theta & 0 & S\theta & 0 \\ 0 & 1 & 0 & 0 \\ -S\theta & 0 & C\theta & 0 \\ 0 & 0 & 0 & 1 \end{bmatrix}, \tag{2.5.12}$$

where $\cos\theta \equiv C\theta$ and $\sin\theta \equiv S\theta$. Similarly,

$$\mathbf{Rot}(x, \psi) = \begin{bmatrix} 1 & 0 & 0 & 0 \\ 0 & C\psi & -S\psi & 0 \\ 0 & S\psi & C\psi & 0 \\ 0 & 0 & 0 & 1 \end{bmatrix} \quad \text{and} \quad \mathbf{Rot}(z, \phi) = \begin{bmatrix} C\phi & -S\phi & 0 & 0 \\ S\phi & C\phi & 0 & 0 \\ 0 & 0 & 1 & 0 \\ 0 & 0 & 0 & 1 \end{bmatrix}. \tag{2.5.13}$$

For any translation transformation,

$$\mathbf{Trans}(a, b, c) = \begin{bmatrix} 1 & 0 & 0 & a \\ 0 & 1 & 0 & b \\ 0 & 0 & 1 & c \\ 0 & 0 & 0 & 1 \end{bmatrix}. \tag{2.5.14}$$

The points or vectors are translated along the x, y, and z directions by aw, bw, and cw, where w is the scale factor; that is,

$$\mathbf{Trans}(a, b, c) \begin{bmatrix} x \\ y \\ z \\ w \end{bmatrix} \equiv \begin{bmatrix} x + aw \\ y + bw \\ z + cw \\ w \end{bmatrix}. \tag{2.5.15}$$

Definition. *Robot kinematics* is defined as the systematic study of the geometry of motion of a dynamic robotic system in three-dimensional space as a function of time regardless of the forces and moments producing such motion. As mentioned before, in most robotic applications the interest is in the *direct* as well as the *inverse* kinematics equations and in their solutions either in attitude vectors or joint angle vectors. More interest is focused on the inverse problem for obvious reasons. Four different types of coordinate system are in use in connection with robotic kinematics:

1. World reference coordinate system, which is a right-handed rectangular Cartesian coordinate system with six degrees of freedom, that is, x, y, z and three Euler angles ψ, θ, ϕ.
2. Computer (CPU) coordinates, which are internal digital representations of the joint angular positions (motors), for example, 8 bit DACS, optical encoders, and number of holes.
3. Body coordinate system, which is again a right-handed rectangular Cartesian coordinate system with six degrees of freedom, that is, u, v, w and three Euler angles ψ, θ, ϕ. This coordinate system is embedded in the robot and moves exactly with the robot.
4. Joint angles (θ_i, $i = 1, 2, \ldots, n$), which is an angular coordinate system describing the position of the joint motors according to a set of link coordinate systems. They are known as soon as a set of link coordinate systems is established (Fig. 2.5.4).

As we noted before, **H**, the 4×4 homogeneous transformation that in essence provides a computational algorithm for the computer to command robotic joint rotations, is a matrix with a basic rotation submatrix, that is, 3×3 R, and three

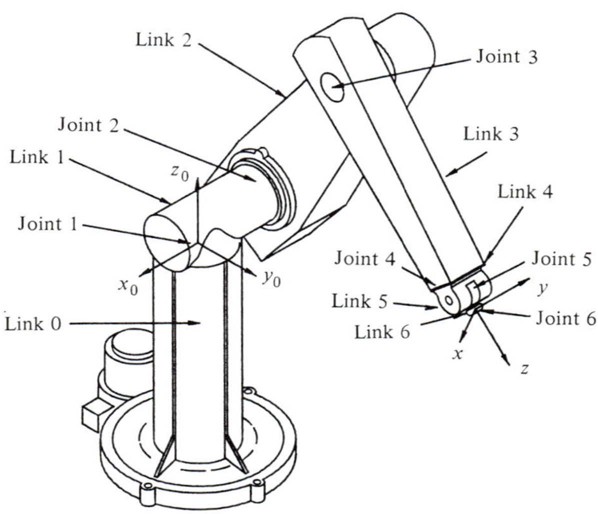

Figure 2.5.4 The PUMA robotic arm and its coordinate systems.

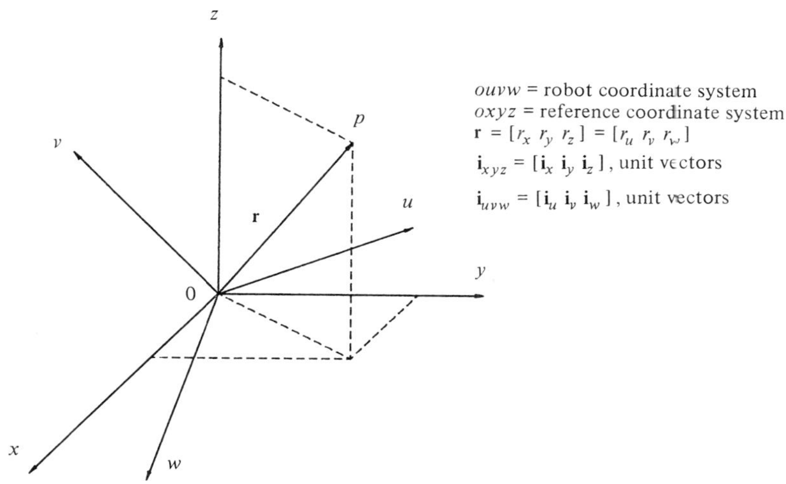

Figure 2.5.5 Reference (x, y, z) and manipulator (u, v, w) coordinate systems.

other basic vectors and matrices. Basic rotation matrices can be obtained by coinciding the coordinates of two rectangular Cartesian coordinate systems (RCCSs)—one fixed and the other moving with the robot (Fig. 2.5.5).

Note that the representations of the position vector **r** cf a point p in space with respect to the xyz and uvw systems will determine the relative rotation of one with respect to the other. In this case it can be shown that

$$\begin{aligned} r_x &= \mathbf{r}\cdot\mathbf{i}_x = (r_u\mathbf{i}_u + r_v\mathbf{i}_v + r_w\mathbf{i}_w)\cdot\mathbf{i}_x \\ &= r_u(\mathbf{i}_u\cdot\mathbf{i}_x) + r_v(\mathbf{i}_v\cdot\mathbf{i}_x) + r_w(\mathbf{i}_w\cdot\mathbf{i}_x). \end{aligned} \tag{2.5.16}$$

Similarly,

$$\begin{aligned} r_y &= \mathbf{r}\cdot\mathbf{i}_y = (r_u\mathbf{i}_u + r_u\mathbf{i}_v + r_w\mathbf{i}_w)\cdot\mathbf{i}_y \\ &= r_u(\mathbf{i}_u\cdot\mathbf{i}_y) + r_v(\mathbf{i}_j\cdot\mathbf{i}_y) + r_w(\mathbf{i}_w\cdot\mathbf{i}_y) \end{aligned} \tag{2.5.17}$$

and

$$r_z = r_u(\mathbf{i}_u\cdot\mathbf{i}_z) + r_v(\mathbf{i}_v\cdot\mathbf{i}_z) + r_w(\mathbf{i}_w\cdot\mathbf{i}_z). \tag{2.5.18}$$

In short,

$$\begin{bmatrix} r_x \\ r_y \\ r_z \end{bmatrix} = \begin{bmatrix} \mathbf{i}_x\cdot\mathbf{i}_u & \mathbf{i}_x\cdot\mathbf{i}_v & \mathbf{i}_x\cdot\mathbf{i}_w \\ \mathbf{i}_y\cdot\mathbf{i}_u & \mathbf{i}_y\cdot\mathbf{i}_v & \mathbf{i}_y\cdot\mathbf{i}_w \\ \mathbf{i}_z\cdot\mathbf{i}_u & \mathbf{i}_z\cdot\mathbf{i}_v & \mathbf{i}_z\cdot\mathbf{i}_w \end{bmatrix} \begin{bmatrix} r_u \\ r_v \\ r_w \end{bmatrix} \tag{2.5.19}$$

or

$$\mathbf{r}_{xyz} = \mathbf{R}\mathbf{r}_{uvw} \tag{2.5.20}$$

Similarly, we can show that

$$\mathbf{r}_{uvw} = \mathbf{B}\mathbf{r}_{xyz}, \tag{2.5.21}$$

where

$$\mathbf{B} = \begin{bmatrix} \mathbf{i}_u \cdot \mathbf{i}_x & \mathbf{i}_u \cdot \mathbf{i}_y & \mathbf{i}_u \cdot \mathbf{i}_z \\ \mathbf{i}_v \cdot \mathbf{i}_x & \mathbf{i}_v \cdot \mathbf{i}_y & \mathbf{i}_v \cdot \mathbf{i}_z \\ \mathbf{i}_w \cdot \mathbf{i}_x & \mathbf{i}_w \cdot \mathbf{i}_y & \mathbf{i}_w \cdot \mathbf{i}_z \end{bmatrix}. \tag{2.5.22}$$

Note that $\mathbf{B} = \mathbf{R}^T$. Furthermore, from Eq. (2.5.20), it is also true that

$$\mathbf{r}_{uvw} = \mathbf{R}^{-1}\mathbf{r}_{xyz}. \tag{2.5.23}$$

Thus, from Eqs. (2.5.21) and (2.5.23) and the fact that $\mathbf{B} = \mathbf{R}^T$ it follows that

$$\mathbf{R}^T = \mathbf{R}^{-1}. \tag{2.5.24}$$

All matrices with such properties are called *orthogonal matrices.* Note that the determinant of such matrices is ± 1.

EXAMPLE 2.5.2 ■

Prove that det $\mathbf{R} = \pm 1$.

Solution. Note that $\mathbf{R}\mathbf{R}^{-1} = \mathbf{R}\mathbf{R}^T = \mathbf{I}$. Therefore,

$$(\det \mathbf{R})^2 = \det \mathbf{I} = 1.$$

$$\therefore \det \mathbf{R} = \pm 1.$$

■ ■

Matrix $\mathbf{R}$ also can be expressed in terms of the direction cosines. For example,

$$(\mathbf{i}_x \cdot \mathbf{i}_u) = \cos(\alpha_{xu}),$$

where α_{xu} is the angle of rotation between the x axis and the u axis. Another symbolic rotation is to write $\mathbf{R}$ as

$$\mathbf{R} = \begin{bmatrix} n_x & o_x & a_x \\ n_y & o_y & a_y \\ n_z & o_z & a_z \end{bmatrix} \tag{2.5.25}$$

or

$$\mathbf{R} = \begin{bmatrix} \cos(\alpha_{xu}) & \cos(\alpha_{xv}) & \cos(\alpha_{xw}) \\ \cos(\alpha_{yu}) & \cos(\alpha_{yv}) & \cos(\alpha_{yw}) \\ \cos(\alpha_{zu}) & \cos(\alpha_{zv}) & \cos(\alpha_{zw}) \end{bmatrix}. \tag{2.5.26}$$

Thus, $\mathbf{R}(x, \alpha)$ is the rotation by α about the x axis which means $\alpha_{yv} = \alpha_{zw} = \alpha$, $\alpha_{zv} = 90° - \alpha$, $\alpha_{zw} = 90° + \alpha$, $\alpha_{yu} = \alpha_{zu} = 90°$, $\alpha_{xu} = 0$, and $\alpha_{xv} = \alpha_{xw} = 90°$. Therefore,

$$\mathbf{R}(x, \alpha) = \begin{bmatrix} 1 & 0 & 0 \\ 0 & \cos\alpha & -\sin\alpha \\ 0 & \sin\alpha & \cos\alpha \end{bmatrix}. \tag{2.5.27}$$

EXAMPLE 2.5.3 ■

Given a vector $\mathbf{r}_{uvw} = [6, 4, 2]^T$ with respect to the $0uvw$ coordinate system, determine $\mathbf{r}_{xyz}$ if $ouvw$ is rotated about the z axis by 60°.

Solution.

$$\mathbf{R}(z, 60) = \begin{bmatrix} \cos 60 & -\sin 60 & 0 \\ \sin 60 & \cos 60 & 0 \\ 0 & 0 & 1 \end{bmatrix} \tag{2.5.28}$$

and

$$\mathbf{r}_{xyz} = \mathbf{R}(z, 60)\mathbf{r}_{uvw} = \begin{bmatrix} 0.5 & -0.866 & 0 \\ 0.866 & 0.5 & 0 \\ 0 & 0 & 1 \end{bmatrix} \begin{bmatrix} 6 \\ 4 \\ 2 \end{bmatrix} = \begin{bmatrix} -0.464 \\ 7.196 \\ 2 \end{bmatrix}. \quad ■■ \tag{2.5.29}$$

Note that the column vectors in the expression for **R** in Eq. (2.5.25) represent the components of the respective unit base vectors of the $0uvw$ coordinate system with respect to the $0xyz$ frame. The row vectors, on the other hand, show the components of the base unit vectors of the $0xyz$ coordinate system with respect to the $0uvw$ frame.

The sequence of rotations may be such that the object may rotate about the principal axis of either the $0xyz$ frame or the $0uvw$ frame. The order of the rotation is always important. For example, a "composite rotation" R may be given by $R(y, \Phi)R(z, \theta)R(y, \psi)$ and another rotation R_2 may be given by $R(x, \psi)R(z, \theta)R(y, \Phi)$. These two rotations produce different results, and thus $R_1 \neq R_2$. Note that, initially, the $0xyz$ and $0uvw$ coordinate frames must coincide for the multiplication rule to apply. In this situation, the rotation matrix is $\mathbf{R} = \mathbf{I}$, the unitary matrix. If, originally, the $0xyz$ and $0uvw$ coordinate frames do not coincide, you must apply the proper transformation to make them coincide. If the body frame $0uvw$ rotates about the frame $0xyz$, then "premultiply" the **I** matrix by the appropriate basic rotation matrix (2.5.26) or its counterpart with respect to the $0uvw$ frame. If the body frame $0uvw$ rotates about itself, that is, the frame $0uvw$, then "postmultiply" the **I** matrix by the appropriate rotation matrices.

EXAMPLE 2.5.4 ■

Find the composite rotation matrix representing the following

(a) a rotation of ϕ about the y axis;
(b) a rotation of θ about the w axis; and
(c) a rotation of α about the u axis.

Solution. Note that the final composite rotation is

$$\mathbf{R} = \mathbf{R}(y, \phi)\mathbf{I}\mathbf{R}(w, \theta)\mathbf{R}(u, \alpha) \tag{2.5.30}$$

$$\mathbf{R} = \begin{bmatrix} C\phi & 0 & S\phi \\ 0 & 1 & 0 \\ -S\phi & 0 & C\phi \end{bmatrix} \begin{bmatrix} 1 & 0 & 0 \\ 0 & 1 & 0 \\ 0 & 0 & 1 \end{bmatrix} \begin{bmatrix} C\theta & -S\theta & 0 \\ S\theta & C\theta & 0 \\ 0 & 0 & 1 \end{bmatrix} \begin{bmatrix} 1 & 0 & 0 \\ 0 & C\alpha & -S\alpha \\ 0 & S\alpha & C\alpha \end{bmatrix}. \tag{2.5.31}$$

The students are urged to carry out the details! ■ ■

EXAMPLE 2.5.5 ■

Find the composite rotation matrix representing the following:

(a) a rotation of α about the u axis;
(b) a rotation of θ about the w axis; and
(c) a rotation of ϕ about the y axis.

This is the reverse sequence of rotations in Example 2.5.4.

Solution. In this case $\mathbf{R} = \mathbf{R}(y, \phi)\mathbf{I}\mathbf{R}(u, \alpha)\mathbf{R}(w, \theta)$.

$$\mathbf{R} = \begin{bmatrix} C\phi & 0 & S\phi \\ 0 & 1 & 0 \\ -S\phi & 0 & C\phi \end{bmatrix} \begin{bmatrix} 1 & 0 & 0 \\ 0 & 1 & 0 \\ 0 & 0 & 1 \end{bmatrix} \begin{bmatrix} 1 & 0 & 0 \\ 0 & C\alpha & -S\alpha \\ 0 & S\alpha & C\alpha \end{bmatrix} \begin{bmatrix} C\theta & -S\theta & 0 \\ S\theta & C\theta & 0 \\ 0 & 0 & 1 \end{bmatrix}. \tag{2.5.32}$$

Multiplying through, you will find that the above composite rotation is different from the composite rotation found in the previous example. ■ ■

EXAMPLE 2.5.6 ■

Find an expression for a general rotational transformation that rotates a body about an arbitrary direction **k** by an angle θ.

Solution. Let us denote the general direction **k** by a unit vector such that

$$\mathbf{k} = k_x\mathbf{i} + k_y\mathbf{u} + k_z\mathbf{k} \tag{2.5.33}$$

is the representation of **k** with respect to a reference frame $0xyz$. We may assume that there exists a rectangular Cartesian frame **K** such that its z axis coincides with the assumed **k** axis and its origin coincides with the origin of $0xyz$. Let us represent **K** by the following homogeneous transformation:

$$\mathbf{K} = \begin{bmatrix} n_x & o_x & a_x & 0 \\ n_y & o_y & a_y & 0 \\ n_z & o_z & a_z & 0 \\ 0 & 0 & 0 & 1 \end{bmatrix}. \tag{2.5.34}$$

It is then clear that

$$a_x = k_x, \qquad a_y = k_y, \qquad a_z = k_z. \tag{2.5.35}$$

Since rotating a body about **k** is equivalent to rotating a body about the z axis of the frame **K**, the following transformation equation then holds true:

$$\mathbf{R}(k, \theta)\mathbf{K} = \mathbf{K}\mathbf{R}(z, \theta), \tag{2.5.36}$$

where $\mathbf{R}(k, \theta)$ is the general rotation by θ about **k**.

$$\mathbf{R}(k, \theta) = \mathbf{K}\mathbf{R}(z, \theta)\mathbf{K}^{-1}. \tag{2.5.37}$$

Thus,

$$\mathbf{R}(\mathbf{k}, \theta) = \begin{bmatrix} n_x & o_x & a_x & 0 \\ n_y & o_y & a_y & 0 \\ n_z & o_z & a_z & 0 \\ 0 & 0 & 0 & 1 \end{bmatrix} \begin{bmatrix} C\theta & -S\theta & 0 & 0 \\ S\theta & C\theta & 0 & 0 \\ 0 & 0 & 1 & 0 \\ 0 & 0 & 0 & 1 \end{bmatrix} \begin{bmatrix} n_x & n_y & n_z & 0 \\ o_x & o_y & o_z & 0 \\ a_x & a_y & a_z & 0 \\ 0 & 0 & 0 & 1 \end{bmatrix} \tag{2.5.38}$$

Multiplying the matrices in Eq. (2.5.38) and using Eq. (2.5.36) and the fact that

$$a_x = k_x = n_y a_z - n_z o_y, \tag{2.5.39}$$

$$a_y = k_y = n_z o_x - n_x o_z, \tag{2.5.40}$$

$$a_z = k_z = n_x o_y - n_y o_x, \tag{2.5.41}$$

we obtain the following expression for $\mathbf{R}(\mathbf{k}, \theta)$:

$$\mathbf{R}(\mathbf{k}, \theta) = \begin{bmatrix} k_x k_x \operatorname{vers}\theta + \cos\theta & k_y k_x \operatorname{vers}\theta - k_z \sin\theta & k_z k_x \operatorname{vers}\theta + k_y \sin\theta & 0 \\ k_x k_y \operatorname{vers}\theta + k_z \sin\theta & k_y k_y \operatorname{vers}\theta + \cos\theta & k_z k_y \operatorname{vers}\theta - k_x \sin\theta & 0 \\ k_x k_z \operatorname{vers}\theta - k_y \sin\theta & k_y k_z \operatorname{vers}\theta + k_x \sin\theta & k_z k_z \operatorname{vers}\theta + \cos\theta & 0 \\ 0 & 0 & 0 & 1 \end{bmatrix} \tag{2.5.42}$$

where $\operatorname{vers}\theta \equiv (1 - \cos\theta)$. In order to find a desired direction **k**, we note that a rotation of θ about **k** yields a transformation **R** such that

$$\mathbf{R} = \begin{bmatrix} n_x & o_x & a_x & 0 \\ n_y & o_y & a_y & 0 \\ n_z & o_z & a_z & 0 \\ 0 & 0 & 0 & 1 \end{bmatrix}. \tag{2.5.43}$$

One then obtains nine equations for the four unknowns k_x, k_y, k_z, and θ. ■■

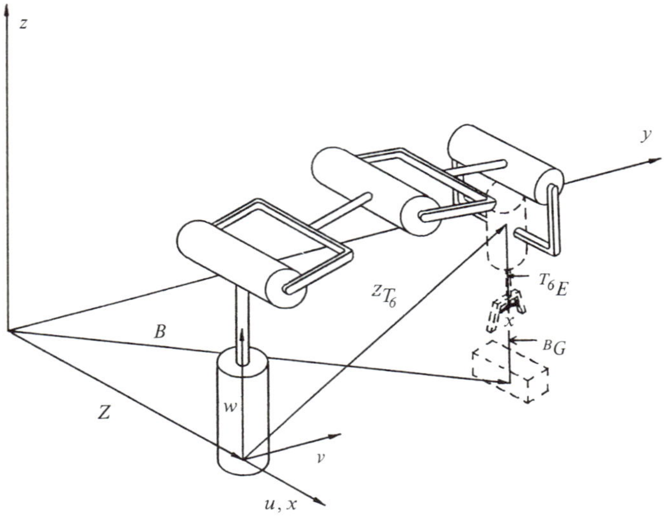

Figure 2.5.6 Manipulator positioning over an object.

EXAMPLE 2.5.7 ■

Referring to Figure 2.5.6, determine the transform $^Z\mathbf{T}_6$ with respect to the base coordinate of the robot which is located from the reference coordinate by a transform **Z**. The object is located at a point with a transform **B** and the position of the end effector with respect to **B** is determined from a transform $^B\mathbf{G}$. The position of the end effector with respect to $^ZT_6$ (the tip of the robot) is $^{T_6}\mathbf{E}$.

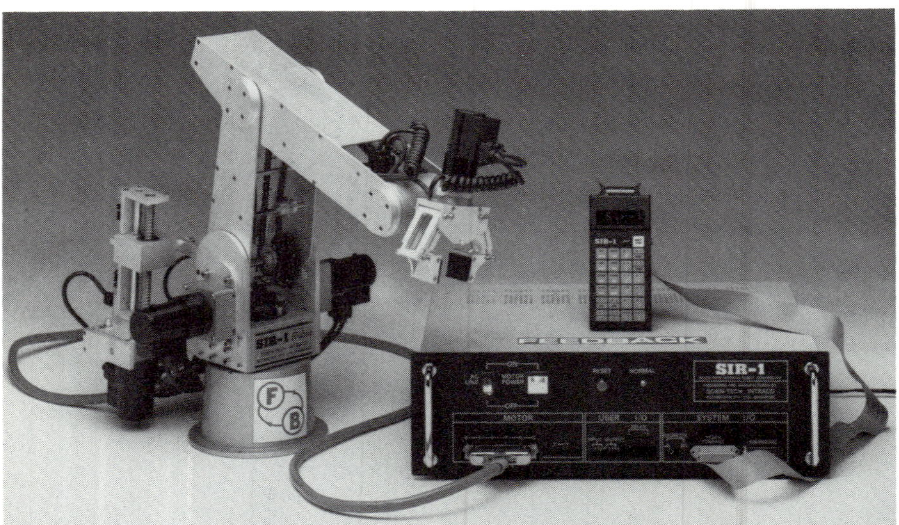

Figure 2.5.7 SIR-1 Robot. (Courtesy of the Feedback Corp.)

Solution. The position of the end effector must be the same in both descriptions (Fig. 2.5.6). Obviously,

$$\mathbf{B}\,{}^{B}\mathbf{G} = \mathbf{Z}\,{}^{Z}\mathbf{T}_6\,{}^{T_6}\mathbf{E}. \tag{2.5.44}$$

Thus,

$${}^{Z}\mathbf{T}_6 = \mathbf{Z}^{-1}\mathbf{B}\,{}^{B}\mathbf{G}\,{}^{T_6}\mathbf{E}^{-1}. \quad ■■ \tag{2.5.45}$$

EXAMPLE 2.5.8 ■

In the previous example, let the position vectors corresponding to $\mathbf{Z}$, ${}^{Z}\mathbf{T}_6$, ${}^{T_6}\mathbf{E}$, $\mathbf{B}$, and ${}^{B}\mathbf{G}$ be $[a\ 0\ 0\ 1]^T$, $[0\ b\ c\ 1]^T$, $[0\ 0\ -d\ 1]$, $[e\ d\ 0\ 1]^T$, and $[h_1\ h_2\ h_3\ 1]^T$, respectively. Find an expression for ${}^{Z}\mathbf{T}_6$ and ${}^{B}\mathbf{G}$. (Fig. 2.5.6).

Solution. The transform corresponding to $\mathbf{Z}$ is

$$\mathbf{Z} = \begin{bmatrix} 1 & 0 & 0 & a \\ 0 & 1 & 0 & 0 \\ 0 & 0 & 1 & 0 \\ 0 & 0 & 0 & 1 \end{bmatrix}, \qquad {}^{Z}\mathbf{T}_6 = \begin{bmatrix} 1 & 0 & 0 & 0 \\ 0 & 1 & 0 & b \\ 0 & 0 & 1 & c \\ 0 & 0 & 0 & 1 \end{bmatrix}; \tag{2.5.46}$$

$${}^{T_6}\mathbf{E} = \begin{bmatrix} 1 & 0 & 0 & 0 \\ 0 & 1 & 0 & 0 \\ 0 & 0 & 1 & -d \\ 0 & 0 & 0 & 1 \end{bmatrix}, \qquad \mathbf{B} = \begin{bmatrix} 1 & 0 & 0 & e \\ 0 & 1 & 0 & f \\ 0 & 0 & 1 & 0 \\ 0 & 0 & 0 & 1 \end{bmatrix}; \tag{2.5.47}$$

$${}^{B}\mathbf{G} = \mathbf{B}^{-1}\mathbf{Z}\,{}^{Z}\mathbf{T}_6\,{}^{T_6}\mathbf{E} = \overset{\mathbf{B}^{-1}}{\begin{bmatrix} 1 & 0 & 0 & -e \\ 0 & 1 & 0 & -f \\ 0 & 0 & 1 & 1 \\ 0 & 0 & 0 & 0 \end{bmatrix}} \overset{\mathbf{Z}}{\begin{bmatrix} 1 & 0 & 0 & a \\ 0 & 1 & 0 & 0 \\ 0 & 0 & 1 & 0 \\ 0 & 0 & 0 & 1 \end{bmatrix}} {}^{Z}\mathbf{T}_6\,{}^{T_6}\mathbf{E}. \tag{2.5.48}$$

Finally,

$${}^{B}\mathbf{G} = \mathbf{B}^{-1}\mathbf{Z} \overset{{}^{Z}\mathbf{T}_6}{\begin{bmatrix} 1 & 0 & 0 & 0 \\ 0 & 1 & 0 & b \\ 0 & 0 & 1 & c \\ 0 & 0 & 0 & 1 \end{bmatrix}} \overset{{}^{T_6}\mathbf{E}}{\begin{bmatrix} 1 & 0 & 0 & 0 \\ 0 & 1 & 0 & 0 \\ 0 & 0 & 1 & -d \\ 0 & 0 & 0 & 1 \end{bmatrix}}$$

$$= \overset{\mathbf{B}^{-1}\mathbf{Z}}{\begin{bmatrix} 1 & 0 & 0 & a-e \\ 0 & 1 & 0 & -f \\ 0 & 0 & 1 & 0 \\ 0 & 0 & 0 & 1 \end{bmatrix}} \overset{{}^{Z}\mathbf{T}_6\,{}^{T_6}\mathbf{E}}{\begin{bmatrix} 1 & 0 & 0 & 0 \\ 0 & 1 & 0 & b \\ 0 & 0 & 1 & c-d \\ 0 & 0 & 0 & 1 \end{bmatrix}}$$

$$= \begin{bmatrix} 1 & 0 & 0 & a-e \\ 0 & 1 & 0 & b-f \\ 0 & 0 & 1 & c-d \\ 0 & 0 & 0 & 1 \end{bmatrix} = {}^{B}\mathbf{G}. \quad ■■ \tag{2.5.49}$$

EXAMPLE 2.5.9 ■

Show that the inverse of a generalized transform matrix with no perspective projection, that is

$$\mathbf{T} = \begin{bmatrix} n_x & n_y & n_z & p_x \\ n_y & o_y & a_y & p_y \\ n_z & o_z & a_z & p_z \\ 0 & 0 & 0 & 1 \end{bmatrix}, \tag{2.5.50}$$

is given by

$$\mathbf{T}^{-1} = \begin{bmatrix} n_x & n_y & n_z & -\mathbf{p}\cdot\mathbf{n} \\ o_x & o_y & o_z & -\mathbf{p}\cdot\mathbf{o} \\ a_x & a_y & a_z & -\mathbf{p}\cdot\mathbf{a} \\ 0 & 0 & 0 & 1 \end{bmatrix}, \tag{2.5.51}$$

where $\mathbf{p}\cdot\mathbf{n} = p_x n_x + p_y n_y + p_z n_z$ is a vector dot product.

Solution. Form the product $\mathbf{T}^{-1}\mathbf{T}$ to obtain

$$\mathbf{T}^{-1}\mathbf{T} = \begin{bmatrix} n_x & n_y & n_z & -\mathbf{p}\cdot\mathbf{n} \\ o_x & o_y & o_x & -\mathbf{p}\cdot\mathbf{o} \\ a_x & a_y & a_z & -\mathbf{p}\cdot\mathbf{a} \\ 0 & 0 & 0 & 1 \end{bmatrix} \begin{bmatrix} n_x & o_x & a_x & p_x \\ n_y & o_y & a_y & p_y \\ n_z & o_z & a_z & p_z \\ 0 & 0 & 0 & 1 \end{bmatrix}. \tag{2.5.52}$$

$$\mathbf{T}^{-1}\mathbf{T} = \begin{bmatrix} \mathbf{n}\cdot\mathbf{n} & \mathbf{n}\cdot\mathbf{o} & \mathbf{n}\cdot\mathbf{a} & \mathbf{n}\cdot\mathbf{p} - \mathbf{p}\cdot\mathbf{n} \\ \mathbf{o}\cdot\mathbf{n} & \mathbf{o}\cdot\mathbf{o} & \mathbf{o}\cdot\mathbf{a} & \mathbf{o}\cdot\mathbf{p} - \mathbf{p}\cdot\mathbf{o} \\ \mathbf{a}\cdot\mathbf{a} & \mathbf{a}\cdot\mathbf{o} & \mathbf{a}\cdot\mathbf{a} & \mathbf{a}\cdot\mathbf{p} - \mathbf{p}\cdot\mathbf{a} \\ 0 & 0 & 0 & 1 \end{bmatrix} = \begin{bmatrix} 1 & 0 & 0 & 0 \\ 0 & 1 & 0 & 0 \\ 0 & 0 & 1 & 0 \\ 0 & 0 & 0 & 1 \end{bmatrix}$$

checks. ■ ■

Stretching transformations generally are of the form

$$\mathbf{H} = \begin{bmatrix} a & 0 & 0 & 0 \\ 0 & b & 0 & 0 \\ 0 & 0 & c & 0 \\ 0 & 0 & 0 & 1 \end{bmatrix}, \tag{2.5.53}$$

which causes a vector to stretch its coordinates or to shrink them. For example,

$$\mathbf{Hx} = \begin{bmatrix} a & 0 & 0 & 0 \\ 0 & b & 0 & 0 \\ 0 & 0 & c & 0 \\ 0 & 0 & 0 & 1 \end{bmatrix} \begin{bmatrix} x \\ y \\ z \\ 1 \end{bmatrix} \equiv \begin{bmatrix} ax \\ by \\ cz \\ 1 \end{bmatrix}. \tag{2.5.54}$$

Note that in this case a cube may become a rectangular parallelepiped.

Scaling transforms, on the other hand, shrink or expand coordinates in a uniform manner. For

$$\mathbf{H} \equiv \begin{bmatrix} 1 & 0 & 0 & 0 \\ 0 & 1 & 0 & 0 \\ 0 & 0 & 1 & 0 \\ 0 & 0 & 0 & s \end{bmatrix}, \tag{2.5.55}$$

$$\mathbf{H}\mathbf{x} = \begin{bmatrix} 1 & 0 & 0 & 0 \\ 0 & 1 & 0 & 0 \\ 0 & 0 & 1 & 0 \\ 0 & 0 & 0 & s \end{bmatrix} \begin{bmatrix} x \\ y \\ z \\ 1 \end{bmatrix} = \begin{bmatrix} x \\ y \\ z \\ s \end{bmatrix} \leftrightarrow \begin{bmatrix} x/s \\ y/s \\ z/s \\ 1 \end{bmatrix}. \tag{2.5.56}$$

Thus, in this case, a cube becomes another cube either larger ($0 < s < 1$) or smaller ($s > 1$) than the original cube.

2.6 ROBOTIC MANIPULATOR JOINT COORDINATE SYSTEMS

A manipulator consists of a series of links connected by a series of joints. Homogeneous transformations with respect to local joint coordinates are called **A** matrices (i.e., Denavit–Hartenberg transformations). Let $\mathbf{A}_1$ describe the position and orientation of the first link, $\mathbf{A}_2$ that of the second link, $\mathbf{A}_3$ that of the third link, and so on with respect to the base coordinate frame. Then the position and orientation of the nth link with respect to the base coordinate will be

$$\mathbf{T}_n = \mathbf{A}_1 \mathbf{A}_2 \mathbf{A}_3 \cdots \mathbf{A}_{n-1} \mathbf{A}_n \tag{2.6.1}$$

The products of **A** matrices are called **T** matrices. $\mathbf{T}_6$ is designated the hand or gripper transformation, such that the hand origin has a position vector $\mathbf{p} \equiv [p_x \; p_y \; p_z]^T$. Attached to the origin is a rectangular Cartesian coordinate system whose orientation designates the orientation of the hand (Fig. 2.6.1). Note that **a**, **o**, and **n** form a right-handed coordinate system; that is, $\mathbf{a} \times \mathbf{o} = \mathbf{n}$, $\mathbf{o} \times \mathbf{n} = \mathbf{a}$, and $\mathbf{n} \times \mathbf{a} = \mathbf{o}$. Note that in this case

$$\mathbf{T}_6 \equiv \begin{bmatrix} n_x & o_x & a_x & p_x \\ n_y & o_y & a_y & p_y \\ n_z & o_z & a_z & p_z \\ 0 & 0 & 0 & 1 \end{bmatrix}. \tag{2.6.2}$$

Similarly, for any link joint coordinate i, the link direction at the joint is in the approach vector direction ($\mathbf{a}_i$), the direction of the link-to-link rotation is in the $\mathbf{o}_i$ direction (the orientation vector), and the normal unit vector $\mathbf{n}_i$ forms a right-handed system with $\mathbf{o}_i$ and $\mathbf{a}_i$ (see Fig. 2.6.2). Again, the position vector of

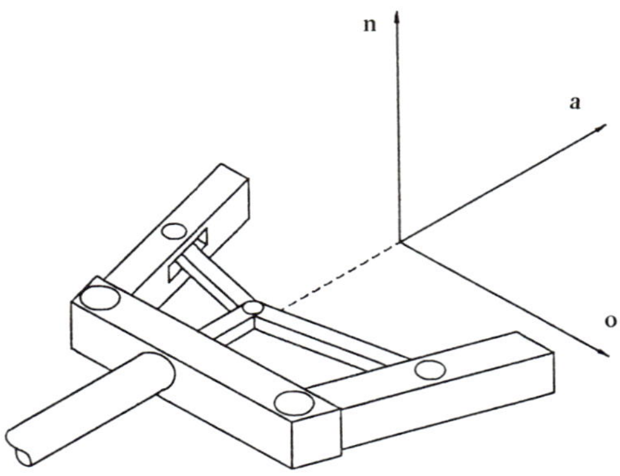

a = approach vector
o = orientation vector of fingertips
n = normal vector

Figure 2.6.1 Orientation of a robotic gripper and its coordinate frame.

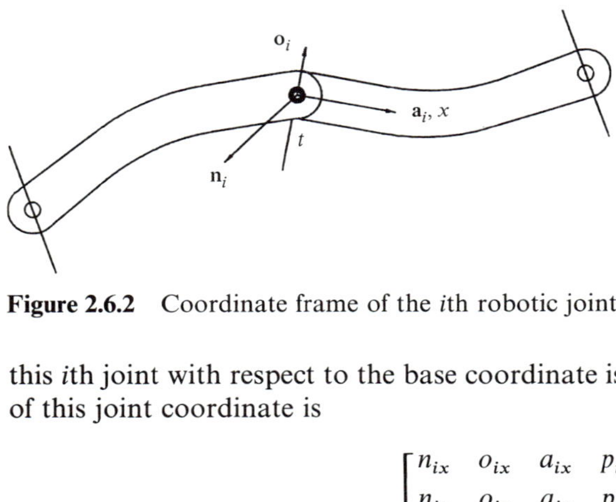

Figure 2.6.2 Coordinate frame of the *i*th robotic joint.

this *i*th joint with respect to the base coordinate is $\mathbf{p}_i$ and the transformation $\mathbf{T}_i$ of this joint coordinate is

$$\mathbf{T}_i = \begin{bmatrix} n_{ix} & o_{ix} & a_{ix} & p_{ix} \\ n_{iy} & o_{iy} & a_{iy} & p_{iy} \\ n_{iz} & o_{iz} & a_{iz} & p_{iz} \\ 0 & 0 & 0 & 1 \end{bmatrix}. \tag{2.6.3}$$

2.7 EULER ANGLES AND EULER TRANSFORMATIONS

Given *xyz* and *uvw* coordinate systems, first rotate about the *z* or *w* axis by ϕ (see Fig. 2.7.1). Then rotate about the *v* axis by θ and about the *x* axis by ψ to get a combined transform $\mathbf{Euler}(\phi, \theta, \psi)$; that is,

$$\mathbf{Euler}(\phi, \theta, \psi) = \mathbf{R}(z, \phi)\mathbf{R}(y, \theta)\mathbf{R}(x, \psi) \tag{2.7.1}$$

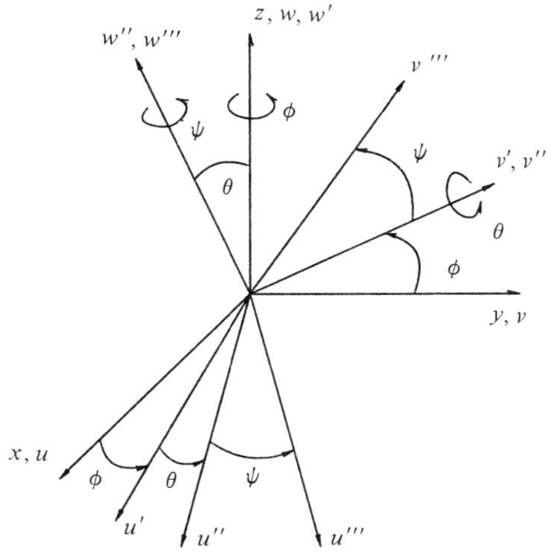

Figure 2.7.1 Gecmetrical description of Euler angles.

or

$$\mathbf{Euler}(\phi, \theta, \psi) = \begin{bmatrix} C\phi & -S\phi & 0 & 0 \\ S\phi & C\phi & 0 & 0 \\ 0 & 0 & 1 & 0 \\ 0 & 0 & 0 & 1 \end{bmatrix} \begin{bmatrix} C\theta & 0 & S\theta & 0 \\ 0 & 1 & 0 & 0 \\ -S\theta & 0 & C\theta & 0 \\ 0 & 0 & 0 & 1 \end{bmatrix}$$

$$\times \begin{bmatrix} 1 & 0 & 0 & 0 \\ 0 & C\psi & -S\psi & 0 \\ 0 & S\psi & C\psi & 0 \\ 0 & 0 & 0 & 1 \end{bmatrix}. \tag{2.7.2}$$

Finally,

$$\mathbf{Euler}(\phi, \theta, \psi) = \begin{bmatrix} C\phi & C\phi S\theta S\psi - S\phi C\psi & C\phi S\theta C\psi + S\theta S\psi & 0 \\ S\phi C\theta & S\phi S\theta S\psi + C\phi C\psi & S\phi S\theta C\psi - C\theta S\psi & 0 \\ -S\theta & C\theta S\psi & C\theta C\psi & 0 \\ 0 & 0 & 0 & 1 \end{bmatrix}. \tag{2.7.3}$$

2.8 ROLL–PITCH–YAW (RPY) TRANSFORMATIONS

Given an xyz coordinate frame (Fig. 2.8.1) attached to the gripper, first rotate with respect to z by ϕ, then with respect to y by θ, and finally with respect to x by ψ. Thus, the composite RPY transformation will be

$$\mathbf{RPY}(\phi, \theta, \psi) = \mathbf{R}(z, \phi)\mathbf{R}(y, \theta)\mathbf{R}(x, \psi) \tag{2.8.1}$$

Therefore, we basically obtain an expression similar to Eq. (2.7.2).

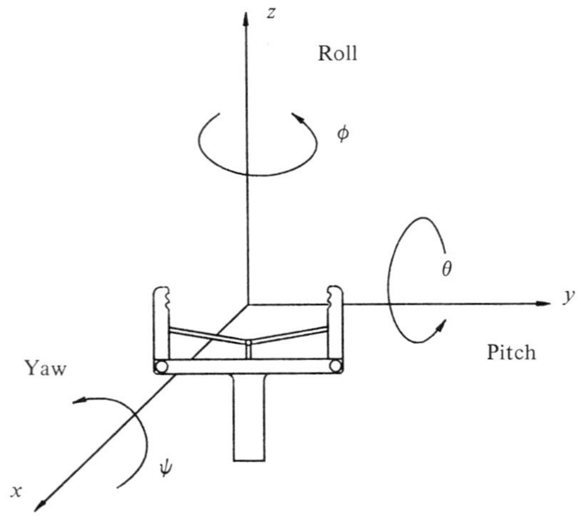

Figure 2.8.1 Roll, pitch, and yaw (RPY) angles.

2.9 SOME EXAMPLES OF BASIC HOMOGENEOUS TRANSFORMATIONS

The basic translation matrix is

$$\mathbf{T}_{\text{translation}} = \begin{bmatrix} 1 & 0 & 0 & dx \\ 0 & 1 & 0 & dy \\ 0 & 0 & 1 & dz \\ 0 & 0 & 0 & 1 \end{bmatrix}, \tag{2.9.1}$$

because

$$\mathbf{T}_{\text{translation}} \cdot \mathbf{x} = (\mathbf{x} + dx) = \begin{bmatrix} 1 & 0 & 0 & dx \\ 0 & 1 & 0 & dy \\ 0 & 0 & 1 & dz \\ 0 & 0 & 0 & 1 \end{bmatrix} \begin{bmatrix} x \\ y \\ z \\ 1 \end{bmatrix} = \begin{bmatrix} x + dx \\ y + dy \\ z + dz \\ 1 \end{bmatrix}. \tag{2.9.2}$$

The basic scaling transformation is

$$\mathbf{T}_{\text{scaling}} = \begin{bmatrix} 1 & 0 & 0 & 0 \\ 0 & 1 & 0 & 0 \\ 0 & 0 & 1 & 0 \\ 0 & 0 & 0 & S \end{bmatrix}, \tag{2.9.3}$$

such that $\mathbf{T}_{\text{scaling}}\mathbf{x} = (1/S)\mathbf{x} = S^{-1}\mathbf{x}$; that is,

$$\begin{bmatrix} 1 & 0 & 0 & 0 \\ 0 & 1 & 0 & 0 \\ 0 & 0 & 1 & 0 \\ 0 & 0 & 0 & S \end{bmatrix} \begin{bmatrix} x \\ y \\ z \\ 1 \end{bmatrix} = \begin{bmatrix} x \\ y \\ z \\ S \end{bmatrix} \equiv \begin{bmatrix} S^{-1}x \\ S^{-1}y \\ S^{-1}z \\ 1 \end{bmatrix}.$$

The *basic perspective transformation* is

$$\mathbf{T}_{\text{perspective}} = \begin{bmatrix} 1 & 0 & 0 & 0 \\ 0 & 1 & 0 & 0 \\ 0 & 0 & 1 & 0 \\ a & b & c & 1 \end{bmatrix}, \tag{2.9.4}$$

such that

$$\mathbf{T}_{\text{perspective}}\mathbf{x} = (\mathbf{f} \cdot \mathbf{x})^{-1}\mathbf{x}; \tag{2.9.5}$$

that is,

$$\begin{bmatrix} 1 & 0 & 0 & 0 \\ 0 & 1 & 0 & 0 \\ 0 & 0 & 1 & 0 \\ a & b & c & 1 \end{bmatrix} \begin{bmatrix} x \\ y \\ z \\ 1 \end{bmatrix} = \begin{bmatrix} x \\ y \\ z \\ 1 + ax + by + cz \end{bmatrix} = \begin{bmatrix} x \\ y \\ z \\ (\mathbf{f} \cdot \mathbf{x}) \end{bmatrix}, \quad \text{Q.E.D.} \tag{2.9.6}$$

where $\mathbf{f}$ is the perspective row vector, $\mathbf{f} \equiv [a\ b\ c\ 1]$.

EXAMPLE 2.9.1 ■

A pair of points $p_1 \equiv [1\ 2\ 1]^T$ and $p_2 \equiv [2\ 3\ 4]^T$ are to be translated along the x and z axes by 3 and -2, respectively. Determine their new positions with respect to the xyz coordinate frame.

Solution. The homogeneous transformation that corresponds to this is

$$\mathbf{H} = \begin{bmatrix} 1 & 0 & 0 & 3 \\ 0 & 1 & 0 & 0 \\ 0 & 0 & 1 & -2 \\ 0 & 0 & 0 & 1 \end{bmatrix}, \tag{2.9.7}$$

and the new locations of p_1 and p_2 are, respectively,

$$p_{1xyz} = \mathbf{H}p_1 = \begin{bmatrix} 1 & 0 & 0 & 3 \\ 0 & 1 & 0 & 0 \\ 0 & 0 & 1 & -2 \\ 0 & 0 & 0 & 1 \end{bmatrix} \begin{bmatrix} 1 \\ 2 \\ 1 \\ 1 \end{bmatrix} = \begin{bmatrix} 4 \\ 2 \\ -1 \\ 1 \end{bmatrix}, \qquad p_{2xyz} = \begin{bmatrix} 5 \\ 3 \\ 2 \\ 1 \end{bmatrix}. \quad ■■ \tag{2.9.8}$$

EXAMPLE 2.9.2 ■

Determine a homogeneous transformation corresponding to a rotation of θ about the ox axis, followed by a translation of a units along the ov axis (Fig. 2.9.1).

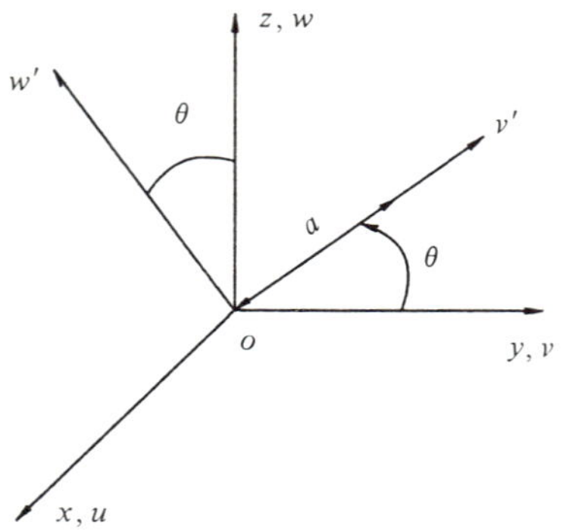

Figure 2.9.1 Rotation–translation transformation.

Solution. For the *first method*,

$$\mathbf{T} = T(v, a)R(x, \theta) \tag{2.9.9}$$

$$\mathbf{T} = T(y, aC\theta)T(z, aS\theta)R(x, \theta) \tag{2.9.10}$$

and

$$\mathbf{T} = \begin{bmatrix} 1 & 0 & 0 & 0 \\ 0 & 1 & 1 & aS\theta \\ 0 & 0 & 1 & aS\theta \\ 0 & 0 & 0 & 1 \end{bmatrix} \begin{bmatrix} 1 & 0 & 0 & 0 \\ 0 & C\theta & -S\theta & 0 \\ 0 & S\theta & C\theta & 0 \\ 0 & 0 & 0 & 1 \end{bmatrix} = \begin{bmatrix} 1 & 0 & 0 & 0 \\ 0 & C\theta & -S\theta & aC\theta \\ 0 & S\theta & C\theta & aS\theta \\ 0 & 0 & 0 & 1 \end{bmatrix}. \tag{2.9.11}$$

For the *second method*, $\mathbf{T} = \mathbf{R}(x, \theta)\mathbf{T}(v, a)$.

$$\therefore \quad \mathbf{T} = \begin{bmatrix} 1 & 0 & 0 & 0 \\ 0 & C\theta & S\theta & 0 \\ 0 & S\theta & C\theta & 0 \\ 0 & 0 & 0 & 1 \end{bmatrix} \begin{bmatrix} 1 & 0 & 0 & 0 \\ 0 & 1 & 0 & a \\ 0 & 0 & 1 & 0 \\ 0 & 0 & 0 & 1 \end{bmatrix}$$

$$= \begin{bmatrix} 1 & 0 & 0 & 0 \\ 0 & C\theta & -S\theta & aC\theta \\ 0 & S\theta & C\theta & aS\theta \\ 0 & 0 & 0 & 1 \end{bmatrix} \quad ■■ \tag{2.9.12}$$

EXAMPLE 2.9.3 ■

Determine the homogeneous transformation corresponding to a rotation of ψ about the x axis, followed by a translation of a units along the x axis, followed by a translation of b units along the z axis, followed by a rotation of θ about the z axis.

Solution. In this case,

$$\mathbf{T} = \mathbf{R}(z, \theta)\mathbf{T}(z, b)\mathbf{T}(x, a)\mathbf{R}(x, \psi). \tag{2.9.13}$$

For the second method, $\mathbf{T} = R(x, \theta)T(v, b)$.

$$\therefore \quad \mathbf{T} = \begin{bmatrix} 1 & 0 & 0 \\ 0 & C\theta & -S\theta \\ 0 & S\theta & C\theta \\ 0 & 0 & 0 \end{bmatrix} \begin{bmatrix} 0 & 1 & 0 & 0 & 0 \\ 0 & 0 & 1 & 0 & 0 \\ 0 & 0 & 0 & 1 & b \\ 1 & 0 & 0 & 0 & 1 \end{bmatrix}$$

$$\times \begin{bmatrix} 1 & 0 & 0 & a \\ 0 & 1 & 0 & 0 \\ 0 & 0 & 1 & 0 \\ 0 & 0 & 0 & 1 \end{bmatrix} \begin{bmatrix} 1 & 0 & 0 & 0 \\ 0 & C\psi & -S\psi & 0 \\ 0 & S\psi & C\psi & 0 \\ 0 & 0 & 0 & 0 \end{bmatrix}$$

$$= \begin{bmatrix} C\theta & -C\psi S\theta & S\psi S\theta & aC\theta \\ S\theta & C\psi C\theta & -S\psi C\theta & aS\theta \\ 0 & S\psi & C\psi & b \\ 0 & 0 & 0 & 1 \end{bmatrix} \quad \blacksquare\blacksquare \tag{2.9.14}$$

2.10 DENAVIT–HARTENBERG (D-H) REPRESENTATIONS

As defined previously, a generalized robotic arm may be considered a kinematic chain interconnected at joints whose motions can be servo-controlled by proper actuators. The interconnections are either revolute (rotary) or prismatic (telescopic) or a combination of the two. Figure 2.10.1 depicts a number of common robotic joints.

Denavit and Hartenberg (1955) defined the concept of a "lower pair" joint as the one created by contacting surfaces. In this sense, "upper pair" joints were the ones created by either point or line contacts between two bodies. Thus, in Fig. 2.10.1 the joints (a), (b), (c), (d), (e), and (f) are lower pairs while the joints (g), (h), (i), and (j) are upper pairs. Now consider a pair of adjacent robotic links interconnected by lower pairs capable of rotation and telescopic motion, that is, case (a) in Figure 2.10.1 (see Fig. 2.10.2).

The coordinates are chosen as follows:

1. The z_i coordinate is along the axis of motion or rotation of the $(i + 1)$th joint.
2. The x_i axis is in the direction normal to both the z_i and z_{i-1} axes, pointing away from the z_{i-1} axis (such as to make a right-handed coordinate with either z_i and z_{i-1} or z_{i-1} and z_i).
3. The y_i coordinate is chosen so as to make the $x_i y_i z_i$ coordinate system a right-handed coordinate system.
4. The distance between the two common normals a_i and a_{i-1} is called the link distance d_i.

5. The length of the normal a_i is the minimum distance between the z_i and z_{i-1} axes. If this is zero, the direction of a_i becomes arbitrary. Note that a_i is also referred to as length of the ith link.
6. The relative orientation of the x_i axis with respect to the x_{i-1} axis is called θ or the vector of the joint angles.
7. The relative orientation of the z_i axis with respect to the z_{i-1} axis is called α_i or the vector of the twist angles.

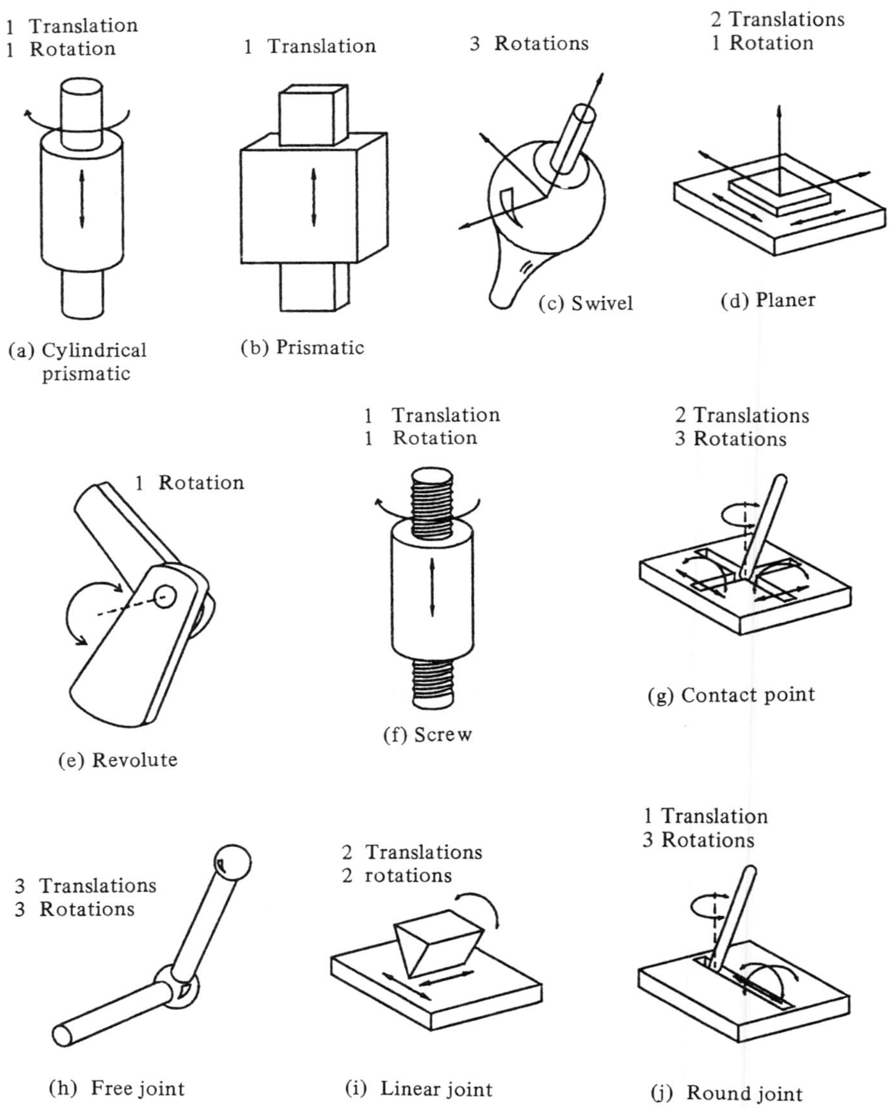

Figure 2.10.1 General description of robotic joints.

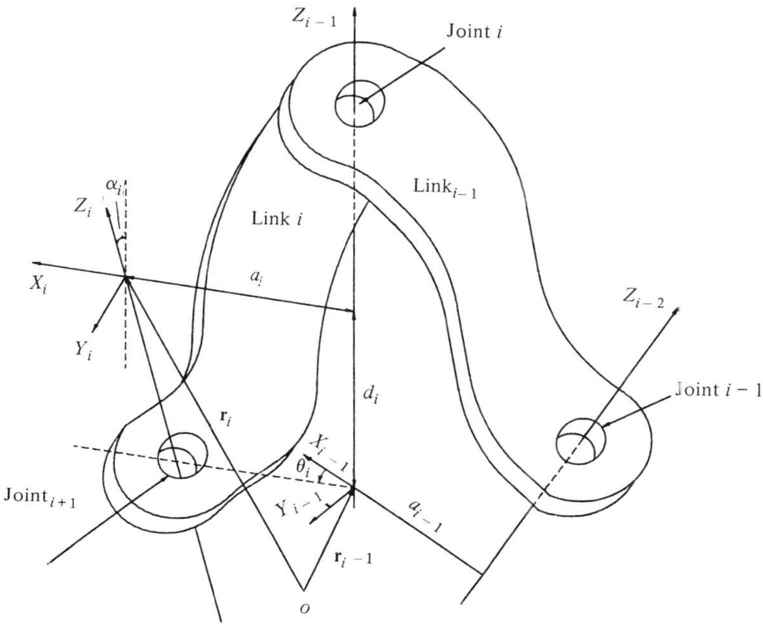

Figure 2.10.2 Generalized robotic link coordinate system.

The four parameters d_i, θ_i, a_i, and α_i constitute a minimally sufficient set to determine the kinematic configuration of each link of the robotic arm. Note that d_i is the *joint distance*, θ_i is the *joint angle*, a_i is the *link length*, and α_i is the *link twist angle*. For a plane revolute joint, generally d_i, a_i, and α_i are all constant while θ_i varies as link i rotates about the axis of joint i. On the other hand, for a prismatic joint, θ_i, a_i, and α_i are constant while d_i varies as link i slides along the axis of joint i. Thus, for both cases, a_i and α_i are generally constant and depend on the design of the robot. For example, when the joint axis of adjacent joints intersect, then the link length a is also zero (such is the case with prismatic joints).

Once the D-H coordinate system for each link is established, a homogeneous transformation matrix can be developed easily relating the ith coordinate frame. Referring to Fig. 2.10.2, we can see that a point $\mathbf{r}_i$ expressed in the ith coordinate system may be expressed in the $(i-1)$th coordinate system as $\mathbf{r}_{i-1}$ by performing the following successive transformations:

1. Rotate about the z_{i-1} axis by an angle θ_i to align the x_{i-1} and x_i axes: $\mathbf{R}(z_{i-1}, \theta_i)$.
2. Translate along the z_{i-1} axis a distance d_i to bring the x_{i-1} and x_i axis: $\mathbf{T}(z_{i-1}, d_i)$.
3. Translate along the x_i axis a distance a_i to bring the two origins into coincidence: $\mathbf{T}(x_i, a_i)$.
4. Rotate about the x_i axis an angle α_i to make the two coordinate systems completely coincide: $\mathbf{R}(x_i, \alpha_i)$.

Thus, the complete transformation of link i with respect to link $i-1$ or joint i with respect to joint $i-1$ is

$$\mathbf{H}_{i-1}^{i} = \mathbf{T}_{i-1}^{i} = \mathbf{A}_{i-1}^{i} = \mathbf{R}(z_{i-1}, \theta_i)\mathbf{T}(z_{i-1}, d_i)\mathbf{T}(x_i, a_i)\mathbf{R}(x_i, \alpha_i) \tag{2.10.1}$$

or

$$\mathbf{A}_{i-1}^{i} = \begin{bmatrix} C\theta_i & -S\theta_i & 0 \\ S\theta_i & C\theta_i & 0 \\ 0 & 0 & 1 \\ 0 & 0 & 0 \end{bmatrix} \begin{bmatrix} 0 & 1 & 0 & 0 & 0 \\ 0 & 0 & 1 & 0 & 0 \\ 0 & 0 & 0 & 1 & d_i \\ 1 & 0 & 0 & 0 & 1 \end{bmatrix}$$

$$\times \begin{bmatrix} 1 & 0 & 0 & a_i \\ 0 & 1 & 0 & 0 \\ 0 & 0 & 1 & 0 \\ 0 & 0 & 0 & 1 \end{bmatrix} \begin{bmatrix} 1 & 0 & 0 & 0 \\ 0 & C\alpha_i & -S\alpha_i & 0 \\ 0 & S\alpha_i & C\alpha_i & 0 \\ 0 & 0 & 0 & 1 \end{bmatrix}$$

$$= \begin{bmatrix} C\theta_i & -C\alpha_i S\theta_i & S\alpha_i S\theta_i & a_i C\theta_i \\ S\theta_i & C\alpha_i C\theta_i & -S\alpha_i C\theta_i & a_i S\theta_i \\ 0 & S\alpha_i & C\alpha_i & d_i \\ 0 & 0 & 0 & 1 \end{bmatrix} \tag{2.10.2}$$

We frequently omit the subscript $i-1$ and call $\mathbf{A}_{i-1}^{i}$ simply $\mathbf{A}_i$ matrices pertaining to the ith joint or link. Note that $\mathbf{A}_i$ generally has the form

$$\mathbf{A}_{i-1}^{i} = \mathbf{A}_i = \begin{bmatrix} n_{x_i} & o_{x_i} & a_{x_i} & p_{x_i} \\ n_{y_i} & o_{y_i} & a_{y_i} & p_{y_i} \\ n_{z_i} & o_{z_i} & a_{z_i} & p_{z_i} \\ 0 & 0 & 0 & 1 \end{bmatrix}, \tag{2.10.3}$$

Thus, its inverse is simply $\mathbf{A}_i^{-1}$ such that

$$\mathbf{A}_i^{i-1} = \mathbf{A}_i^{-1} = \begin{bmatrix} n_{x_i} & n_{y_i} & n_{z_i} & -\mathbf{n}_i \cdot \mathbf{p} \\ o_{x_i} & o_{y_i} & o_{z_i} & -\mathbf{o}_i \cdot \mathbf{p} \\ a_{x_i} & a_{y_i} & a_{z_i} & -\mathbf{a}_i \cdot \mathbf{p} \\ 0 & 0 & 0 & 1 \end{bmatrix}. \tag{2.10.4}$$

Furthermore, if $\mathbf{r}_i$ is the homogeneous coordinate of the ith joint and $\mathbf{r}_{i-1}$ is the homogeneous coordinate of the $(i-1)$th joint, it is clear that

$$\mathbf{r}_{i-1} = \mathbf{A}_{i-1}^{i}\mathbf{r}_i, \tag{2.10.5}$$

where

$$\mathbf{r}_i \equiv [x_i \; y_i \; z_i \; 1]^T \tag{2.10.6}$$

and

$$\mathbf{r}_{i-1} \equiv [x_{i-1} \; y_{i-1} \; z_{i-1} \; 1]^T. \tag{2.10.7}$$

The following algorithm can be used to establish the link coordinate system.

Algorithm 2.10.1. This procedure completely assigns and establishes link coordinate systems compatible with D-H transformations for a given n-degree of freedom robotic arm. Adjacent links can be related to each other kinematically through a set of 4×4 homogeneous transformation matrices. The first coordinate system pertaining to the base is denoted by a homogeneous system $[x_0\ y_0\ z_0\ 1]^T$ at the supporting base. The origin of this system is denoted as the 0th joint.

1. Establish the base coordinate $[x_0\ y_0\ z_0]$ such that the z_0 axis lies along the axis of motion of joint 1.
2. Initialize and loop for each $i = 1, 2, \ldots, n$.
3. Align all z_i with the axis of motion or rotation of joint $i + 1$.
4. Establish the origin of the ith coordinate system either at the intersection of the z_i and z_{i-1} axes or at the intersection of the common normal between the z_i and z_{i-1} axes.
5. Establish the x_i at each ith joint, either by $\mathbf{x}_i = \pm(\mathbf{z}_{i-1} \times \mathbf{z}_i)$ or along the common normal between $\mathbf{z}_{i-1}$ and $\mathbf{z}_i$ when they are parallel.
6. Establish the y_i axis by $\mathbf{y}_i = +(\mathbf{z}_i \times \mathbf{x}_i)$ to complete the right-handed coordinate system.
7. Find d_i as the distance from the origin of the $(i - 1)$th coordinate system to the intersection of the z_{i-1} axis and the x_i axis.
8. Find a_i as the distance from the intersection of the z_{i-1} axis to the origin of the ith coordinate system.
9. Find θ_i as the angle of rotation from the x_{i-1} axis to the x axis.
10. Find α_i as the angle of rotation from the z_{i-1} axis to the z_i axis about the x_i axis.

Specification of $\mathbf{T}_n^6$ Matrices The description of the end of the manipulator, in terms of homogeneous transformations, is given by

$$\mathbf{T}_{n-1}^6 = \mathbf{A}_n \mathbf{A}_{n+1} \cdots \mathbf{A}_6 \tag{2.10.8}$$

with respect to the link coordinate frame $n - 1$. The description of the end point of any link i with respect to the link coordinate frame j is

$$\mathbf{T}_j^i = \mathbf{A}_{j+1} \mathbf{A}_{j+2} \cdots \mathbf{A}_i, \qquad j < i. \tag{2.10.9}$$

Commonly, the end of the manipulator, link coordinate frame 6, with respect to the base coordinate is given by the following transformation:

$$\mathbf{T}_6 = \mathbf{A}_1 \mathbf{A}_2 \mathbf{A}_3 \mathbf{A}_4 \mathbf{A}_5 \mathbf{A}_6. \tag{2.10.10}$$

If the manipulator is described with respect to a reference coordinate frame by a transform $\mathbf{Z}$ and has a tool attached to its end whose attachment is described by transform $\mathbf{E}$ with respect to the hand coordinate, then from Fig. 2.10.3 it is clear that the position and orientation of the tool with respect to the base coordinate is given by $\mathbf{X} = \mathbf{Z}\mathbf{T}_6\mathbf{E}$. Thus,

$$\mathbf{T}_6 = \mathbf{Z}^{-1}\mathbf{X}\mathbf{E}^{-1}.$$

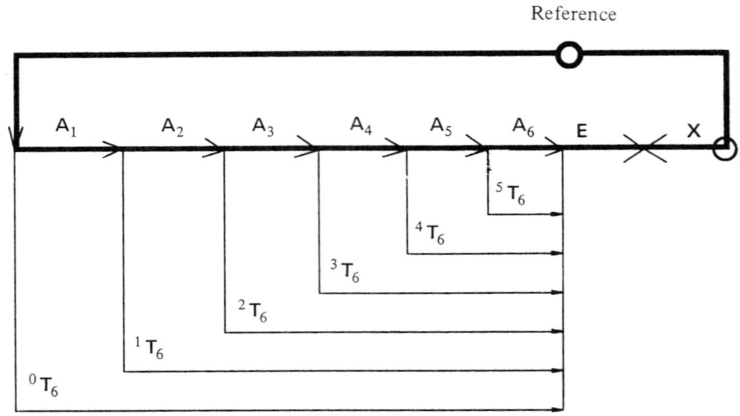

Figure 2.10.3 Transformation graph for a tool.

2.11 DIRECT KINEMATICS IN ROBOTICS

Equating Eqs. (2.10.10) and (2.11.1), we obtain an equation for the tool kinematics in robotic manipulation with six degrees of freedom:

$$\mathbf{A}_1\mathbf{A}_2\mathbf{A}_3\mathbf{A}_4\mathbf{A}_5\mathbf{A}_6 = \mathbf{A}^{-1}\mathbf{X}\mathbf{E}^{-1} \tag{2.11.1}$$

If **Z**, **X**, and **E** are given, then one can set up 16 equations with n or less unknowns to solve for certain joint parameters. Since the last row of all D-H transformations of interest has known values, one is dealing with 12 equations in n unknowns. Also, note that the following equations hold:

$$\mathbf{n}\cdot\mathbf{n} = \mathbf{o}\cdot\mathbf{o} = \mathbf{a}\cdot\mathbf{a} = 1 \tag{2.11.2}$$

and

$$\mathbf{n}\times\mathbf{o} = \mathbf{a} \tag{2.11.3}$$

Thus, one is actually dealing with six independent equations in n unknowns with n independent joint motion variables.

In direct kinematics, one calculates the D-H matrices and finally comes up with an expression for the tip transformation matrix $\mathbf{T}_o^n$. In what follows, we elaborate on some common robot geometries and derive the pertinent equations for the direct kinematics. The derived equations are used later for the solutions to the inverse kinematics problem.

Before proceeding further, let us restate the direct kinematics problem in robotics:

Given the n-dimensional vector of the joint variables, such as the angles or prismatic displacements in the form $\boldsymbol{\theta}_i = [\theta_1\ \theta_2\ \theta_3 \cdots \theta_n]$, *what is the corresponding kinematic attitude matrix* $\mathbf{T}_o^n$ *pertaining to the position and the orientation of the robot end effector?*

EXAMPLE 2.11.1 ▪

Derive an expression for the direct kinematics of a simple cylindrical robot (Fig. 2.11.1).

Solution. Note that in this case

$$\mathbf{T}_0^3 = \mathbf{Trans}(z_0 z_1)\mathbf{Rot}(z_0, \theta)\mathbf{Trans}(u, r)$$

$$= \begin{bmatrix} 1 & 0 & 0 & 0 \\ 0 & 1 & 0 & 0 \\ 0 & 0 & 1 & Z \\ 0 & 0 & 0 & 1 \end{bmatrix} \begin{bmatrix} C\theta & -S\theta & 0 & 0 \\ S\theta & C\theta & 0 & 0 \\ 0 & 0 & 1 & 0 \\ 0 & 0 & 0 & 1 \end{bmatrix} \begin{bmatrix} 1 & 0 & 0 & rC\theta \\ 0 & 1 & 0 & 0 \\ 0 & 0 & 1 & d_i \\ 0 & 0 & 0 & 1 \end{bmatrix} \qquad (2.11.4)$$

or, finally,

$$\mathbf{T}_0^3 = \begin{bmatrix} C\theta & -S\theta & 0 & rC\theta \\ S\theta & C\theta & 0 & rS\theta \\ 0 & 0 & 1 & Z \\ 0 & 0 & 0 & 1 \end{bmatrix}. \qquad \blacksquare\blacksquare \qquad (2.11.5)$$

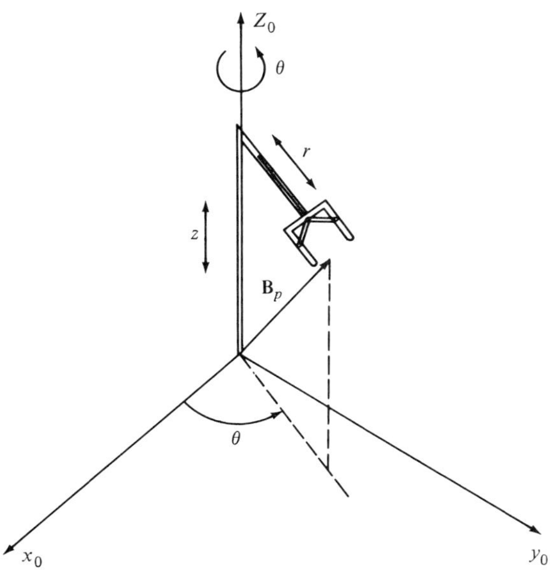

Figure 2.11.1 Simple cylindrical polar robot.

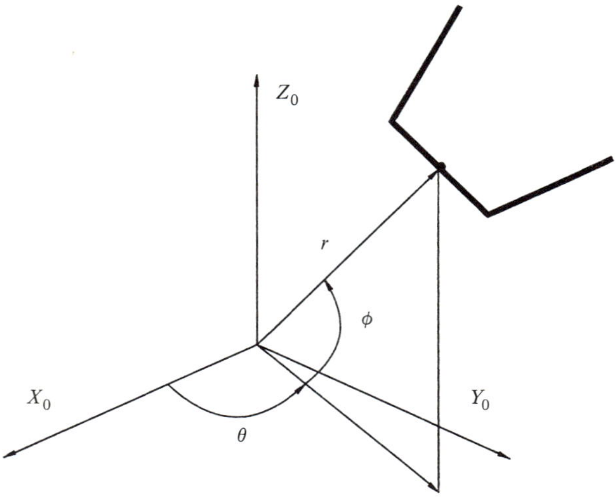

Figure 2.11.2 Simple spherical robot.

EXAMPLE 2.11.2 ■

Derive an expression for the direct kinematics of a simple spherical robot (Fig. 2.11.2).

Solution. Note that in this case

$$\mathbf{T}_0^3 = \mathbf{Rot}(z_0, \theta)\,\mathbf{Rot}(v, -\phi)\,\mathbf{Trans}(u, r)$$

$$= \begin{bmatrix} C\theta & -S\theta & 0 & 0 \\ S\theta & C\theta & 0 & 0 \\ 0 & 0 & 1 & 0 \\ 0 & 0 & 0 & 1 \end{bmatrix} \begin{bmatrix} C\phi & 0 & -S\phi & 0 \\ 0 & 1 & 0 & 0 \\ S\phi & 0 & C\phi & 0 \\ 0 & 0 & 0 & 1 \end{bmatrix} \begin{bmatrix} 1 & 0 & 0 & r \\ 0 & 1 & 0 & 0 \\ 0 & 0 & 1 & 0 \\ 0 & 0 & 0 & 1 \end{bmatrix} \tag{2.11.6}$$

or, finally,

$$\mathbf{T}_0^3 = \begin{bmatrix} C\theta C\phi & -S\theta & -C\theta S\phi & rC\theta C\phi \\ S\theta C\phi & C\theta & -S\theta S\phi & rS\theta C\phi \\ S\phi & 0 & C\phi & rS\phi \end{bmatrix}. \quad ■■ \tag{2.11.7}$$

To arrive at $\mathbf{A}_i$ and $\mathbf{T}$ matrices, we first construct Table 2.11.1 for joint parameters.

TABLE 2.11.1 JOINT AND LINK PARAMETERS FOR A POLAR CYLINDRICAL ROBOT

Link	Variable	α_i	a_i	d_i	$S\alpha$	$C\alpha$
1	θ_1	$\alpha_1 = -90°$	$a_1 = 0$	$d_1 = 0$	-1	0
2	θ_2	$\alpha_2 = 90°$	$a_2 = 0$	d_2	1	0
3	d_3	$\alpha_3 = 0°$	$a_3 = 0$	d_3	0	1
4	θ_3	$\alpha_3 = 90°$	$a_4 = 0$	$d_4 = 0$	-1	0
5	θ_5	$\alpha_5 = 90°$	$a_5 = 0$	$d_5 = 0$	1	0
6	θ_6	$\alpha_6 = 0°$	$a_6 = 0$	$d_6 = 0$	0	1

EXAMPLE 2.11.3 ■

Derive expressions for the direct kinematics of a RPY (roll–pitch–yaw) and Euler type robotic hand.

Solution. Refer to Figures 2.7.1 and 2.8.1. The exprssions obtained are the same, namely,

$$\mathbf{RPY}(\phi, \theta, \psi) = \mathbf{Euler}(\phi, \theta, \psi)$$

$$= \begin{bmatrix} C\phi C\theta & C\phi S\theta S\psi - S\phi C\psi & C\phi S\theta C\psi + S\theta S\psi & 0 \\ S\psi C\theta & S\phi S\theta S\psi + C\phi C\psi & S\phi S\theta C\psi - C\theta S\psi & 0 \\ -S\theta & C\theta S\psi & C\theta C\psi & 0 \\ 0 & 0 & 0 & 1 \end{bmatrix}. \qquad (2.11.8)$$

■ ■

EXAMPLE 2.11.4 ■

Completely derive expressions for the transformation matrices of a cylindrical polar manipulator shown in Figure 2.11.3.

Solution. The cylindrical polar robotic manipulators (Stanford arm) may be represented by the joint coordinate systems shown in Figure 2.11.3. Note that there are one telescopic motion and five rotary motions involved with this robot. Furthermore, there are two rolling motions involved before and after the pitch motion to simulate the yaw motion.

$$\mathbf{A}_0^1 = \begin{bmatrix} C_1 & 0 & -S_1 & 0 \\ S_1 & 0 & C_1 & 0 \\ 0 & -1 & 0 & 0 \\ 0 & 0 & 0 & 1 \end{bmatrix}, \qquad (2.11.9)$$

$$\mathbf{A}_1^2 = \begin{bmatrix} C_2 & 0 & S_2 & 0 \\ S_2 & 0 & -C_2 & 0 \\ 0 & 1 & 0 & d_2 \\ 0 & 0 & 0 & 1 \end{bmatrix}, \qquad (2.11.10)$$

$$\mathbf{A}_2^3 = \begin{bmatrix} 1 & 0 & 0 & 0 \\ 0 & 1 & 0 & 0 \\ 0 & 0 & 1 & d_3 \\ 0 & 0 & 0 & 1 \end{bmatrix}, \qquad (2.11.11)$$

$$\mathbf{A}_3^4 = \begin{bmatrix} C_4 & 0 & -S_4 & 0 \\ S_4 & 0 & C_4 & 0 \\ 0 & -1 & 0 & 0 \\ 0 & 0 & 0 & 1 \end{bmatrix}, \qquad (2.11.12)$$

$$\mathbf{A}_4^5 = \begin{bmatrix} C_5 & 0 & S_5 & 0 \\ S_5 & 0 & -C_5 & 0 \\ 0 & 1 & 0 & 0 \\ 0 & 0 & 0 & 1 \end{bmatrix}, \qquad (2.11.13)$$

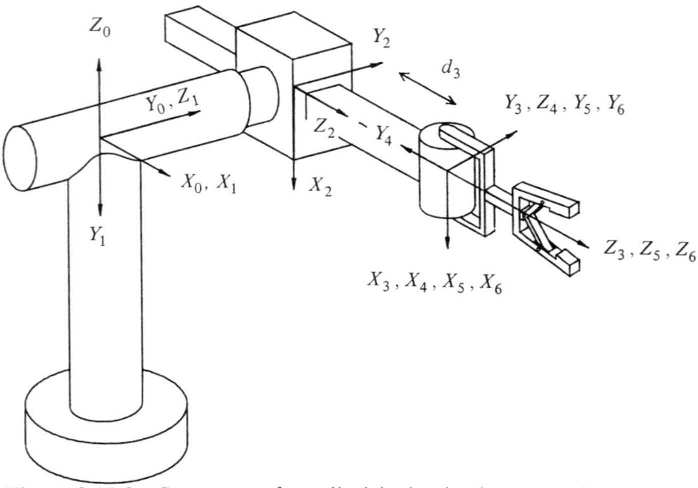

Figure 2.11.3 Structure of a cylindrical robotic manipulator.

and

$$\mathbf{A}_5^6 = \begin{bmatrix} C_6 & -S_6 & 0 & 0 \\ S_6 & C_6 & 0 & 0 \\ 0 & 0 & 1 & 0 \\ 0 & 0 & 0 & 1 \end{bmatrix}. \tag{2.11.14}$$

To find the **T** matrices we must find the products of **A** matrices such that

$$\mathbf{T}_i^j = \mathbf{A}_i \mathbf{A}_{i+1} \cdots \mathbf{A}_j. \tag{2.11.15}$$

$$\mathbf{T}_5^6 = \mathbf{A}_5^6 = \begin{bmatrix} C_6 & -S_6 & 0 & 0 \\ S_6 & C_6 & 0 & 0 \\ 0 & 0 & 1 & 0 \\ 0 & 0 & 0 & 1 \end{bmatrix}, \tag{2.11.16}$$

$$\mathbf{T}_4^6 = \mathbf{A}_4^6 \mathbf{A}_5^6 = \begin{bmatrix} C_5 C_6 & -C_5 S_6 & S_5 & 0 \\ S_5 C_6 & -S_5 S_6 & -C_5 & 0 \\ S_6 & C_6 & 0 & 0 \\ 0 & 0 & 0 & 1 \end{bmatrix}, \tag{2.11.17}$$

$$\mathbf{T}_3^6 = \mathbf{A}_3^4 \mathbf{A}_4^5 \mathbf{A}_5^6 = \begin{bmatrix} C_4 C_5 C_6 - S_4 S_6 & -C_4 C_5 S_6 - S_4 C_6 & C_4 S_5 & 0 \\ S_4 C_5 C_6 + C_4 S_6 & -S_4 C_5 S_6 + C_4 C_6 & S_4 S_5 & 0 \\ -S_5 C_6 & S_5 S_6 & C_5 & 0 \\ 0 & 0 & 0 & 1 \end{bmatrix}, \tag{2.11.18}$$

$$\mathbf{T}_2^6 = \mathbf{A}_2^3 \mathbf{A}_3^4 \mathbf{A}_4^5 \mathbf{A}_5^6 = \begin{bmatrix} C_4 C_5 C_6 - S_4 S_6 & -C_4 C_5 S_6 - S_4 C_6 & C_4 S_5 & 0 \\ S_4 C_5 C_6 + C_4 S_6 & -S_4 C_5 S_6 + C_4 C_6 & S_4 S_5 & 0 \\ -S_5 C_6 & S_5 S_6 & C_5 & d_3 \\ 0 & 0 & 0 & 1 \end{bmatrix}, \tag{2.11.19}$$

$$
\mathbf{T}_1^6 = \mathbf{A}_1^2\mathbf{A}_2^3\mathbf{A}_3^4\mathbf{A}_4^5\mathbf{A}_5^6
$$

$$
= \begin{bmatrix}
C_2(C_4C_5C_6 - S_4S_6) - S_2S_5C_6 & -C_2(C_4C_5S_6 + S_4C_6) + S_2S_5S_6 & C_2C_4S_5 + S_2C_5 & S_2d_3 \\
S_2(C_4C_5C_6 - S_4S_6) + C_2S_5C_6 & -S_2(C_4C_5S_6 + S_4C_6) - C_2S_5S_6 & S_2C_4S_5 - C_2C_5 & -C_2d_3 \\
S_4C_5C_6 + C_4S_6 & -S_4C_5S_6 + C_4C_6 & S_4S_5 & d_2 \\
0 & 0 & 0 & 1
\end{bmatrix}, \qquad (2.11.20)
$$

and, finally,

$$
\mathbf{T}_0^6 = \mathbf{A}_0^1\mathbf{A}_1^2\mathbf{A}_2^3\mathbf{A}_3^4\mathbf{A}_4^5\mathbf{A}_5^6 = \begin{bmatrix} n_x & o_x & a_x & p_x \\ n_y & o_y & a_y & p_y \\ n_z & o_z & a_z & p_z \\ 0 & 0 & 0 & 1 \end{bmatrix}, \qquad (2.11.21)
$$

where

$$
n_x = C_1[C_2(C_4C_5C_6 - S_4S_6) - S_2S_5C_6] - S_1(S_4C_5C_6 + C_4C_6), \qquad (2.11.22)
$$

$$
n_y = S_1[C_2(C_4C_5C_6 - S_4S_6) - S_2S_5C_6] + C_1(S_4C_5C_6 + C_4S_6), \qquad (2.11.23)
$$

$$
n_z = -S_2(C_4C_5C_6 - S_4S_6) - C_2S_5C_6, \qquad (2.11.24)
$$

$$
o_x = C_1[-C_2(C_4C_5S_6 - S_4C_6) - S_2S_5S_6] - S_1(-S_4C_5S_6 + C_4C_6), \qquad (2.11.25)
$$

$$
o_y = S_1[-C_2(C_4C_5S_6 - S_4C_6) - S_2S_5S_6] + C_1(-S_4C_5S_6 + C_4C_6), \qquad (2.11.26)
$$

$$
o_z = S_2(C_4C_5S_6 + S_4C_6) + C_2S_5S_6, \qquad (2.11.27)
$$

$$
a_x = C_1(C_2C_4S_5 + S_2C_5) - S_1S_4S_5, \qquad (2.11.28)
$$

$$
a_y = S_1(C_2C_4S_5 + S_2C_5) + C_1S_4C_5, \qquad (2.11.29)
$$

$$
a_z = -S_2C_4S_5 + C_2C_5, \qquad (2.11.30)
$$

$$
p_x = C_1S_2d_3 - S_1d_2, \qquad (2.11.31)
$$

$$
p_y = S_1S_2d_3 + C_1d_2, \qquad (2.11.32)
$$

and

$$
p_z = C_2d_3. \quad \blacksquare\blacksquare \qquad (2.11.33)
$$

We shall use the above equations for the construction of the inverse kinematics solutions for a cylindrical polar manipulator later on in this chapter.

EXAMPLE 2.11.5 ■

For the MiniMover Microbot shown in Figure 2.11.4, construct (a) a set of robotic coordinate frames, (b) a table for joint parameters, and (c) the components of the $\mathbf{T}_0^5$ matrix.

Solution. (a) The structure of the robotic coordinate frames is shown in Figure 2.11.5.

(b) Table 2.11.2 presents the corresponding joint parameters.

(c) Thus, the D-H represent $\mathbf{A}_{i-1}^i$ takes the form

$$\mathbf{A}_{i-1}^i = \begin{bmatrix} C\theta_i & -C\alpha_i S\theta_i & S\alpha_i S\theta_i & a_i C\theta_i \\ S\theta_i & C\alpha_i C\theta_i & -S\alpha_i C\theta_i & a_i S\theta_i \\ 0 & S\alpha_i & C\alpha_i & d_i \\ 0 & 0 & 0 & 1 \end{bmatrix}, \tag{2.11.34}$$

such that in expanded form

$$\mathbf{A}_0^1 = \begin{bmatrix} C_1 & 0 & -S_1 & 0 \\ S_1 & 0 & C_1 & 0 \\ 0 & -1 & 0 & 8.1 \\ 0 & 0 & 0 & 1 \end{bmatrix}, \quad \mathbf{A}_1^2 = \begin{bmatrix} C_2 & -S_2 & 0 & 7C_2 \\ S_2 & C_2 & 0 & 7S_2 \\ 0 & 0 & 1 & 0 \\ 0 & 0 & 0 & 1 \end{bmatrix}; \tag{2.11.35}$$

$$\mathbf{A}_2^3 = \begin{bmatrix} C_3 & -S_3 & 0 & 7C_3 \\ S_3 & C_3 & 0 & 7S_3 \\ 0 & 0 & 1 & 0 \\ 0 & 0 & 0 & 1 \end{bmatrix}, \quad \mathbf{A}_3^4 = \begin{bmatrix} C_4 & 0 & S_4 & 0 \\ S_4 & 0 & -C_4 & 0 \\ 0 & 1 & 0 & 0 \\ 0 & 0 & 0 & 1 \end{bmatrix}; \tag{2.11.36}$$

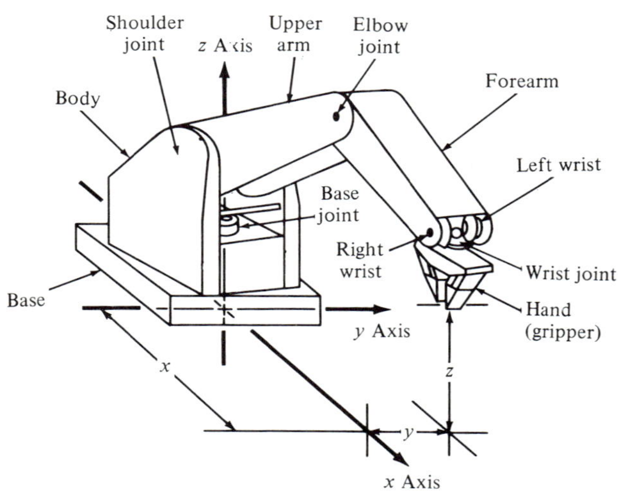

Figure 2.11.4 Configuration of MiniMover Microbot.

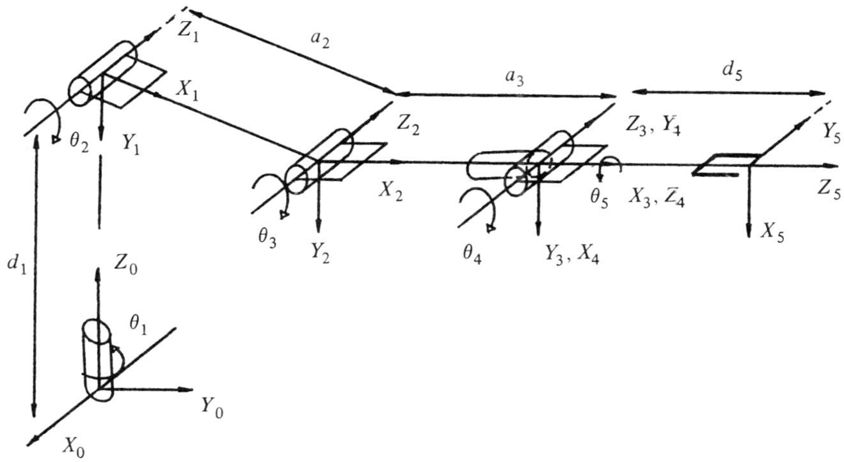

Figure 2.11.5 Structure of coordinate frames.

TABLE 2.11.2 JOINT COORDINATE PARAMETERS FOR MINIMOVER MICROBOT

Joint	θ_i	α_i	a_i (in.)	d_i (in.)
$i = 1$	θ_1	$-90°$	0	8.1
$i = 2$	θ_2	$0°$	7	0
$i = 3$	θ_3	$0°$	7	0
$i = 4$	θ_4	$90°$	0	0
$i = 5$	θ_5	0	0	3.8

and

$$\mathbf{A}_4^5 = \begin{bmatrix} C_5 & -S_5 & 0 & 0 \\ S_5 & C_5 & 0 & 0 \\ 0 & 0 & 1 & 3.8 \\ 0 & 0 & 0 & 1 \end{bmatrix}; \qquad \mathbf{T}_0^5 = \mathbf{A}_0^1\mathbf{A}_1^2\mathbf{A}_2^3\mathbf{A}_3^4\mathbf{A}_4^5. \tag{2.11.37}$$

Thus, the Microbot's arm matrix $\mathbf{T}_0^5$ finally reduces to having the following components:

$$(T_0^5)_{11} = C_1C_{23}C_4C_5 - C_1S_{23}S_4C_5 - S_1S_5, \tag{2.11.38}$$

$$(T_0^5)_{12} = -C_1C_{23}C_4S_5 + C_1S_{23}S_4S_5 - S_1C_5, \tag{2.11.39}$$

$$(T_0^5)_{13} = C_1C_{23}S_4 + C_1S_{23}C_4, \tag{2.11.40}$$

$$(T_0^5)_{14} = 3.8C_1C_{23}S_4 + 3.8C_1S_{23}C_4 + 7C_1(C_{23} + C_2), \tag{2.11.41}$$

$$(T_0^5)_{21} = S_1C_{23}C_4C_5 - S_1S_{23}S_4S_5 + C_1S_5, \tag{2.11.42}$$

$$(T_0^5)_{22} = -S_1C_{23}C_4S_5 + S_1S_{23}S_4S_5 + C_1C_5, \tag{2.11.43}$$

$$(T_0^5)_{23} = S_1C_{23}S_4 + S_1S_{23}C_4, \tag{2.11.44}$$

$$(T_0^5)_{24} = 3.8S_1C_{23}S_4 + 3.8S_1S_{23}C_4 + 7S_1(C_{23} + C_2), \tag{2.11.45}$$

$$(T_0^5)_{31} = -S_{23}C_4C_5 - C_{23}S_4C_5, \tag{2.11.46}$$

$$(T_0^5)_{32} = S_{23}C_4S_5 + C_{23}S_4S_5, \tag{2.11.47}$$

$$(T_0^5)_{33} = -S_{23}S_4 + C_{23}C_4, \tag{2.11.48}$$

$$(T_0^5)_{34} = -3.8S_{23}S_4 + 3.8C_{23}C_4 + 8.1 - 7(S_{23} + S_2), \tag{2.11.49}$$

and

$$(T_0^5)_{41} = (T_0^5)_{42} = (T_0^5)_{43} = 0, \qquad (T_0^5)_{44} = 1. \quad ■■ \tag{2.11.50}$$

EXAMPLE 2.11.6 ■

Consider a standard six-axis robotic manipulator, as depicted in Figure 2.11.6, having general dimensions as well as general motions of the base, shoulder, elbow, wrist pitch, wrist yaw, and wrist roll.

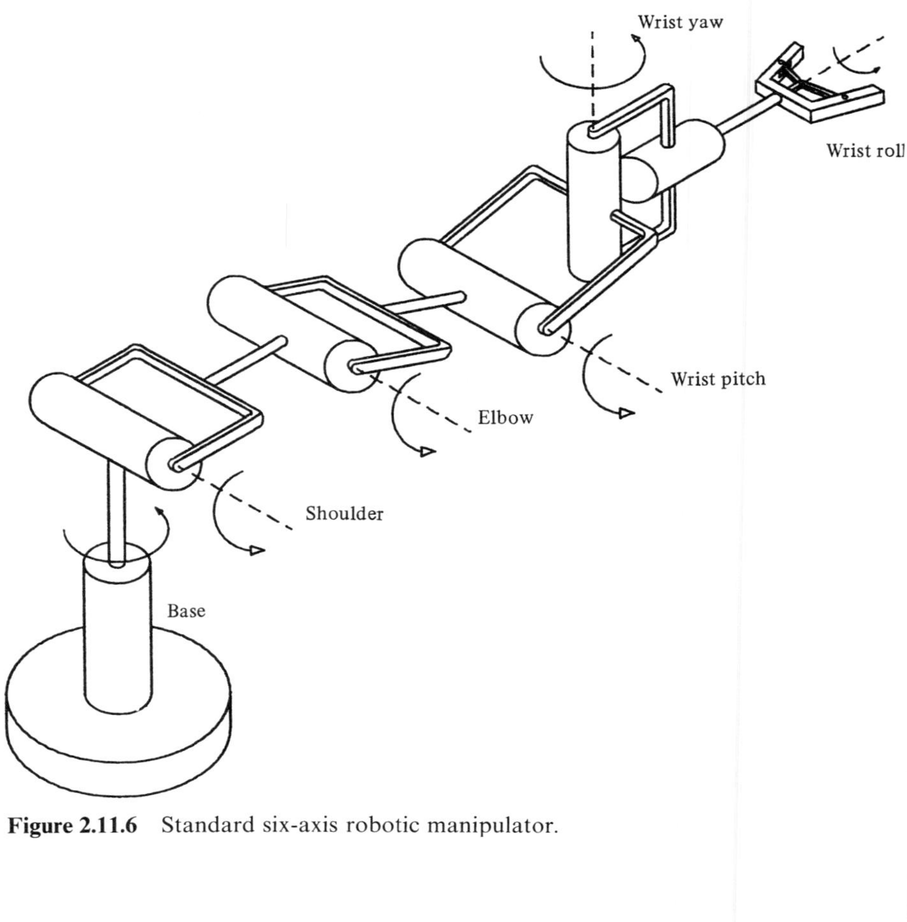

Figure 2.11.6 Standard six-axis robotic manipulator.

For the first version of this robotic manipulator, consider the gripper coordinate frame such that the axis of motion is co-directional with the axis of wrist roll (Fig. 2.11.7). For the second version of this same robotic manipulator, we consider the gripper axis of motion to be perpendicular to the axis of wrist roll (Fig. 2.11.8). For both cases, find the Denavit–Hartenberg transformations as well as the **T** matrices.

Solution. Considering Figure 2.11.7 and Algorithm 2.10.1, one can construct Table 2.11.3 for the joint parameters.

Now, based on the generalized Denavit–Hartenberg transformations, the following $\mathbf{A}_{i-1}^{i} = \mathbf{A}_i$ matrices are derived for the ith joint given that

$$\mathbf{A}_i = \begin{bmatrix} C\theta_i & -S\theta_i C\alpha_i & S\theta_i S\alpha_i & a_i C\theta_i \\ S\theta_i & C\theta_i C\alpha_i & -C\theta_i S\alpha_i & a_i S\theta_i \\ 0 & S\alpha_i & C\alpha_i & d_i \\ 0 & 0 & 0 & 1 \end{bmatrix}, \qquad (2.11.51)$$

$$\mathbf{A}_1 = \begin{bmatrix} C\theta_1 & 0 & -S\theta_1 & 0 \\ S\theta_1 & 0 & C\theta_1 & 0 \\ 0 & -1 & 0 & d_1 \\ 0 & 0 & 0 & 1 \end{bmatrix}, \qquad (2.11.52)$$

$$\mathbf{A}_2 = \begin{bmatrix} C\theta_2 & -S\theta_2 & 0 & a_2 C\theta_2 \\ S\theta_2 & C\theta_2 & 0 & a_2 S\theta_2 \\ 0 & 0 & 1 & 0 \\ 0 & 0 & 0 & 1 \end{bmatrix}, \qquad (2.11.53)$$

$$\mathbf{A}_3 = \begin{bmatrix} C\theta_3 & -S\theta_3 & 0 & a_3 C\theta_3 \\ S\theta_3 & C\theta_3 & 0 & a_3 S\theta_3 \\ 0 & 0 & 1 & 0 \\ 0 & 0 & 0 & 1 \end{bmatrix}, \qquad (2.11.54)$$

$$\mathbf{A}_4 = \begin{bmatrix} C\theta_4 & 0 & S\theta_4 & a_4 C\theta_4 \\ S\theta_4 & 0 & -C\theta_4 & a_4 S\theta_4 \\ 0 & 1 & 0 & 0 \\ 0 & 0 & 0 & 1 \end{bmatrix}, \qquad (2.11.55)$$

$$\mathbf{A}_5 = \begin{bmatrix} C\theta_5 & 0 & -S\theta_5 & 0 \\ S\theta_5 & 0 & C\theta_5 & 0 \\ 0 & -1 & 0 & 0 \\ 0 & 0 & 0 & 1 \end{bmatrix}, \qquad (2.11.56)$$

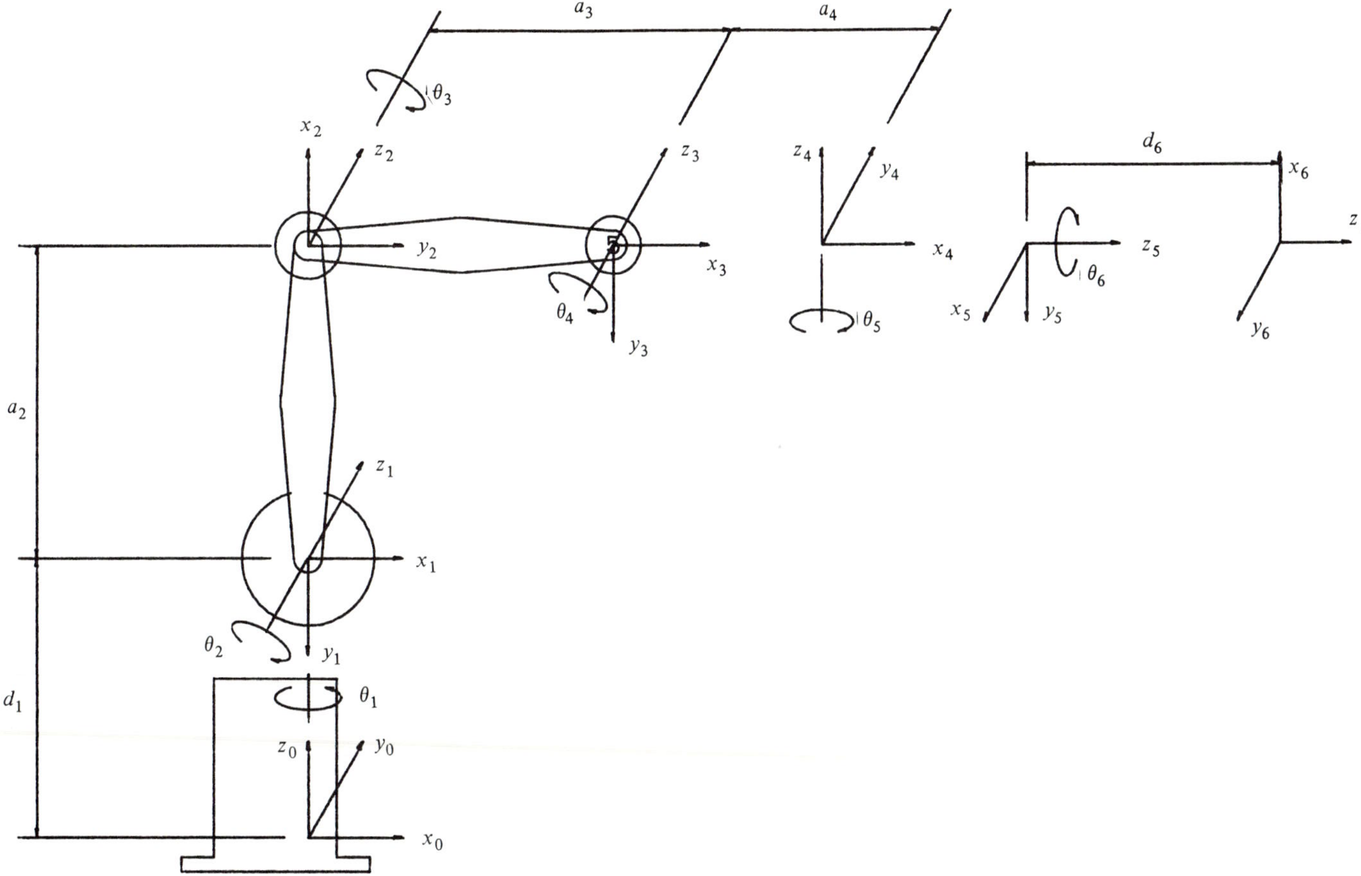

Figure 2.11.7 Standard six-axis revolute robot with a first-version gripper.

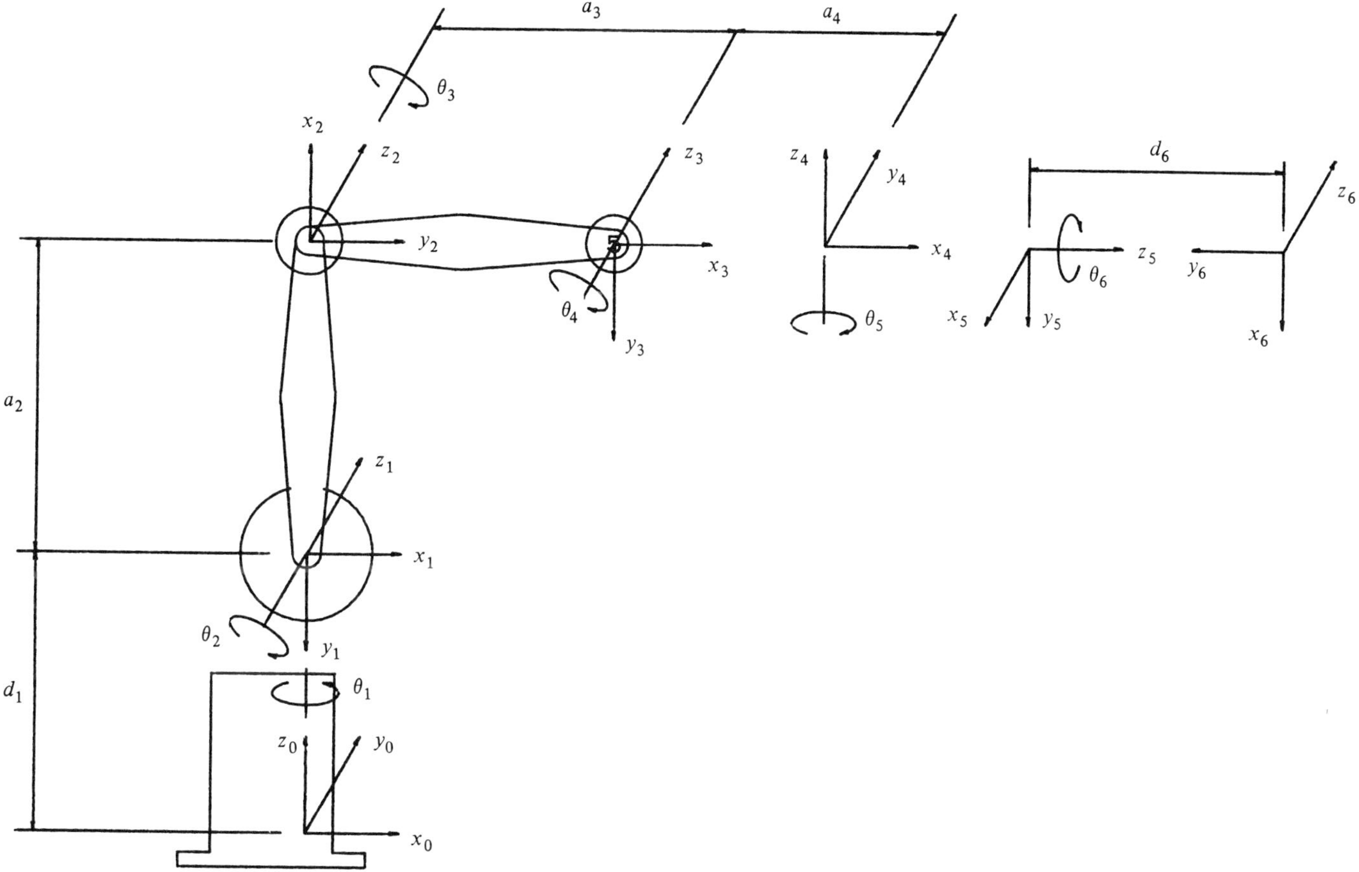

Figure 2.11.8 Standard six-axis revolute robot with a second-version gripper.

TABLE 2.11.3 JOINT PARAMETERS FOR A STANDARD SIX-AXIS ROBOT WITH A GRIPPER AXIS OF MOTION IN THE SAME DIRECTION AS THE ROLL AXIS OF MOTION

Joint	θ_i	α_i	a_i	d_i	$C\alpha_i$	$S\alpha_i$
1	θ_1	$-90°$	0	d_1	0	-1
2	θ_2	$0°$	a_2	0	1	0
3	θ_3	$0°$	a_3	0	1	0
4	θ_4	$90°$	a_4	0	0	1
5	θ_5	$-90°$	0	0	0	-1
6	θ_6	$0°$	0	d_6	1	0

and

$$\mathbf{A}_6 = \begin{bmatrix} C\theta_6 & -S\theta_6 & 0 & 0 \\ S\theta_6 & C\theta_6 & 0 & 0 \\ 0 & 0 & 1 & d_6 \\ 0 & 0 & 0 & 1 \end{bmatrix}. \tag{2.11.57}$$

Note that for this six-axis robot the **T** transformation is given by

$$\mathbf{T}_0^6 = \mathbf{A}_1\mathbf{A}_2\mathbf{A}_3\mathbf{A}_4\mathbf{A}_5\mathbf{A}_6, \tag{2.11.58}$$

where

$$\mathbf{T}_0^6 = \begin{bmatrix} n_x & o_x & a_x & p_x \\ n_y & o_y & a_y & p_y \\ n_z & o_z & a_z & p_z \\ 0 & 0 & 0 & 1 \end{bmatrix} \tag{2.11.59}$$

is completely known while the six angles θ_1, θ_2, θ_3, θ_4, θ_5, and θ_6 are the unknowns to be found.

Multiplying the matrices $\mathbf{A}_1$ through $\mathbf{A}_6$ and setting the result equal to $\mathbf{T}_0^6$ yields the following 12 equations for the determination of the vector of the joint angle $\boldsymbol{\theta}_i \equiv [\theta_1\ \theta_2\ \theta_3\ \theta_4\ \theta_5\ \theta_6]^T$:

$$n_x = C_1(C_5C_6C_{234} - S_6S_{234}) - S_1S_5C_6, \tag{2.11.60}$$

$$n_y = S_1(C_5C_6C_{234} - S_6S_{234}) + C_1S_5C_6, \tag{2.11.61}$$

$$n_z = -C_5C_6C_{234} - S_6C_{234}, \tag{2.11.62}$$

$$o_x = -C_1(C_5S_6C_{234} + C_6S_{234}) + S_1S_5S_6, \tag{2.11.63}$$

$$o_y = -S_1(C_5S_6C_{234} + C_6S_{234}) - C_1S_5S_6, \tag{2.11.64}$$

$$o_z = S_{234}C_5S_6 - C_6C_{234}, \tag{2.11.65}$$

$$a_x = -C_1S_5C_{234} - S_1C_5, \tag{2.11.66}$$

$$a_y = -S_1S_5C_{234} + C_1C_5, \tag{2.11.67}$$

$$a_z = S_{234}S_5, \tag{2.11.68}$$

$$p_x = -C_1(d_6S_5C_{234} - a_4C_{234} - a_3C_{23} - a_2C_2) - d_6S_1C_5 \tag{2.11.69}$$

$$p_y = -S_1(d_6S_5C_{234} - a_4C_{234} - a_3C_{23} - a_2C_2) + d_6C_1C_5, \tag{2.11.70}$$

and

$$p_z = d_6S_5S_{234} - a_4S_{234} - a_3S_{23} - a_2S_2 + d_1, \tag{2.11.71}$$

where for brevity we have used the following compact notations:

$$C\theta_i = C_i, \qquad S\theta_i = S_i; \tag{2.11.72}$$

$$C(\theta_i + \theta_j) = C_{ij}, \qquad S(\theta_i + \theta_j) = S_{ij}; \tag{2.11.73}$$

and

$$C(\theta_i + \theta_j + \theta_k) = C_{ijk}, \qquad S(\theta_i + \theta_j + \theta_k) = S_{ijk}. \tag{2.11.74}$$

Now consider the gripper axis of motion to be perpendicular to the axis of wrist roll. Using Figure 2.11.8 and Algorithm 2.10.1, one can construct Table 2.11.4 for the joint parameters.

Based on Table 2.11.4, the generalized Denavit–Hartenberg transformations or the **A** matrices will be the same as before, except for $\mathbf{A}_6$ which becomes

$$\mathbf{A}_6 = \begin{bmatrix} C_6 & 0 & -S_6 & 0 \\ S_6 & 0 & C_6 & 0 \\ 0 & -1 & 0 & d_6 \\ 0 & 0 & 0 & 1 \end{bmatrix}. \tag{2.11.75}$$

Now consider the inverse kinematics problem,

$$\mathbf{T}_0^6 = \mathbf{A}_1\mathbf{A}_2\mathbf{A}_3\mathbf{A}_4\mathbf{A}_5\mathbf{A}_6, \tag{2.11.76}$$

where the first five **A** matrices are given by Eqs. (2.11.52)-(2.11.57) and the last matrix is given by Eq. (2.11.75). The $\mathbf{T}_0^6$ matrix is given by Eq. (2.11.59) and all elements of $\mathbf{T}_0^6$ are known completely, as before.

TABLE 2.11.4 JOINT PARAMETERS FOR A SECOND FIVE-AXIS ROBOT WITH A GRIPPER AXIS OF MOTION PERPENDICULAR TO THE ROLL AXIS OF MOTION

Joint	θ_i	α_i	a_i	d_i	$C\alpha_i$	$S\alpha_i$
1	θ_1	$-90°$	0	d_1	0	-1
2	θ_2	$0°$	a_2	0	1	0
3	θ_3	$0°$	a_3	0	1	0
4	θ_4	$90°$	a_4	0	0	-1
5	θ_5	$-90°$	0	0	0	-1
6	θ_6	$-90°$	0	d_6	0	-1

Multiplying the above six matrices and setting the results equal to $\mathbf{T}_0^6$ yields the following 12 equations for the determination of the vector of the joint angles:

$$n_x = C_1(C_5C_6C_{234} - S_6S_{234}) - S_1S_5C_6, \tag{2.11.77}$$

$$n_y = S_1(C_5C_6C_{234} - S_6S_{234}) + C_1S_5C_6, \tag{2.11.78}$$

$$n_z = -C_5C_6C_{234} - S_6C_{234}, \tag{2.11.79}$$

$$o_x = C_1S_5C_{234} + S_1C_5, \tag{2.11.80}$$

$$o_y = S_1S_5C_{234} - C_1C_5, \tag{2.11.81}$$

$$o_z = -S_5S_{234}, \tag{2.11.82}$$

$$a_x = -C_1(C_5S_6C_{234} + C_6S_{234}) + S_1S_5S_6, \tag{2.11.83}$$

$$a_y = -S_1(C_5S_6C_{234} + C_6S_{234}) - C_1S_5S_6, \tag{2.11.84}$$

$$a_z = C_5S_6S_{234} - C_6C_{234}, \tag{2.11.85}$$

$$p_x = -C_1(d_6S_5C_{234} - a_4C_{234} - a_3C_{23} - a_2C_2) - d_6S_1C_5, \tag{2.11.86}$$

$$p_y = -S_1(d_6S_5C_{234} - a_4C_{234} - a_3C_{23} - a_2C_2) + d_6C_1C_5, \tag{2.11.87}$$

$$p_z = d_6S_5S_{234} - a_4S_{234} - a_3S_{23} - a_2S_2 + d_1. \tag{2.11.88}$$

Note that again

$$\mathbf{T}_0^6 = \begin{bmatrix} n_x & o_x & a_x & p_x \\ n_y & o_y & a_y & p_y \\ n_z & o_z & a_z & p_z \\ 0 & 0 & 0 & 1 \end{bmatrix}. \quad ■■ \tag{2.11.89}$$

EXAMPLE 2.11.7 ■

The robot shown in Figure 2.11.9 is in its home position. The axes of motion are as shown in the figure.

(a) Complete the description of all joint coordinate systems by assigning the proper set of x and y axes.
(b) Construct a table for joint parameters α_i, θ_i, a_i, and d_i for this robot.
(c) For $\boldsymbol{\theta} \equiv [90° \; 0° \; 90° \; 0° \; 90°]$ determine the matrix $\mathbf{T}_5^0$.

Solution. (a) To construct the complete set of coordinate frames, let us apply Algorithm 2.10.1 and construct a right-handed coordinate system beginning with the base coordinate system $x_0y_0z_0$. The completed description of all coordinate systems are then as shown in Figure 2.11.10.

(b) Table 2.11.5 for joint parameters can then be constructed.

(c) For $\boldsymbol{\theta} \equiv [90° \; 0° \; 90° \; 0° \; 90°]$ the matrix $\mathbf{T}_0^5$ is given by

$$\mathbf{T}_0^5 = \mathbf{A}_0^1\mathbf{A}_1^2\mathbf{A}_2^3\mathbf{A}_3^4\mathbf{A}_4^5, \tag{2.11.90}$$

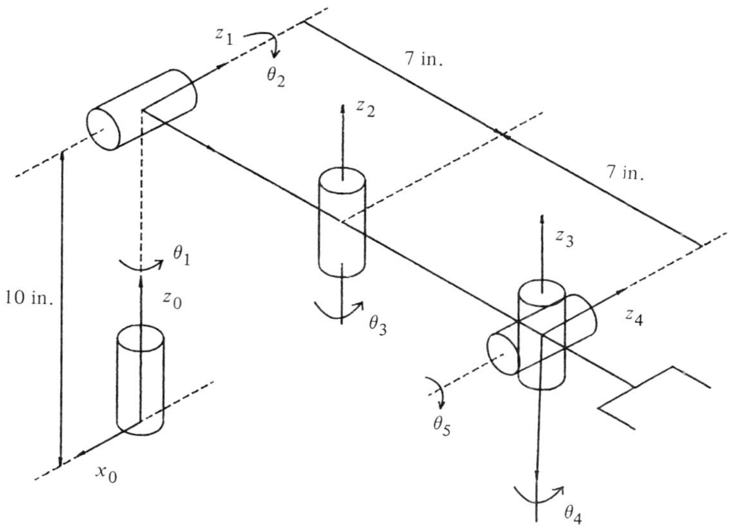

Figure 2.11.9 Robot set of coordinate frames showing the axes of motions.

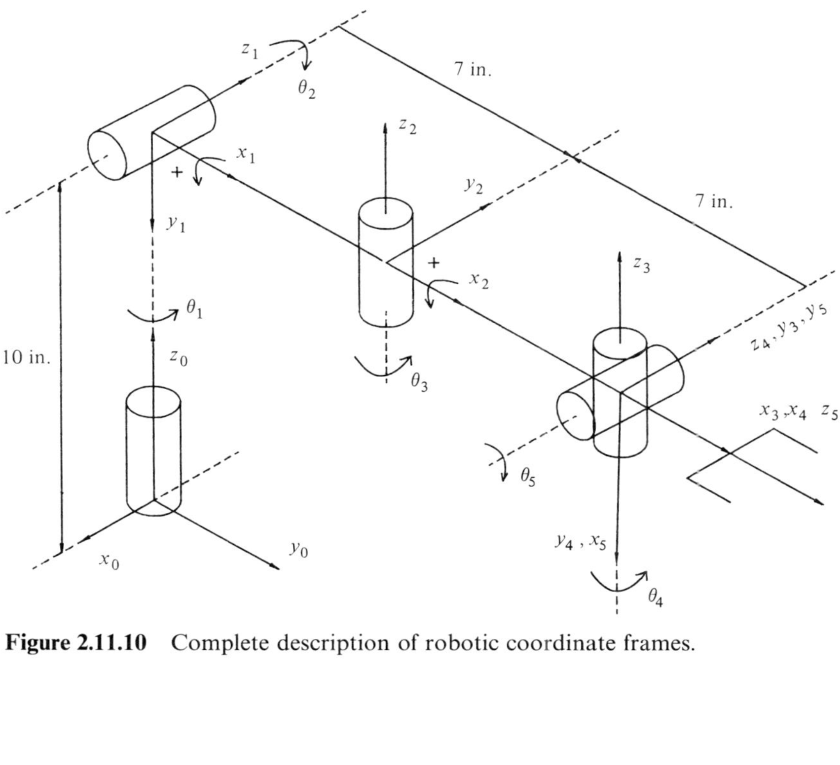

Figure 2.11.10 Complete description of robotic coordinate frames.

TABLE 2.11.5 JOINT PARAMETERS

Joint	θ_h	θ	α_i	a_i (in.)	d_i (in.)
1	90°	θ_1	−90°	0	10
2	0°	θ_2	90°	7	0
3	0°	θ_3	0°	7	0
4	0°	θ_4	−90°	0	0
5	90°	θ_5	90°	0	0

where, as usual,

$$\mathbf{A}_{i-1}^{i} = \begin{bmatrix} C\theta_i & -C\alpha_i S\theta_i & S\alpha_i S\theta_i & a_i C\theta_i \\ S\theta_i & C\alpha_i C\theta_i & -S\alpha_i C\theta_i & a_i S\theta_i \\ 0 & S\alpha_i & C\alpha_i & d_i \\ 0 & 0 & 0 & 1 \end{bmatrix}. \tag{2.11.91}$$

Thus,

$$\mathbf{A}_0^1 = \begin{bmatrix} 0 & 0 & -1 & 0 \\ 1 & 0 & 0 & 0 \\ 0 & -1 & 0 & 10 \\ 0 & 0 & 0 & 1 \end{bmatrix}, \qquad \mathbf{A}_1^2 = \begin{bmatrix} 1 & 0 & 0 & 7 \\ 0 & 0 & -1 & 0 \\ 0 & 1 & 0 & 0 \\ 0 & 0 & 0 & 1 \end{bmatrix}, \tag{2.11.92}$$

$$\mathbf{A}_2^3 = \begin{bmatrix} 0 & -1 & 0 & 0 \\ 1 & 0 & 0 & 7 \\ 0 & 0 & 1 & 0 \\ 0 & 0 & 0 & 1 \end{bmatrix}, \qquad \mathbf{A}_3^4 = \begin{bmatrix} 1 & 0 & 0 & 0 \\ 0 & 0 & 1 & 0 \\ 0 & -1 & 0 & 0 \\ 0 & 0 & 0 & 1 \end{bmatrix}, \tag{2.11.93}$$

$$\mathbf{A}_4^5 = \begin{bmatrix} 0 & 0 & 1 & 0 \\ 1 & 0 & 0 & 0 \\ 0 & 1 & 0 & 0 \\ 0 & 0 & 0 & 1 \end{bmatrix}, \qquad \mathbf{T}_0^5 = \begin{bmatrix} 0 & 0 & -1 & -7 \\ 0 & -1 & 0 & 7 \\ -1 & 0 & 0 & 10 \\ 0 & 0 & 0 & 1 \end{bmatrix}. \tag{2.11.94}$$

Note that all α_i, a_i, and d_i values are assumed to be known. Find as many solutions as you can for the above problem. ■ ■

EXAMPLE 2.11.8 ■

Consider the PUMA 600 series robotic arm (Fig. 2.11.11). Establish the joint coordinate systems for the given robot configuration. Find the $\mathbf{A}_i$ and the $\mathbf{T}_0^6$ matrices.

Solution. Recalling Algorithm 2.10.1, we can establish a set of link coordinates as shown in Table 2.11.6. The parameter a_3 is rather small

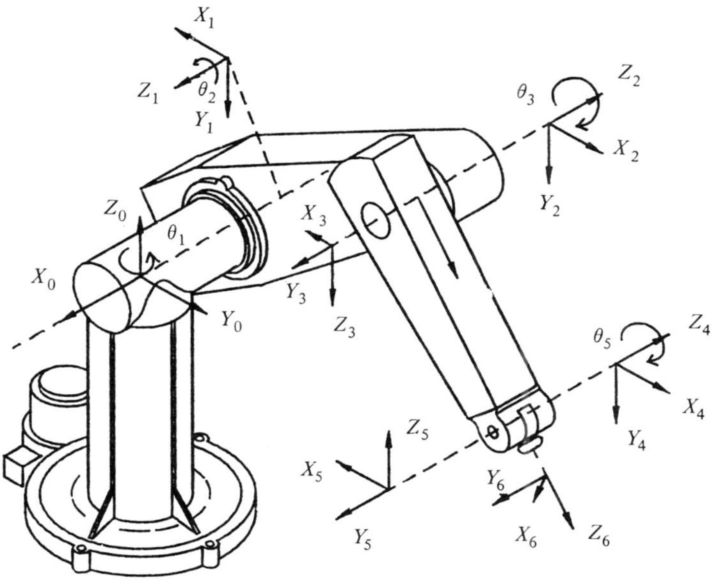

Figure 2.11.11 Given configuration of PUMA 600.

compared to the other parameters such as a_2, d_2, d_4, and d_6. For example, $a_2 = 431.8$ mm, $d_2 = 149.1$ mm, $a_3 = 20.3$ mm, $d_4 = 433$ mm, and $d_6 = 56.5$ mm. Therefore, in the following discussion we neglect a_3.

Using Eq. (2.11.51) and the abbreviations $\cos\theta_i \equiv C\theta_i \equiv C_i$ and $\sin\theta_i \equiv S\theta_i \equiv S_i$, we obtain the following D-H representations:

$$\mathbf{A}_0^1 = \begin{bmatrix} C_1 & 0 & -S_1 & 0 \\ S_1 & 0 & C_1 & 0 \\ 0 & -1 & 0 & 0 \\ 0 & 0 & 0 & 1 \end{bmatrix}, \qquad \mathbf{A}_1^2 = \begin{bmatrix} C_2 & S_2 & 0 & +432\,C_2 \\ S_2 & -C_2 & 0 & +432\,S_2 \\ 0 & 0 & -1 & +149.5 \\ 0 & 0 & 0 & 1 \end{bmatrix}, \tag{2.11.95}$$

TABLE 2.11.6 JOINT PARAMETERS

Joint	θ_i	α_i	a_i	d_i	$S\alpha_i$	$C\alpha_i$
1	θ_1	$-90°$	0	0	-1	0
2	θ_2	$180°$	a_2	d_2	0	-1
3	θ_3	$90°$	a_3	0	1	0
4	θ_4	$90°$	0	d_4	1	0
5	θ_5	$-90°$	0	0	-1	0
6	θ_6	$-180°$	0	d_6	0	-1

$$\mathbf{A}_2^3 = \begin{bmatrix} C_3 & 0 & S_3 & 0 \\ S_3 & 0 & -C_3 & 0 \\ 0 & 1 & 0 & 0 \\ 0 & 0 & 0 & 1 \end{bmatrix}, \qquad \mathbf{A}_3^4 = \begin{bmatrix} C_4 & 0 & +S_4 & 0 \\ S_4 & 0 & -C_4 & 0 \\ 0 & +1 & 0 & 432 \\ 0 & 0 & 0 & 1 \end{bmatrix}, \tag{2.11.96}$$

$$\mathbf{A}_4^5 = \begin{bmatrix} C_5 & 0 & -S_5 & 0 \\ S_5 & 0 & +C_5 & 0 \\ 0 & -1 & 0 & 0 \\ 0 & 0 & 0 & 1 \end{bmatrix}, \qquad \mathbf{A}_5^6 = \begin{bmatrix} C_6 & -S_6 & 0 & 0 \\ S_6 & C_6 & 0 & 0 \\ 0 & 0 & 1 & 56.5 \\ 0 & 0 & 0 & 1 \end{bmatrix}. \tag{2.11.97}$$

For the direct solution we have to evaluate $\mathbf{T}_0^6 \equiv \mathbf{A}_0^1\mathbf{A}_1^2\mathbf{A}_2^3\mathbf{A}_3^4\mathbf{A}_4^5\mathbf{A}_5^6$. Based on the values given in Table 2.11.6, this can be calculated! The result is

$$\mathbf{T}_0^6 \equiv \begin{bmatrix} n_x & o_x & a_x & p_x \\ n_y & o_y & a_y & p_y \\ n_z & o_z & a_z & p_z \\ 0 & 0 & 0 & 1 \end{bmatrix}, \tag{2.11.98}$$

where $S_{ij} \equiv \sin(\theta_i + \theta_j)$ and $C_{ij} = \cos(\theta_i + \theta_j)$, and

$$n_x = C_1[C_{23}(C_4C_5C_6 - S_4S_6) - S_{23}S_5C_6] - S_1(S_4C_5C_6 + C_4S_6), \tag{2.11.99}$$

$$n_y = C_1[C_{23}(C_4C_5C_6 - S_4S_6) - S_{23}S_5C_6] - S_1(S_4C_5C_6 + C_4S_6), \tag{2.11.100}$$

$$n_z = -S_{23}(C_4C_5C_6 - S_4S_5) - C_{23}S_5C_6, \tag{2.11.101}$$

$$o_x = C_1[-C_{23}(C_4C_5S_6 + S_4C_6) + S_{23}S_5S_6] - S_1(-S_4C_5S_6 + C_4C_6), \tag{2.11.102}$$

$$o_y = S_1[-C_{23}(C_4C_5S_6 + S_4C_6) + S_{23}S_5S_6] - S_1(-S_4C_5S_6 + C_4C_6), \tag{2.11.103}$$

$$o_z = S_{23}(C_4C_5S_6 + S_4C_6) + C_{23}S_5S_6, \tag{2.11.104}$$

$$a_x = C_1(C_{23}C_4S_5 + S_{23}C_5) - S_1S_4S_5, \tag{2.11.105}$$

$$a_x = S_1(C_{23}C_4S_5 + S_{23}C_5) + C_1S_4S_5, \tag{2.11.106}$$

$$a_z = -S_{23}C_4S_5 + C_{23}C_5, \tag{2.11.107}$$

$$p_x = C_1[56.5(C_{23}C_4S_5 + S_{23}C_5) + 432S_{23} + 432C_2] - S_1(56.5S_4S_5 + 149.5), \qquad (2.11.108)$$

$$p_x = S_1[56.5(C_{23}C_4S_5 + S_{23}C_5) + 432S_{23} + 432C_2] - C_1(56.5S_4S_5 + 149.5), \qquad (2.11.109)$$

$$p_z = 56.5(C_{23}C_5 - S_{23}C_4S_5) + 432C_{32} - 432S_2. \qquad (2.11.110)$$

Quiz 2.11.1 ■

Show that for the PUMA 600 series if $\theta_1 = 90°$, $\theta_2 = 0°$, $\theta_3 = 90°$, $\theta_4 = 0°$, $\theta_5 = 0°$, and $\theta_6 = 0°$ then

$$^0\mathbf{T}_6 = \begin{bmatrix} 0 & -1 & 0 & +149.5 \\ 0 & 0 & 0 & 920.5 \\ -1 & 0 & 0 & 0 \\ 0 & 0 & 0 & 1 \end{bmatrix}. \quad ■■ \qquad (2.11.111)$$

EXAMPLE 2.11.9 ■

A gripper originally lies along the z axis such that the w axis is coincident with the z axis. The gripper then rolls 90°, then pitches 180°, and finally yaws −90°. Find the resulting transformation pertaining to this sequence of rotations (see Fig. 2.11.12).

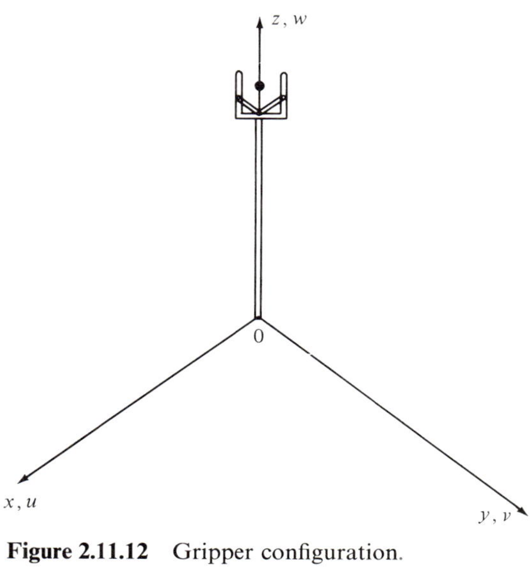

Figure 2.11.12 Gripper configuration.

Figure 2.11.13 Rhino XR-2 robotic workstation at the University of New Mexico Mechanical Engineering Department's instructional laboratories.

Solution. Referring to Eq. (2.8.1), we see that the problem involves a RPY transformation given by $\mathbf{RPY}(\phi, \theta, \psi) = \mathbf{R}(z, \phi)\mathbf{R}(v, \theta)\mathbf{R}(u, \psi)$.

$$\therefore \quad \mathbf{RPY}(-90, 180, -90) = \mathbf{R}(z, -90)\mathbf{R}(v, 180)\mathbf{R}(u, -90)$$

$$= \begin{bmatrix} 0 & 1 & 0 & 0 \\ -1 & 0 & 0 & 0 \\ 0 & 0 & 1 & 0 \\ 0 & 0 & 0 & 1 \end{bmatrix} \begin{bmatrix} -1 & 0 & 0 & 0 \\ 0 & 1 & 0 & 0 \\ 0 & 0 & -1 & 0 \\ 0 & 0 & 0 & 1 \end{bmatrix} \begin{bmatrix} 1 & 0 & 0 & 0 \\ 0 & 0 & 1 & 0 \\ 0 & -1 & 0 & 0 \\ 0 & 0 & 0 & 1 \end{bmatrix}$$

$$= \begin{bmatrix} 0 & 0 & 1 & 0 \\ 1 & 0 & 0 & 0 \\ 0 & 1 & 0 & 0 \\ 0 & 0 & 0 & 1 \end{bmatrix}. \quad \blacksquare\blacksquare \qquad (2.11.112)$$

In the next section we elaborate on the solutions to the inverse kinematics problem in robotics.

2.12 INVERSE KINEMATICS SOLUTIONS

Generally, for a six-jointed robot we have six joint angles θ_i, where $i = 1, 2, 3, 4, 5, 6$. In terms of $\mathbf{A}_j^i$ matrices this can be written

$$\mathbf{T}_0^n = \mathbf{A}_0^1 \mathbf{A}_1^2 \mathbf{A}_2^3 \cdots \mathbf{A}_{n-1}^n, \tag{2.12.1}$$

where $\mathbf{T}_0^2$ is the nth joint matrix with respect to the base coordinate. This joint matrix is represented by

$$\mathbf{T}_0^n = \begin{bmatrix} n_x & o_x & a_x & p_x \\ n_y & o_y & a_y & p_y \\ n_z & o_z & a_z & p_z \\ 0 & 0 & 0 & 1 \end{bmatrix}. \tag{2.12.2}$$

Note that we may write Eq. (2.12.1) in the following compact form:

$$\mathbf{T}_0^n = \prod_{i=1}^{n} \mathbf{A}_{i-1}^i, \qquad i = 1, 2, 3, \ldots, n, \tag{2.12.3}$$

or

$$\begin{bmatrix} \mathbf{n} & \mathbf{o} & \mathbf{a} & \mathbf{p} \\ 0 & 0 & 0 & 1 \end{bmatrix} = \prod_{i=1}^{n} \begin{bmatrix} C\theta_i & -C\alpha_i S\theta_i & S\alpha_i S\theta_i & a_i C\theta_i \\ S\theta_i & C\alpha_i C\theta_i & -S\alpha_i C\theta_i & a_i S\theta_i \\ 0 & S\alpha_i & C\alpha_i & d_i \\ 0 & 0 & 0 & 1 \end{bmatrix}, \tag{2.12.4}$$

where **n** is the normal vector, **o** is the sliding vector, and **a** is the approach vector of the hand.

If the manipulator is related to a reference coordinate frame by a transformation $\mathbf{B}$ and has a tool attached to its last joint mounting plate described by $\mathbf{H}$, then the end joint of the tool can be related to the reference coordinate frame by multiplying matrices $\mathbf{B}$, $\mathbf{T}_0^n$, and $\mathbf{H}$ together as

$$\mathbf{T}_{\text{ref}}^{\text{tool}} = \mathbf{B}\mathbf{T}_0^n\mathbf{H}. \tag{2.12.5}$$

Clearly, $\mathbf{H} \equiv \mathbf{A}_n^{\text{tool}}$ and $\mathbf{B} \equiv \mathbf{A}_{\text{ref}}^n$.

As discussed before, the direct kinematics solution is obtained by simply calculating $\mathbf{T} = \mathbf{A}_0^n$ by the chain multiplication of n matrices $\mathbf{A}_{i-1}^i$, $i = 1, 2, \ldots, n$. For any given robot configuration, a complete table for joint coordinates θ_i, a_i, α_i, and d_i can be constructed such that one can compute all $\mathbf{A}_{i-1}^i$ matrices and eventually the orientation matrix $\mathbf{T}_o^n$ of the nth joint. This procedure is demonstrated for the PUMA 600 series robot in Example 2.11.8.

EXAMPLE 2.12.1 ■

Consider a six-jointed robotic manipulator equipped with a digital TV camera capable of continuously monitoring the positions and movements of an object (a tea cup). The position and orientation of the cup with respect to the camera is expressed by a matrix $\mathbf{T}_1$ and the origin of the robot base coordinate system with respect to the camera is given by $\mathbf{T}_2$ such that

$$\mathbf{T}_1 = \begin{bmatrix} 0 & 1 & 0 & 5 \\ 1 & 0 & 0 & 6 \\ 0 & 0 & -1 & 10 \\ 0 & 0 & 0 & 1 \end{bmatrix} \quad \text{and} \quad \mathbf{T}_2 = \begin{bmatrix} 1 & 0 & 0 & -20 \\ 0 & -1 & 0 & 10 \\ 0 & 0 & -1 & 12 \\ 0 & 0 & 0 & 1 \end{bmatrix}. \tag{2.12.6}$$

Determine the position and orientation of the cup with respect to the base corrdinate (Fig. 2.12.1).

Solution. Note that

$$\mathbf{T}_1 = \mathbf{T}_{\text{camera}}^{\text{cup}} = \begin{bmatrix} 0 & 1 & 0 & 5 \\ 1 & 0 & 0 & 6 \\ 0 & 0 & -1 & 10 \\ 0 & 0 & 0 & 1 \end{bmatrix} \quad \text{and} \quad \mathbf{T}_2 = \begin{bmatrix} 1 & 0 & 0 & -20 \\ 0 & -1 & 0 & 10 \\ 0 & 0 & -1 & 12 \\ 0 & 0 & 0 & 1 \end{bmatrix}. \tag{2.12.7}$$

Clearly,

$$\mathbf{T}_{\text{base}}^{\text{cup}} = \mathbf{T}_{\text{base}}^{\text{camera}} \mathbf{T}_{\text{camera}}^{\text{cup}} = \mathbf{T}_2^{-1} \mathbf{T}_1. \tag{2.12.8}$$

Thus,

$$\mathbf{T}_{\text{base}}^{\text{cup}} = \begin{bmatrix} 1 & 0 & 0 & 20 & 0 & 1 & 0 & 5 \\ 0 & -1 & 0 & 10 & 1 & 0 & 0 & 6 \\ 0 & 0 & -1 & 12 & 0 & 0 & -1 & 10 \\ 0 & 0 & 0 & 1 & 0 & 0 & 0 & 1 \end{bmatrix} \tag{2.12.9}$$

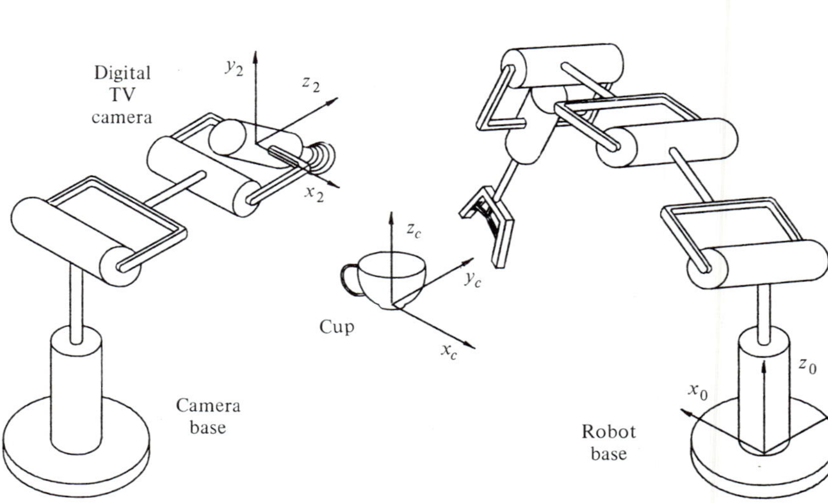

Figure 2.12.1 Camera-based robotic work station.

and

$$\therefore \quad \mathbf{T}_{\text{base}}^{\text{cup}} = \begin{bmatrix} 0 & 1 & 0 & 25 \\ -1 & 0 & 0 & 4 \\ 0 & 0 & 1 & 2 \\ 0 & 0 & 0 & 1 \end{bmatrix}. \quad \blacksquare\blacksquare \tag{2.12.10}$$

EXAMPLE 2.12.2 ■

Referring to Figure 2.12.1 and Example 2.12.1, determine the position and orientation of the cup with respect to the gripper if the position and orientation of the gripper with respect to the base coordinate frame is given by $\mathbf{T}_3$ such that

$$\mathbf{T}_3 = \mathbf{T}_{\text{base}}^{\text{gripper}} = \begin{bmatrix} 1 & 0 & 0 & 8 \\ 0 & 1 & 0 & 6 \\ 0 & 0 & 1 & 6 \\ 0 & 0 & 0 & 1 \end{bmatrix}. \tag{2.12.11}$$

Solution. Note that

$$\mathbf{T}_{\text{gripper}}^{\text{cup}} = \mathbf{T}_{\text{gripper}}^{\text{base}} \mathbf{T}_{\text{base}}^{\text{cup}} = \mathbf{T}_3^{-1} \mathbf{T}_{\text{base}}^{\text{cup}}. \tag{2.12.12}$$

$$\therefore \quad \mathbf{T}_{\text{gripper}}^{\text{cup}} = \begin{bmatrix} 1 & 0 & 0 & -8 \\ 0 & 1 & 0 & -6 \\ 0 & 0 & 1 & -6 \\ 0 & 0 & 0 & 1 \end{bmatrix} \begin{bmatrix} 0 & 1 & 0 & 25 \\ -1 & 0 & 0 & 4 \\ 0 & 0 & 1 & 2 \\ 0 & 0 & 0 & 1 \end{bmatrix} \tag{2.12.13}$$

and

$$\therefore \quad \mathbf{T}_{\text{gripper}}^{\text{cup}} = \begin{bmatrix} 0 & 1 & 0 & 17 \\ -1 & 0 & 0 & -2 \\ 0 & 0 & 1 & -4 \\ 0 & 0 & 0 & 1 \end{bmatrix}. \quad \blacksquare\blacksquare \tag{2.12.14}$$

Inverse Kinematics Solutions for Six-Axis Manipulators In this case, Eq. (2.91) reduces to

$$\begin{bmatrix} \mathbf{n} & \mathbf{o} & \mathbf{a} & \mathbf{p} \\ 0 & 0 & 0 & 1 \end{bmatrix} = \prod_{i=1}^{6} \begin{bmatrix} C\theta_i & -C\alpha_i S\theta_i & S\alpha_i S\theta_i & a_i C\theta_i \\ S\theta_i & C\alpha_i C\theta_i & -S\alpha_i C\theta_i & a_i S\theta_i \\ 0 & S\alpha_i & C\alpha_i & d_i \\ 0 & 0 & 0 & 1 \end{bmatrix} \tag{2.12.15}$$

or, alternatively,

$$\mathbf{T}_0^6 = \mathbf{A}_0^1 \mathbf{A}_1^2 \mathbf{A}_2^3 \mathbf{A}_3^4 \mathbf{A}_4^5 \mathbf{A}_5^6. \tag{2.12.16}$$

It is clear that the θ_i are the variables to be determined while normally the α_i, d_i, and a_i values are all known for a given manipulator. Of course, the situation is different with prismatic manipulators for which some a_i and d_i

are variables similar to joint angles θ_i. The two treatments are basically similar.

Once the link parameters are established, the matrices $\mathbf{A}_{i-1}^{i}$ can easily be constructed and chain multiplied. It is clear that in these cases we generally have 12 equations in 6 unknowns, and therefore finding the right solutions requires great care! In the absence of any translatory action, we can find general inverse solutions in terms of Euler angles (ϕ, θ, ψ) or RPY (roll, pitch, yaw) angles (ϕ, θ, ψ).

EXAMPLE 2.12.3 ■

Find the general inverse kinematics solutions for the Euler angles ϕ, θ, and ψ.

Solution. Recalling Eq. (2.11.8), we note that

$$\mathbf{T}_0^3 = \begin{bmatrix} n_x & o_x & a_x & p_x \\ n_y & o_y & a_y & p_y \\ n_z & o_z & a_z & p_z \\ 0 & 0 & 0 & 1 \end{bmatrix} = \mathbf{Euler}(\phi, \theta, \psi). \tag{2.12.17}$$

Thus, $p_x = p_y = p_z = 0$, and

$$n_x = C\phi C\theta, \tag{2.12.18}$$

$$n_y = S\phi C\theta, \tag{2.12.19}$$

$$n_z = -S\theta, \tag{2.12.20}$$

$$o_x = C\phi S\theta S\psi - S\phi C\psi, \tag{2.12.21}$$

$$o_y = S\phi S\theta S\psi + C\phi C\psi, \tag{2.12.22}$$

$$o_z = C\theta S\psi, \tag{2.12.23}$$

$$a_x = C\phi S\theta C\psi + S\theta S\psi, \tag{2.12.24}$$

$$a_y = S\phi S\theta C\psi - C\theta S\psi, \tag{2.12.25}$$

$$a_z = C\theta C\psi. \tag{2.12.26}$$

From Eqs. (2.12.18) and (2.12.19) one obtains

$$\phi = \tan^{-1}\left(\frac{n_y}{n_x}\right) + n\pi. \tag{2.12.27}$$

From Eqs. (2.12.18), (2.12.19), and (2.12.20) one obtains

$$\theta = \tan^{-1}\left(-\frac{nz}{(n_x^2 + n_y^2)^{1/2}}\right) + n\pi, \tag{2.12.28}$$

and from Eqs. (2.12.23) and (2.12.26) one obtains

$$\psi = \tan^{-1}\left(\frac{o_z}{a_z}\right) + n\pi. \tag{2.12.29}$$

To obtain useful solutions, the computer function ATAN2 should be used such that

$$\phi = \text{ATAN2}(n_y, n_x), \tag{2.12.30}$$

$$\theta = \text{ATAN2}[-n_z, (n_x^2 + n_y^2)^{1/2}], \tag{2.12.31}$$

and

$$\psi = \text{ATAN2}(o_z, a_z). \quad \blacksquare\blacksquare \tag{2.12.32}$$

EXAMPLE 2.12.4 ■

Find the general inverse kinematics solutions for the RPY transformations in terms of ϕ, θ, and ψ.

Solution. Recalling Eq. (2.11.8), note that

$$\mathbf{T}_0^3 = \begin{bmatrix} n_x & o_x & a_x & p_x \\ n_y & o_y & a_y & p_y \\ n_z & o_z & a_z & p_z \\ 0 & 0 & 0 & 1 \end{bmatrix} = \mathbf{RPY}(\phi, \theta, \psi). \tag{2.12.33}$$

Thus, $p_x = p_y = p_z = 0$ and the same equations (2.12.18)–(2.12.26) apply to this case as well. Thus, the inverse kinematics solutions are

$$\phi = \text{ATAN2}(n_y, n_x), \tag{2.12.34}$$

$$\theta = \text{ATAN2}[-n_z, (n_x^2 + n_y^2)^{1/2}], \tag{2.12.35}$$

and

$$\psi = \text{ATAN2}(o_z, a_z). \quad \blacksquare\blacksquare \tag{2.12.36}$$

EXAMPLE 2.12.5 ■

Find the general inverse kinematics solutions for the spherical robot transform **SPH**(θ, ϕ, r). This is applicable to the Unimate 2100G (Fig. 2.12.2).

Solution. Referring to Eq. (2.11.7), one notes that

$$\mathbf{T}_0^3 = \begin{bmatrix} n_x & o_x & a_x & p_x \\ n_y & o_y & a_y & p_y \\ n_z & o_z & a_z & p_z \\ 0 & 0 & 0 & 1 \end{bmatrix} = \begin{bmatrix} C\theta C\phi & -S\theta & -C\theta S\phi & rC\theta C\phi \\ S\theta C\phi & C\theta & -S\theta S\phi & rS\theta C\phi \\ S\phi & 0 & C\phi & rS\phi \\ 0 & 0 & 0 & 1 \end{bmatrix}. \tag{2.12.37}$$

Figure 2.12.2 Unimate robot. (Courtesy of Unimation Robots.)

Thus,

$$\theta \pm n\pi = \tan^{-1}\left(\frac{n_y}{n_x}\right) = \tan^{-1}\left(-\frac{o_x}{o_y}\right) = \tan^{-1}\left(\frac{a_y}{a_x}\right) = \tan^{-1}\left(\frac{p_y}{p_x}\right), \tag{2.12.38}$$

$$\phi \pm n\pi = \tan^{-1}\left(\frac{(a_x^2 + a_y^2)^{1/2}}{(n_x^2 + n_y^2)^{1/2}}\right) = \tan^{-1}\left(\frac{p_z}{(p_x^2 + p_y^2)^{1/2}}\right) = \tan^{-1}\left(\frac{n_z}{a_z}\right), \tag{2.12.39}$$

and

$$r = (p_x^2 + p_y^2 + p_z^2)^{1/2}. \tag{2.12.40}$$

For computer applications θ and ϕ are given by

$$\theta = \text{ATAN2}(p_y, p_x) \tag{2.12.41}$$

and

$$\phi = \text{ATAN2}[p_z, (p_x^2 + p_y^2)^{1/2}]. \quad \blacksquare\blacksquare \tag{2.12.42}$$

EXAMPLE 2.12.6 ■

Find the inverse kinematics solutions for the cylindrical robotic manipulator described in Example 2.11.4.

Solution. The governing equations for the inverse kinematics are given by Eqs. (2.11.22)–(2.11.33). Let us now form the following product by multiplying $\mathbf{A}_1^{-1}$ by $\mathbf{T}_0^6$ such that

$$\mathbf{A}_1^{-1}\mathbf{T}_6 = \begin{bmatrix} C_1 & S_1 & 0 & 0 \\ 0 & 0 & -1 & 0 \\ -S_1 & C_1 & 0 & 0 \\ 0 & 0 & 0 & 1 \end{bmatrix} \begin{bmatrix} n_x & o_x & a_x & p_x \\ n_y & o_y & a_y & p_y \\ n_z & o_z & a_z & p_z \\ 0 & 0 & 0 & 1 \end{bmatrix} \tag{2.12.43}$$

or

$$\mathbf{A}_1^{-1}\mathbf{T}_6 = \begin{bmatrix} F_1(\mathbf{n}) & F_1(\mathbf{o}) & F_1(\mathbf{a}) & F_1(\mathbf{p}) \\ F_2(\mathbf{n}) & F_2(\mathbf{o}) & F_2(\mathbf{a}) & F_2(\mathbf{p}) \\ F_3(\mathbf{n}) & F_3(\mathbf{o}) & F_3(\mathbf{a}) & F_3(\mathbf{p}) \\ 0 & 0 & 0 & 1 \end{bmatrix}, \tag{2.12.44}$$

where $F_i(\mathbf{u})$, $i = 1, 2, 3$, of any vector $\mathbf{u}$ is given by

$$F_1(\mathbf{u}) = C_1 u_x + S_1 u_y, \tag{2.12.45}$$

$$F_2(\mathbf{u}) = -u_z, \tag{2.12.46}$$

and

$$F_3(\mathbf{u}) = -S_1 u_x + C_1 u_y. \tag{2.12.47}$$

Now it is clear that $\mathbf{A}_1^{-1}\mathbf{T}_6 = \mathbf{T}_1^6$, and therefore, based on matrix equation (2.11.20), we have

$$F_1(\mathbf{n}) = C_2(C_4C_5C_6 - S_4S_6) - S_2S_5C_6, \tag{2.12.48}$$

$$F_1(\mathbf{o}) = -C_2(C_4C_5C_6 + S_4C_6) + S_2S_5S_6, \tag{2.12.49}$$

$$F_1(\mathbf{a}) = C_2C_4S_5 + S_2C_5, \tag{2.12.50}$$

$$F_1(\mathbf{p}) = S_2d_3, \qquad F_2(\mathbf{p}) = -C_2d_3, \qquad F_3(\mathbf{p}) = d_2, \tag{2.12.51}$$

$$F_2(\mathbf{n}) = S_2(C_4C_5C_6 - S_4S_6) + C_2S_5C_6, \tag{2.12.52}$$

$$F_2(\mathbf{o}) = -S_2(C_4C_5C_6 + S_4C_6) - C_2S_5S_6, \tag{2.12.53}$$

$$F_2(\mathbf{a}) = S_2C_4S_5 - C_2C_5, \tag{2.12.54}$$

$$F_3(\mathbf{n}) = S_4C_5C_6 + C_4S_6, \tag{2.12.55}$$

$$F_3(\mathbf{o}) = -S_4C_5S_6 + C_4C_6, \tag{2.12.56}$$

and

$$F_3(\mathbf{a}) = S_4S_5. \tag{2.12.57}$$

From Eq. (2.12.51) one finds that

$$-S_1p_x + C_1p_y = d_2, \tag{2.12.58}$$

which can easily be solved for θ_1 to yield a unique solution:

$$\theta_1 = \tan^{-1}\left(\frac{p_y}{p_x}\right) - \tan^{-1}\left(\frac{d_2}{\pm (r^2 - d_2^2)^{1/2}}\right). \tag{2.12.59}$$

Now from Eqs. (2.12.51) one finds that

$$S_2d_3 = C_1p_x + S_1p_y \quad \text{and} \quad -C_2d_3 = -p_z. \tag{2.12.60}$$

Thus, a unique solution for θ_2 exists because $d_3 > 0$ in general:

$$\theta_2 = \tan^{-1}\left(\frac{C_1p_x + S_1p_y}{p_z}\right) \tag{2.12.61}$$

and, obviously,

$$d_3 = S_2(C_1p_x + S_1p_y) + C_2p_z. \tag{2.12.62}$$

From Eqs. (2.12.33), (2.12.34), and (2.12.35) one obtains the following equations:

$$-S_1a_x + C_1a_y = S_4S_5, \tag{2.12.63}$$

$$(C_1a_x + S_1a_y) = C_2C_4S_5 + S_2C_5, \tag{2.12.64}$$

and

$$C_2(C_1a_x + S_1a_y) - S_1a_z = C_4S_5. \tag{2.12.65}$$

Now dividing Eq. (2.12.63) by (2.12.64) yields a nonunique solution for θ_4 such that

$$\theta_4 = \tan^{-1}\left(\frac{C_1a_y - S_1a_x}{C_2(C_1a_x + S_1a_y) - S_2a_z}\right) \quad \text{if} \quad \theta_5 > 0 \tag{2.12.66a}$$

and

$$\theta_4 = \theta_4 + \pi \quad \text{if } \theta_5 < 0. \tag{2.12.66b}$$

If both the numerator and denominator in Eq. (2.12.65) go to zero, it would imply that $\theta_5 = 0$ or $n\pi$, and this would be a "degenerate" case deserving special attention. In this case the axis of joint 4 and joint 6 are aligned and only the sum of $\theta_4 + \theta_6$ becomes significant; basically, θ_4 can be picked arbitrarily, that is, the current value should suffice.

Again, multiplying Eq. (2.11.63) by S_4 and Eq. (2.11.64) by C_4 and adding, we obtain

$$S_5 = C_4[C_2(C_1a_x + S_1a_y) - S_2a_z] + S_4(-S_1a_x + C_1a_y). \tag{2.12.67}$$

Now dividing Eq. (2.12.67) by (2.12.64) yields

$$\theta_5 = \tan^{-1}\left(\frac{C_4[C_2(C_1a_x + S_1a_y) - S_2a_z] + S_4(-S_1a_x + C_1a_y)}{S_2(C_1a_x + S_1a_y) + C_2a_z}\right). \tag{2.12.68}$$

Now from Eqs. (2.11.25), (2.11.26), and (2.11.27) we obtain the following relation:

$$C_1o_x + S_1o_y = -C_2(C_4C_5S_6 + S_4C_6) + S_2S_5S_6, \tag{2.12.69}$$

$$-S_1a_x + C_1o_y = -S_4C_5S_6 + C_4C_6, \tag{2.12.70}$$

$$C_4[C_2(C_1o_x + S_1o_y) - S_2o_z] = -(C_4C_5S_6 + S_4C_6)C_4, \tag{2.12.71}$$

$$S_4(-S_1o_x + C_1o_y) = -S_4^2C_5S_6 + S_4C_4C_6, \tag{2.12.72}$$

and

$$C_4(-S_1o_x + C_1o_y) = -S_4C_4C_5S_6 + C_4^2C_6. \tag{2.12.73}$$

From Eqs. (2.12.69)–(2.12.73) it is clear that

$$S_6 = -C_5\{C_4[C_2(C_1o_x + S_1o_y) - S_2o_z] + S_4(-S_1o_x + C_1o_y)]$$
$$+ S_5[S_2(C_1o_x + S_1o_y) + C_2o_z] = a^* \tag{2.12.74}$$

and

$$C_6 = -S_4[C_2(C_1o_x + S_1o_y) - S_2o_z] + C_4[-S_1o_x + C_1o_y] = b^*. \tag{2.12.75}$$

Thus, θ_6 can be obtained as

$$\theta_6 = \tan^{-1}\left(\frac{a^*}{b^*}\right). \quad \blacksquare\blacksquare \tag{2.12.76}$$

EXAMPLE 2.12.7 ■

Referring to Example 2.11.5 for the Minimover Microbot, construct a set of inverse kinematics solutions.

Solution. Recalling Eqs. (2.11.38)–(2.11.50), we note that

$$(T_0^5)_{32}C_5 + (T_0^5)_{31}S_5 = 0. \tag{2.12.77}$$

Thus,

$$\frac{S_5}{C_5} = \tan\theta_5 = \frac{-o_z}{n_z} \tag{2.12.78}$$

or

$$\theta_5 = \tan^{-1}\left(\frac{-o_z}{n_z}\right), \tag{2.12.79}$$

where $(T_0^5)_{32} = o_z$ and $(T_0^5)_{31} = n_z$. Furthermore, note that

$$(T_0^5)_{11}S_5 + (T_0^5)_{12}C_5 = -S_1. \tag{2.12.80}$$

Thus,

$$\theta_1 = \sin^{-1}(n_x \sin\theta_5 + o_x \cos\theta_5), \tag{2.12.81}$$

where, generally,

$$\mathbf{T}_0^5 = \begin{bmatrix} n_x & o_x & a_x & p_x \\ n_y & o_y & a_y & p_y \\ n_z & o_z & a_z & p_z \\ 0 & 0 & 0 & 1 \end{bmatrix}. \tag{2.12.82}$$

Now set up

$$(\mathbf{A}_1^2)^{-1}(\mathbf{A}_0^1)^{-1}\mathbf{T}_0^5 = \mathbf{A}_2^3\mathbf{A}_3^4\mathbf{A}_4^5 = \mathbf{A}^* \tag{2.12.83}$$

and obtain

$$A_{11}^* = n_xC_1C_2 + n_yC_2S_1 - n_zS_2 = C_5C_{34}, \tag{2.12.84}$$

$$A_{12}^* = o_xC_1C_2 + o_yC_2S_1 - o_zS_2 = -S_5C_{34}, \tag{2.12.85}$$

$$A_{13}^* = a_xC_1C_2 + a_yS_1C_2 - a_zS_2 = S_{34}, \tag{2.12.86}$$

$$A_{14}^* = p_xC_1C_2 + p_yC_2S_1 - p_zS_2 + 8.1S_2 - 7 = 3.2S_{34} + 7C_3, \tag{2.12.87}$$

$$A_{21}^* = -n_xS_2C_1 - n_yS_1S_2 - n_zC_2 = C_5S_{34}, \tag{2.12.88}$$

$$A_{22}^* = -o_xS_2C_1 - o_yS_1S_2 - o_zC_2 = -S_5S_{34}, \tag{2.12.89}$$

$$A_{23}^* = a_xS_2C_1 - a_yS_1S_2 - a_zC_2 = -C_{34}, \tag{2.12.90}$$

and

$$A_{24}^* = -p_xS_2C_1 - p_yS_1S_2 - p_zC_2 + 8.1C_2 = -3.8C_{34} + 7S_3. \tag{2.12.91}$$

From Eqs. (2.12.84) and (2.12.88) we obtain

$$n_xC_1 + n_yS_1 = C_5\cos(\theta_2 + \theta_3 + \theta_4) \tag{2.12.92}$$

or

$$-o_z = -S_S\sin(\theta_2 + \theta_3 + \theta_4). \tag{2.12.93}$$

Thus,

$$\theta_2 + \theta_3 + \theta_4 = \tan^{-1}\left(\frac{o_z}{\tan\theta_5(n_xC_1 + n_yS_1)}\right). \tag{2.12.94}$$

From Eqs. (2.12.84) and (2.12.85) we obtain

$$-\tan\theta_5(n_xC_1C_2 + n_yC_2S_1 - n_zS_2) = o_xC_1C_2 + o_yC_2S_1 - o_zS_2, \tag{2.12.95}$$

or

$$-\tan\theta_5(n_xC_1 + n_yS_1 - n_z\tan\theta_2) = o_xC_1 + o_yS_1 - o_z\tan\theta_2. \tag{2.12.96}$$

$$\therefore \quad \theta_2 = \tan^{-1}\left(\frac{o_xC_1 + o_yS_1 + \tan\theta_5(n_xC_1 + n_yS_1)}{n_z\tan\theta_5 + o_z}\right). \tag{2.12.97}$$

From Eq. (2.12.86)

$$\sin(\theta_3 + \theta_4) = a_xC_1C_2 + a_yS_1C_2 - a_zS_2, \tag{2.12.98}$$

or

$$\theta_3 + \theta_4 = \sin^{-1}(a_xC_2C_1 + a_yS_1C_2 - a_zS_2). \tag{2.12.99}$$

From Eq. (2.12.87)

$$7\cos\theta_3 = p_xC_1C_2 + p_yC_2S_1 - p_zS_2 + 8.1S_2 - 7 - 3.8S_{34}, \tag{2.12.100}$$

or

$$\theta_3 = \cos^{-1}[\tfrac{1}{7}(p_xC_1C_2 + p_yC_2S_1 - p_zS_2 + 8.1S_2 - 7 - 3.8S_{34})]. \tag{2.12.101}$$

Finally,

$$\begin{aligned}\theta_4 &= \theta_3 + \theta_4 - \theta_3 \\ &= -\cos^{-1}[\tfrac{1}{7}(p_xC_2C_1 + p_yC_2S_1 - p_zS_2 + 8.1S_2 - 7 - 3.8S_{34})] \\ &\quad + \sin^{-1}(a_xC_1C_2 + a_yS_1C_2 - a_zS_2). \quad ■■\end{aligned} \tag{2.12.102}$$

EXAMPLE 2.12.8 ■

Find the inverse kinematics solutions for the standard six-axis robotic manipulator described in Example 2.11.6 and Figures 2.11.6, 2.11.7, and 2.11.8.

Solution. The governing equations for the two cases were given by Eqs. (2.11.60)–(2.11.71) and (2.11.77)–(2.11.88), respectively. To construct the pertinent solutions for both cases we proceed as follows: Multiply Eq. (2.11.67) by C_1 and Eq. (2.11.66) by S_1 and subtract the resulting equations to obtain

$$a_yC_1 - a_xS_1 = C_5. \tag{2.12.103}$$

Multiply Eq. (2.11.70) by C_1 and Eq. (2.11.69) by S_1 and subtract the resulting equations to obtain

$$p_y C_1 - p_x S_1 = d_6 C_5. \tag{2.12.104}$$

From Eqs. (2.12.103) and (2.12.104) one obtains

$$d_6 a_y C_1 - d_6 a_x S_1 = p_y C_1 - p_x S_1. \tag{2.12.105}$$

Therefore,

$$\tan \theta_1 = \left(\frac{p_y - d_6 a_y}{p_x - d_6 a_x}\right), \tag{2.12.106}$$

or

$$\theta_1 = \tan^{-1}\left(\frac{p_y - d_6 a_y}{p_x - d_6 a_x}\right) \tag{2.12.107}$$

and

$$\theta_1 = \theta_1 + \pi.$$

From Eqs. (2.12.103) and (2.12.104) one can show that

$$\tan \theta_5 = \frac{\pm d_6[1 - (a_y C_1 - a_x S_1)^2]^{1/2}}{p_y C_1 - p_x S_1}, \tag{2.12.108}$$

or

$$\theta_5 = \tan^{-1}\left(\frac{\pm d_6[1 - (a_y C_1 - a_x S_1)^2]^{1/2}}{p_y C_1 - p_x S_1}\right). \tag{2.12.109}$$

Now multiplying Eq. (2.11.65) by C_1 and Eq. (2.11.66) by S_1 and add the resulting equation to obtain

$$-(a_x C_1 + a_y S_1) = S_5 C_{234}. \tag{2.12.110}$$

Considering Eq. (2.11.68), one notes that if $S_5 \neq 0$, then from Eqs. (2.11.67) and (2.12.110) one finds that

$$\theta_2 + \theta_3 + \theta_4 = \theta_{234} = \tan^{-1}\left(\frac{-a_z}{a_x C_1 + a_1 S_1}\right) \quad \text{for} \quad \theta_5 > 0 \tag{2.12.111}$$

and

$$\theta_{234} = \theta_{234} + \pi \quad \text{for} \quad \theta_5 < 0. \tag{2.12.112}$$

Multiplying Eq. (2.11.60) by C_1 and Eq. (2.11.59) by S_1 and subtracting the resulting equations, we obtain

$$S_5 C_6 = n_y C_1 - n_x S_1. \tag{2.12.113}$$

Multiplying Eq. (2.11.62) by S_1 and Eq. (2.11.63) by C_1 and subtracting the resulting equations, we obtain

$$S_5 S_6 = o_x S_1 - o_y C_1. \tag{2.12.114}$$

Now if $S_5 \neq 0$, we can divide Eq. (2.12.114) by Eq. (2.12.113) to obtain

$$\theta_6 = \tan^{-1}\left(\frac{o_x S_1 - o_y C_1}{n_y C_1 - n_x S_1}\right) \quad \text{for} \quad \theta_5 > 0 \tag{2.12.115}$$

and

$$\theta_6 = \theta_6 + \pi \quad \text{for} \quad \theta_5 < 0. \tag{2.12.116}$$

It should be noted that in actual implementation of these solutions one must incorporate in the pertinent robotic inverse kinematics software the expression ATAN2 instead of $\tan^{-1}$. This is the computer ATAN2 function such that

$$\tan^{-1}\left(\frac{a}{b}\right) \rightarrow \text{ATAN2}(a, b). \tag{2.12.117}$$

In this sense, the solution for θ_6 as given by expression (2.12.115) for $\theta_5 > 0$ should read

$$\theta_6 = \text{ATAN2}[(o_x S_1 - o_y C_1), (n_y C_1 - n_x S_1)], \tag{2.12.118}$$

so that in the pertinent software the correct value of θ_6 is implemented. Multiplying Eq. (2.11.69) by C_1 and Eq. (2.11.70) by S_1 and adding and rearranging, we obtain the following:

$$a_2 C_2 + a_3 C_{23} = C_1 p_x + S_1 p_y + d_6 S_5 C_{234} - a_4 C_{234} = t. \tag{2.12.119}$$

In the above equation, we have set the result equal to t for simplicity. Furthermore, from Eq. (2.11.71), one obtains the following:

$$a_2 S_2 + a_3 S_{23} = -p_z + d_1 - a_4 S_{234} + d_6 S_5 S_{234} = u. \tag{2.12.120}$$

Again, for simplicity, we have set the result equal to u. Thus, from Eqs. (2.12.119) and (2.12.120),

$$a_3 C_{23} = t - a_2 C_2 \tag{2.12.121}$$

and

$$a_3 S_{23} = u - a_2 S_2, \tag{2.12.122}$$

where both t and u are known quantities, thus far. Hence, from Eqs. (2.12.121) and (2.12.122),

$$a_3^2 = (t - a_2 C_2)^2 + (u - a_2 S_2)^2, \tag{2.12.123}$$

which yields the following: which is set equal to w (known)

$$t C_2 + u S_2 = \frac{-a_3^2 + t^2 + u^2 + a_2^2}{2a_2} = w. \tag{2.12.124}$$

For simplicity, we have set the result equal to w. It then follows that

$$\theta_2 = -\tan^{-1}\left(\frac{\pm[1-(w/q)^2]^{1/2}}{w/q}\right) + \tan^{-1}\left(\frac{u}{t}\right), \qquad (2.12.125)$$

where

$$q = (t^2 + u^2)^{1/2}. \qquad (2.12.126)$$

Consequently, from Eqs. (2.12.121) and (2.12.122),

$$\tan(\theta_2 + \theta_3) = \left(\frac{u - a_2 S_2}{t - a_2 C_2}\right), \qquad (2.12.127)$$

or

$$\theta_3 = \tan^{-1}\left(\frac{u - a_2 S_2}{t - a_2 C_2}\right) - \theta_2 \qquad (2.12.128)$$

and, finally,

$$\theta_4 = \theta_{234} - \theta_2 - \theta_3 \quad \text{and} \quad \theta_{234} = \theta_2 + \theta_3 + \theta_4 \qquad (2.12.129)$$

are now known completely.

This completes the solution for the inverse kinematics problem of a standard robot when $\theta_4 \neq 0$ (nondegenerate case). ■ ■

The Degenerate Case: $\theta_5 = 0 \pm n\pi$ If $S_5 = 0$ ($\theta_5 = 0 \pm n\pi$), the arm geometry becomes one of a degenerate kind because θ_{234} becomes essentially arbitrary based on Eq. (2.11.68). Frequently, one would use the current value for θ_{234}. For the sake of calculations, let us assume that $\theta_{234} = k$ (arbitrary for $\theta_5 = 0$ or $\pm n\pi$) and Eqs. (2.11.62) and (2.11.65) yield

$$n_z = -C_5 C_6 S_{234} - S_6 C_{234} \qquad (2.12.130)$$

and

$$o_z = -C_6 C_{234} + S_6 C_5 S_{234}, \qquad (2.12.131)$$

which further yield

$$C_6 = -n_z S_{234} C_5 - o_z C_{234} \qquad (2.12.132)$$

and

$$S_6 = C_5 o_z S_{234} - n_z C_{234}, \qquad (2.12.133)$$

so that

$$\theta_6 = \tan^{-1}\left(\frac{C_5 o_z S_{234} - n_z C_{234}}{-C_5 n_z S_{234} - o_z C_{234}}\right). \qquad (2.12.14)$$

Now again multiplying Eq. (2.11.69) by C_1 and Eq. (2.11.70) by S_1 and adding the resulting equations with S_5 set to zero, we obtain

$$p_x C_1 + p_y S_1 = a_4 C_{234} + a_3 C_{23} + a_2 C_2. \qquad (2.12.135)$$

Equation (2.11.71) for the degenerate case yields

$$p_z = -a_3 S_{23} - a_2 S_2 + d_1 - a_4 S_{234}. \qquad (2.12.136)$$

Thus, from Eqs. (2.12.134) and (2.12.135), one obtains the following equation, which is set equal to X (a known value):

$$a_3 C_{23} + a_2 C_2 + a_2 C_2 = p_x C_1 + p_y S_1 - a_4 C_{234} = X. \qquad (2.12.137)$$

From Eq. (2.12.135), one obtains the following equations, which is set equal to Y:

$$a_3 S_{23} + a_2 S_2 = -p_z + d_1 - a_4 S_{234} = Y. \qquad (2.12.138)$$

Thus, from Eqs. (2.12.136) and (2.12.138), one finally finds

$$\theta_2 = \tan^{-1}\left(\frac{Z/R}{\pm[1 - (Z/R)^2]^{1/2}}\right) - \tan^{-1}\left(\frac{X}{Y}\right), \qquad (2.12.139)$$

where

$$Z = \frac{a_2^2 + X^2 + Y^2 - a_3^2}{2a_2} \qquad (2.12.140)$$

and

$$R = (X^2 + Y^2)^{1/2}. \qquad (2.12.141)$$

Now from Eqs. (2.12.136) and (2.12.137), one easily can obtain

$$\theta_{23} = \theta_2 + \theta_3 = \tan^{-1}\left(\frac{Y - a_2 S_2}{X - a_2 C_2}\right) \qquad (2.12.142)$$

and thus $\theta_3 + \theta_{23} - \theta_2$ and

$$\theta_4 = k - \theta_{23}. \qquad (2.12.143)$$

This completes the solution for one of the degenerate cases, and obviously there is a multitude of such degenerate solutions that can be obtained.

Inverse Kinematics Solutions for the Second Case The governing equations for this case (Fig. 2.11.8) are Eqs. (2.11.77)–(2.11.88). Note that the structure of these equations is similar to the structure of previous equations (2.11.60)–(2.11.71) except that o_x, o_y, and o_z are replaced by a_x, a_y, and a_z, respectively. Thus, it is clear that in the previous solutions for the nondegenerate and the degenerate cases if we had replaced o_x by a_x, a_x by $-o_x$, o_y by a_y, a_y by

$-o_y$, o_z by a_z, and a_z by $-o_z$ we should find the exact solutions for the second case. For this case the nondegenerate solutions are

$$\theta_1 = \tan^{-1}\left(\frac{p_y + d_6 o_y}{p_y + d_6 o_x}\right) \quad \text{and} \quad \theta_1 = \theta_1 + \pi, \tag{2.12.144}$$

$$\theta_5 = \tan^{-1}\left(\frac{\pm d_6[1 - (o_x S_1 - o_y C_2)^2]^{1/2}}{p_y C_1 - p_x S_1}\right), \tag{2.12.145}$$

$$\theta_{234} = \tan^{-1}\left(\frac{-o_z}{o_x C_1 + o_y S_1}\right) \quad \text{for} \quad \theta_5 > 0, \tag{2.12.146}$$

$$\theta_{234} = \theta_{234} + \pi \quad \text{for} \quad \theta_5 < 0, \tag{2.12.147}$$

$$\theta_6 = \tan^{-1}\left(\frac{a_x S_1 - a_y C_1}{n_y C_1 - n_x S_1}\right) \quad \text{for} \quad \theta_5 > 0, \tag{2.12.148}$$

$$\theta_6 = \theta_6 + \pi \quad \text{for} \quad \theta_5 > 0, \tag{2.12.149}$$

$$\theta_2 = -\tan^{-1}\left(\frac{\pm [1 - (3/q)^{1/2}}{w/q}\right) + \tan^{-1}\left(\frac{u}{t}\right), \tag{2.12.150}$$

$$\theta_3 = \tan^{-1}\left(\frac{u - a_2 S_2}{t - a_2 C_2}\right) - \theta_2, \tag{2.12.151}$$

and, finally,

$$\theta_4 = \theta_{234} - \theta_{23}. \tag{2.12.152}$$

This completes the solution for the inverse kinematics problem of a standard six-axis robot having the second kind of gripper when $\theta_5 \neq 0$.

For the degenerate case when $\theta_5 = 0$, one similarly finds that

$$\theta_1 = \tan^{-1}\left(\frac{p_y + d_6 o_y}{p_y + d_6 o_x}\right) \quad \text{and} \quad \theta_1 = \theta_1 + \pi, \tag{2.12.153}$$

$$\theta_5 = 0 \pm n\pi \quad \text{and} \quad \theta_{234} = k \text{ (arbitrary)}, \tag{2.12.154}$$

$$\theta_6 = \tan^{-1}\left(\frac{C_5 a_z S_{234} - n_z C_{234}}{-C_5 n_z S_{234} - a_z C_{234}}\right), \tag{2.12.155}$$

$$\theta_2 = \tan^{-1}\left(\frac{Z/R}{\pm [1 - (Z/R)^2]^{1/2}}\right) - \tan^{-1}\left(\frac{X}{Y}\right), \tag{2.12.156}$$

$$\theta_3 = \tan^{-1}\left(\frac{Y - a_2 S_2}{X - a_2 C_2}\right) - \theta_2, \tag{2.12.157}$$

and

$$\theta_4 = k - \tan^{-1}\left(\frac{Y - a_2 S_2}{X - a_2 C_2}\right). \tag{2.12.158}$$

This completes the solution for one of the degenerate cases of the standard six-axis robot with a gripper of the second kind.

EXAMPLE 2.12.9 ■

For a Rhino-XR2 robotic manipulator shown schematically in Figure 2.12.3, determine the exact inverse kinematics solutions.

Solution. Referring to Figure 2.12.4, which shows a complete set of coordinate frames, we can construct Table 2.12.1 for the joint parameters. (See also Fig. 2.12.5.) The D-H attitude matrices can now be derived based on Eq. (2.12.4) and Table 2.12.1 such that

$$\mathbf{A}_0^1 = \begin{bmatrix} C_1 & 0 & -S_1 & 0 \\ S_1 & 0 & C_1 & 0 \\ 0 & -1 & 0 & d_1 \\ 0 & 0 & 0 & 1 \end{bmatrix}, \tag{2.12.159}$$

$$\mathbf{A}_1^2 = \begin{bmatrix} C_2 & -S_2 & 0 & a_2 C_2 \\ S_2 & C_2 & 0 & a_2 S_2 \\ 0 & 0 & 1 & 0 \\ 0 & 0 & 0 & 1 \end{bmatrix}, \tag{2.12.160}$$

$$\mathbf{A}_2^3 = \begin{bmatrix} C_3 & -S_3 & 0 & a_3 C_3 \\ S_3 & C_3 & 0 & a_3 S_3 \\ 0 & 0 & 1 & 0 \\ 0 & 0 & 0 & 1 \end{bmatrix}, \tag{2.12.161}$$

$$\mathbf{A}_3^4 = \begin{bmatrix} C_4 & 0 & -S_4 & a_4 C_4 \\ S_4 & 0 & C_4 & a_4 S_2 \\ 0 & -1 & 0 & 0 \\ 0 & 0 & 0 & 1 \end{bmatrix}, \tag{2.12.162}$$

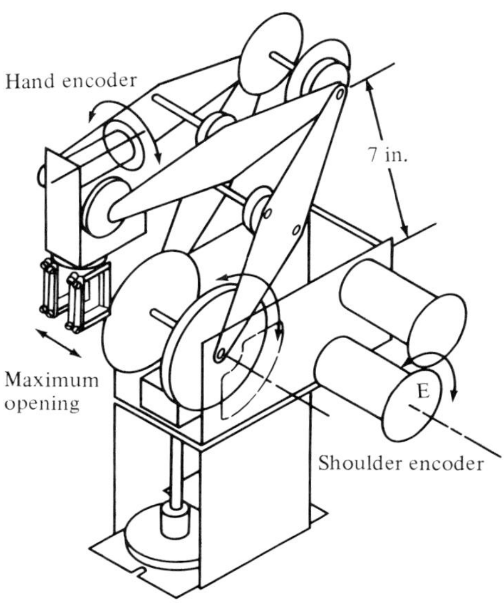

Figure 2.12.3 A PUMA 600 robot. (Courtesy of Unimation Robots, Inc.).

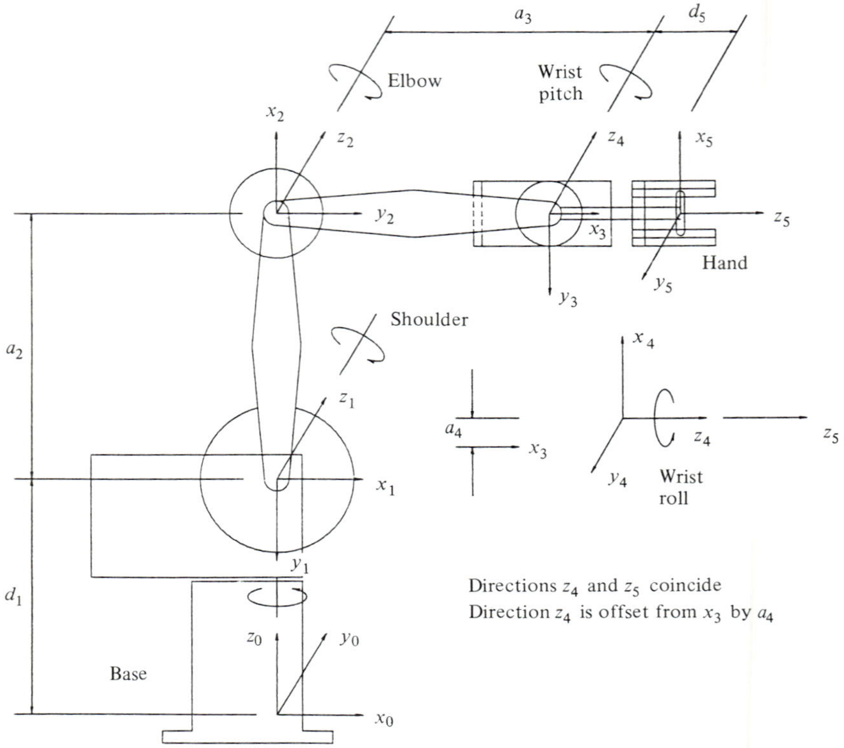

Figure 2.12.4 Schematic for Rhino XR-2 robotic manipulator, depicting the set of joint coordinate frames.

TABLE 2.12.1 GENERAL TABLE FOR JOINT PARAMETERS OF RHINO XR-2 ROBOTIC MANIPULATOR

Joint	θ_i	α_i	a_i	d_i
1	θ_1	$-90°$	0	d_i
2	θ_2	$0°$	a_2	0
3	θ_3	$0°$	a_3	0
4	θ_4	$-90°$	a_4	0
5	θ_5	$0°$	0	d_5

and

$$\mathbf{A}_4^5 = \begin{bmatrix} C_5 & -S_5 & 0 & 0 \\ S_5 & C_5 & 0 & 0 \\ 0 & 0 & 1 & d_5 \\ 0 & 0 & 0 & 1 \end{bmatrix}. \tag{2.12.163}$$

Now the inverse kinematics equation is

$$\mathbf{T}_0^5 = \mathbf{A}_0^1 \mathbf{A}_1^2 \mathbf{A}_2^3 \mathbf{A}_3^4 \mathbf{A}_4^5, \tag{2.12.164}$$

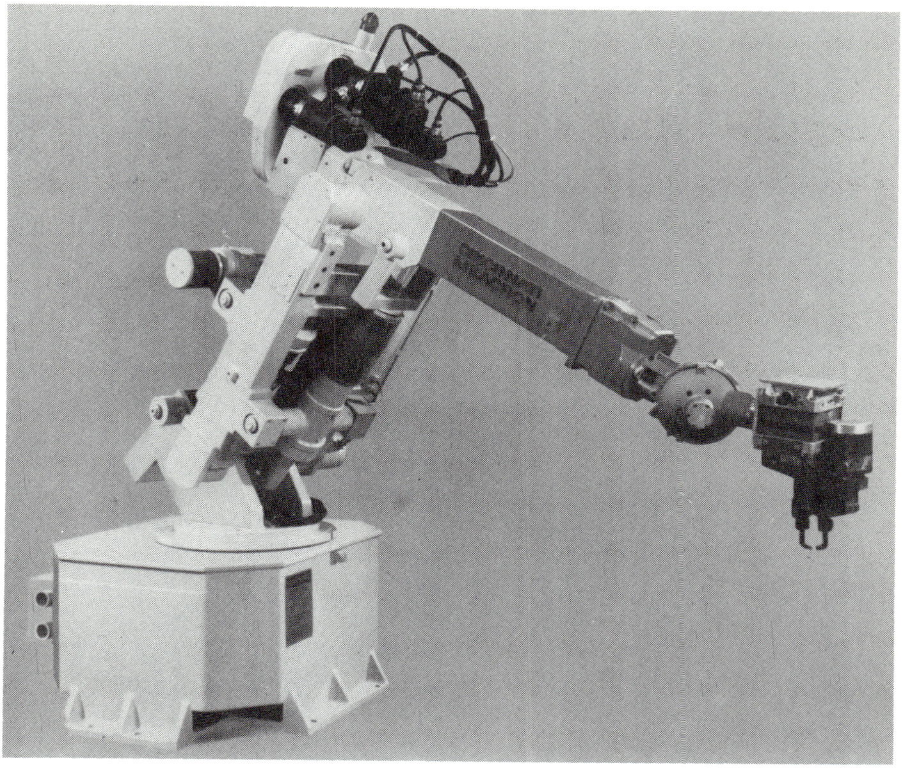

Figure 2.12.5 A T[3]-766 Robot. (Courtesy of Cincinnati Milacron.)

where the $\mathbf{T}_0^5$ matrix is known and given by

$$\mathbf{T}_0^5 = \begin{bmatrix} n_x & o_x & a_x & p_x \\ n_y & o_y & a_y & p_y \\ n_z & o_z & a_z & p_z \\ 0 & 0 & 0 & 1 \end{bmatrix}. \tag{2.12.165}$$

Now the task is to determine the five unknowns, that is $\theta_1, \theta_2, \theta_3, \theta_4$, and θ_5, in terms of the components of the attitude matrix $\mathbf{T}_0^5$ and other structural parameters. We may solve the inverse kinematics problem by simply expanding Eq. (2.12.164). Thus, one finds the following expanded expressions:

$$n_x = C_1 C_{234} C_5 + S_1 S_5, \tag{2.12.166}$$

$$o_x = -C_1 C_{234} S_5 + S_1 C_5, \tag{2.12.167}$$

$$a_x = -C_1 S_{234}, \tag{2.12.168}$$

$$p_x = -d_5 C_1 S_{234} + a_4 C_1 C_{234} + a_3 C_1 C_{23} + a_2 C_1 C_2, \tag{2.12.169}$$

$$n_y = S_1 C_{234} C_5 - C_1 S_5, \tag{2.12.170}$$

$$o_y = -S_1 C_{234} S_5 - C_1 C_5, \tag{2.12.171}$$

$$a_y = S_1 S_{234}, \tag{2.12.172}$$

$$p_y = -d_5 S_1 S_{234} + a_4 S_1 C_{234} + a_3 S_1 C_{23} + a_2 S_1 C_2, \tag{2.12.173}$$

$$n_z = -S_{234} C_5, \tag{2.12.174}$$

$$o_z = S_{234} S_5, \tag{2.12.175}$$

$$a_z = -C_{234}, \tag{2.12.176}$$

and

$$p_z = -d_5 C_{234} - a_4 S_{234} - a_3 S_{23} - a_2 S_2 + d_1. \tag{2.12.177}$$

From Eq. (2.12.176),

$$\theta_2 + \theta_3 + \theta_4 = \cos^{-1}(-a_z). \tag{2.12.178}$$

Multiplying Eq. (2.12.169) by S_1 and Eq. (2.12.173) by C_1 and subtracting the results, that is, $S_1 p_x - C_1 p_y$, we obtain

$$S_1 p_x - C_1 p_y = 0 \quad \text{or} \quad \theta_1 = \tan^{-1}\left(\frac{p_y}{p_x}\right). \tag{2.12.179}$$

Now if $S_{234} \neq 0$, then from Eqs. (2.12.174) and (2.12.175) one finds that

$$\theta_5 = \tan^{-1}\left(\frac{-o_z}{n_z}\right) \quad \text{and} \quad S_{234} \neq 0. \tag{2.12.180}$$

But if $S_{234} = 0$ (degenerate case), then $\theta_2 + \theta_3 + \theta_4 = 0$ (resp., $\pm n\pi$) (vertical approach for Rhino). Thus, $C_{234} = 1$ (resp., ± 1). From Eqs. (2.12.166) and (2.12.167),

$$n_x = \pm C_1 C_5 + S_1 S_5 = \cos(\theta_1 \mp \theta_5) \tag{2.12.181}$$

and

$$o_x = \mp C_1 S_5 + S_1 C_5 = \sin(\theta_1 \mp \theta_5). \tag{2.12.182}$$

Therefore,

$$\theta_1 \mp \theta_5 = \tan^{-1}\left(\frac{o_x}{n_x}\right) \quad \text{if} \quad S_{234} = 0. \tag{2.12.183}$$

From Eqs. (2.12.169) and (2.12.173), one may show that

$$C_1 p_x + S_1 p_y = -d_5 S_{234} + a_4 C_{234} + a_2 C_2 + a_3 C_{23}, \tag{2.12.184}$$

and from Eq. (2.12.177) and (2.12.184), one may show that

$$a_2 S_2 + a_3 S_{23} = d_1 - d_5 C_{234} - a_4 S_{234} - p_z = \alpha \tag{2.12.185}$$

and

$$a_2 C_2 + a_3 C_{23} = C_1 p_x + S_1 p_y + d_5 S_{234} - a_4 C_{234} = \beta, \tag{2.12.186}$$

where α and β are known quantities because $\theta_2 + \theta_3 + \theta_4$ is known already. From Eqs. (2.12.185) and (2.12.186) one may show that

$$\alpha^2 + \beta^2 = a_2^2 + a_3^2 + 2a_2 a_3 C_3, \tag{2.12.187}$$

such that

$$\theta_3 = \cos^{-1}\left(\frac{\alpha^2 + \beta^2 - a_2^2 - a_3^2}{2a_2 a_3}\right). \tag{2.12.188}$$

Here the arccosine solution is acceptable because it can easily be shown that the argument cannot exceed ± 1. Generally, speaking, arctangent solutions are preferred because they cover the whole domain from $-\infty$ to $+\infty$ for their argument. Expanding Eqs. (2.12.185) and (2.12.186), one obtains

$$a_3(S_2 C_3 + C_2 S_3) + a_2 S_2 = \alpha \tag{2.12.189}$$

and

$$a_3(C_2 C_3 - S_2 S_3) + a_2 C_2 = \beta. \tag{2.12.190}$$

Multiply Eq. (2.12.189) by S_2 and Eq. (2.12.190) by C_2 and add to obtain

$$a_3 C_3 + a_2 = \alpha S_2 + \beta C_2. \tag{2.12.191}$$

Now multiply Eq. (2.12.189) by C_2 and Eq. (2.12.190) by S_2 and subtract to obtain

$$a_3 S_3 = \alpha C_2 - \beta S_2. \tag{2.12.192}$$

Multiplying Eq. (2.12.191) by β and Eq. (2.12.192) by α and adding, one obtains

$$\beta(a_3 C_3 + a_2) + \alpha a_3 S_3 = (\alpha^2 + \beta^2) C_2 \tag{2.12.193}$$

Multiplying Eq. (2.12.191) by α and Eq. (2.12.192) by β and subtracting, one obtains

$$\alpha(a_3 C_3 + a_2) - \beta a_3 S_3 = (\alpha^2 + \beta^2) S_2. \tag{2.12.194}$$

Now from Eqs. (2.12.193) and (2.12.194), one finds that

$$\theta_2 = \tan^{-1}\left(\frac{\alpha(a_3 C_3 + a_2) - \beta a_3 S_3}{\beta(a_3 C_3 + a_2) + \alpha a_3 S_3}\right), \tag{2.12.195}$$

so that from Eqs. (2.12.178), (2.12.188), and (2.12.195),

$$\theta_4 = \cos^{-1}(-a_z) - \cos^{-1}\left(\frac{\alpha^2 + \beta^2 - a_2^2 - a_3^2}{2a_2 a_3}\right) - \tan^{-1}\left(\frac{\alpha(a_3 C_3 + a_2) - \beta a_3 S_3}{\beta(a_3 C_3 + a_2) + \alpha a_3 S_3}\right). \tag{2.12.196}$$

This completes the inverse kinematics solutions for Rhino XR-2. ■ ■

TABLE 2.12.2 INVERSE KINEMATICS SOLUTIONS FOR THE RHINO XR-2 ROBOT

Joint	Solutions	Valid range
1	$\theta_1 = \tan^{-1}\left(\frac{p_y}{p_x}\right)$	$-180^\circ \leq \theta_1 \leq 180^\circ$
2	$\theta_2 = \tan^{-1}\left(\frac{\alpha(a_3C_3 + a_2) - \beta a_3S_3}{\beta(a_3C_3 + a_2) + \alpha a_3S_3}\right)$	$-90^\circ \leq \theta_2 \leq 90^\circ$
3	$\theta_3 = \cos^{-1}\left(\frac{\alpha^2 + \beta^2 - a_2^2 - a_3^2}{2a_2a_3}\right)$	$0^\circ \leq \theta_3 \leq 180^\circ$
4	$\theta_4 = \cos^{-1}(-a_z) - \cos^{-1}\left(\frac{\alpha^2 + \beta^2 - a_2^2 - a_2^2}{2a_2a_3}\right) - \tan^{-1}\left(\frac{\alpha(a_3C_3 + a_2) - \beta a_3S_3}{\beta(a_3C_3 + a_2) + \alpha a_3S_3}\right)$	$-270^\circ \leq \theta_4 \leq 90^\circ$
5	$\theta_5 = \tan^{-1}\left(\frac{-o_z}{n_z}\right)$ if $S_{234} \neq 0$	$-180^\circ \leq \theta_5 \leq 180^\circ$
	$\pm\theta_5 = \tan^{-1}\left(\frac{p_y}{p_x}\right) - \tan^{-1}\left(\frac{o_x}{n_x}\right)$ if $S_{234} = 0$	

Table 2.12.2 summarizes the inverse kinematics solutions as well as their valid ranges for all motions of the Rhino XR-2 robot.

The Rhino XR-2 hardware is designed such that when the shoulder motor is activated to produce a rotation $\theta_2 = \delta$, the elbow motor is activated simultaneously to produce a rotation of $\theta_3 = -\delta$. Thus, the Rhino operation is such that $\theta_2 + \theta_3 + \theta_4$ remains constant when either the shoulder motor or elbow motor is activated.

This means that the orientation of the wrist relative to the base is unaffected by either the shoulder motor or elbow motor. Thus, the wrist pitch motor may be used to set the sum $\theta_2 + \theta_3 + \theta_4$, the elbow motor can be used to set $\theta_2 + \theta_3$, and, finally, the shoulder motor may be used to set θ_2. Table 2.12.3 displays the angular control ranges of the Rhino XR-2 motors.

TABLE 2.12.3 ANGULAR CONTROL OF ROBOT MOTORS

Motor	Function	Angular control
A	Fingers	—
B	Wrist roll	θ_5
C	Wrist pitch	$\theta_2 + \theta_3 + \theta_4$
D	Elbow	$\theta_2 + \theta_3$
E	Shoulder	θ_2
F	Body (base)	θ_1
G	Carousel	—
H	Spare	—

Now consider the home joint vector $\boldsymbol{\theta}^h = [0^\circ \ -90^\circ \ 90^\circ \ -90^\circ \ 0^\circ]^T$. Let the motor angles be denoted by $\phi^m \equiv [\phi_1 \ \phi_2 \ \phi_3 \ \phi_4 \ \phi_5]^T$. Note that $\phi_1 = \theta_1$, $\phi_2 = \theta_2$, $\phi_3 = \theta_3 + (\theta_2 - \theta_2^h) = \theta_2 + \theta_3 + 90^\circ$, $\phi_4 = \theta_4 + (\theta_2 - \theta_2^h) + (\theta_3 - \theta_3^h) = \theta_2 + \theta_3 + \theta_4 = \theta_{234}$, and $\phi_5 = \theta_5$. It is clear that

$$\boldsymbol{\phi}^h = \begin{bmatrix} 0 \\ -90 \\ 90 \\ -90 \\ 0 \end{bmatrix} = \boldsymbol{\theta}^h. \tag{2.12.197}$$

Furthermore,

$$\Delta\boldsymbol{\phi} = \begin{bmatrix} 1 & 0 & 0 & 0 & 0 \\ 0 & 1 & 0 & 0 & 0 \\ 0 & 1 & 1 & 0 & 0 \\ 0 & 1 & 1 & 1 & 0 \\ 0 & 0 & 0 & 0 & 1 \end{bmatrix} \Delta\boldsymbol{\theta} + \begin{bmatrix} 0 \\ 0 \\ 90 \\ 0 \\ 0 \end{bmatrix}, \tag{2.12.198}$$

where $\Delta\boldsymbol{\phi}$ and $\Delta\boldsymbol{\theta}$ are incremental changes in joint motor angles and the joint angles, respectively.

The following algorithm can be used for the inverse problem:

Algorithm 2.12.1

1. Calculate θ_1, θ_2, θ_3, θ_4, and θ_5 using Table 2.12.2.
2. Drive the wrist roll motor B to the position $\phi_5 = \theta_5$.
3. Drive the wrist pitch motor C to the position $\phi_4 = \theta_2 + \theta_3 + \theta_4 = 0$.
4. Drive the elbow motor D to the position $\phi_3 = \theta_2 + \theta_3$.
5. Drive the shoulder motor E to the position $\theta_2 = \theta_2$.
6. Drive the base motor F to the position $\phi_1 = \theta_1$.

The above procedure can be employed to calculate the desired solutions for the joint vector $\boldsymbol{\theta}$ and translate that in terms of the motor angles. The motor angles can then be activated by appropriate joint commands through the controller. Figure 2.12.6 presents a flowchart for this purpose.

Based on the algorithm given in Fig. 2.12.6, a computer program can be established to implement the inverse kinematics solutions for the motion of the Rhino XR-2 robot. This program is discussed in Chapter 9 on robot programming. Other motion programs are also discussed in Chapter 9.

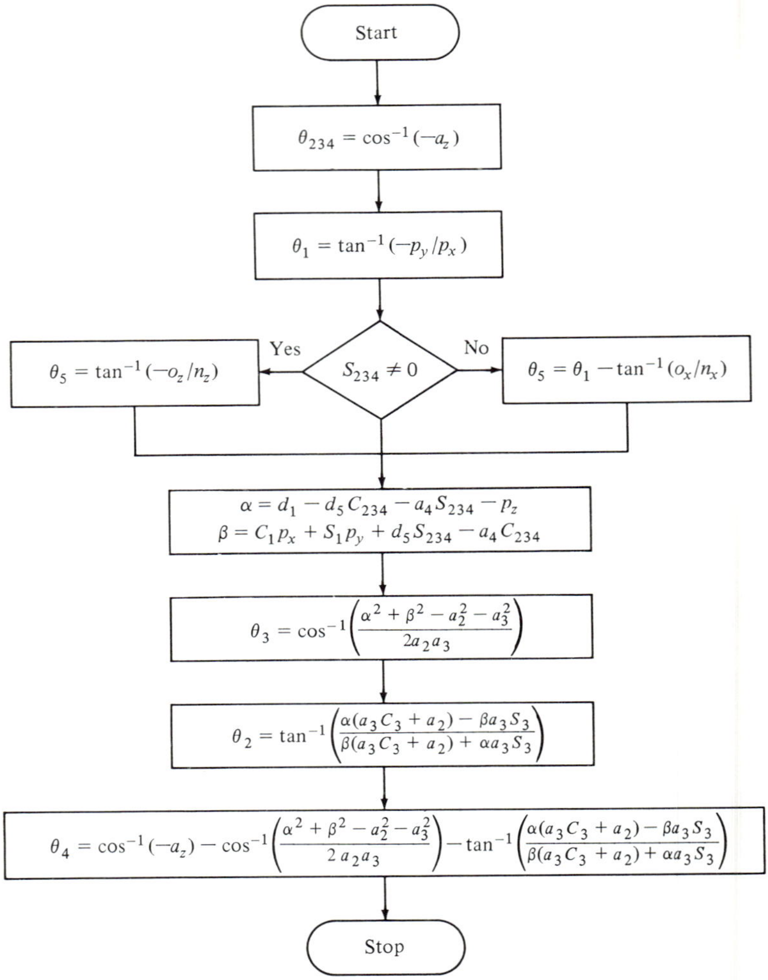

Figure 2.12.6 Inverse kinematics solution algorithm for Rhino XR-2.

2.13 GEOMETRICAL APPROACH IN INVERSE KINEMATICS

Sometimes, employing certain geometrical considerations on robotic manipulators greatly simplifies the procedure for obtaining the solutions to their inverse kinematics problems. The geometrical approach can be generalized to solve the inverse kinematics problem for a great variety of industrial robotic manipulators. There are a number of geometrical techniques available for this purpose. However, the most common one is the one that calculates the solution to the inverse kinematics problem in two stages. In the first stage a position vector is

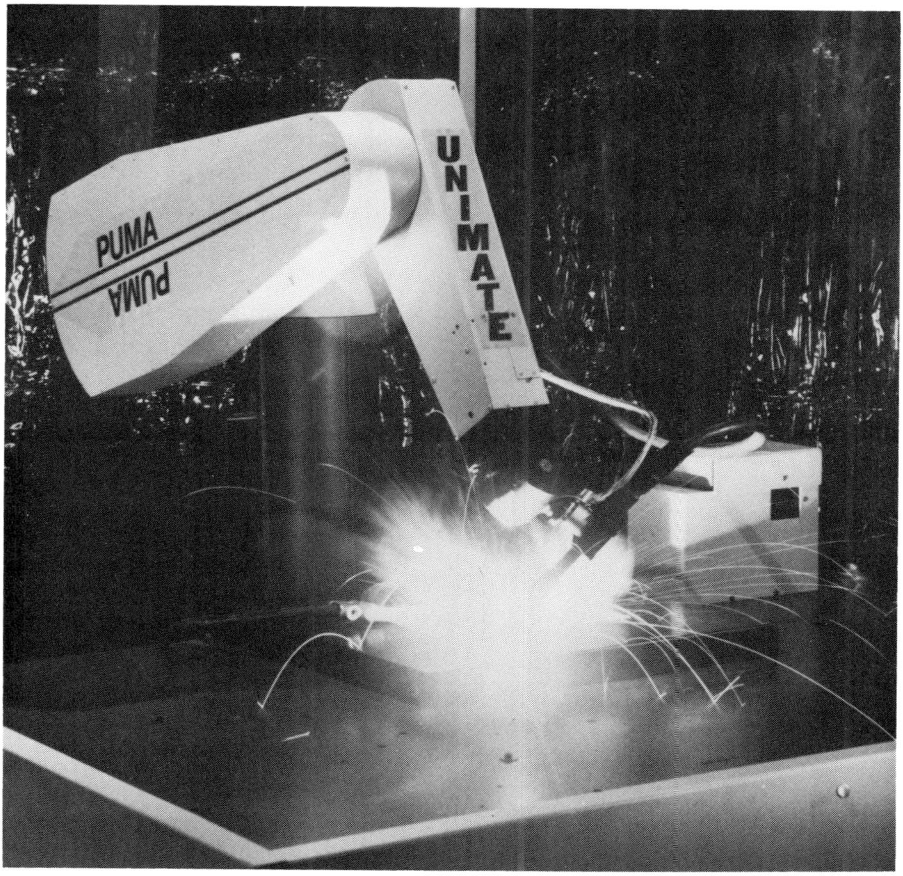

Figure 2.13.1 PUMA 600 Robot. (Courtesy of Unimation Robots.)

calculated for the robotic wrist and used to find the solution to the first three joints. In the second stage the solutions obtained in the first stage are used to calculate the inverse kinematics solutions for the last three joints. In the following example this techiques is used for a PUMA 600 robotic manipulator (Fig. 2.13.1).

EXAMPLE 2.13.1 ■

Referring to Figure 2.13.2 for a PUMA 600 series robotic manipulator, construct the inverse kinematics solutions considering a_3 to be negligible. (Refer to Homework Problem 25 for the case where $a_3 \neq 0$; a_3 is the offset between the y_3 and the z_2 coordinate directions and is approximately equal to 20 mm.)

Solution. Based on Figure 2.13.2 and Algorithm 2.10.1 we can construct Table 2.13.1 for the joint parameters.

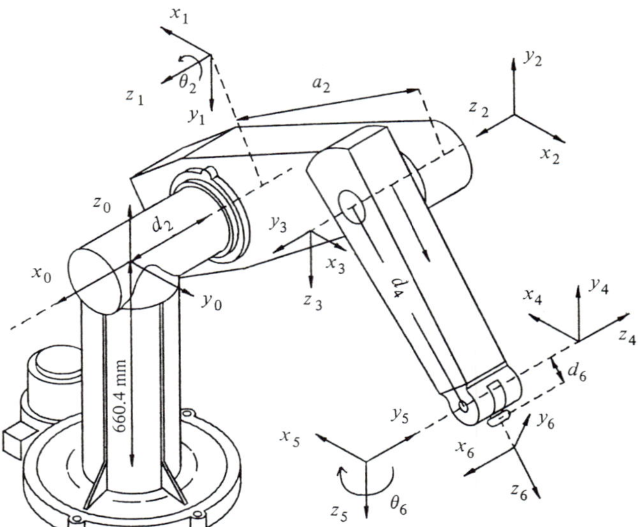

Figure 2.13.2 New coordinate configuration for PUMA 600.

Based on Table 2.13.1 for the joint parameters, the following **A** matrices are obtained:

$$\mathbf{A}_0^1 = \begin{bmatrix} C_1 & 0 & -S_1 & 0 \\ S_1 & 0 & C_1 & 0 \\ 0 & -1 & 0 & 0 \\ 0 & 0 & 0 & 1 \end{bmatrix}, \qquad \mathbf{A}_1^2 = \begin{bmatrix} C_2 & -S_2 & 0 & a_2 C_2 \\ S_2 & C_2 & 0 & a_2 S_2 \\ 0 & 0 & 1 & d_2 \\ 0 & 0 & 0 & 1 \end{bmatrix}, \tag{2.13.1}$$

$$\mathbf{A}_2^3 = \begin{bmatrix} C_3 & 0 & S_4 & 0 \\ S_4 & 0 & -C_3 & 0 \\ 0 & 1 & 0 & 0 \\ 0 & 0 & 0 & 1 \end{bmatrix}, \qquad \mathbf{A}_3^4 = \begin{bmatrix} C_4 & 0 & -S_4 & 0 \\ S_4 & 0 & C_4 & 0 \\ 0 & -1 & 0 & d_4 \\ 0 & 0 & 0 & 1 \end{bmatrix}, \tag{2.13.2}$$

TABLE 2.13.1 A DIFFERENT PUMA 600 ROBOT: LINK AND JOINT PARAMETERS

Joint i	θ_i	α_i	a_i (mm)	d_i (mm)	Range
1	θ_1	$-90°$	0	0	$-160°$–$160°$
2	θ_2	$0°$	432	149.5	$-225°$– $45°$
3	θ_3	$90°$	0	0	$-45°$–$225°$
4	θ_4	$-90°$	0	432.0	$-110°$–$170°$
5	θ_5	$90°$	0	0	$-100°$–$100°$
6	0	$0°$	0	56.5	$-266°$–$266°$

$$\mathbf{A}_2^3 = \begin{bmatrix} C_5 & 0 & S_5 & 0 \\ S_5 & 0 & -C_5 & 0 \\ 0 & 1 & 0 & 0 \\ 0 & 0 & 0 & 1 \end{bmatrix}, \qquad \mathbf{A}_5^6 = \begin{bmatrix} C_6 & -S_6 & 0 & 0 \\ S_6 & C_6 & 0 & 0 \\ 0 & 0 & 1 & d_6 \\ 0 & 0 & 0 & 1 \end{bmatrix}. \tag{2.13.3}$$

Now $\mathbf{T}_0^6 = \mathbf{A}_0^1\mathbf{A}_1^2\mathbf{A}_2^3\mathbf{A}_3^4\mathbf{A}_4^5\mathbf{A}_5^6$ can be obtained since

$$A_{11} = n_x = C_1[C_{23}(C_4C_5C_6 - S_4S_6) - S_{23}S_5C_6] - S_1(S_4C_5C_6 + C_4S_6), \tag{2.13.4}$$

$$A_{21} = n_y = S_1[C_{23}(C_4C_5C_6 - S_4S_6) - S_{23}S_5C_6] + C_1(S_4C_5C_6 + C_4S_6), \tag{2.13.5}$$

$$A_{31} = n_z = -S_{23}(C_4C_5C_6 - S_4S_6) - C_{23}S_5C_6, \tag{2.13.6}$$

$$A_{12} = o_x = C_1[-C_{23}(C_4C_5C_6 - S_4S_6) - S_{23}S_5C_6] - S_1(-S_4C_5C_6 + C_4S_6), \tag{2.13.7}$$

$$A_{22} = o_y = S_1[-C_{23}(C_4C_5C_6 - S_4S_6) - S_{23}S_5C_6] + C_1(-S_4C_5C_6 + C_4S_6), \tag{2.13.8}$$

$$A_{32} = o_z = S_{23}(C_4C_5S_6 - S_4C_6) - C_{23}S_5S_6, \tag{2.13.9}$$

$$A_{13} = a_x = C_1(C_{23}C_4S_5 + S_{23}C_5) - S_1S_4S_5, \tag{2.13.10}$$

$$A_{23} = a_y = S_1(C_{23}C_4S_5 + S_{23}C_5) + C_1S_4S_5, \tag{2.13.11}$$

$$A_{33} = a_z = -S_{23}C_4S_5 + C_{23}C_5, \tag{2.13.12}$$

$$A_{14} = P_x = C_1[56.5(C_{23}C_4S_5 + S_{23}C_5) + 432S_{23} + 432C_2] - S_1(56.5S_4S_5 + 149.5), \tag{2.13.13}$$

$$A_{24} = p_y = S_1[56.5(C_{23}C_4S_5 + S_{23}C_5) + 432S_{23} + 432C_2] + C_1(56.5S_4S_5 + 149.5), \tag{2.13.14}$$

and

$$A_{34} = p_z = 56.5(C_{23}C_5 - S_{23}C_4S_5) + 432C_{23} - 432S_2. \tag{2.13.15}$$

Here, we employ the geometrical approach in obtaining the inverse kinematics solutions. First, we define a position vector pointing from the shoulder to the wrist, by means of which we find θ_1, θ_2, and θ_3. Then, we use the results obtained in finding θ_4, θ_5, and θ_6 for the rest of the robotic arm, that is, the gripper. Note that the position vector from the base origin of the wrist coordinate can be defined as $\mathbf{p}^*$ such that

$$\mathbf{p}^* = \mathbf{p} - d_6\mathbf{a}, \tag{2.13.16}$$

which is obviously a known quantity. Observe that $\mathbf{p}^*$ forms the last column in the $\mathbf{T}_0^4 = \mathbf{A}_0^1\mathbf{A}_1^2\mathbf{A}_2^3\mathbf{A}_3^4$ matrix. Then, one can show easily that

$$p_x^* = C_1(a_2C_2 + d_4S_{23}) - d_2S_1, \tag{2.13.17}$$

$$p_y^* = S_1(a_2C_2 + d_4S_{23}) + d_2C_1, \tag{2.13.18}$$

and

$$p_z^* = d_4C_{23} - a_2S_2. \tag{2.13.19}$$

From Eq. (2.13.17) and (2.13.18), it can be shown that

$$\theta_1 = \tan^{-1}\left(\frac{\pm p_y(\mathbf{p}^* \cdot \mathbf{p}^* - p_x^{*2} - d_2^2)^{1/2} - d_2p_x^*}{\pm p_x(\mathbf{p}^* \cdot \mathbf{p}^* - p^{*2} - d_2^2)^{1/2} + d_2p_y^*}\right), \tag{2.13.20}$$

$$\theta_2 = \tan^{-1}\left(\frac{-[p_z^*(a_2 + d_4S_3) + (d_4C_3) \pm (\mathbf{p}^* \cdot \mathbf{p}^* - p_z^{*^2} - d_2^2)^{1/2}]}{p_z(d_4C_3) - (a_2 + d_4S_3)[\pm (\mathbf{p}^* \cdot \mathbf{p}^* - p_z^{*^2} - d_2^2)^{1/2}]}\right), \tag{2.13.21}$$

and

$$\theta_3 = \tan^{-1}\left(\frac{\mathbf{p}^* \cdot \mathbf{p}^* - d_4^2 - a_2^2 + d_2^2}{\pm [4d_4^2a_2^2 - (\mathbf{p}^* \cdot \mathbf{p}^* - d_4^2 - a_2^2 - d_2^2)^2]^{1/2}}\right). \tag{2.13.22}$$

Various combinations of signs produce various possible solution configurations. Note from Eqs. (2.13.10) and (2.13.11) that

$$C_1a_y - S_1a_x = S_4S_5 \tag{2.13.23}$$

and

$$C_1C_{23}a_x + S_2C_{23}a_y = C_{23}C_4C_5 + C_{23}S_{23}C_5, \tag{2.13.24}$$

and from Eqs. (2.13.24) and (2.13.12) that

$$C_1C_{23}a_x + S_2C_{23}a_y - S_{23}a_z = C_4S_5. \tag{2.13.25}$$

Thus, from Eqs. (2.13.23) and (2.13.25), one finds that

$$\theta_4 = \tan^{-1}\left(\frac{C_1a_y - S_1a_x}{C_1C_{23}a_x + S_1C_{23}a_y - S_{23}a_z}\right). \tag{2.13.26}$$

Similarly, from Eqs. (2.13.10), (2.13.11), and (2.13.13) for a_x, a_y, and a_z, one can show that

$$(C_1C_{23}C_4 - S_1S_4)a_x + (S_1C_{23}C_4 + C_1S_4)a_y - C_4S_{23}a_z = S_5 \tag{2.13.27}$$

and

$$C_1S_{23}a_x + S_1S_{23}a_y + C_{23}a_z = C_5, \tag{2.13.28}$$

so that

$$\theta_5 = \tan^{-1}\left(\frac{(C_1C_{23}C_4 - S_1S_4)a_x + (S_1C_{23}C_4 + C_1S_4)a_y - C_4S_{23}a_z}{C_1S_{23}a_x + S_1S_{23}a_y + C_{23}a_z}\right). \quad (2.13.29)$$

We note that for $\theta_5 = 0$, the solution is degenerate; that is, the robot becomes a five-axis arm and needs a special treatment as follows. ■ ■

The Degenerate Case: $\theta_5 = 0$ In this case, the value of the first wrist roll θ_4 essentially becomes arbitrary and so we choose the current values; set $\theta_4 = 0$. The sum of the first and second wrist rolls θ_4 and θ_6 then determines the direction of the sliding vector **o**. Other solutions essentially stay the same. In particular, the solution for θ_6 can be obtained from Eqs. (2.13.4), (2.13.5), and (2.13.6):

$$(-S_1C_4 - C_1C_{23}S_4)n_x + (C_1C_4 - S_1C_{23}S_4)n_y + S_4S_{23}n_z = S_6 \quad (2.13.30$$

and

$$(-S_1C_4 - C_1C_{23}S_4)o_x + (C_1C_4 - S_1C_{23}S_4)o_y + S_4S_{23}o_z = C_6, \quad (2.13.31)$$

such that

$$\theta_6 = \tan^{-1}\left(\frac{(-S_1C_4 - C_1C_{23}S_4)n_x + (C_1C_4 - S_1C_{23}S_4)n_y + S_4S_{23}n_z}{(-S_1C_4 - C_1C_{23}S_4)o_x + (C_1C_4 - S_1C_{23}S_4)o_y + S_4S_{23}o_z}\right), \quad (2.13.32)$$

This completes the inverse kinematics solution for the vector of the joint angles of a PUMA arm for a given kinematic attitude of the gripper frame.

Finally, it must be mentioned that certain generalized algorithms are presently available that enable the user to generate fairly accurate solutions to the inverse kinematics problem of a general robotic arm. However, the nature of such algorithms is essentially iterative and real-time inverse solutions are not possible for a general robotic manipulator at this time.

BIBLIOGRAPHY

Campbell, D. (1984). "An Iterative Algorithm for the Inverse Kinematics Solution of a General n-axis Robot Manipulator Using Powell's Optimization Technique," M.Sc. Thesis, Department of Mechanical Engineering, Clarkson University, Potsdam, NY (Aug.).

Denavit, J., and Hartenberg, R. S. (1955). A Kinematic Notation for Lower-Pair Mechanisms Based on Matrices, *ASME J. Appl. Mech.*, **22**, 215–221.

Duffy, J. (1980). *Analysis of Mechanisms and Robot Manipulators*, Wiley, New York.

Duffy, J., and Rooney, J. (1975). A Foundation for a Unified Theory of Analysis of Spatial Mechanisms, *J. Eng. Indus. Trans. ASME, Ser. B*, **97**(4), 1159–1164.

Goldenberg, A. A., and Lawrence, D. L. (1985). A Generalized Solution to the Inverse Kinematics of Robot Manipulators, *J. Dyn. Syst. Meas. Control*, **107**, 103–106.

Goldenberg, A. A., Benhabib, B., and Fenton, R. G. (1985). A Complete Generalized Solution to the Inverse Kinematics of Robots, *IEEE J. Robotics Automation*, **RA-1**(1), 14–20.

Hollerbach J. M., and Gideon, Sahar (1983). Wrist-Partitioned, Inverse Kinematic Accelerations and Manipulator Dynamics, *Int. J. Robotics Res.*, **4**, 61–76.

Lee, C. S. G., and Ziegler, M. (1984). Geometric Approach in Solving Inverse Kinematics of PUMA Robots, *IEEE Trans. Aerospace & Electron. Syst.* **AES-20**(6), 695–706.

Lumelsky, V. J. (1983). "Iterative Procedure for Inverse Coordinate Transformation for One Class of Robots," Rept. No. 82CRD332, General Electric Co. (Feb.).

Paul, R. P. (1981). Kinematics Control Equations for Simple Manipulators, *IEEE Trans. Syst. Man & Cyberne.*, **SMC-11**(6), 449–455.

Paul, R. P. (1981). *Robot Manipulators: Mathematics, Programming and Control*, M.I.T. Press, Cambridge, MA.

Pieper, D. L. (1968). "The Kinematics of Manipulators Under Computer Control," Ph.D. Dissertation, Mechanical Engineering Department, Stanford University, 1969; also AIM72.

Pieper, D. L., and Roth, B. (1969). The Kinematics of Manipulators Under Computer Control, in *Proceedings of the Second International Congress on the Theory of Machines and Mechanisms*, Zakopane, Poland, vol. 2, pp. 159–168.

Powell, M. J. D. (1964). "An Efficient Method for Finding the Minimum of a Function of Several Variables Without Calculating Derivatives, *Comput. J.*, 155–164.

Shahinpoor, M. (1985). The Exact Inverse Kinematics Solutions for the Rhino XR-2 Robot Manipulators, *Robotics Age J. Intelligent Machines*, **7**(8), 6–14.

Tsai, L. W., and Morgan, A. P. (1984). "Solving the Kinematics of the Most General Six- and Five-Degree of Freedom Manipulators by Continuation Methods," General Motors Res. Rept. No. GMR-4631 (Oct.).

HOMEWORK PROBLEMS

1. Given the attitude matrix $\mathbf{T}_0^n$, such that

$$\mathbf{T}_0^n = \begin{bmatrix} n_x & o_x & a_x & p_a \\ n_x & o_y & a_y & p_y \\ n_z & o_z & a_z & p_y \\ 0 & 0 & 0 & 1 \end{bmatrix},$$

find the inverse of $\mathbf{T}_0^n$ by the conventional techniques discussed in Section 2.4.

2. Describe a set of transformations that creates a mirror image of an object at a distance d from a mirror.

3. Find a composite homogeneous transformation matrix representing the following:

(a) A translation of a along the w axis.

(b) A rotation of θ about the x axis.

(c) A rotation of ψ about the u axis.
(d) A translation of b along the y axis.
(e) A rotation of ϕ about the z axis.
Answer: $\mathbf{H} = R(z, \phi)T(y, b)R(x, \theta)T(w, a)R(u, \psi)$.

4. A robotic hand initially lies along the y axis such that the direction of gripping is in the z direction which is along the u axis of the robotic hand as shown in Figure P2.4. The hand is then translated by 4 units along the y axis, rotated about the z axis by $-90°$, rotated about the v axis by $150°$, and finally rotated about the u axis by $-90°$. Find the following:
(a) The overall homogeneous transformation $\mathbf{H}$.
(b) The final location of the hand.

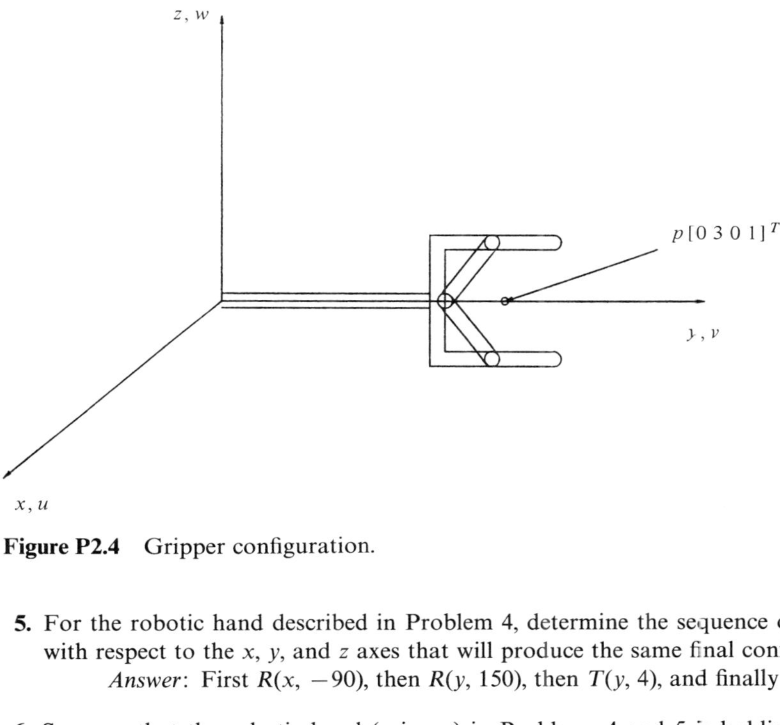

Figure P2.4 Gripper configuration.

5. For the robotic hand described in Problem 4, determine the sequence of rotations with respect to the x, y, and z axes that will produce the same final configuration.
Answer: First $R(x, -90)$, then $R(y, 150)$, then $T(y, 4)$, and finally $R(z, -90)$.

6. Suppose that the robotic hand (gripper) in Problems 4 and 5 is holding a box of dimensions 6, 4, and 2 whose center of volume is at $\mathbf{p} \equiv [0\ 3\ 0\ 1]^T$ as shown in Figure P2.6. Determine the following:
(a) The initial configuration (attitude) matrix for the box.
(b) The final configuration matrix for the box after it undergoes the homogeneous transformation $\mathbf{H}$ of Problem 1.
Answers:

$$\text{(a)}\ (\mathbf{Box})_{\text{initial}} \equiv \begin{bmatrix} 3 & 3 & -3 & -3 & -3 & -3 & 3 & 3 \\ 5 & 5 & 5 & 5 & 1 & 1 & 1 & 1 \\ 1 & -1 & -1 & 1 & 1 & -1 & -1 & 1 \\ 1 & 1 & 1 & 1 & 1 & 1 & 1 & 1 \end{bmatrix}.$$

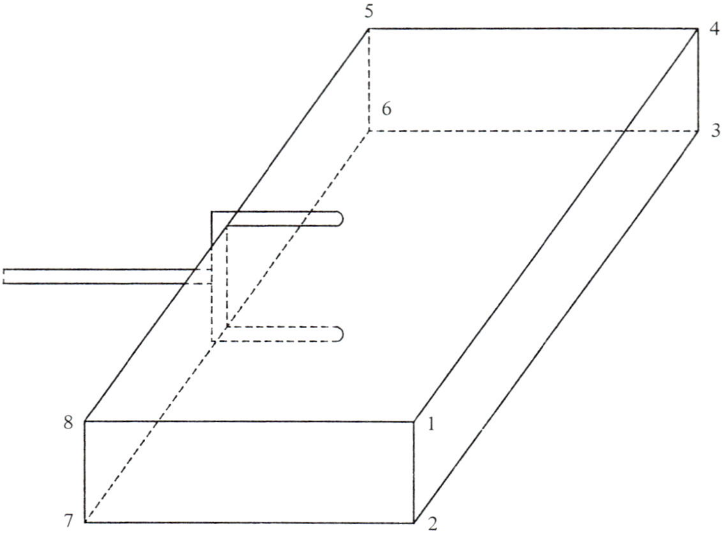

Figure P2.6

(b) $(\mathbf{Box})_{\text{final}} = \mathbf{H}(\mathbf{Box})_{\text{initial}}$

$$= \begin{bmatrix} 5 & 3 & 3 & 5 & 5 & 3 & 3 & 5 \\ 5.1 & 5.1 & -0.1 & -0.1 & -2.1 & -2.1 & -3.1 & 3.1 \\ 2.8 & 2.8 & 5.8 & 5.8 & 2.36 & 2.36 & -0.63 & -0.63 \\ 1 & 1 & 1 & 1 & 1 & 1 & 1 & 1 \end{bmatrix}.$$

7. Determine the new location of the robotic hand as well as the box configuration matrix if the hand rotates 50° about an axis whose unit vector is $\mathbf{k} = [(2)^{1/2}/2 \; 0 \; (2)^{1/2}/2 \;\; 1]^T$.

Answer: $\mathbf{p} \equiv [-1.625 \;\; 1.929 \;\; 1.625 \;\; 1]^T$ and

$$(\mathbf{Box})_{\text{final}} = \mathbf{H}(\mathbf{Box})_{\text{initial}},$$

$$\mathbf{H} = \begin{bmatrix} 0.8215 & -0.5416 & 0.1785 & 0 \\ 0.5416 & 0.643 & -0.5416 & 0 \\ 0.1785 & 0.5416 & 0.8215 & 0 \\ 0 & 0 & 0 & 1 \end{bmatrix}.$$

8. Determine the direction $\mathbf{k}$ and the angle of rotation θ about that direction which cause a gripper to relocate from an initial point $\mathbf{p}_0 \equiv [0\,0\,4\,1]^T$ to a final location $\mathbf{p} \equiv [2\;2\;(8)^{1/2}\;\;1]^T$.

9. A robotic hand is initially at a point $\mathbf{p} \equiv [1\;\;1\;\;3\;\;1]^T$. It is then rotated about the z axis by $-90°$, then around its own v axis by 150°, and finally about its own u axis by $-90°$ (Fig. P2.9).

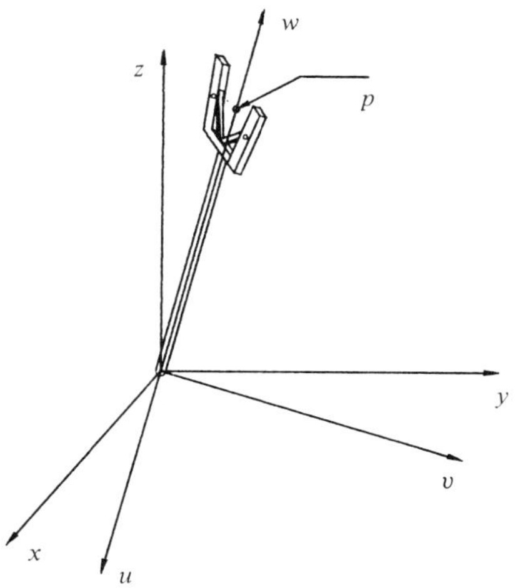

Figure P2.9

(a) Find the resulting homogeneous transformation **H**.
(b) Find the final location of the robotic hand.

Answers:

(a) $\mathbf{H} = \begin{bmatrix} 0 & 0 & 1 & 0 \\ 0.866 & 0.5 & 0 & 0 \\ -0.5 & 0.866 & 0 & 0 \\ 0 & 0 & 0 & 1 \end{bmatrix}.$

(b) $\mathbf{p} = [3\ 1.366\ 0.366\ 1]^T$.

10. For the Rhino XR-2 robot in the home position (Fig. P2.10), use the D-H procedure to construct an appropriate table for joint parameters d_i, a_i, α_i, and θ_i, $i = 1, 2, 3, 4, 5$, by means of an appropriate set of joint coordinate frames.

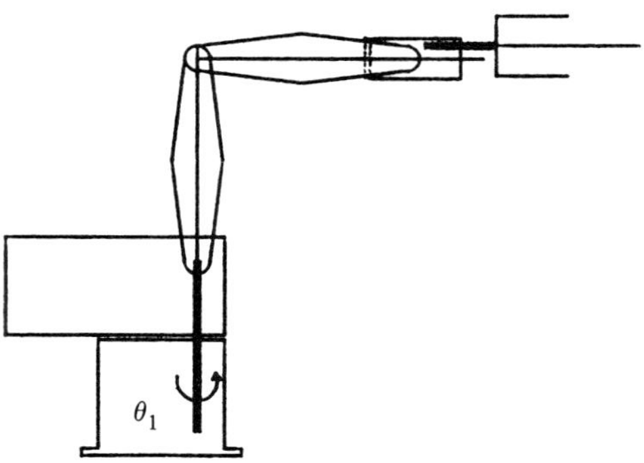

Figure P2.10

11. The robot shown in Figure P2.11 is in its home position. The axes of motion are as shown.

(a) Complete the description of all joint coordinate systems by assigning the proper set of x and y axes.

(b) Construct a table for the joint parameters α_i, θ_i, a_i, and d_i for this robot.

(c) For $\theta = [90° \; 0° \; 90° \; 0° \; 90°]$, determine the matrix $\mathbf{T}_0^5$.

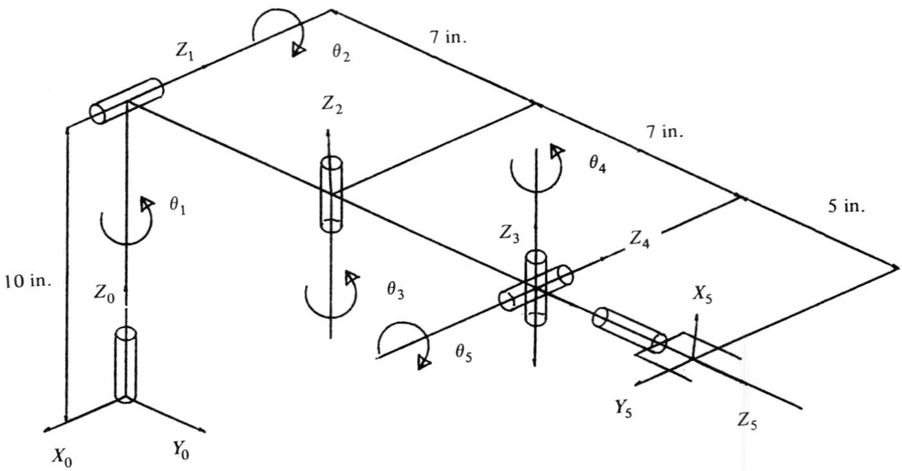

Figure P2.11

12. A six-axis robotic manipulator is described in Table P2.12.1, which gives the joint coordinate parameters. Find the inverse kinematics solution for the joint angles $\boldsymbol{\theta}_i \equiv [\theta_1 \; \theta_2 \; \theta_3 \; \theta_4 \; \theta_5 \; \theta_6]^T$. Assume that the position and orientation of the gripper, that is, T_0^6, with respect to the base coordinate is completely known:

$$\mathbf{T}_0^6 \equiv \begin{bmatrix} n_x & o_x & a_x & p_x \\ n_y & o_y & a_y & p_y \\ n_z & o_z & a_z & p_z \\ 0 & 0 & 0 & 1 \end{bmatrix}.$$

TABLE P2.12.1

Joint i	θ_i	α_i	a_i	d_i
1	θ_1	$-90°$	0	0
2	θ_2	$0°$	a_2	d_2
3	θ_3	$90°$	0	0
4	θ_4	$-90°$	0	d_3
5	θ_5	$90°$	0	0
6	θ_6	$0°$	0	d_6

13. For the robotic hand in Problem 9, determine the sequence of rotations with respect to the x, y, and z axes that will produce the same final configuration.

Answer: First $R(x, -90)$, then $R(y, 150)$, and finally $R(z, -90)$.

14. Assume that a robotic hand that is initially located at $\mathbf{p} \equiv [1\ 1\ 3\ 1]^T$ is holding an ellipsoid whose long axis is 6 units in length and aligned parallel to the u axis and whose short axis is 3 units in length and aligned parallel to the v axis (see Fig. P2.6). The origin of the ellipsoid is at p. Determine the final configuration matrix of the ellipsoid after the homogeneous transformation of Problem 6. (Hint: First describe the initial configuration by a 4×6 matrix, then multiply by $\mathbf{H}$ to get the final configuration.)

15. A gripper initially lies along the z axis which is coincident with the w axis of the gripper (Fig. P2.15). The $0xyz$ and $0uvw$ frames originally coincide. Now the gripper is first rotated about the z axis by $-90°$, then around the v axis by $180°$, and finally around u axis by $-90°$. Find the resulting RPY transformation. Note that

$$\mathbf{RPY}(\phi, \theta, \psi) = R(z, -90)R(v, 180)R(u, -90).$$

Answer

$$\mathbf{RPY} = \begin{bmatrix} 0 & 0 & 1 & 0 \\ 1 & 0 & 0 & 0 \\ 0 & 1 & 0 & 0 \\ 0 & 0 & 0 & 1 \end{bmatrix}.$$

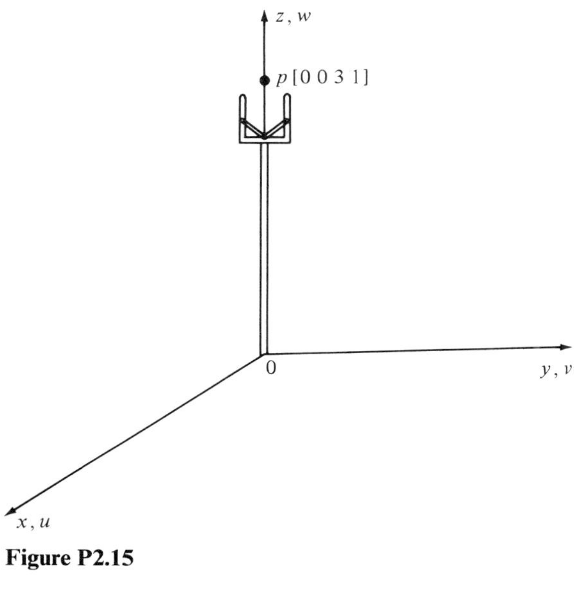

Figure P2.15

16. What sequence of rotations produces the same final configuration as in Problem 15 if the rotations are with respect to the x, y, and z axes?

Answer: First $R(x, -90)$, then $R(y, 180)$, and finally $R(z, -90)$.

17. Suppose that the gripper in the previous problems is holding a cube of unit length and the cube is initially aligned with the xyz frame with its center of volume located at $\mathbf{p} \equiv [0\ 0\ 3\ 1]^T$, which is also the origin of the gripper (Fig. P2.17). Determine the final configuration matrix of the cube after the RPY transformation of Problem 1.

Answer

$$(\mathbf{Cube})_{\text{final}} = \begin{bmatrix} 3.5 & 2.5 & 2.5 & 3.5 & 3.5 & 2.5 & 2.5 & 3.5 \\ 0.5 & 0.5 & -0.5 & -0.5 & -0.5 & -0.5 & 0.5 & 0.5 \\ 0.5 & 0.5 & 0.5 & 0.5 & -0.5 & -0.5 & -0.5 & -0.5 \\ 1 & 1 & 1 & 1 & 1 & 1 & 1 & 1 \end{bmatrix}.$$

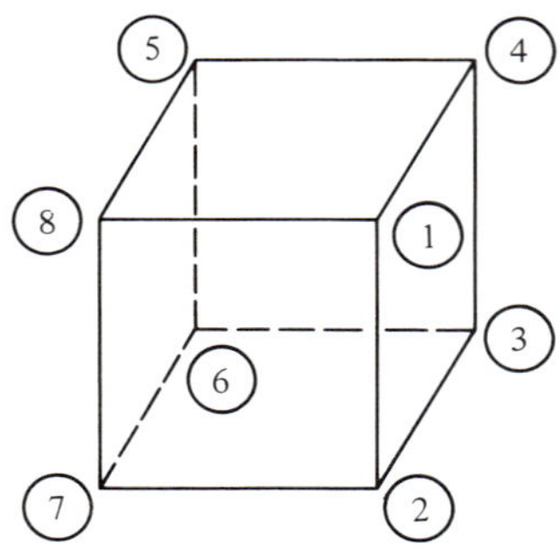

Figure P2.17

18. Find the new location of the gripper point p as well as the cube configuration matrix if the gripper rotates 45° around an axis whose unit vector is $\mathbf{k} \equiv [1/\sqrt{3}\ 1/\sqrt{3}\ 1/\sqrt{3}\ 1]^T$.

Answers: $\mathbf{p}_{\text{final}} = [1.52\ -0.98\ 2.42\ 1]^T$

$$(\mathbf{Cube})_{\text{final}} = \begin{bmatrix} 2.0177 & 1.5118 & 0.707 & 1.213 & 1.524 & 1.018 & 1.822 & 2.328 \\ -0.432 & -0.1212 & -0.627 & -0.938 & -1.74 & -1.43 & -0.926 & -1.34 \\ 2.914 & 2.1094 & 2.421 & 3.225 & 2.719 & 1.914 & 1.603 & 2.408 \\ 1 & 1 & 1 & 1 & 1 & 1 & 1 & 1 \end{bmatrix}.$$

19. Write a program in BASIC to multiply a set of 4 × 4 matrices of the form:

$$\mathbf{A} = \begin{bmatrix} n_x & o_x & a_x & p_x \\ n_y & o_y & a_y & p_y \\ n_z & o_z & a_z & p_z \\ 0 & 0 & 0 & 1 \end{bmatrix}.$$

20. For a general six-axis robotic arm, the inverse kinematics equations for the six unknowns θ_i, $i = 1, 2, 3, 4, 5, 6$, of the joint vector are the following 12 equations:

$$\begin{bmatrix} n_x & o_x & a_x & p_x \\ n_y & o_y & a_y & p_y \\ n_z & o_z & a_z & p_z \\ 0 & 0 & 0 & 1 \end{bmatrix} = \prod_{i=1}^{6} \begin{bmatrix} C\theta_i & -C\alpha_i S\theta_i & S\alpha_i S\theta_i & a_i C\theta_i \\ S\theta_i & C\alpha_i C\theta_i & -S\alpha_i C\theta_i & a_i S\theta_i \\ 0 & S\alpha_i & C\alpha_i & d_i \\ 0 & 0 & 0 & 1 \end{bmatrix}.$$

Table P2.20.1 gives the joint coordinate parameters. Note that all values of α_i, a_i, and d_i are assumed to be known. Find as many solutions as you can for the problem.

TABLE P2.20.1

Joint	θ_i	α_i	a_i	d_i
1	θ_1	α_1	a_1	d_1
2	θ_2	α_2	a_2	d_2
3	θ_3	α_3	a_3	d_3
4	θ_4	α_4	a_4	d_4
5	θ_5	α_5	a_5	d_5
6	θ_6	α_6	a_6	d_6

21. For a simple elbow manipulator, Table P2.21.1 presents the joint parameters. Find the complete set of inverse kinematics solutions.

TABLE P2.21.1

Joint	θ_i	α_i	d_i	a_i	$C\alpha_i$	$S\alpha_i$
1	θ_1	90°	0	0	0	1
2	θ_2	0°	0	a_2	1	0
3	θ_3	0°	0	a_3	1	0
4	θ_4	−90°	0	a_4	0	−1
5	θ_5	90°	0	0	0	1
6	θ_6	0°	0	0	1	0

22. Solve the inverse kinematics problem for Rhino XR-2 ($a_4 \neq 0$) if $\theta_2 + \theta_3 = 0$. Note that this allows the robotic hand to be oriented in any arbitrary direction and not necessarily perpendicular to the work surface.

23. A six-axis robotic manipulator is described by the joint coordinate parameters given in Table P2.23.1. Find the inverse kinematics solutions for the joint angles $\boldsymbol{\theta}_i \equiv [\theta_1\ \theta_2\ \theta_3\ \theta_4\ \theta_5\ \theta_6]^T$. Assume that the position and orientation of the gripper, that is, $\mathbf{T}_0^6$, with respect to the base coordinate is known completely:

$$\mathbf{T}_0^6 \equiv \begin{bmatrix} n_x & o_x & a_x & p_x \\ n_y & o_y & a_y & p_y \\ n_z & o_z & a_z & p_z \\ 0 & 0 & 0 & 1 \end{bmatrix}.$$

TABLE P2.23.1

Joint i	θ_i	α_i	a_i	d_i
1	θ_1	−90°	0	d_1
2	θ_2	0°	a_2	0
3	θ_3	0°	a_3	0
4	θ_4	90°	0	0
5	θ_5	0°	0	d_5
6	$\theta_6 = \pi$	0°	0	0

24. A robotic manipulator is described by the joint parameters given in Table P2.24.1.

(a) Completely set up and draw the coordinate frames with a proper scale.

(b) Solve the inverse kinematics problem for the joint angles θ_1, θ_2, θ_3, θ_4, θ_5, and θ_6.

TABLE P2.24.1

Joint	θ_i	α_i	a_i	d_i
1	θ_1	$-90°$	0	d_1
2	θ_2	$0°$	a_2	0
3	θ_3	$0°$	a_3	0
4	θ_4	$-90°$	a_4	0
5	θ_5	$90°$	0	d_5
6	θ_6	$0°$	0	d_6

25. Referring to Example 2.11.8 and Table 2.11.6, find the inverse kinematics solutions for the PUMA 600 series when $a_3 \neq 0$.

Chapter 3

Homogeneous Robotic Differential Transformations

3.1 INTRODUCTION

There are many important applications in robot engineering that require some knowledge of homogeneous differential transformations. Basically, when small (differential) changes in the attitude matrix (position and orientation) is of any importance, the knowledge of homogeneous differential transformations become essential—the small or differential motion of objects, robotic grippers, robotic joints, cameras, or any other auxiliary equipment associated with a robotic work station. Of particular importance are the motions of accommodators and the velocity and acceleration transformations. As discussed before, there are two distinct kinematic spaces at work in robotic manipulators. One is the joint space and the other is the Cartesian tip or gripper space.

It is always of interest to robotic applications to devise transformations that bridge the kinematic and dynamic knowledge of the joint space to that of the Cartesian gripper space. For example, given the joint velocities in the form of a vector of the joint velocities, what would be the generalized vector of the tip velocities? Similarly, given the vector of the joint accelerations, what would be the generalized vector of the tip accelerations? These questions will be answered in this chapter and in Chapters 5 and 7.

In many robotic applications a vision system such as a TV camera (digitizable) keeps track of both the object motion and manipulator motion. Incremental or minute changes in the positions of both object and manipulator must be fed back continuously to a microcomputer in order to perform a variety of assembly tasks such as insert and place. Thus, one would like to be able to compute differential changes via suitable differential transformations. Differential changes are also important because they can be related readily to the corresponding differential changes in the joint coordinate vector $\boldsymbol{\theta}_i \equiv [\theta_1, \theta_2, \cdots, \theta_n]$. Such joint vector differential changes can be related to both the angular velocities and the angular accelerations. For example, consider the generalized D-H matrix $\mathbf{A}_{i-1}^i$ which is given by

$$\mathbf{A}_{i-1}^i = \begin{bmatrix} C\theta_i & -C\alpha_i S\theta_i & S\alpha_i S\theta_i & a_i C\theta_i \\ S\theta_i & C\alpha_i C\theta_i & -S\alpha_i C\theta_i & a_i S\theta_i \\ 0 & S\alpha_i & C\alpha_i & d_i \\ 0 & 0 & 0 & 1 \end{bmatrix}. \tag{3.1.1}$$

Note that

$$\frac{d\mathbf{A}_{i-1}^i}{d\theta_i} = \begin{bmatrix} -S\theta_i & -C\alpha_i C\theta_i & S\alpha_i C\theta_i & -a_i S\theta_i \\ C\theta_i & -C\alpha_i S\theta_i & +S\alpha_i S\theta_i & a_i C\theta_i \\ 0 & 0 & 0 & 0 \\ 0 & 0 & 0 & 0 \end{bmatrix} = \mathbf{B}_{i-1}^i. \tag{3.1.2}$$

Furthermore, it is also true that

$$\frac{d\mathbf{A}_{i-1}^i}{dt} = \dot{\mathbf{A}}_{i-1}^i = \begin{bmatrix} -S\theta_i & -C\alpha_i C\theta_i & S\alpha_i C\theta_i & -a_i S\theta_i \\ C\theta_i & -C\alpha_i S\theta_i & +S\alpha_i S\theta_i & a_i C\theta_i \\ 0 & 0 & 0 & 0 \\ 0 & 0 & 0 & 0 \end{bmatrix} \frac{d\theta_i}{dt}, \tag{3.1.3}$$

where, for brevity, we may put $d\theta_i/dt = \dot{\theta}_i$.

$$\frac{d^2\mathbf{A}_{i-1}^i}{dt^2} = \ddot{\mathbf{A}}_{i-1}^i = \mathbf{B}_{i-1}^i \ddot{\theta}_i + \begin{bmatrix} -C\theta_i & +C\alpha_i S\theta_i & -S\alpha_i S\theta_i & -a_i C\theta_i \\ -S\theta_i & -C\alpha_i C\theta_i & S\alpha_i C\theta_i & -a_i S\theta_i \\ 0 & 0 & 0 & 0 \\ 0 & 0 & 0 & 0 \end{bmatrix} \dot{\theta}_i. \tag{3.1.4}$$

From Eqs. (3.1.1) and (3.1.2) we see that

$$\mathbf{A}_{i-1}^i + d\mathbf{A}_{i-1} = \mathbf{Trans}(d\mathbf{r})\mathbf{R}(\mathbf{k}, d\theta)\mathbf{A}_{i-1}^i. \tag{3.1.5}$$

$$\therefore \quad d\mathbf{A}_{i-1}^i = [\mathbf{Trans}(d\mathbf{r})\mathbf{R}(\mathbf{k}, d\theta) - \mathbf{I}]\mathbf{A}_{i-1}^i. \tag{3.1.6}$$

The expression in brackets is denoted by $\boldsymbol{\Delta}_{i-1}^i$ and is called the differential transformation, that is, the differential translation–rotation matrix. Thus,

$$d\mathbf{A}_{i-1}^i = \boldsymbol{\Delta}_{i-1}^i \, \mathbf{A}_{i-1}^i. \tag{3.1.7}$$

The multiplication rule still applies! Namely, if $\boldsymbol{\Delta}_{i-1}^i$ is "premultiplied" then the change is with respect to the base or $(i-1)$th coordinate frame. However, it is also true that

$$d\mathbf{A}_{i-1}^i = \mathbf{A}_{i-1}^i \, {}^A\Delta_i^{i-1}, \tag{3.1.8}$$

where, in this case, ${}^A\Delta_i^{i-1}$ is "postmultiplied" and it is a transformation with respect to the **A** coordinate frame.

The differential transformation Δ can be analyzed further by noting that

$$\textbf{Trans}(d\mathbf{r}) = \begin{bmatrix} 1 & 0 & 0 & dx \\ 0 & 1 & 0 & dy \\ 0 & 0 & 1 & dz \\ 0 & 0 & 0 & 1 \end{bmatrix}, \tag{3.1.9}$$

and the $\mathbf{R}(\mathbf{k}, d\theta)$ can be shown to be

$$\mathbf{R}(\mathbf{k}, d\theta) = \begin{bmatrix} 1 & -k_z d\theta & k_y d\theta & 0 \\ k_z d\theta & 1 & -k_x d\theta & 0 \\ -k_y d\theta & k_x d\theta & 1 & 0 \\ 0 & 0 & 0 & 1 \end{bmatrix}. \tag{3.1.10}$$

EXAMPLE 3.1.1 ■

Show that $\mathbf{R}(\mathbf{k}, d\theta)$ is given by Eq. (3.1.10).

Solution. Recall from Eq. (2.4.40) that

$$\mathbf{R}(\mathbf{k}, \theta) = \begin{bmatrix} k_x^2(1 - C\theta) + C\theta & k_y k_x(1 - C\theta) - k_z S\theta & k_z k_x(1 - C\theta) + k_y S\theta & 0 \\ k_x k_y(1 - C\theta) + k_z S\theta & k_y^2(1 - C\theta) + C\theta & k_z k_y(1 - C\theta) - k_x S\theta & 0 \\ k_x k_z(1 - C\theta) - k_y S\theta & k_y k_z(1 - C\theta) + k_x S\theta & k_z^2(1 - C\theta) + C\theta & 0 \\ 0 & 0 & 0 & 1 \end{bmatrix} \tag{3.1.10}$$

Now as $\theta \to d\theta$ the $\lim(\sin d\theta) \to d\theta$, the $\lim(\cos \theta) \to 1$, and the $\lim(1 - C\theta) \to 0$; thus, Eq. (3.1.11) reduces to expression (3.1.10). Q.E.D. ■ ■

Note that the differential transformation Δ can then be defined as

$$\Delta = \textbf{Trans}(d\mathbf{r})\mathbf{R}(\mathbf{k}, d\theta) - \mathbf{I}, \tag{3.1.12}$$

or

$$\Delta = \begin{bmatrix} 1 & 0 & 0 & dx \\ 0 & 1 & 0 & dy \\ 0 & 0 & 1 & dz \\ 0 & 0 & 0 & 1 \end{bmatrix} \begin{bmatrix} 1 & -k_z d\theta & k_y d\theta & 0 \\ k_z d\theta & 1 & -k_x d\theta & 0 \\ -k_y d\theta & k_x d\theta & 1 & 0 \\ 0 & 0 & 0 & 1 \end{bmatrix} - \begin{bmatrix} 1 & 0 & 0 & 0 \\ 0 & 1 & 0 & 0 \\ 0 & 0 & 1 & 0 \\ 0 & 0 & 0 & 1 \end{bmatrix}. \tag{3.1.13}$$

Finally, we obtain the following expression for the differential transformation:

$$\Delta = \begin{bmatrix} 0 & -k_z d\theta & k_y d\theta & dx \\ k_z d\theta & 0 & -k_x d\theta & dy \\ -k_y d\theta & k_x d\theta & 0 & dz \\ 0 & 0 & 0 & 1 \end{bmatrix}. \tag{3.1.14}$$

This is a generalized differential transformation. This generalized differential transformation can be written for the ith link such that

$$\Delta_{i-1}^{i} = \begin{bmatrix} 0 & -k_{iz} d\theta_i & k_{iy} d\theta_i & dx_i \\ k_{iz} d\theta_i & 0 & -k_{ix} d\theta_i & dy_i \\ -k_{iy} d\theta_i & k_{ix} d\theta_i & 0 & dz_i \\ 0 & 0 & 0 & 0 \end{bmatrix}, \tag{3.1.15}$$

which is a differential transformation with respect to the $(i-1)$th coordinate frame. It can be shown that, based on Eqs. (2.4.10) and (2.4.11), one has the following:

$$\mathbf{R}(x, \delta_x) = \begin{bmatrix} 1 & 0 & 0 & 0 \\ 0 & 1 & -\delta_x & 0 \\ 0 & \delta_x & 1 & 0 \\ 0 & 0 & 0 & 1 \end{bmatrix}, \tag{3.1.16}$$

$$\mathbf{R}(y, \delta_y) = \begin{bmatrix} 1 & 0 & \delta_y & 0 \\ 0 & 1 & 0 & 0 \\ -\delta_y & 0 & 1 & 0 \\ 0 & 0 & 0 & 1 \end{bmatrix}, \tag{3.1.17}$$

and

$$\mathbf{R}(z, \delta_z) = \begin{bmatrix} 1 & -\delta_z & 0 & 0 \\ \delta_z & 1 & 0 & 0 \\ 0 & 0 & 1 & 0 \\ 0 & 0 & 0 & 1 \end{bmatrix}. \tag{3.1.18}$$

We can compute Δ by noting that without the translation effects $\Delta = \Delta^*$, so that

$$\Delta^* = \mathbf{R}(x, \delta_x)\mathbf{R}(y, \delta_y)\mathbf{R}(z, \delta_z) = \mathbf{R}(z, \delta_z)\mathbf{R}(y, \delta_y)\mathbf{R}(x, \delta_x) \tag{3.1.19}$$

is independent of the order in which the rotations are made. Thus, we find that $\Delta = \text{Trans}(d\mathbf{r})\Delta^* - \mathbf{I})$ reduces to the same expression if we set $k_x d\theta = \delta_x$, $k_y d\theta = \delta_y$, and $k_z d\theta = \delta_z$:

$$\Delta = \begin{bmatrix} 0 & -\delta_z & \delta_y & dx \\ \delta_z & 0 & -\delta_x & dy \\ -\delta_y & \delta_x & 0 & dz \\ 0 & 0 & 0 & 1 \end{bmatrix}. \tag{3.1.20}$$

The two vectors $d\mathbf{r} \equiv [dx\,dy\,dz]^T$ and $\delta \equiv [\delta_x\,\delta_y\,\delta_z]^T$ are called differential translation and rotation vectors. Sometimes we refer to them in a combined fashion such that

$$\mathbf{D} = [dx \quad dy \quad dz \quad \delta_x \quad \delta_y \quad \delta_z]^T \tag{3.1.21}$$

is called the generalized differential displacement vector.

EXAMPLE 3.1.2 ■

Show that ${}^A\Delta^i_{i-1} = \mathbf{A}^i_{i-1}\,\Delta^i_{i-1}\,\mathbf{A}^i_{i-1}$.

Solution. Recalling expressions (2.13.7) and (2.13.8) we find that

$$\Delta\mathbf{A}^i_{i-1} = \mathbf{A}^i_{i-1}\,{}^A\Delta. \tag{3.1.22}$$

Note that $\Delta = \Delta^i_{i-1}$ and ${}^A\Delta = \Delta^{i-1}_i$. Thus, premultiplying both sides of Eq. (3.1.22) by $(\mathbf{A}^i_{i-1})^{-1}$, we obtain

$${}^A\Delta = (\mathbf{A}^i_{i-1})^{-1}\Delta\mathbf{A}^i_{i-1} \quad \text{or} \quad \Delta^{i-1}_i = \mathbf{A}^{i-1}_i\,\Delta^i_{i-1}\,\mathbf{A}^i_{i-1}, \tag{3.1.23}$$

which is the desired proof. ■ ■

Note that ${}^A\Delta$ is a generalized differential transform with respect to the ith frame. Equation (3.1.15) can now be simplified:

$${}^A\Delta^i_{i-1} = \begin{bmatrix} 0 & -\delta_{iz} & \delta_{iy} & dx_i \\ \delta_{iz} & 0 & -\delta_{ix} & dy_i \\ -\delta_{iy} & \delta_{ix} & 0 & dz_i \\ 0 & 0 & 0 & 0 \end{bmatrix}, \tag{3.1.24}$$

where $\delta_{ix} = k_{ix}d\theta_i$, $\delta_{iy} = k_{iy}d\theta_i$, and $\delta_{iz} = k_{iz}d\theta_i$. The ith joint differential rotation vector $\boldsymbol{\delta}_i \equiv [\delta_{ix}\,\delta_{iy}\,\delta_{iz}]^T$ can be calculated with respect to joint differential rotation vector $\delta \equiv [\delta_x\,\delta_y\,\delta_z]^T$ with respect to the base coordinate such that

$$\delta_{ix} = \boldsymbol{\delta}\cdot\mathbf{n}, \qquad \delta_{iy} = \boldsymbol{\delta}\cdot\mathbf{o}, \qquad \delta_{iz} = \boldsymbol{\delta}\cdot\mathbf{a}, \tag{3.1.25}$$

where **n**, **o**, and **a** pertain to the **A** or **T** matrices. Furthermore,

$$dx_i = \boldsymbol{\delta}\cdot(\mathbf{p}\times\mathbf{n}) + \mathbf{d}\cdot\mathbf{n}, \tag{3.1.26}$$

$$dy_i = \boldsymbol{\delta}\cdot(\mathbf{p}\times\mathbf{o}) + \mathbf{d}\cdot\mathbf{0}, \tag{3.1.27}$$

$$dz_i = \boldsymbol{\delta}\cdot(\mathbf{p}\times\mathbf{a}) + \mathbf{d}\cdot\mathbf{a}. \quad ■■ \tag{3.1.28}$$

EXAMPLE 3.1.3 ■

Derive the relationships (3.1.25)–(3.1.28).

Solution. To derive Eqs. (3.1.25)–(3.1.28), we note that

$${}^A\Delta^{i-1}_i = [\mathbf{A}^i_{i-1}]^{-1}\,\Delta^i_{i-1}\,[\mathbf{A}^i_{i-1}], \tag{3.1.29}$$

where

$$\mathbf{A}^{i}_{i-1} = \begin{bmatrix} \mathbf{n} & \mathbf{o} & \mathbf{a} & \mathbf{p} \\ 0 & 0 & 0 & 1 \end{bmatrix} = \begin{bmatrix} A_{11} & p \\ 0 & 1 \end{bmatrix}. \tag{3.1.30}$$

$$\mathbf{n} \times \mathbf{o} = \mathbf{a}, \qquad \mathbf{o} \times \mathbf{a} = \mathbf{n}, \tag{3.1.31}$$

$$\mathbf{A}_{ii}\mathbf{A}_{ii}^{T} \equiv \mathbf{I}, \tag{3.1.32}$$

and

$$\boldsymbol{\Delta}^{i}_{i-1} = \left[\begin{array}{ccc|c} 0 & -\delta_z & \delta_y & dx \\ \delta_z & 0 & -\delta_x & dy \\ -\delta_y & \delta_x & 0 & dz \\ \hline 0 & 0 & 0 & 0 \end{array}\right] = \left[\begin{array}{c|c} \mathbf{R} & \mathbf{d} \\ \hline 0 & 0 \end{array}\right]. \tag{3.1.33}$$

One can see easily that

$${}^{A}\boldsymbol{\Delta}^{i-1}_{i} = \begin{bmatrix} \mathbf{A}^{T}_{11} & -\mathbf{A}^{T}_{11}\mathbf{p} \\ 0 & \mathbf{1} \end{bmatrix} \begin{bmatrix} \mathbf{R} & \mathbf{d} \\ 0 & 0 \end{bmatrix} \begin{bmatrix} \mathbf{A}_{11} & \mathbf{p} \\ 0 & \mathbf{1} \end{bmatrix}, \tag{3.1.34}$$

or

$${}^{A}\boldsymbol{\Delta}^{i-1}_{i} = \begin{bmatrix} \mathbf{A}^{T}_{11} & -\mathbf{A}^{T}_{11}\mathbf{p} \\ 0 & \mathbf{1} \end{bmatrix} \begin{bmatrix} \mathbf{R}\mathbf{A}_{11} & \mathbf{R}\mathbf{p} + \mathbf{d} \\ 0 & \mathbf{0} \end{bmatrix} \tag{3.1.35}$$

and

$${}^{A}\boldsymbol{\Delta}^{i-1}_{i} = \begin{bmatrix} \mathbf{A}^{T}_{11}\mathbf{R}\mathbf{A}_{11} & \mathbf{A}_{11}[\mathbf{R}\mathbf{p} + \mathbf{d}] \\ 0 & \mathbf{0} \end{bmatrix}. \tag{3.1.36}$$

Furthermore,

$$\mathbf{A}^{T}_{11}\mathbf{R}\mathbf{A}_{11} = \begin{bmatrix} n_x & n_y & n_z \\ o_x & o_y & o_x \\ a_x & a_y & a_z \end{bmatrix} \begin{bmatrix} 0 & -\delta_z & \delta_y \\ \delta_z & 0 & -\delta_x \\ -\delta_y & \delta_x & 0 \end{bmatrix} \times \begin{bmatrix} n_x & o_x & a_x \\ n_y & o_y & a_y \\ n_z & o_z & a_z \end{bmatrix}, \tag{3.1.37}$$

or

$$\mathbf{A}^{T}_{11}\mathbf{R}\mathbf{A}_{11} = \begin{bmatrix} \mathbf{n}^T \\ \mathbf{o}^T \\ \mathbf{a}^T \end{bmatrix} \begin{bmatrix} (\delta_y n_z - n_y\delta_z) & (\delta_y o_z - o_y\delta_z) & (\delta_y a_z - a_y\delta_z) \\ (\delta_z n_x - n_z\delta_x) & (\delta_z o_x - o_z\delta_x) & (\delta_z a_x - a_z\delta_x) \\ (\delta_x n_y - n_x\delta_y) & (\delta_x o_y - o_x\delta_y) & (\delta_x a_y - a_x\delta_y) \end{bmatrix}$$

$$= \begin{bmatrix} \mathbf{N}^T \\ \mathbf{o}^T \\ \mathbf{a}^T \end{bmatrix} [\mathbf{S}_1 \quad \mathbf{S}_2 \quad \mathbf{S}_3], \tag{3.1.38}$$

where

$$\mathbf{S}_1 = \begin{bmatrix} \mathbf{i} & \mathbf{j} & \mathbf{k} \\ \delta_x & \delta_y & \delta_z \\ n_x & n_y & n_z \end{bmatrix} = \begin{bmatrix} (\delta_y n_z - n_y \delta_z) \\ (\delta_z n_x - n_z \delta_x) \\ (\delta_x n_y - n_x \delta_y) \end{bmatrix} = \boldsymbol{\delta} \times \mathbf{n}. \tag{3.1.39}$$

Similarly,

$$\mathbf{S}_2 = \boldsymbol{\delta} \times \mathbf{o} \quad \text{and} \quad \mathbf{S}_3 = \boldsymbol{\delta} \times \mathbf{a}. \tag{3.1.40}$$

Then

$$\mathbf{A}_{11}^T \mathbf{R} \mathbf{A}_{11} = \begin{bmatrix} \mathbf{n} \cdot (\boldsymbol{\delta} \times \mathbf{n}) & \mathbf{n} \cdot (\boldsymbol{\delta} \times \mathbf{o}) & \mathbf{n} \cdot (\boldsymbol{\delta} \times \mathbf{a}) \\ \mathbf{o} \cdot (\boldsymbol{\delta} \times \mathbf{n}) & \mathbf{o} \cdot (\boldsymbol{\delta} \times \mathbf{o}) & \mathbf{o} \cdot (\boldsymbol{\delta} \times \mathbf{a}) \\ \mathbf{a} \cdot (\boldsymbol{\delta} \times \mathbf{n}) & \mathbf{a} \cdot (\boldsymbol{\delta} \times \mathbf{o}) & \mathbf{a} \cdot (\boldsymbol{\delta} \times \mathbf{a}) \end{bmatrix} \tag{3.1.41}$$

Finally,

$$\mathbf{A}_{11}^T \mathbf{R} \mathbf{A}_{11} = \begin{bmatrix} 0 & -\boldsymbol{\delta} \cdot \mathbf{a} & \boldsymbol{\delta} \cdot \mathbf{o} \\ \boldsymbol{\delta} \cdot \mathbf{a} & 0 & -\boldsymbol{\delta} \cdot \mathbf{n} \\ -\boldsymbol{\delta} \cdot \mathbf{o} & \boldsymbol{\delta} \cdot \mathbf{n} & 0 \end{bmatrix}. \tag{3.1.42}$$

Since it is true that

$$\mathbf{a} \cdot (\mathbf{b} \times \mathbf{c}) = \mathbf{b} \cdot (\mathbf{c} \times \mathbf{a}) = -\mathbf{b}(\mathbf{a} \times \mathbf{c}), \tag{3.1.43}$$

$$\mathbf{n} \times \mathbf{o} = \mathbf{a}, \qquad \mathbf{o} \times \mathbf{a} = \mathbf{n}, \qquad \mathbf{a} \times \mathbf{n} = \mathbf{o}, \tag{3.1.44}$$

and

$$\mathbf{u} \times \mathbf{u} = \mathbf{o} \times \mathbf{o} = \mathbf{a} \times \mathbf{a} = 0, \tag{3.1.45}$$

one can conclude that

$$\mathbf{A}_{11}^T [\mathbf{R}\mathbf{p} + \mathbf{d}] = \mathbf{A}_{11}^T \mathbf{R}\mathbf{p} + \mathbf{A}_{11}^T \mathbf{d}. \tag{3.1.46}$$

Furthermore, since

$$\mathbf{A}_{11}^T \mathbf{d} = \begin{bmatrix} \mathbf{n}^T \\ \mathbf{o}^T \\ \mathbf{a}^T \end{bmatrix} [\mathbf{d}] = \begin{bmatrix} \mathbf{n} \cdot \mathbf{d} \\ \mathbf{o} \cdot \mathbf{d} \\ \mathbf{a} \cdot \mathbf{d} \end{bmatrix} \tag{3.1.47}$$

and

$$\mathbf{A}_{11}^T \mathbf{R}\mathbf{p} = \begin{bmatrix} \mathbf{n}^T \\ \mathbf{o}^T \\ \mathbf{a}^T \end{bmatrix} \begin{bmatrix} 0 & -\delta_z & +\delta_y \\ +\delta_z & 0 & -\delta_x \\ -\delta_y & +\delta_x & 0 \end{bmatrix} \begin{bmatrix} p_x \\ p_y \\ p_z \end{bmatrix}, \tag{3.1.48}$$

which means that

$$\mathbf{A}_{11}^T \mathbf{R}\mathbf{p} = \begin{bmatrix} \mathbf{n}^T \\ \mathbf{o}^T \\ \mathbf{a}^T \end{bmatrix} \begin{bmatrix} \delta_y p_z - \delta_z p_y \\ \delta_z p_x - \delta_x p_z \\ \delta_x p_y - \delta_y p_x \end{bmatrix} = \begin{bmatrix} \mathbf{n}^T \\ \mathbf{o}^T \\ \mathbf{a}^T \end{bmatrix} \begin{bmatrix} g_1 \\ g_2 \\ g_3 \end{bmatrix} \tag{3.1.49}$$

where

$$\begin{bmatrix} g_1 \\ g_2 \\ g_3 \end{bmatrix} = \begin{bmatrix} \delta_y p_z - \delta_z p_y \\ \delta_z p_x - \delta_x p_z \\ \delta_x p_y - \delta_y p_x \end{bmatrix} = \begin{bmatrix} \mathbf{i} & \mathbf{j} & \mathbf{k} \\ \delta_x & \delta_y & \delta_z \\ p_x & p_y & p_z \end{bmatrix} = \boldsymbol{\delta} \times \mathbf{p}, \tag{3.1.50}$$

then

$$\mathbf{A}_{11}^T \mathbf{Rp} = \begin{bmatrix} \mathbf{n}^T \\ \mathbf{o}^T \\ \mathbf{a}^T \end{bmatrix} [\boldsymbol{\delta} \times \mathbf{p}] = \begin{bmatrix} \mathbf{n} \cdot (\boldsymbol{\delta} \times \mathbf{p}) \\ \mathbf{o} \cdot (\boldsymbol{\delta} \times \mathbf{p}) \\ \mathbf{a} \cdot (\boldsymbol{\delta} \times \mathbf{p}) \end{bmatrix} = \begin{bmatrix} \boldsymbol{\delta} \cdot (\mathbf{p} \times \mathbf{n}) \\ \boldsymbol{\delta} \cdot (\mathbf{p} \times \mathbf{o}) \\ \boldsymbol{\delta} \cdot (\mathbf{p} \times \mathbf{a}) \end{bmatrix}, \tag{3.1.51}$$

$$\mathbf{A}_{ii}^T [\mathbf{Rp} + \mathbf{d}] = \begin{bmatrix} \boldsymbol{\delta} \cdot (\mathbf{p} \times \mathbf{n}) + \mathbf{n} \cdot \mathbf{d} \\ \boldsymbol{\delta} \cdot (\mathbf{p} \times \mathbf{o}) + \mathbf{o} \cdot \mathbf{d} \\ \boldsymbol{\delta} \cdot (\mathbf{p} \times \mathbf{a}) + \mathbf{a} \cdot \mathbf{d} \end{bmatrix}, \tag{3.1.52}$$

and, finally,

$$^{A}\Delta^{i-1} = \begin{bmatrix} 0 & -\boldsymbol{\delta} \cdot \mathbf{a} & \boldsymbol{\delta} \cdot \mathbf{o} & [\boldsymbol{\delta} \cdot (\mathbf{p} \times \mathbf{n}) + \mathbf{n} \cdot \mathbf{d}] \\ \boldsymbol{\delta} \cdot \mathbf{a} & 0 & -\boldsymbol{\delta} \cdot \mathbf{n} & [\boldsymbol{\delta} \cdot (\mathbf{p} \times \mathbf{o}) + \mathbf{o} \cdot \mathbf{d}] \\ -\boldsymbol{\delta} \cdot \mathbf{o} & \boldsymbol{\delta} \cdot \mathbf{n} & 0 & [\boldsymbol{\delta} \cdot (\mathbf{p} \times \mathbf{a}) + \mathbf{a} \cdot \mathbf{d}] \\ 0 & 0 & 0 & 0 \end{bmatrix}, \tag{3.1.53}$$

which is the desired proof. ■■

EXAMPLE 3.1.4 ■

Suppose that one of the **A** matrices is given by

$$\mathbf{A}_{i-1}^{i} = \mathbf{A} = \begin{bmatrix} 0 & 0 & 1 & 5 \\ 0 & 1 & 0 & 6 \\ -1 & 0 & 0 & 1 \\ 0 & 0 & 0 & 1 \end{bmatrix}, \tag{3.1.54}$$

and, furthermore, that $\mathbf{d} = 0.5\mathbf{i} + 0\mathbf{j} + \mathbf{k}$ and $\boldsymbol{\delta} = 0\mathbf{i} + 0.1\mathbf{j} + 0\mathbf{k}$. Find the equivalent differential translation and rotation with respect to the coordinate frame **A**.

Solution.

$$\mathbf{A}_{i-1}^{i} = \mathbf{A} = \begin{bmatrix} n_x & o_x & a_x & p_x \\ n_y & o_y & a_y & p_y \\ n_z & o_z & a_z & p_z \\ 0 & 0 & 0 & 1 \end{bmatrix}. \tag{3.1.55}$$

Therefore, $\mathbf{n} \equiv 0\mathbf{i} + 0\mathbf{j} - \mathbf{k}$, $\mathbf{o} \equiv 0\mathbf{i} + \mathbf{j} + \mathbf{k}$, $\mathbf{a} \equiv \mathbf{i} + 0\mathbf{j} + 0\mathbf{k}$, and $\mathbf{p} = 5\mathbf{i} + 6\mathbf{j} + \mathbf{k}$. Now

$$d\mathbf{A}_{i-1}^{i} = \Delta_{i-1}^{i} \, \mathbf{A}_{i-1}^{i} = \mathbf{A}^{i}{}_{i-1} \, {}^{A}\Delta_{i}^{i-1} \tag{3.1.56}$$

and

$$\Delta_{i-1}^{i} = \begin{bmatrix} 0 & 0 & 0.1 & dx \\ 0 & 0 & 0 & dy \\ -0.1 & +0 & 0 & dz \\ 0 & 0 & 0 & 0 \end{bmatrix}, \tag{3.1.57}$$

Thus,

$$\Delta_{i-1}^{i} = \begin{bmatrix} 0 & 0 & 0.1 & 0.5 \\ 0 & 0 & 0 & 0 \\ -0.1 & 0 & 0 & 1 \\ 0 & 0 & 0 & 0 \end{bmatrix}, \tag{3.1.58}$$

and

$$\begin{aligned} d\mathbf{A}_{i-1}^{i} &= \begin{bmatrix} 0 & 0 & 0.1 & 0.5 \\ 0 & 0 & 0 & 0 \\ -0.1 & 0 & 0 & 1 \\ 0 & 0 & 0 & 0 \end{bmatrix} \begin{bmatrix} 0 & 0 & 1 & 5 \\ +0 & 1 & 0 & 6 \\ -1 & 0 & 0 & 1 \\ 0 & 0 & 0 & 1 \end{bmatrix} \\ &= \begin{bmatrix} -0.1 & 0 & 0 & 0.6 \\ 0 & 0 & 0 & 0 \\ 0 & 0 & -0.1 & 0.5 \\ 0 & 0 & 0 & 0 \end{bmatrix}. \end{aligned} \tag{3.1.59}$$

Now consider $d\mathbf{A}_{i-1}^{i} = \mathbf{A}_{i-1}^{i}\, {}^{A}\Delta_{i-1}^{i}$. Here

$${}^{A}\Delta^{i}_{i-1} = \begin{bmatrix} 0 & -\boldsymbol{\delta}\cdot\mathbf{a} & \boldsymbol{\delta}\cdot\mathbf{o} & [\boldsymbol{\delta}\cdot(\mathbf{p}\times\mathbf{n}) + \mathbf{d}\cdot\mathbf{n}] \\ \boldsymbol{\delta}\cdot\mathbf{a} & 0 & -\boldsymbol{\delta}\cdot\mathbf{n} & [\boldsymbol{\delta}\cdot(\mathbf{p}\times\mathbf{o}) + \mathbf{d}\cdot\mathbf{o}] \\ -\boldsymbol{\delta}\cdot\mathbf{o} & \boldsymbol{\delta}\cdot\mathbf{n} & 0 & [\boldsymbol{\delta}\cdot(\mathbf{p}\times\mathbf{a}) + \mathbf{d}\cdot\mathbf{a}] \\ 0 & 0 & 0 & 0 \end{bmatrix}. \tag{3.1.60}$$

We now have $\boldsymbol{\delta}\cdot\mathbf{a} = 0$, $\boldsymbol{\delta}\cdot\mathbf{o} = 0.1$, and $\boldsymbol{\delta}\cdot\mathbf{n} = 0.2$. Furthermore,

$$\mathbf{p}\times\mathbf{n} = \begin{bmatrix} \mathbf{i} & \mathbf{j} & \mathbf{k} \\ 5 & 6 & 1 \\ 0 & 0 & -1 \end{bmatrix} = -6\mathbf{i} + 5\mathbf{j} + 0\mathbf{k}, \tag{3.1.61}$$

$$\mathbf{p}\times\mathbf{o} = \begin{bmatrix} \mathbf{i} & \mathbf{j} & \mathbf{k} \\ 5 & 6 & 1 \\ 0 & 1 & 0 \end{bmatrix} = -1\mathbf{i} - 0\mathbf{j} + 5\mathbf{k}, \tag{3.1.62}$$

and

$$\mathbf{p}\times\mathbf{a} = \begin{bmatrix} \mathbf{i} & \mathbf{j} & \mathbf{k} \\ 5 & 6 & 1 \\ 1 & 0 & 0 \end{bmatrix} = 0\mathbf{i} + \mathbf{j} - 6\mathbf{k}. \tag{3.1.63}$$

Thus,

$$\boldsymbol{\delta}\cdot(\mathbf{p}\times\mathbf{n}) = 0.5, \qquad \boldsymbol{\delta}\cdot(\mathbf{p}\times\mathbf{o}) = -0, \qquad \mathbf{d}\cdot\mathbf{n} = -1,$$
$$\mathbf{d}\cdot\mathbf{o} = 0, \qquad \mathbf{d}\cdot\mathbf{a} = 0.5, \quad \text{and} \quad \boldsymbol{\delta}\cdot(\mathbf{p}\times\mathbf{a}) = 0. \tag{3.1.64}$$

Equation (3.1.60) further reduces to

$$^{A}\boldsymbol{\Delta}^{i}_{i-1} = \begin{bmatrix} 0 & 0 & 0.1 & -0.5 \\ 0 & 0 & -0.0 & -0.0 \\ -0.1 & +0 & 0 & 0.6 \\ 0 & 0 & 0 & 0 \end{bmatrix}. \tag{3.1.65}$$

Now $d\mathbf{A}^{i}_{i-1} = \mathbf{A}^{i}_{i-1}\,{}^{A}\boldsymbol{\Delta}^{i}_{i-1}$ and therefore, based on Eqs. (3.1.54) and (3.1.60),

$$d\mathbf{A}^{i}_{i-1} = \begin{bmatrix} 0 & 0 & 1 & 5 \\ 0 & 1 & 0 & 6 \\ -1 & 0 & 0 & 1 \\ 0 & 0 & 0 & 1 \end{bmatrix} \begin{bmatrix} 0 & 0 & 0.1 & -0.5 \\ 0 & 0 & 0.0 & 0 \\ -0.1 & 0 & 0 & 0.6 \\ 0 & 0 & 0 & 0 \end{bmatrix}. \quad \blacksquare\,\blacksquare \tag{3.1.66}$$

Finally,

$$d\mathbf{A}^{i}_{i-1} = \begin{bmatrix} -0.1 & 0 & 0 & 0.5 \\ 0 & 0 & 0 & 0 \\ 0 & 0 & -0.1 & 0.5 \\ 0 & 0 & 0 & 0 \end{bmatrix}. \tag{3.1.67}$$

EXAMPLE 3.1.5 ■

A robotic vision system is tracking the motion of an object by means of a camera attached to the robot's link 5 of the manipulator (Figs. 3.1.1 and 3.1.2). The transform describing the camera with respect to link 5 is given by $^{T_5}\mathbf{CAM}$ and is expressed numerically as

$$^{T_5}\mathbf{CAM} = \begin{bmatrix} 0 & 0 & 1 & 6 \\ 0 & +1 & 0 & 0 \\ -1 & 0 & 0 & 10 \\ 0 & 0 & 0 & 1 \end{bmatrix}. \tag{3.1.68}$$

The gripper is described by $\mathbf{A}^{6}_{5}$ such that

$$\mathbf{A}^{6}_{5} = \begin{bmatrix} 0 & -1 & 0 & 0 \\ 1 & 0 & 0 & 0 \\ 0 & 0 & 1 & 8 \\ 0 & 0 & 0 & 1 \end{bmatrix}. \tag{3.1.69}$$

The object is continuously observed by the camera and the object's position is expressed by a transform $(^{\text{Cam}}\mathbf{Object}) \equiv {}^{\text{CAM}}\mathbf{O}$. The differential changes in the motion of the object are translated into differential changes

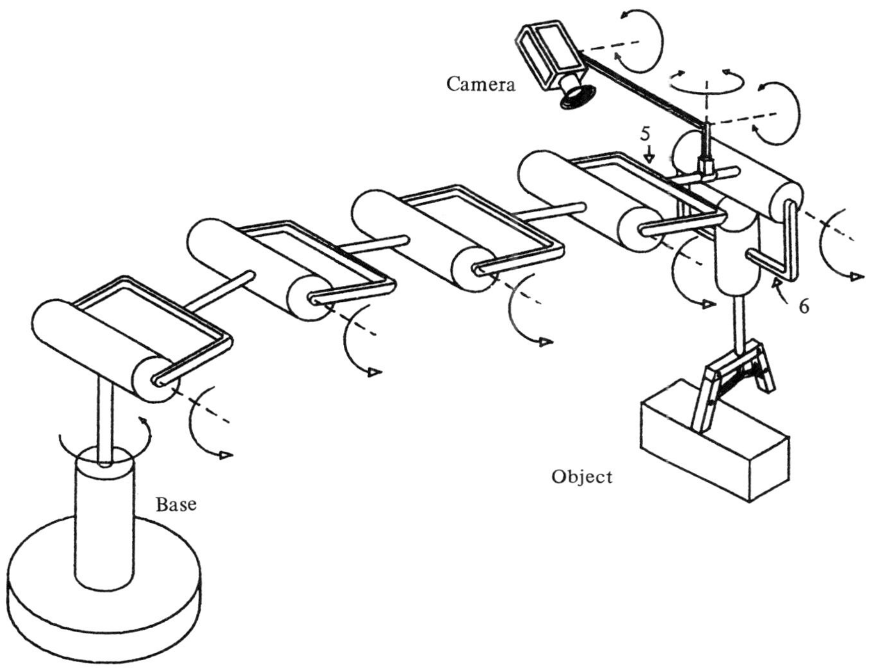

Figure 3.1.1 TV-camera-equipped robotic manipulator performing differential chase of an object.

in the camera coordinate. Suppose these are given by

$$^{\mathrm{CAM}}\mathbf{d} = -0.1\mathbf{i} + 0.2\mathbf{j} + 0\mathbf{k} \tag{3.1.70}$$

and

$$^{\mathrm{CAM}}\boldsymbol{\delta} = 0\mathbf{i} + 0\mathbf{j} + 0.2\mathbf{k}. \tag{3.1.71}$$

Determine the corresponding differential change $d\mathbf{T}_\epsilon$ for the gripper that eventually will bring the gripper into contact with the object.

Solution. Referring to Fig. 3.1.3, one notes that

$$\mathbf{T}_0^5\,\mathbf{A}_5^6\mathbf{X} = \mathbf{T}_0^5\,{}^{T_5}\mathbf{CAM}\;{}^{\mathrm{CAM}}\mathbf{O}, \tag{3.1.72}$$

where $\mathbf{X}$ is a transform describing the object with respect to the gripper, that is,

$$\mathbf{X} \equiv {}^{T_0^6}\mathbf{Object} \equiv {}^{T_6}\mathbf{O}. \tag{3.1.73}$$

Thus, the gripper coordinate frame is at a transform $^{C}\mathbf{G}$ such that

$$^{C}\mathbf{G} = \mathbf{CAM}^{-1}\,\mathbf{A}_5^6. \tag{3.1.74}$$

Figure 3.1.2 Typical robot chase situation in auto factory.

Equation (3.1.74) for ${}^{C}\mathbf{G}$ now reduces to

$$
{}^{C}\mathbf{G} = \mathbf{CAM}^{-1}\mathbf{A}_6 = \begin{bmatrix} 0 & 0 & -1 & 10 \\ 0 & +1 & 0 & 0 \\ 1 & 0 & 0 & -6 \\ 0 & 0 & 0 & 1 \end{bmatrix} \begin{bmatrix} 0 & -1 & 0 & 0 \\ 1 & 0 & 0 & 0 \\ 0 & 0 & 1 & 8 \\ 0 & 0 & 0 & 1 \end{bmatrix}, \tag{3.1.75}
$$

or

$$
{}^{C}\mathbf{G} = \begin{bmatrix} n_x & o_x & a_x & p_x \\ n_y & o_y & a_y & p_y \\ n_z & o_z & a_z & p_z \\ 0 & 0 & 0 & 1 \end{bmatrix} = \begin{bmatrix} 0 & 0 & -1 & 2 \\ 1 & 0 & 0 & 0 \\ 0 & -1 & 0 & -6 \\ 0 & 0 & 0 & 1 \end{bmatrix}. \tag{3.1.76}
$$

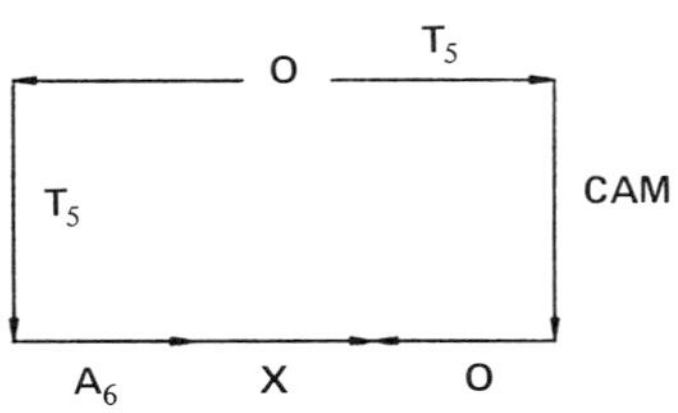

Figure 3.1.3 Transform graph for Example 3.1.5.

Thus, $\mathbf{n} = 0\mathbf{i} + 1\mathbf{j} + 0\mathbf{k}$, $\mathbf{o} = 0\mathbf{i} + 0\mathbf{j} - 1\mathbf{k}$, $\mathbf{a} = -\mathbf{i} + 0\mathbf{j} + 0\mathbf{k}$, and $\mathbf{p} = 2\mathbf{i} + 0\mathbf{j} - 6\mathbf{k}$. One easily can obtain

$$^{T}d_x = \boldsymbol{\delta}\cdot(\mathbf{p}\times\mathbf{n}) + \mathbf{d}\cdot\mathbf{n} = \mathbf{n}\cdot[(\boldsymbol{\delta}\times\mathbf{p}) + \mathbf{d}], \tag{3.1.77}$$

$$^{T}d_y = \boldsymbol{\delta}\cdot(\mathbf{p}\times\mathbf{o}) + \mathbf{d}\cdot\mathbf{o} = \mathbf{o}\cdot[(\boldsymbol{\delta}\times\mathbf{p}) + \mathbf{d}], \tag{3.1.78}$$

$$^{T}d_z = \boldsymbol{\delta}\cdot(\mathbf{p}\times\mathbf{a}) + \mathbf{d}\cdot\mathbf{a} = \mathbf{a}\cdot[(\boldsymbol{\delta}\times\mathbf{p}) + \mathbf{d}], \tag{3.1.79}$$

and

$$^{T}\delta_x = \mathbf{n}\cdot\boldsymbol{\delta}, \qquad ^{T}\delta_y = \mathbf{o}\cdot\boldsymbol{\delta}, \qquad ^{T}\delta_z = \mathbf{a}\cdot\boldsymbol{\delta}. \tag{3.1.80}$$

So that

$$\boldsymbol{\delta}\times\mathbf{p} = \begin{bmatrix} \mathbf{i} & \mathbf{j} & \mathbf{k} \\ 0 & 0 & 0.2 \\ 2 & 0 & -6 \end{bmatrix} = 0\mathbf{i} + 0.4\mathbf{j} + 0\mathbf{k}, \tag{3.1.81}$$

and therefore

$$\boldsymbol{\delta}\times\mathbf{p} + \mathbf{d} = -0.1\mathbf{i} - 0.2\mathbf{j} + 0\mathbf{k}. \tag{3.1.82}$$

Finally,

$$^{T_6}\mathbf{d} = -0.2\mathbf{i} + 0\mathbf{j} + 0.1\mathbf{k} \tag{3.1.83}$$

and

$$^{T_6}\boldsymbol{\delta} = 0\mathbf{i} - 0.2\mathbf{j} + 0\mathbf{k}. \tag{3.1.84}$$

3.2 JACOBIAN OF TRANSFORMATION IN ROBOTIC MANIPULATION

The Jacobian of transformation in robotics is a matrix that relates the differential changes from one frame to another frame. It also relates such differential changes from the joint space to the tip space.

Let us recall the generalized differential displacement $^{A}\mathbf{D}$ measured with respect to a frame A such that

$$^{A}\mathbf{D} = [^{A}dx \quad ^{A}dy \quad ^{A}dz \quad ^{A}\delta_x \quad ^{A}\delta_y \quad ^{A}\delta_z]^T. \tag{3.2.1}$$

Now ${}^A\mathbf{D}$ is related to ${}^B\mathbf{D}$, the differential displacement of another frame B of the same robot, by means of a Jacobian ${}^B\mathbf{J}_A$ such that

$$ {}^B\mathbf{D} = {}^B\mathbf{J}_A {}^A\mathbf{D}, \tag{3.2.2} $$

where ${}^B\mathbf{J}_A$ is the Jacobian of transformation from frame A to frame B. To clarify the situation, let us assign y_i, $i = 1, 2, 3, 4, 5, 6$, as the general coordinates of frame B and x_i, $i = 1, 2, 3, 4, 5, 6$, as the general coordinates of frame A, such that

$$ dx_1 = {}^A dx, \qquad dx_2 = {}^A dy, \qquad dx_3 = {}^A dz, \tag{3.2.3} $$

$$ dx_4 = {}^A\delta x, \qquad dx_5 = {}^A\delta y, \qquad dx_6 = {}^A\delta z; \tag{3.2.4} $$

$$ dy_1 = {}^B dx, \qquad dy_2 = {}^B dy, \qquad dy_3 = {}^B dz, \tag{3.2.5} $$

$$ dy_4 = {}^B\delta x, \qquad dy_5 = {}^B\delta y, \qquad dy_6 = {}^B\delta z. \tag{3.2.6} $$

Note that there exist functional relationships between the y_i and x_i coordinates, such that

$$ y_i = f_i(x_1, x_2, x_3, x_4, x_5, x_6), \qquad i = 1, 2, 3, 4, 5, 6, \tag{3.2.7} $$

and therefore

$$ dy_i = \sum_{j=1}^{6} \frac{\partial f_i}{\partial x_j} dx_i, \qquad i, j = 1, 2, 3, 4, 5, 6. \tag{3.2.8} $$

Note that

$$ dy_i = \begin{bmatrix} dy_1 \\ dy_2 \\ dy_3 \\ dy_4 \\ dy_5 \\ dy_6 \end{bmatrix} = \begin{bmatrix} {}^B dx \\ {}^B dy \\ {}^B dz \\ {}^B\delta x \\ {}^B\delta y \\ {}^B\delta z \end{bmatrix} = {}^B\mathbf{D}, \tag{3.2.9} $$

and

$$ dx_i = \begin{bmatrix} dx_1 \\ dx_2 \\ dx_2 \\ dx_4 \\ dx_5 \\ dx_6 \end{bmatrix} = \begin{bmatrix} {}^A dx \\ {}^A dy \\ {}^A dz \\ {}^A\delta x \\ {}^A\delta y \\ {}^A\delta z \end{bmatrix} = {}^A\mathbf{D}. \tag{3.2.10} $$

Therefore, based on Eqs. (3.2.2), (3.2.8), (3.2.9), and (3.2.10), it is clear that

$$
{}^{B}\mathbf{J}_{A} = \begin{bmatrix}
\frac{\delta f_1}{\partial x_1} & \frac{\delta f_1}{\partial x_2} & \frac{\delta f_1}{\partial x_3} & \frac{\delta f_1}{\partial x_4} & \frac{\delta f_1}{\partial x_5} & \frac{\delta f_1}{\partial x_6} \\
\frac{\delta f_2}{\partial x_1} & \frac{\delta f_2}{\partial x_2} & \frac{\delta f_2}{\partial x_3} & \frac{\delta f_2}{\partial x_4} & \frac{\delta f_2}{\partial x_5} & \frac{\delta f_2}{\partial x_6} \\
\frac{\delta f_3}{\partial x_1} & \frac{\delta f_3}{\partial x_2} & \frac{\delta f_3}{\partial x_3} & \frac{\delta f_3}{\partial x_4} & \frac{\delta f_3}{\partial x_5} & \frac{\delta f_3}{\partial x_6} \\
\frac{\delta f_4}{\partial x_1} & \frac{\delta f_4}{\partial x_2} & \frac{\delta f_4}{\partial x_3} & \frac{\delta f_4}{\partial x_4} & \frac{\delta f_4}{\partial x_5} & \frac{\delta f_4}{\partial x_6} \\
\frac{\delta f_5}{\partial x_1} & \frac{\delta f_5}{\partial x_2} & \frac{\delta f_5}{\partial x_3} & \frac{\delta f_5}{\partial x_4} & \frac{\delta f_5}{\partial x_5} & \frac{\delta f_5}{\partial x_6} \\
\frac{\delta f_6}{\partial x_1} & \frac{\delta f_6}{\partial x_2} & \frac{\delta f_6}{\partial x_3} & \frac{\delta f_6}{\partial x_4} & \frac{\delta f_6}{\partial x_5} & \frac{\delta f_6}{\partial x_6}
\end{bmatrix}, \tag{3.2.11}
$$

or, in general, the components of the ${}^{B}\mathbf{J}_{A}$ matrix are simply $\delta f_i/\delta x_i$ for $i, j = 1, 2, 3, 4, 5, 6$. Dividing Eq. (3.2.2) by the time differential dt, we obtain a generalized expression for velocities, such that

$$
{}^{A}\mathbf{V} = \begin{bmatrix} \frac{{}^{A}dx}{dt} & \frac{{}^{A}dy}{dt} & \frac{{}^{A}dz}{dt} & \frac{{}^{A}\delta x}{dt} & \frac{{}^{A}\delta y}{dt} & \frac{{}^{A}\delta z}{dt} \end{bmatrix}^{T} \tag{3.2.12}
$$

or

$$
{}^{A}\mathbf{V} = [{}^{A}v_x \quad {}^{A}v_y \quad {}^{A}v_z \quad {}^{A}\omega_x \quad {}^{A}\omega_y \quad {}^{A}\omega_z]^{T}, \tag{3.2.13}
$$

where ${}^{A}\omega_x$, ${}^{A}\omega_y$, and ${}^{A}\omega_z$ are the angular velocities of the A frame about the x, y, and z coordinates, respectively. Thus, it is also true that

$$
{}^{B}\mathbf{V} = {}^{B}\mathbf{J}_{A}\,{}^{A}\mathbf{V}, \tag{3.2.14}
$$

where the generalized six-dimensional velocities are defined such that

$$
\mathbf{V} = \begin{bmatrix} \mathbf{v} \\ \boldsymbol{\omega} \end{bmatrix}. \tag{3.2.15}
$$

In a similar manner, one may relate the generalized tip velocity ${}^{0}\mathbf{V}$ to the vector of the joint velocities $\dot{\boldsymbol{\theta}}$ such that

$$
\dot{\boldsymbol{\theta}} = [\dot{\theta}_1 \quad \dot{\theta}_2 \quad \dot{\theta}_3 \quad \dot{\theta}_4 \quad \dot{\theta}_5 \quad \dot{\theta}_6] \tag{3.2.16}
$$

for a six-axis robotic manipulator. Under these circumstances,

$$
{}^{0}\mathbf{V} = {}^{0}\mathbf{J}(\boldsymbol{\theta})\dot{\boldsymbol{\theta}}, \tag{3.2.17}
$$

where ${}^{0}\mathbf{J}(\boldsymbol{\theta})$ is the Jacobian of transformation relating the generalized differential displacement ${}^{0}\mathbf{D}$ at the tip with respect to the base frame to the vector of the joint displacement, $d\boldsymbol{\theta} = [d\theta_1\ d\theta_2\ d\theta_3\ d\theta_4\ d\theta_5\ d\theta_6]^{T}$, such that

$$
{}^{0}\mathbf{D} = {}^{0}\mathbf{J}(\boldsymbol{\theta})d\boldsymbol{\theta}. \quad ■■ \tag{3.2.18}
$$

EXAMPLE 3.2.1. ■

For a two-link planar robotic manipulator (Fig. 3.2.1), derive expressions for the Jacobians of transformation between frames as well as between the tip and the joint spaces.

Solution. Let us write out the D-H transformations $\mathbf{A}_i^j$ between the coordinate frames in the following form:

$$\mathbf{A}_0^1 = \begin{bmatrix} C_1 & -S_1 & 0 & 0 \\ S_1 & C_1 & 0 & 0 \\ 0 & 0 & 0 & 0 \\ 0 & 0 & 0 & 1 \end{bmatrix}, \qquad \mathbf{A}_1^2 = \begin{bmatrix} 1 & 0 & 0 & l_1 \\ 0 & 1 & 0 & 0 \\ 0 & 0 & 1 & 0 \\ 0 & 0 & 0 & 1 \end{bmatrix}, \tag{3.2.19}$$

$$\mathbf{A}_0^2 = \begin{bmatrix} C_1 & -S_1 & 0 & l_1C_1 \\ S_1 & C_1 & 0 & l_1S_1 \\ 0 & 0 & 1 & 0 \\ 0 & 0 & 0 & 1 \end{bmatrix}, \qquad \mathbf{A}_1^3 = \begin{bmatrix} C_2 & -S_2 & 0 & l_1 \\ S_2 & C_2 & 0 & 0 \\ 0 & 0 & 1 & 0 \\ 0 & 0 & 0 & 1 \end{bmatrix}, \tag{3.2.20}$$

$$\mathbf{A}_0^3 = \begin{bmatrix} C_{12} & -S_{12} & 0 & l_1C_1 \\ S_{12} & C_{12} & 0 & l_1S_1 \\ 0 & 0 & 1 & 0 \\ 0 & 0 & 0 & 1 \end{bmatrix}, \qquad \mathbf{A}_2^3 = \begin{bmatrix} C_2 & -S_2 & 0 & 0 \\ S_2 & C_2 & 0 & 0 \\ 0 & 0 & 1 & 0 \\ 0 & 0 & 0 & 1 \end{bmatrix}, \tag{3.2.21}$$

$$\mathbf{A}_0^4 = \begin{bmatrix} C_{12} & -S_{12} & 0 & l_1C_1 + l_2C_{12} \\ S_{12} & C_{12} & 0 & l_1S_1 + l_2S_{12} \\ 0 & 0 & 1 & 0 \\ 0 & 0 & 0 & 1 \end{bmatrix},$$

$$\mathbf{A}_1^4 = \begin{bmatrix} C_2 & -S_2 & 0 & l_1 + l_2C_2 \\ S_2 & C_2 & 0 & l_2S_2 \\ 0 & 0 & 1 & 0 \\ 0 & 0 & 0 & 0 \end{bmatrix}, \tag{3.2.22}$$

$$\mathbf{A}_2^4 = \begin{bmatrix} C_2 & -S_2 & 0 & l_2C_2 \\ S_2 & C_2 & 0 & l_2S_2 \\ 0 & 0 & 1 & 0 \\ 0 & 0 & 0 & 1 \end{bmatrix}, \qquad \mathbf{A}_3^4 = \begin{bmatrix} 1 & 0 & 0 & l_2 \\ 0 & 1 & 0 & 0 \\ 0 & 0 & 1 & 0 \\ 0 & 0 & 0 & 1 \end{bmatrix}. \tag{3.2.23}$$

The generalized velocities ${}^j\mathbf{V}_i$ of the ith frame with respect to the jth frame can easily be found as

$${}^0\mathbf{V}_1 = \begin{bmatrix} 0 \\ 0 \\ \dot{\theta}_1 \end{bmatrix}, \qquad {}^1\mathbf{V}_2 = \begin{bmatrix} 0 \\ 0 \\ 0 \end{bmatrix}, \qquad {}^0\mathbf{V}_2 = \begin{bmatrix} -l_1S_1\dot{\theta}_1 \\ l_1C_1\dot{\theta}_1 \\ \dot{\theta}_1 \end{bmatrix}, \tag{3.2.24}$$

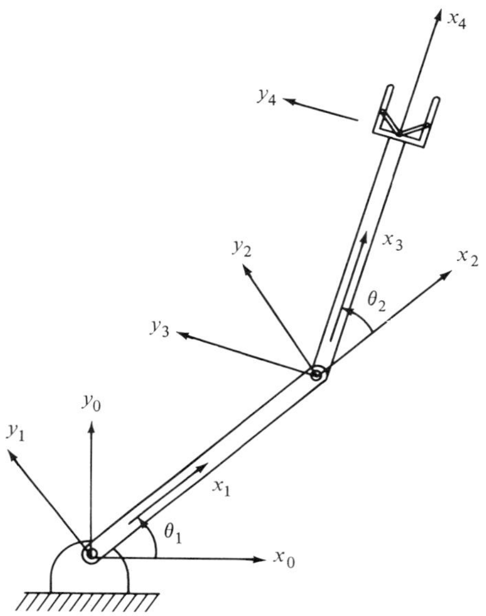

Figure 3.2.1 Geometry of a two-link manipulator.

$$
{}^{1}\mathbf{V}_3 = \begin{bmatrix} 0 \\ 0 \\ \dot{\theta}_2 \end{bmatrix}, \qquad {}^{0}\mathbf{V}_3 = \begin{bmatrix} -l_1 S_1 \dot{\theta}_2 \\ l_1 C_1 \dot{\theta}_2 \\ \dot{\theta}_1 + \dot{\theta}_2 \end{bmatrix}, \qquad {}^{2}\mathbf{V}_3 = \begin{bmatrix} 0 \\ 0 \\ \dot{\theta}_2 \end{bmatrix}, \tag{3.2.25}
$$

$$
{}^{0}\mathbf{V}_4 = \begin{bmatrix} -l_1 S_1 \dot{\theta}_1 - l S_{12}(\dot{\theta}_1 + \dot{\theta}_2) \\ l_1 C_1 \dot{\theta}_1 + l C_{12}(\dot{\theta}_1 + \dot{\theta}_2) \\ \dot{\theta}_1 + \dot{\theta}_2 \end{bmatrix}, \qquad {}^{1}\mathbf{V}_4 = \begin{bmatrix} -l_2 S_2 \dot{\theta}_2 \\ l_2 C_1 \dot{\theta}_2 \\ \dot{\theta}_2 \end{bmatrix}, \tag{3.2.26}
$$

$$
{}^{2}\mathbf{V}_4 = \begin{bmatrix} -l_2 S_2 \dot{\theta}_2 \\ l_2 C_1 \dot{\theta}_2 \\ \dot{\theta}_2 \end{bmatrix}, \qquad {}^{3}\mathbf{V}_4 = \begin{bmatrix} 0 \\ 0 \\ 0 \end{bmatrix}. \tag{3.2.27}
$$

Now the corresponding Jacobians of transformation can be derived, based on Eq. (3.2.14), such that

$$
{}^{1}\mathbf{J}_2 = \begin{bmatrix} 0 & 0 & -l_1 S_1 \\ 0 & 0 & l_1 C_1 \\ 0 & 0 & 1 \end{bmatrix}, \qquad {}^{0}\mathbf{V}_2 = {}^{1}\mathbf{J}_2 \, {}^{0}\mathbf{V}_1; \tag{3.2.28}
$$

$$
{}^{2}\mathbf{J}_3 = \begin{bmatrix} 1 & 0 & 0 \\ 0 & 1 & 0 \\ 0 & 0 & 1 + \dot{\theta}_2/\dot{\theta}_1 \end{bmatrix}, \qquad {}^{0}\mathbf{V}_3 = {}^{2}\mathbf{J}_3 \, {}^{0}\mathbf{V}_2; \tag{3.2.29}
$$

$$ {}^{1}\mathbf{J}_3 = \begin{bmatrix} 0 & 0 & -l_1 S_1 \\ 0 & 0 & l_1 C_1 \\ 0 & 0 & 1 + \dot{\theta}_2/\dot{\theta}_1 \end{bmatrix}, \qquad {}^{0}\mathbf{V}_3 = {}^{1}\mathbf{J}_3 {}^{0}\mathbf{V}_2; \tag{3.2.30} $$

$$ {}^{1}\mathbf{J}_4 = \begin{bmatrix} 0 & 0 & -l_1 S_1 - l_2 S_{12}(\dot{\theta}_1 + \dot{\theta}_2)/\dot{\theta}_1 \\ 0 & 0 & l_1 C_1 + l_2 C_{12}(\dot{\theta}_1 + \dot{\theta}_2)/\dot{\theta}_1 \\ 0 & 0 & 1 + \dot{\theta}_2/\dot{\theta}_1 \end{bmatrix}, \qquad {}^{0}\mathbf{V}_4 = {}^{1}\mathbf{J}_4 {}^{0}\mathbf{V}_1; \tag{3.2.31} $$

$$ {}^{2}\mathbf{J}_4 = \begin{bmatrix} 1 & 0 & -l_2 S_{12}(\dot{\theta}_1 + \dot{\theta}_2)/\dot{\theta}_1 \\ 0 & 1 & l_2 C_{12}(\dot{\theta}_1 + \dot{\theta}_2)/\dot{\theta}_1 \\ 0 & 0 & 1 + \dot{\theta}_2/\dot{\theta}_1 \end{bmatrix}, \qquad {}^{0}\mathbf{V}_4 = {}^{2}\mathbf{J}_4 {}^{0}\mathbf{V}_2; \tag{3.2.32} $$

$$ {}^{2}\mathbf{J}_4 = \begin{bmatrix} 0 & 0 & [l_1 S_1 \dot{\theta}_1 - l_2 S_{12}(\dot{\theta}_1 + \dot{\theta}_2)/\dot{\theta}_1] \\ 0 & 0 & l_1 C_1 \dot{\theta}_1 + l_2 C_{12}(\dot{\theta}_1 + \dot{\theta}_2)/\dot{\theta}_1 \\ 0 & 0 & 1 + \dot{\theta}_2/\dot{\theta}_1 \end{bmatrix}, \qquad {}^{0}\mathbf{V}_4 = {}^{3}\mathbf{J}_4 {}^{0}\mathbf{V}_3. \tag{3.2.33} $$

The relation between the tip velocity ${}^{0}\mathbf{V}_4$ and the vector of the joint velocities ${}^{0}\boldsymbol{\theta} = [{}^{0}\theta_1 \; {}^{0}\theta_1]^T$ is through a Jacobian of transformation $\mathbf{J}(\boldsymbol{\theta})$, such that

$$ {}^{0}\mathbf{V}_4 = \mathbf{J}(\boldsymbol{\theta})\dot{\theta}, \tag{3.2.34} $$

where

$$ \mathbf{J}(\boldsymbol{\theta}) = \begin{bmatrix} -l_1 S_1 - l_2 S_{12} & -l_2 S_{12} \\ l_1 C_1 + l_2 C_{12} & l_1 C_{12} \\ 1 & 1 \end{bmatrix}. \tag{3.2.35} $$

Generally speaking, in terms of the set of generalized coordinates q_i, $i = 1, 2, \ldots, n$, we may write

$$ \frac{\delta T_0^n}{\delta q_i} = \mathbf{T}_0^n \, {}^{T_0^n}\Delta_i \tag{3.2.36} $$

or, for $n = 6$, the above relation can be written

$$ d\mathbf{T}_0^6 = d\mathbf{T}_6 = \mathbf{T}_6 \, {}^{T_6}\Delta_i \, dq_i. \tag{3.2.37} $$

This may be written in terms of nine 6×6 Jacobian of transformation matrices, such that

$$ d\mathbf{T}_6 = \mathbf{J} \, d\mathbf{q}, \tag{3.2.38} $$

or, in expanded form,

$$ d\mathbf{T}_6 = \begin{bmatrix} {}^{T_6}d_x \\ {}^{T_6}d_y \\ {}^{T_6}d_z \\ {}^{T_6}\delta_x \\ {}^{T_6}\delta_y \\ {}^{T_6}\delta_z \end{bmatrix} = \begin{bmatrix} d_{1x} & d_{2x} & d_{3x} & d_{4x} & d_{5x} & d_{6x} \\ d_{1y} & d_{2y} & d_{3y} & d_{4y} & d_{5y} & d_{6y} \\ d_{1z} & d_{2z} & d_{3z} & d_{4z} & d_{5z} & d_{6z} \\ \delta_{1x} & \delta_{2x} & \delta_{3x} & \delta_{4x} & \delta_{5x} & \delta_{6x} \\ \delta_{1y} & \delta_{2y} & \delta_{3y} & \delta_{4y} & \delta_{5y} & \delta_{6y} \\ \delta_{1z} & \delta_{2z} & \delta_{3z} & \delta_{4z} & \delta_{5z} & \delta_{6z} \end{bmatrix} \begin{bmatrix} dq_1 \\ dq_2 \\ dq_3 \\ dq_4 \\ dq_5 \\ dq_6 \end{bmatrix}. \tag{3.2.39} $$

It can easily be shown that

$$d_{ix} = (-n_x p_y + n_y p_x)\mathbf{i}, \qquad d_{iy} = (o_y p_x - o_x p_y)\mathbf{j}, \qquad d_{iz} = (a_y p_x - a_x p_y)\mathbf{k}; \tag{3.2.40}$$

$$\delta_{iz} = n_z\mathbf{i} + o_z\mathbf{j} + a_z\mathbf{k}, \qquad \delta_{ix} = n_x\mathbf{i} + o_x\mathbf{j} - a_x\mathbf{k},$$
$$\delta_{iy} = n_y\mathbf{i} + o_y\mathbf{j} + a_y\mathbf{k}. \quad ■■ \tag{3.2.41}$$

BIBLIOGRAPHY

Featherstone, R. (1983). Position and Velocity Transformations Between Robot End-Effector Coordinates and Joint Angles, *Int. J. Robotics Res.*, **2**(2), 35–45.

Groom, R. C., Jr. (1972). "Force Feedback Steering of a Teleoperator System," M.Sc. Thesis, M.I.T., Cambridge, MA.

Hollerbach, J. M., and Gideon, Sahar (1983). Wrist-Partitioned, Inverse Kinematic Accelerations and Manipulator Dynamics, *Int. J. Robotics Res.*, **2**(4), 61–76.

Luh, J. Y. S., Walker, M. W., and Paul, R. P. (1980). Resolved Acceleration Control of Mechanical Manipulators, *IEEE Trans. Autom. Control*, **25**(3), 468–474.

Paul, R. P. (1983). *Robot Manipulators: Mathematics, Programming, and Control*, M.I.T. Press, Cambridge, MA.

Whitney, D. E. (1969). Resolved Motion Rate Control of Manipulators and Human Prostheses, *IEEE Trans. Man, Machine, Syst.*, **10**(2), 47–53.

Whitney, D. E. (1972). The Mathematics of Coordinated Control of Prosthetic Arms and Manipulators, *J. Dyn. Syst. Meas. Control*, 303–309 (Dec.).

HOMEWORK PROBLEMS

1. If an **A** matrix is given by

$$\mathbf{A} = \begin{bmatrix} 0 & 0 & -1 & 1 \\ -1 & 0 & 0 & 2 \\ 0 & 1 & 0 & 0 \\ 0 & 0 & 0 & 0 \end{bmatrix},$$

where **A** describes a coordinate frame with respect to the base coordinate of a robot, determine the differential transformation $d\mathbf{A}$ that results if the coordinate frame is translated by **d** and then differentially rotated by **δ** such that

$$\mathbf{d} = 0.1\mathbf{i} + 0.2\mathbf{j} + 0.3\mathbf{k}$$

and

$$\boldsymbol{\delta} = 0.1\mathbf{i} + 0.2\mathbf{j} + 0.3\mathbf{k}.$$

Find $d\mathbf{A}$ with respect to both the base coordinate frame and the **A** frame itself.

2. For the robotic vision system described in Example 3.1.5, the differential changes in the motion and rotation of the cup with respect to the camera are given by

$$^{\text{CAM}}\mathbf{d} = 0.1\mathbf{i} + 0.2\mathbf{j} + 0.3\mathbf{k}$$

and

$$^{\text{CAM}}\mathbf{d} = 0.1\mathbf{i} + 0.2\mathbf{j} + 0.3\mathbf{k}.$$

Determine the corresponding differential changes that the gripper must perform in order to move toward the cup and eventually come into contact with the moving cup.

3. If an **A** matrix is represented in the form

$$\mathbf{A} = \begin{bmatrix} 1 & 0 & 0 & 1 \\ 0 & -1 & 0 & 2 \\ 0 & 0 & -1 & 3 \\ 0 & 0 & 0 & 1 \end{bmatrix}$$

and describes a coordinate frame with respect to the base coordinate frame of a robot, what is the differential transformation $d\mathbf{A}$ that results if the coordinate frame is translated by $\mathbf{d}$ such that

$$\mathbf{d} = 0.1\mathbf{i} + 0.1\mathbf{j} - 0.1\mathbf{k}$$

and

$$\boldsymbol{\delta} = -0.1\mathbf{i} - 0.1\mathbf{j} + 0.1\mathbf{k},$$

both with respect to the base coordinate?

4. Using the information in Problem 3, compute the corresponding differential transform $^{A}\Delta$ with respect to the **A** coordinate frame as well as the corresponding differential translation $^{A}\mathbf{d}$ and differential rotation $^{A}\boldsymbol{\delta}$ with respect to the **A** frame.

5. For the robotic differential chase system (Fig. 3.1.2), the differential changes in the motion and rotation of the object with respect to the camera are

$$^{\text{CAM}}\mathbf{d} = -0.1\mathbf{i} - 0.1\mathbf{j} - 0.1\mathbf{k} \quad \text{and} \quad ^{\text{CAM}}\boldsymbol{\delta} = 0.1\mathbf{i} - 0.1\mathbf{j} + 0.1\mathbf{k}.$$

Determine the corresponding differential changes $^{T_6}\mathbf{d}$ and $^{T_6}\boldsymbol{\delta}$ of the gripper in order to keep track of the motion of the cup and eventually come into contact with it.

6. Find the Jacobian of transformations between the generalized tip and joint velocities for a PUMA 600 robotic manipulator as described in Example 3.2.1.

7. Find the Jacobian of differential transformation between the generalized joint and tip differential displacements.

Chapter 4

Robotic Workspace and Motion Trajectory Design

4.1 INTRODUCTION

The workspace of a robotic manipulator is defined as the set of all three-dimensional points that can be reached by a reference point located on the robotic hand. A restricted version of a robotic workspace is also defined as the set of all points that can be reached by a fixed orientation of the robotic hand. The knowledge of robotic workspaces is important in arranging the associated flexible manufacturing cell of a robot and assessing the efficiency of a production line.

In general, the boundary of a robotic workspace, which is called the robotic work envelope, is a complex surface difficult to represent explicitly by geometrical equations. To derive mathematical expressions for robotic work surfaces, we customarily employ the Denavit–Hartenberg (D-H) homogeneous transformations to describe the locus of the trajectories produced by the robotic hand.

Generally, there are two issues involved in analyzing robotic workspaces:

1. Given the robotic structure, that is, a table for robotic joint parameters, what is the geometric structure of the workspace?
2. Given a desired geometrical description of a workspace for robotic applications, what is the necessary robotic structure?

To answer the above two questions we begin by elaborating on the general structure of robotic workspaces.

4.2 GENERAL STRUCTURE OF ROBOTIC WORKSPACES

Since a robotic workspace is the aggregate of all possible positions of a point attached to a point on the robotic hand, we begin by analyzing the workspace of a point. Let us consider a manipulator with n revolute joints as shown in Figure 4.2.1. Let P be a point attached to the origin of the gripper coordinate frame. The workspace of the point P with respect to the end frame n is denoted by $W_n(P)$, where n is the closest axis of rotation to P. It is then clear that (see Fig. 4.2.2)

$$W_n(P) = \text{circle}(n). \tag{4.2.1}$$

Consequently, the workspace of P with respect to the $(n-1)$th axis of rotation would be a torus such that

$$W_n(P) = R_{n-1} W_n(P) = R_{n-1}\ \text{circle}(n) = \text{torus}(n-1), \tag{4.2.2}$$

where R_{n-1} is a rotation operator capable of rotating a workspace about the axis $n-1$. Repeating the operation, we obtain regions that are solid rings with a central void such that

$$\begin{aligned} W_{n-1}(P) &= R_{n-2} W_{n-1}(P) = R_{n-2}[\text{torus}(n-1)] \\ &= \text{SHR}(n-2) \\ &\ \vdots \end{aligned} \tag{4.2.3}$$

$$\text{W}_1(\text{P}) = \text{R}_1[\text{W}_2(\text{P})] = \text{R}_1[\text{SHR}(2)] = \text{SHR}(1), \tag{4.2.4}$$

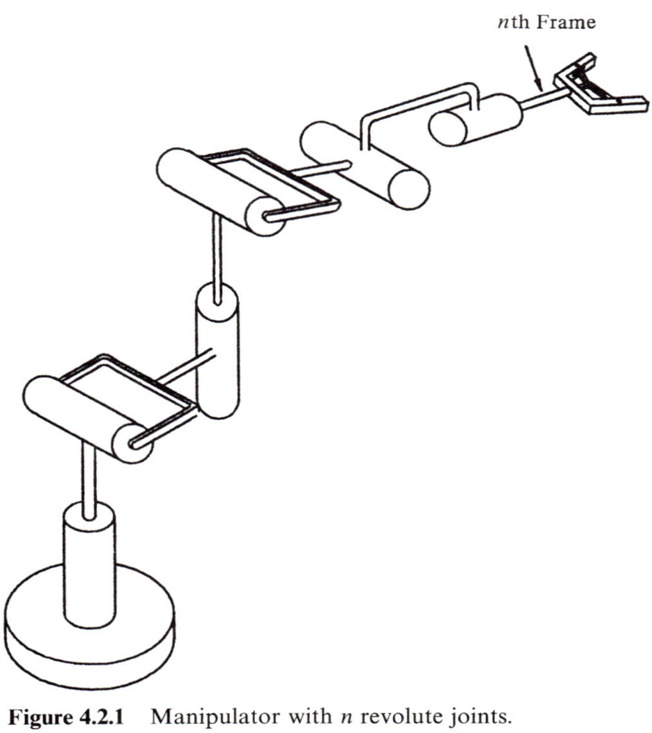

Figure 4.2.1 Manipulator with n revolute joints.

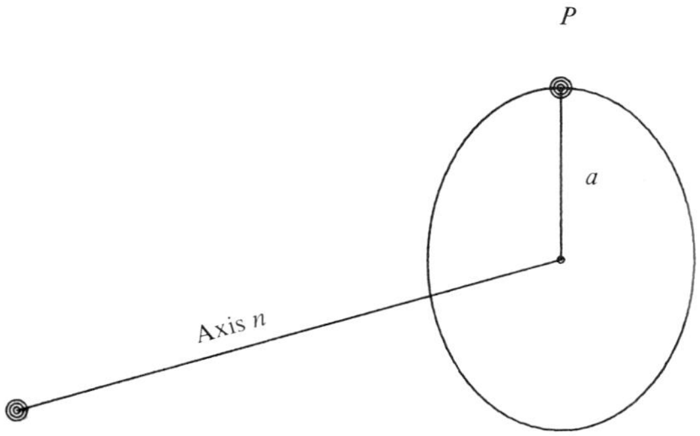

Figure 4.2.2 Workspace $W_n(P)$ = circle (n).

where SHR—*solid hollow ring*—is a notation originally proposed by Gupta and Roth in 1982.

Referring to Figure 4.2.3, we can show that the equation for the torus $(i-1)$ in the x_{i-1}, y_{i-1}, z_{i-1} coordinate system is

$$\{x_{i-1}^2 + y_{i-1}^2 + (z_{i-1} - d_{i-1})^2 - [a_{i-1}^2 + a_i^2 + (d_{i+1} + d_i)^2]\}^2$$
$$= 4a_{i-1}^2\left\{a_i^2 - \frac{[(z_{i-1} - d_{i-1}) - (d_{i+1} + d_i)\cos \alpha_{i-1}]^2}{\sin \alpha_{i-1}}\right\}. \tag{4.2.5}$$

Since P is an end point on the manipulator, the above equation can be transformed easily into an equation describing the torus $(n-1)$ in the $(n-1)$th coordinate frame.

Note that if the ratio $a/a_{n-1} \ll 1$, we have a hollow inaccessible region (hole) around the z_{n-1} axis. If a is large enough, the hole disappears because the

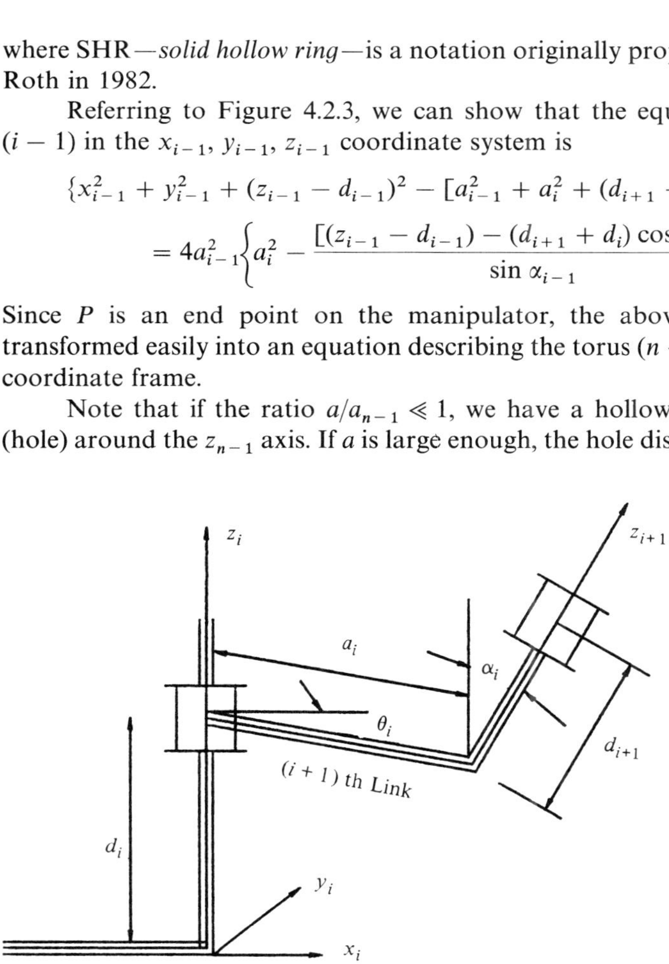

Figure 4.2.3 Representation of link parameters: d_i the link distance, a_i the link length, θ_i the link rotation, and α_i the link twist.

torus intersects itself. This occurs whenever the $n-1$ axis passes through the circle (n). Thus, under these circumstances,

$$a^2 \geq a_{n-1}^2 + (d + d_n)^2 \tan^2 \alpha_{n-1}. \tag{4.2.6}$$

EXAMPLE 4.2.1 ■

Discuss conditions pertaining to the existence or nonexistence of a hole in a robotic workspace.

Solution. Referring to Equation (4.2.6), we see that if either $d + d_n$ or α_{n-1} is zero, there is no hole unless $a \leq a_{n-1}$. However, if $\alpha_{n-1} = \pm 90$ and $d + d_n \neq 0$, there is always a hole because a is finite. Under these conditions, if $d = -d_n$, a hole may be avoided. ■ ■

EXAMPLE 4.2.2 ■

Refer to Figures 4.2.3 and 4.2.4. Let $d_1 = d_2 = d = 0$, $\alpha_1 = 90°$, $a_1 = c$, $a_2 = b$, and $a \leq b \leq c$. Determine the workspaces $W_3(P)$, $W_2(P)$, and $W_1(P)$.

Solution. The workspace $W_3(P)$ is obviously a circle(S), $W_2(P)$ is a torus(2), and $W_1(P)$ is a SHR(1) as shown in Figure 4.2.5. ■ ■

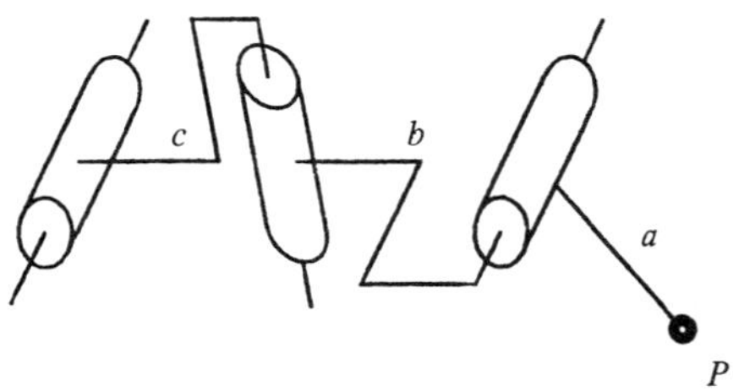

Figure 4.2.4 Special manipulator.

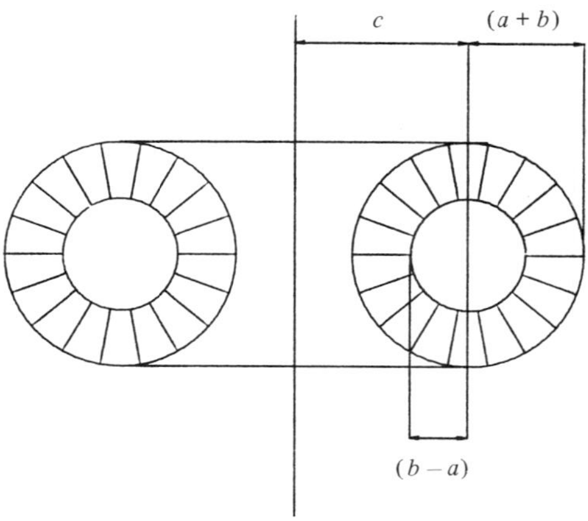

Figure 4.2.5 Workspace $W_1(P)$ for the manipulator shown in Fig. 4.2.4.

4.3 MANIPULATORS WITH n REVOLUTE JOINTS

In a manipulator with n revolute joints in series, there are n coordinate frames to specify the geometrical configuration of the robot. In general, in the ith frame, the Z_i axis is the axis of motion while the R_i axis is the axis normal to both the Z_i and the Z_{i+1} axes.

As we discussed in Chapter 2, there are four parameters—θ_i, α_i, a_i, and d_i—needed for the complete description of the kinematic attitude of one link with respect to the other. Such descriptions led to the definition of the D-H transformations that we are going to use in the complete description of robotic workspaces. Let us consider a general n-jointed manipulator whose nth link is attached to a hand whose center can be considered as a point P (Fig. 4.3.1).

4.3.1 Location of the Robotic Hand on a Plane

Given the location and orientation of the ith and $(i + 1)$th frame, $\mathbf{X}_i$ and $\mathbf{X}_{i+1}$, respectively, it is clear that

$$\mathbf{R}_i = \mathbf{A}_i \mathbf{X}_{i+1}, \tag{4.3.1}$$

where

$$\mathbf{A}_i = \begin{bmatrix} C\theta_i & -S\theta_i C\alpha_i & S\theta_i S\alpha_i & a_i C\theta_i \\ S\theta_i & C\theta_i C\alpha_i & -C\theta_i S\alpha_i & a_i S\theta_i \\ 0 & S\alpha_i & C\alpha_i & S_i \\ 0 & 0 & 0 & 1 \end{bmatrix}$$

and

$$\mathbf{X}_i = (X_i \quad Y_i \quad Z_i \quad 1)^T. \tag{4.3.2}$$

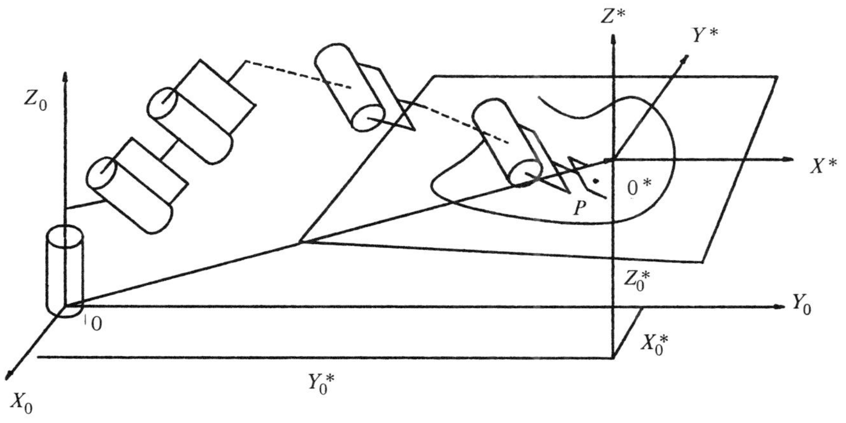

Figure 4.3.1 Workspace on a plane.

Since we are concerned with the tip motion when constructing the workspace, we present some differential relationships. Let there be an incremental change $\delta\theta_i$ on the ith joint displacement such that

$$\theta_i = \theta_{i0} + \delta\theta_i, \tag{4.3.3}$$

where θ_{i0} is the initial value of θ_i before the incremental change. Note that the new value of $\mathbf{A}_i$ is now given by

$$\mathbf{A}_i = \mathbf{A}_{i0} + \delta\theta_i \mathbf{B}_{i0}, \tag{4.3.4}$$

where $\mathbf{A}_{i0}$ is $\mathbf{A}_i$ evaluated at θ_{i0} and

$$\mathbf{B}_{i0} = \begin{bmatrix} -S\theta_{i0} & -C\theta_{i0}C\alpha_i & C\theta_{i0}S\alpha_i & -a_iS\theta_{i0} \\ C\theta_{i0} & -S\theta_{i0}C\alpha_i & S\theta_{i0}S\alpha_i & a_iC\theta_{i0} \\ 0 & 0 & 0 & 0 \\ 0 & 0 & 0 & 0 \end{bmatrix}. \tag{4.3.5}$$

Now the relationship between the coordinate frame $\mathbf{R}$ and the coordinate frame $\mathbf{R}_{n+1}$ attached to the tip of the robotic hand (origin at P) is given by

$$\mathbf{x}^* = \mathbf{A}_n^* \mathbf{x}_n, \tag{4.3.6}$$

where

$$\mathbf{A}_n^* = \mathbf{A}_0 \mathbf{A}_1 \mathbf{A}_2 \cdots \mathbf{A}_n \tag{4.3.7}$$

and

$$\mathbf{A}_0 = \begin{bmatrix} \cos(x_0, x^*) & \cos(x_0, y^*) & \cos(x_0, z^*) & -x_0^* \\ \cos(y_0, x^*) & \cos(y_0, y^*) & \cos(y_0, z^*) & -y_0^* \\ \cos(z_0, x^*) & \cos(z_0, y^*) & \cos(z_0, z^*) & -z_0^* \end{bmatrix}. \tag{4.3.8}$$

The result is the coordinate transformation between the $\mathbf{x}^*$ and the $\mathbf{R}_0$ frames. Now, substituting Equation (4.3.4) in (4.3.7) yields

$$\begin{aligned} \mathbf{A}_n^* = \mathbf{A}_{n0}^* &+ \delta\theta_1(\mathbf{A}_0 \mathbf{B}_{10} \mathbf{A}_{20} \mathbf{A}_{30} \cdots \mathbf{A}_{n0}) \\ &+ \delta\theta_2(\mathbf{A}_0 \mathbf{A}_{10} \mathbf{B}_{20} \mathbf{A}_{30} \mathbf{A}_{40} \cdots \mathbf{A}_{n0}) \\ &+ \cdots \\ &+ \delta\theta_n(\mathbf{A}_0 \mathbf{A}_{10} \mathbf{A}_{20} \mathbf{A}_{30} \cdots \mathbf{B}_{n0}). \end{aligned} \tag{4.3.9}$$

EXAMPLE 4.3.1 ■

Derive Equation (4.3.9).

Solution. Substituting Equation (4.3.4) into (4.3.7), one obtains

$$\mathbf{A}_n^* = (\mathbf{A}_0 + \delta\theta_0 \mathbf{B}_0)(\mathbf{A}_{10} + \delta\theta_1 \mathbf{B}_{10})(\mathbf{A}_{20} + \delta\theta_2 \mathbf{B}_{20}) \cdots (\mathbf{A}_{n0} + \delta\theta_n \mathbf{B}_{n0}). \tag{4.3.10}$$

Now note that $\delta\theta_0 = 0$ and thus Equation (4.3.10) reduces to

$$\begin{aligned} \mathbf{A}_n^* = (\mathbf{A}_0 \mathbf{A}_{10} \mathbf{A}_{20} \cdots \mathbf{A}_{n0}) &+ \delta\theta_1(\mathbf{A}_0 \mathbf{B}_{10} \mathbf{A}_{20} \cdots \mathbf{A}_{n0}) + \cdots \\ &+ \delta\theta_n(\mathbf{A}_0 \mathbf{A}_{10} \mathbf{A}_{20} \cdots \mathbf{A}_{n0}) + \text{higher-order terms in } \delta\theta_i. \end{aligned} \tag{4.3.11}$$

Neglecting the higher-order terms and noting that

$$\mathbf{A}_{n0}^* = \mathbf{A}_0 \mathbf{A}_{10} \mathbf{A}_{20} \cdots \mathbf{A}_{n0}, \tag{4.3.12}$$

we obtain the desired result (4.3.9). ■ ■

Let us now define the difference $\delta \mathbf{A}_n^*$ such that

$$\delta \mathbf{A}_n^* = \mathbf{A}_n^* - \mathbf{A}_{n0}^* = \begin{bmatrix} \delta n_1 & \delta o_1 & \delta a_1 & \delta x_0 \\ \delta n_2 & \delta o_2 & \delta a_2 & \delta y_0 \\ \delta n_3 & \delta o_3 & \delta a_3 & \delta z_0 \end{bmatrix}, \tag{4.3.13}$$

where the submatrix $\delta \mathbf{R}$,

$$\delta \mathbf{R} = \begin{bmatrix} \delta n_1 & \delta o_1 & \delta a_1 \\ \delta n_2 & \delta o_2 & \delta a_2 \\ \delta n_3 & \delta o_3 & \delta a_3 \end{bmatrix}, \tag{4.3.14}$$

is referred to as the change of orientation of the last coordinate frame $\mathbf{x}_n$ with respect to the $\mathbf{R}^*$ frame. Clearly, the components of the vectors $\delta \mathbf{n}$, $\delta \mathbf{o}$, and $\delta \mathbf{a}$ are related such that

$$\delta \mathbf{n} \times \delta \mathbf{o} = \delta \mathbf{a}. \tag{4.3.15}$$

Also, as discussed before,

$$\delta \mathbf{n} \cdot \delta \mathbf{n} = \delta \mathbf{o} \cdot \delta \mathbf{o} = \delta \mathbf{a} \cdot \delta \mathbf{a} = 1. \tag{4.3.16}$$

Therefore, there are only six independent equations obtainable from the matrix equation (4.3.13). Without loss of generality one may take any combination of $\delta n_1, \delta n_2, \delta n_3, \delta o_1, \delta o_2, \delta o_3, \delta a_1, \delta a_2, \delta a_3$ and $\delta x_0, \delta y_0, \delta z_0$. For example, let us consider

$$\delta x_0 = f_1(\delta \theta_1, \delta \theta_2, \ldots, \delta \theta_n) \tag{4.3.17}$$

$$\delta y_0 = f_2(\delta \theta_1, \delta \theta_2, \ldots, \delta \theta_n) \tag{4.3.18}$$

$$\delta z_0 = f_3(\delta \theta_1, \delta \theta_2, \ldots, \delta \theta_n) \tag{4.3.19}$$

$$\delta n_3 = f_4(\delta \theta_1, \delta \theta_2, \ldots, \delta \theta_n) \tag{4.3.20}$$

$$\delta o_1 = f_5(\delta \theta_1, \delta \theta_2, \ldots, \delta \theta_n) \tag{4.3.21}$$

and

$$\delta a_1 = f_6(\delta \theta_1, \delta \theta_2, \ldots, \delta \theta_n) \tag{4.3.22}$$

These equations imply that the robotic hand is perpendicular to the axis of the last link as shown in Figure 4.3.2.

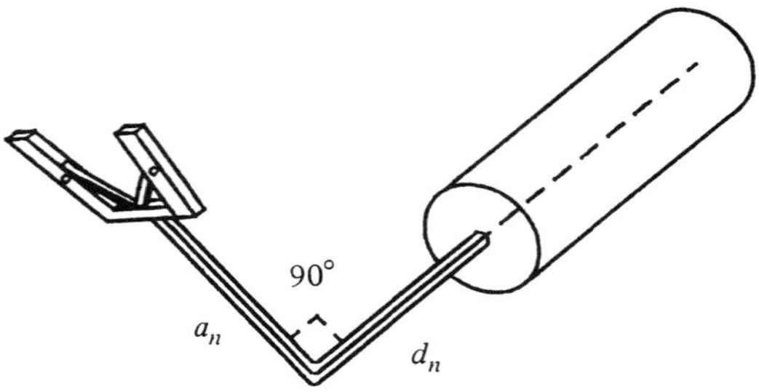

Figure 4.3.2 Hand configuration described by Eqs. (4.3.17)–(4.3.22).

4.3.2 Motion of the Robotic Hand on the Boundary of the Workspace

In this case let us move the robotic hand on a plane along the x_0 axis while $\delta y_0 = \delta z_0 = 0$ and the orientation of the hand is kept unchanged; that is $\delta n_3 = \delta o_1 = \delta a_1 = 0$. To move the hand toward the boundary of the workspace as fast as possible with $\delta\theta_i$ kept small, one should find the extremum of δx_0 subject to all the constraints. For example, one should maximize

$$\delta R_0 = f_1(\theta_1, \theta_2, \ldots, \theta_n) \tag{4.3.23}$$

subject to the following constraints:

$$f_2(\theta_1, \theta_2, \ldots, \theta_n) = 0 \tag{4.3.24}$$

$$f_3(\theta_1, \theta_2, \ldots, \theta_n) = 0 \tag{4.3.25}$$

$$f_4(\theta_1, \theta_2, \ldots, \theta_n) = 0 \tag{4.3.26}$$

$$f_5(\theta_1, \theta_2, \ldots, \theta_n) = 0 \tag{4.3.27}$$

$$f_6(\theta_1, \theta_2, \ldots, \theta_n) = 0 \tag{4.3.28}$$

$$(\delta\theta_i)_{\min} \le (\delta\theta_i) \le (\delta\theta_i)_{\max} \qquad i = 1, 2, \ldots, n. \tag{4.3.29}$$

EXAMPLE 4.3.2 ■

For a simple elbow manipulator described by the joint parameters in Table 4.3.1, find the envelope of the workspace on the horizontal and sagittal planes (Fig. 4.3.3).

Solution. For typical values of $a_2 = a_3 = 5a_4$, the typical work envelopes on the horizontal and sagittal planes of the robot can be obtained by tracing the envelope (contour) of the workspace by means of a computer program. Typical workspaces are sketched in Figures 4.3.4 and 4.3.5 by treating the robotic hand as a point [from Tsai and Soni (1983)]. ■ ■

TABLE 4.3.1 JOINT PARAMETER TABLE

Link	θ_i	α_i	a_i	d_i	Constraints
1	θ_1	90°	0	0	$-120° \leq \theta_1 \leq 120°$
2	θ_2	0°	a_2	0	$0° \leq \theta_2 \leq 90°$
3	θ_3	0°	a_3	0	$-120° \leq \theta_3 \leq 0°$
4	θ_4	−90°	a_4	0	$-120° \leq \theta_4 \leq 120°$
5	θ_5	90°	0	0	$-30° \leq \theta_5 \leq 210°$
6	θ_6	0°	0	0	$-240° \leq \theta_6 \leq 240°$

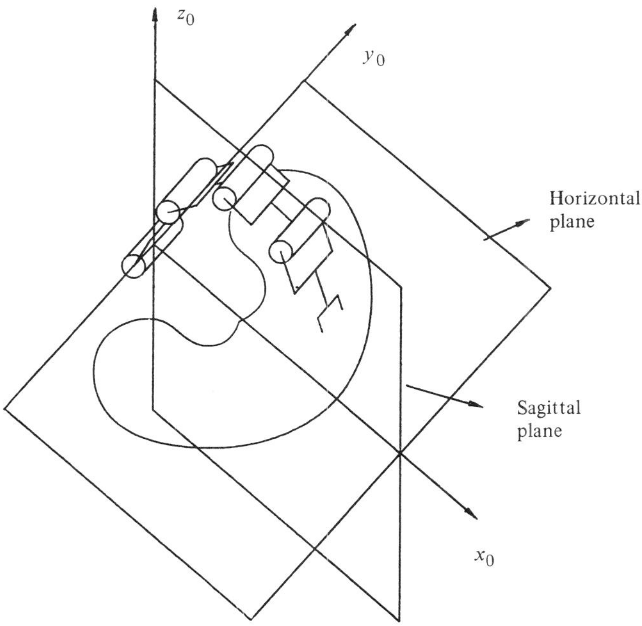

Figure 4.3.3 Robotic hand movement on the horizontal and sagittal planes.

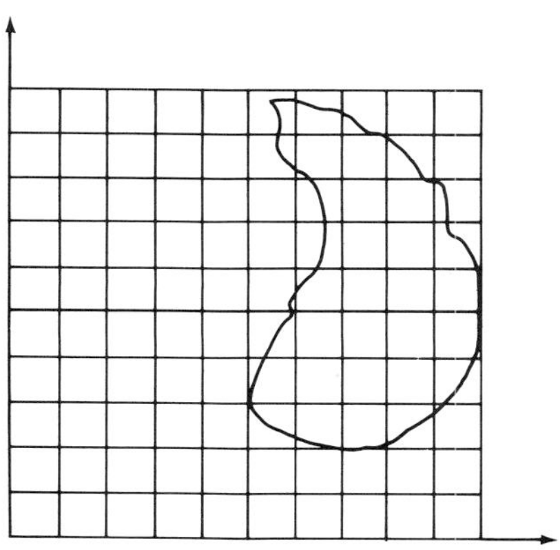

Figure 4.3.4 Workspace of a 6R robot on the sagittal plane.

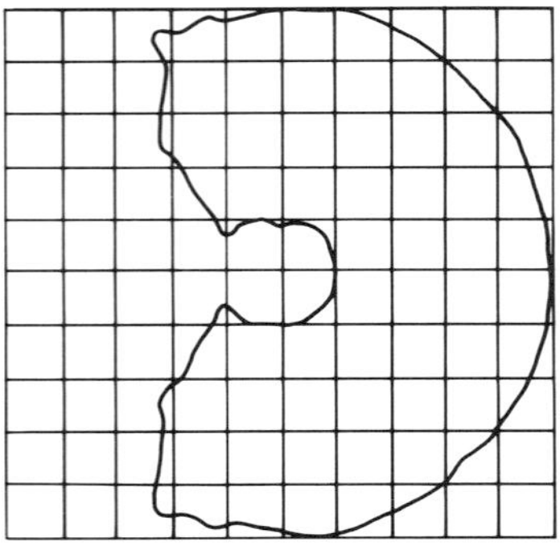

Figure 4.3.5 Workspace of a 6*R* robot on the horizontal plane.

4.4 ROBOTIC WORKSPACE PERFORMANCE INDEX

An important point regarding the robotic workspace is the development of a performance index. As early as 1983 Lee and Yang [6] established a theorem that clearly defines such an index. Before stating the theorem let us consider a general $n - R$ robotic manipulator with n link lengths a_i, $i = 1, 2, \ldots, n$ and n link distances d_i, $i = 1, 2, \ldots, n$. For simplicity let

$$a_i = l_i, \qquad i = 1, 2, 3, \ldots, n; \tag{4.4.1}$$

$$d_i = l_{i+n}, \qquad i = 1, 2, 3, \ldots, n. \tag{4.4.2}$$

Now based on dimensional analysis, the volume of this manipulator can be represented in general by

$$V = \sum_{i,j,k}^{2n} \mu_{ijk} l_i l_j l_k, \tag{4.4.3}$$

where i, j, and k are integers from 1 to $2n$ and μ_{ijk} is the corresponding coefficient of the term $l_i l_j l_k$. We are now ready to state a very important theorem due to Lee and Yang [6].

Lee–Yang Theorem. For a given manipulator (for which the twist angles, rotation limits, location angles, and proportions of link lengths are given) the ratio between the volume of the workspace and the cube of the total length is a constant; that is,

$$\frac{V}{(l_1 + l_2, \ldots, l_{2n})^3} = \frac{V}{L^3} = \text{constant}. \tag{4.4.4}$$

Proof. The lengths and distances of all links can be represented by the axial offset distance of the last link, $d_n = l_{2n}$. Thus,

$$l_i = p_i l_{2n}, \qquad i = 1, 2, \ldots, 2n, \tag{4.4.5}$$

where the p_i are known proportions. Substituting the relationships given by (4.4.5) in Equation (4.4.3) yields

$$V = \sum_{i,j,k}^{2n} \mu_{ijk} l_i l_j l_k = \sum_{i,j,k}^{2n} \mu_{ijk} p_i p_j p_k l_{2n}^3 = P_1 l_{2n}^3, \tag{4.4.6}$$

where $P_1 = \sum_{i,j,k}^{2n} \mu_{ijk} p_i p_j p_k$ is a constant. Now let $L = l_1 + l_2 + \cdots + l_{2n} = \sum_{i=1}^{2n} l_i$ be the total link length of the robot. It is then clear that

$$P_1 = \sum_{i=1}^{2n} l_i = \sum_{i=1}^{2n} p_i l_{2n} = P_2 l_{2n}, \tag{4.4.7}$$

where $P_2 = \sum_{i=1}^{2n} p_i$ is another constant. Thus, a workspace performance index can be defined as the volume index VI such that the workspace is

$$\mathrm{VI} = \frac{V}{L^3} = \frac{P_1 l_{2n}^3}{P_2^3 l_{2n}^3} = P_1 P_2^{-3}, \tag{4.4.8}$$

which proves the theorem.

Clearly, the maximum value of VI is obtained for a highly articulated robot of maximum radius L such that the workspace is a sphere of radius L. In this case,

$$(\mathrm{VI})_{\max} = \frac{4\pi L^3/3}{L^3} = \frac{4\pi}{3} \approx 4.188. \tag{4.4.9}$$

Based on this performance index, one can define a normalized performance index NVI as a kinematic performance index for use in the quantitative evaluation of manipulators based on workspace. Table 4.4.1, from Lee and Yang (1983), depicts the VI and NVI values for a number of commercially available industrial robotic manipulators.

TABLE 4.4.1 PERFORMANCE INDEX FOR SOME INDUSTRIAL ROBOTIC MANIPULATORS

Robot	Degrees of freedom	L (in.)	V (in^3.)	VI	NVI
PUMA600	6	51.0	183986	1.39	0.331
C.M.T^3	6	101.0	2295233	2.23	0.532
Pana-Robo	5	59.84	355937	1.66	0.397
AID-800	5	59.02	267350	1.30	0.310
CT-V30	5	94.49	1299251	1.54	0.368

4.5 EXTREME REACHES OF ROBOTIC HANDS

Let us consider a $2R$ robotic arm with joint distances d_1 and d_2 and link lengths a_1 and a_2 as shown in Figure 4.5.1. Note that in this case

$$R_{02}^2 = d_1^2 + a_1^2 + d_2^2 + a_2^2 + 2d_1d_2 \cos \alpha_1 + 2a_2(d_1 \sin \alpha_1 \sin \theta_2 + a_1 \cos \theta_2). \tag{4.5.1}$$

Thus, the condition for an extreme distance of the hand is

$$\tan \theta_2 = \frac{d_1 \sin \alpha_1}{a_1}, \tag{4.5.2}$$

which clearly yields two extrema, a maximum and a minimum distance.

EXAMPLE 4.5.1 ■

Find the extreme reaches of a general $2R$ robotic manipulator shown in Figure 4.5.1.

Solution. Substituting the value of θ_2 from Equation (4.5.2) into (4.5.1), which gives us the extreme values of Equation (4.5.1), yields

$$(R_{02})^2_{\text{max, min}} = d_1^2 + a_1^2 + d_2^2 + a_2^2 + 2d_1d_2 \cos \alpha_1 + 2a_2\{d_1 \sin \alpha_1 \sin[\tan^{-1}(a_1d_1 \sin \alpha_1)] + a_1 \cos[\tan^{-1}(a_1^{-1}d_1 \sin \alpha_1)]\}. \quad ■■ \tag{4.5.3}$$

To derive a general expression, consider a robotic manipulator with any number of arbitrary joints with axes of motion denoted by the z_i. The hand displacement vector $\mathbf{p}$ is such that

$$\mathbf{p} = p_x\mathbf{i} + p_y\mathbf{j} + p_z\mathbf{k}, \tag{4.5.4}$$

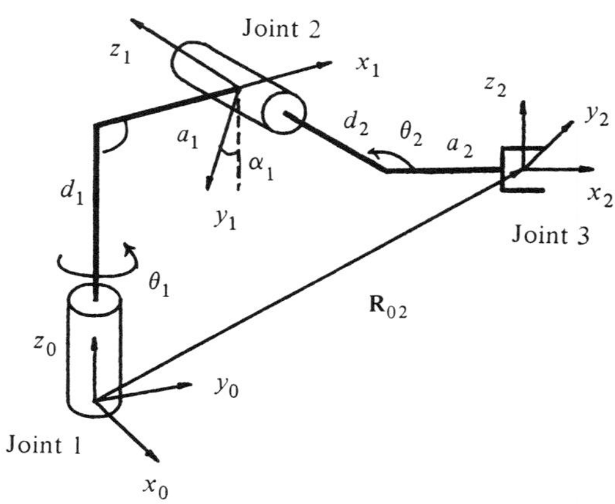

Figure 4.5.1 A $2R$ robotic manipulator.

where clearly (Fig. 4.5.2)

$$p_x = \hat{p}_n(\theta_i), \qquad p_y = \hat{p}_y(\theta_i), \qquad p_z = \hat{p}_z(\theta_i). \tag{4.5.5}$$

Note that

$$p = |p| = (p_x^2 + p_y^2 + p_z^2)^{1/2}. \tag{4.5.6}$$

To find the extreme reaches of this robot we have to find $\partial \mathbf{p}/\partial\theta_i$; that is,

$$\frac{\partial \mathbf{p}}{\partial \theta_i} = p^{-1}(\mathbf{p}\mathbf{T}_i), \tag{4.5.7}$$

where

$$\mathbf{T}_i = \frac{\partial \mathbf{p}}{\partial \theta_i} = \frac{\partial p_x}{\partial \theta_i}\mathbf{i} + \frac{\partial p_y}{\partial \theta_i}\mathbf{j} + \frac{\partial p_z}{\partial \theta_i}\mathbf{k} \tag{4.5.8}$$

and $\mathbf{T}$ is a tangent to the sphere through $\mathbf{p}$ and normal to the axis z_i. To have an extremum (extreme distance) $\partial \mathbf{p}/\partial\theta_i = 0$, and thus

$$\mathbf{p} \cdot \mathbf{T}_i = 0. \tag{4.5.9}$$

Note that, in general, for $\mathbf{T}_i$ to be perpendicular to both $\mathbf{p}$ and axis z_i and tangent to the sphere of radius p and to have its origin at the origin of the base coordinate system, one must have

$$\mathbf{T}_i = \mathbf{z}_i \times (\mathbf{p} - \mathbf{p}_i), \tag{4.5.10}$$

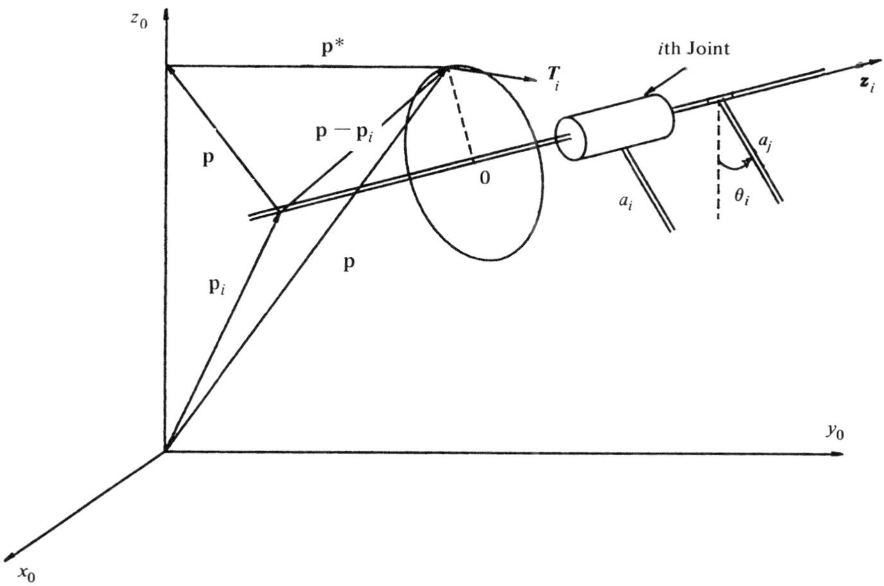

Figure 4.5.2 Generalized robotic joint and its rotation effect on the moment of the gripper in the workspace.

where $\mathbf{p}_i$ is any vector drawn from the origin in the direction of $\mathbf{z}_i$. Thus, the general conditions to have extrema are that

$$\mathbf{p} \cdot (\mathbf{p} \times \mathbf{z}_i) = 0. \tag{4.5.11}$$

Thus, the three vectors $\mathbf{p}$, $\mathbf{p}_i$, and $\mathbf{z}_i$ must lie in the same plane and all intermediate joint axes must intersect an extreme distance line. Thus, we may state the following general theorem by Sugimoto, Duffy, and Thomas (SDT).

Theorem 4.5.1 (SDT-I). For a generalized robotic joint shown in Figure 4.5.2, the three vectors $\mathbf{p}$, $\mathbf{p}_i$, and $\mathbf{z}_i$ must all lie in the same plane and all intermediate joint axes must intersect an extreme distance line.

Proof. Refer to Equations (4.5.4)–(4.5.11). Using Figure 4.5.2 again, consider the vector $\mathbf{p}^*$ drawn from the tangent point A perpendicular to the z_0 axis. Thus, $\mathbf{p}^*$ is represented by

$$\mathbf{p}^* = p_x^*\mathbf{i} + p_y^*\mathbf{j}. \tag{4.5.12}$$

It then follows that

$$\frac{\partial \mathbf{p}^*}{\partial \theta_i} = p^*(\mathbf{p}^* \cdot \mathbf{T}_i), \tag{4.5.13}$$

where

$$p^* = (p_x^{*2} + p_y^{*2})^{1/2}. \tag{4.5.14}$$

Again, for A to be an extreme distance with respect to the z_0 axis we must have

$$\mathbf{p}^* \cdot \mathbf{T}_i = 0, \qquad p \neq 0. \tag{4.5.15}$$

Now from Figure 4.5.2, it is clear that

$$\mathbf{p} = p_z\mathbf{k} + \mathbf{p}^*. \tag{4.5.16}$$

Substituting Equations (4.5.16) into (4.5.10) yields

$$\mathbf{T}_i = \mathbf{z}_i \times (p_z\mathbf{k} + \mathbf{p}^* - \mathbf{p}_i). \tag{4.5.17}$$

Now substituting Equation (4.5.17) into (4.5.15) yields

$$\mathbf{p}^* \cdot [\mathbf{z}_i \times (p_z\mathbf{k} - \mathbf{p}_i)] = 0. \tag{4.5.18}$$

Thus, the following second theorem due to Sugimoto, Duffy, and Thomas (SDT-II) follows.

Theorem 4.5.2 (SDT-II). For a generalized robotic joint shown in Figure 4.5.2, the three vectors $\mathbf{p}$, $\mathbf{z}_i$, and $(p_z\mathbf{k} - \mathbf{p}_i) = \mathbf{p}$ must all be in the same plane and all intermediate joint axes must intersect an extreme distance line.

Proof. Refer to Equations (4.5.12)–(4.5.18) and Figure 4.5.2.

EXAMPLE 4.5.2 ■

Referring to Figure 4.5.3, find the extreme distances (work envelope) of the Rhino XR-2 robotic manipulator.

Solution. Referring to Figure 4.5.3, we see that initially $\psi_1 = -(\theta_2 + \theta_3 + \theta_4)$ and $\psi_2 = \theta_5$ are known. Note that

$$-90^\circ \leq (\theta_2 + \theta_3 + \theta_4) \leq 0^\circ. \tag{4.5.19}$$

Furthermore,

$$-180^\circ \leq \theta_5 \leq 180^\circ, \tag{4.5.20}$$

such that ψ_1 will be the net *wrist pinch angle* with respect to a vertical approach vector directed toward the work surface, and ψ_2 will be the *finger roll angle* measured relative to the home position. It is obvious that ψ_3 does not affect the location of P but it does affect the orientation of the joint coordinate at P.

The objective here is to find the upper and lower bounds on x, y, and z of point P granted that the orientation of the gripper, as described by angles ψ_1 and ψ_2, has already been given. Essentially the bounds will be selected in such a way that a fixed point P can be reached by a set of joint angles θ_i, $i = 1, 2, \ldots, 6$. To find such extrema for x, y, and z, we must

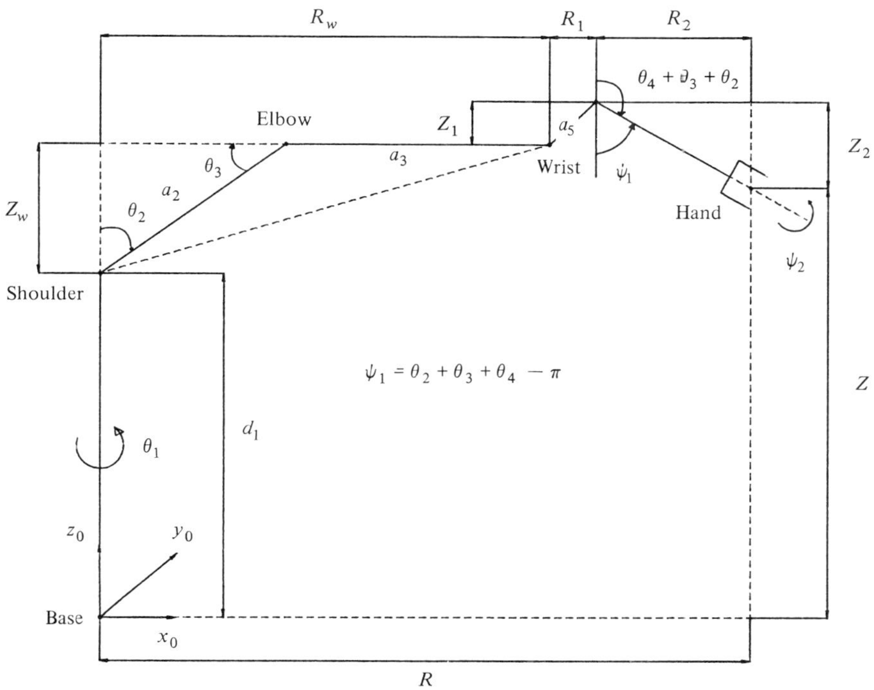

Figure 4.5.3 Typical geometry of the workspace for the Rhino XR-2.

consider constraints for the location of the wrist joint. Thus, from Figure 4.5.3, we find that

$$R_w^2 + Z_w^2 \leq (a_2 + a_3)^2. \tag{4.5.21}$$

Now if one calculates the wrist radius R_w, one finds from Figure 4.5.3 that

$$R_w = R - R_1 - R_2 = (x^2 + y^2)^{1/2} - a_4 \cos \psi_1 - d_5 \sin \psi_i. \tag{4.5.22}$$

The wrist height above the shoulder, Z_w, can be calculated as

$$Z_w = Z + Z_2 - d_1 - Z_1 = Z + d_5 \cos \psi_1 - a_4 \sin \psi_1 - d_1. \tag{4.5.23}$$

Combining Equations (4.5.22) and (4.5.23) with inequality (4.5.2) yields the following expanded inequality with explicit dependence on x, y, and z:

$$[(x^2 + y^2)^{1/2} - \beta_1]^2 + (Z - \beta_2)^2 \leq (a_2 + a_3)^2, \tag{4.5.24}$$

where

$$\beta_1 = a_4 \cos \psi_1 + d_5 \sin \psi_1 \tag{4.5.25}$$

and

$$\beta_2 = a_4 \sin \psi_1 - d_5 \cos \psi_1 + d_1. \tag{4.5.26}$$

Note that inequality (4.5.24) immediately puts an upper bound on x, y, and z as soon as the angle ψ_1 is established. To show these bounds let us consider them one at a time.

Upper and Lower Bounds on x. If we restrict the base joint rotation to $-90°$ to $90°$, $\psi_1 = \theta_1$ satisfies the following bounded inequality:

$$-90° \leq \psi_1 \leq 90°. \tag{4.5.27}$$

This in a way prevents the Rhino XR-2 from wrapping and straining the control cables around the base. From Figure 4.5.3 it is clear that $|\theta_1| \leq 90°$ implies that $x \geq 0$. To get an upper bound on x the inequality (4.3.24) is employed. Thus, setting $y = 0$ and $z = \beta_2$, one obtains from (4.5.24) that

$$0 \leq x \leq a_2 + a_3 + \beta_1 = R_{\max}. \tag{4.5.28}$$

The lower bound may not be realistic owing to structural constraints. Therefore, to get a more realistic lower bound on x, more details pertaining to Rhino XR-2 should be employed.

Upper and Lower Bounds on y. To obtain the appropriate bounds on the y coordinate of the Rhino XR-2, one again must use the inequality (4.5.24) with $Z = \beta_2$ and with the value of x selected according to inequality (4.5.28). This procedure yields

$$|y| \leq [(a_2 + a_3 + \beta_1)^2 - x^2]^{1/2}. \tag{4.5.29}$$

A lower bound on $|y|$ can be established if x is set equal to its maximum value which is R_{max}. This turns out to be equal to zero. On the other hand, a lower bound on $|y|$ can be obtained once a minimum value for R (R_{min}) is established. Obviously, $R_{min} > 0$; otherwise, the Rhino XR-2 fingers would get caught in the base gears (chain and sprocket). Note that the distance of the base corner to the origin of the base coordinate is about 2 in. The hand radius is at most 3 in. and the length of the hand from the back to the fingertips is 12 in. Thus, R_{min} can be defined as

$$R_{min} = 2 + 3 \cos \psi_1 + 12 \sin \psi_1. \tag{4.5.30}$$

Thus, $x^2 + y^2 \geq R_{min}^2$ and one has

$$|y| \geq \max[0, (R_{min}^2 - x^2)^{1/2}]. \tag{4.5.31}$$

The above inequality therefore keeps the entire hand outside a vertical cylinder of radius R_{min} that is centered around the origin's z_0 axis.

Upper and Lower Bounds on z. To obtain the appropriate upper and lower bounds on z, we again employ the inequality (4.5.24). At this stage both x and y have already been established. Thus,

$$z \leq \beta_2 + \{(a_2 + a_3)^2 - [(x^2 + y^2)^{1/2} - \beta_1]^2\}^{1/2} \tag{4.5.32}$$

will establish the upper bound on z. To find the lower bound on z, one must note that the robotic hand should not penetrate the work surface ($z_0 = a$). Essentially, a minimum z (z_{min}) should be established. In this case, one notes that $z_m - 3 \sin \psi_i$ and therefore

$$z \leq \max(z_m, \beta_2 - \{(a_2 + a_3)^2 - [(x^2 + y^2)^{1/2} - \beta_1]^2\}^{1/2}). \tag{4.5.33}$$

This completes the discussion on the Rhino XR-2 workspace. For any other robot the procedure is the same and can be carried out in a straightforward manner. ■ ■

EXAMPLE 4.5.3 ■

For a Microbot Minimover as shown in Figure 4.5.4 determine the upper and lower bounds on x, y, and z.

Solution. First determine an overall inequality similar to (4.5.24) and then analyze the problem as discussed in the preceding paragraphs. From Figure 4.5.4 note that the bounds on x, y, and z are

$$-17.5 \leq x \leq 17.5 \text{ in.}$$
$$-17.5 \leq y \leq 17.5 \text{ in.}$$
$$0 \leq z \leq 25.6 \text{ in.}$$

Table 4.5.1 gives details on the performance of the Microbot Minimover.

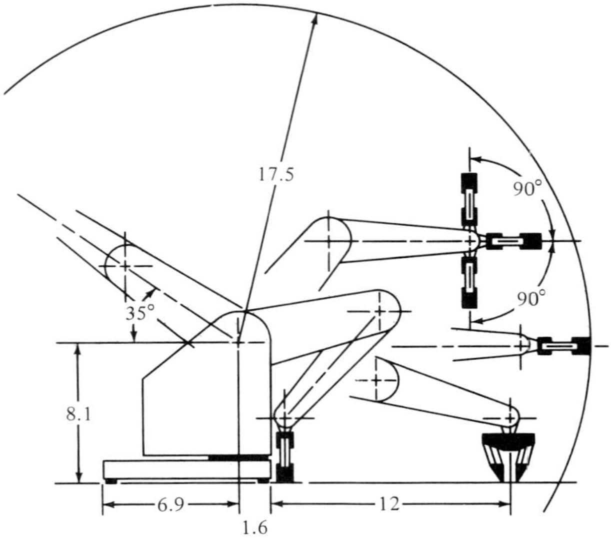

Figure 4.5.4 Side view of a Microrobot Minimover work envelope. (Note: All dimensions are in inches.)

TABLE 4.5.1 DETAILED PERFORMANCE

Motion	Maximum range of motion	Speed (full load)	Speed (no load)
Base	$\pm 90°$	0.37 rad/s	0.42 rad/s
Shoulder	$\pm 144°$, $-35°$	0.15 rad/s	0.36 rad/s
Elbow	$+0°$, $-149°$	0.23 rad/s	0.82 rad/s
Wrist roll	$\pm 360°$	1.31 rad/s	2.02 rad/s
Wrist pitch	$\pm 90°$	1.31 rad/s	2.02 rad/s
Hand	0–3 in.	8 mm/s	(20 mm/s)

4.6 ROBOTIC TASK DESCRIPTION (ROBOT–OBJECT MANIPULATION)

As discussed before, objects can be described by means of $4 \times n$ homogeneous matrices. Moving coordinates, on the other hand, can be described by 4×4 homogeneous matrices. Generally speaking, the 4×4 transformation (specification) of body frames should suffice for the complete description of the object. For complex objects, such as a cup, the 4×4 matrices may not suffice and the use of $4 \times n$ matrices may be more appropriate and convenient.

Consider the cup depicted in Figure 4.6.1. The task may consist of picking up the cup and carrying it over to point A where coffee may be poured into it

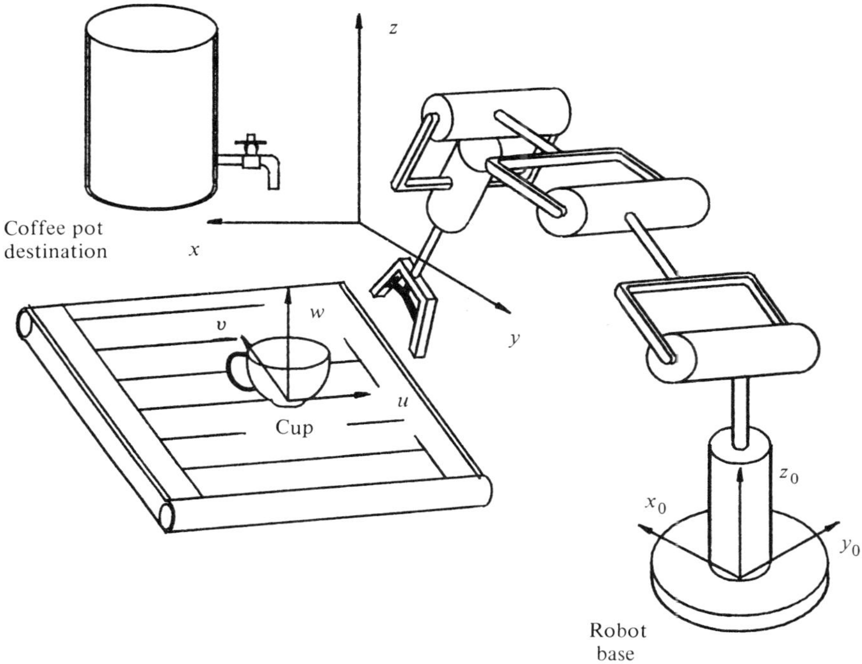

Figure 4.6.1 Typical task description.

from a coffee pot (valve). The robot's base coordinate is at transform **Z** from the reference frame *xyz*. Here **Z** is a general 4 × 4 homogeneous transformation.

The cup is originally located with respect to the base coordinate with a transform $\mathbf{C}_0$. The gripper is at $\mathbf{T}_0^6$ with respect to the base coordinate frame $x_0 y_0 z_0$. Let the final orientation and position of the cup with respect to the base coordinate be denoted by $\mathbf{C}_A$.

The task may be translated in terms of a series of 4 × 4 transformations pertaining to the following:

1. The gripper approaching the cup $T_0^6 \mathbf{C}_0$.
2. The gripper departing from the cup $T_0^6 \mathbf{C}_A$.
3. The gripper holding the cup with respect to the *uvw* coordinate $T_0^6 \mathbf{C}$.
4. The position and orientation of the cup in a sequence of lift-off positions $\mathbf{C}_{il}$, $i = 1, 2, \ldots, L$.
5. The position and orientation of the cup in a sequence of set-down positions $\mathbf{C}_{is}$, $i = 1, 2, \ldots, S$.
6. Transformations pertaining to the operation of the coffee pot valve.

Thus, task description in robotic manipulation consists of prescribing a set of 4 × 4 transformation matrices. Every transformation implies that either the robot or the objects have changed their positions as well as their orientations,

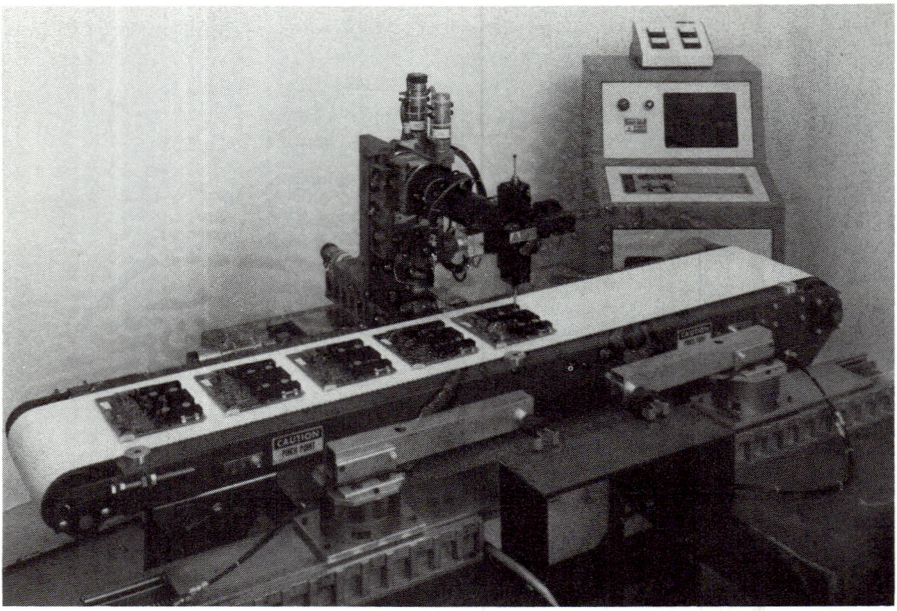

Figure 4.6.2 Robot inspecting printed integrated circuits. (Courtesy of Anorad Robots.)

and subsequent transformations, as they are multiplied through, indicate a sequence of motions and rotations.

Having specified a task by a sequence of homogeneous transformations, we are in a position to communicate directly with a host computer that is serving as the robot's brain and ask that such a sequence of transformations be executed by the robot. At this stage we must design a robotic motion trajectory suitable for the execution of the prescribed task (see Fig. 4.6.2).

EXAMPLE 4.6.1 ▪

Describe the number of ways in which a sequence of transformations can be written and saved in the memory of a host computer.

Solution. There are essentially three distinct ways of teaching the robot how to perform a task. The first technique, which is especially suitable for robots with stepper motors, is to guide the robot manually through a trajectory that connects the various transforms. The second technique is to control and design the motion trajectory manually from a master control station. The third technique is to input numerical data pertaining to various transformations along with the software that instructs the robot to execute such transformations in an orderly fashion. In the next section we elaborate on the details pertaining to the design of motion trajectories once a robotic task is specified completely.

4.7 ROBOTIC MOTION TRAJECTORY DESIGN

4.7.1 Introduction

Now that we have learned how to describe a robot–object manipulation task by means of homogeneous transformations, we must concern ourselves with the techniques of motion trajectory design. We need to describe the desired motion of robotic manipulators in either the joint space, the three-dimensional workspace, or the hand coordinates, inasmuch as the time histories of position, orientation, linear velocity, angular velocity, linear acceleration, and angular acceleration are concerned (see Fig. 4.7.1).

As far as the human task operator is concerned, the robotic system should be capable of taking care of motion trajectory details once the required task transformations are inputted. For example, the operator may just input the desired kinematic attitude of the robotic hand to manipulate an object and leave it to the robotic control system to design the shape of the motion trajectory and the details such as the velocity and acceleration profiles. In a way this is related to what is called the *task-level programming*, which will be discussed in Chapter 9. Let us address the question of how robotic motion trajectories are interpreted by the host computer and how the host computer actually constructs such

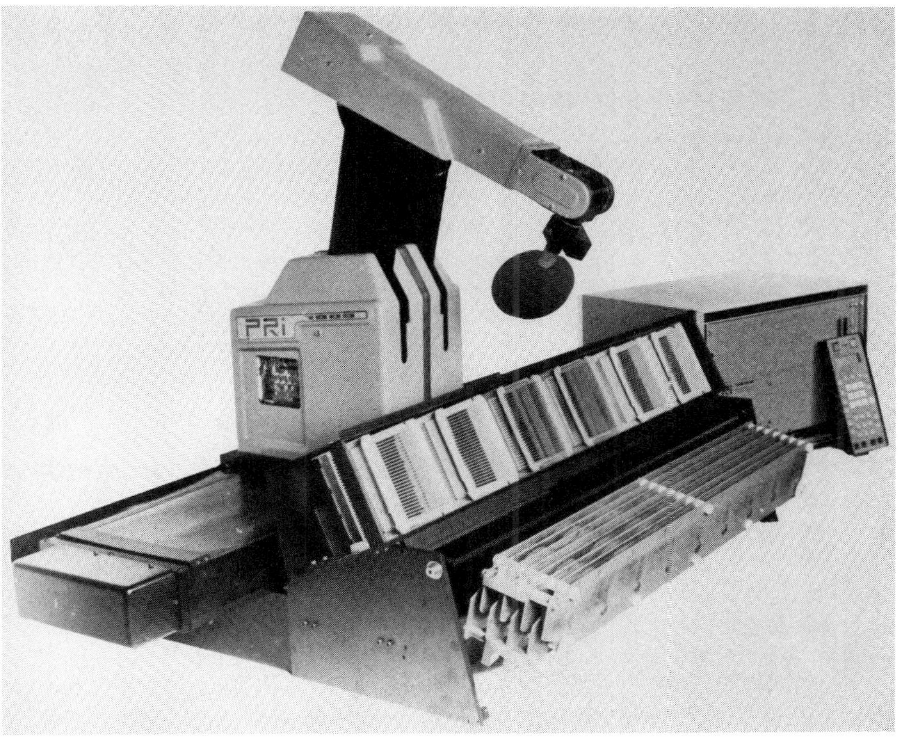

Figure 4.7.1 Robot placing MOS wafers. (Courtesy of PRI.)

trajectories and commands the robot to execute the desired tasks. Because of the digital nature of the host computer, the trajectories are generated in a discrete manner. Thus, the generation of each discrete point of the motion trajectory occurs at what is termed the *trajectory run-time*. They may be generated at the rate of 10–300 Hz depending on how fast the *path update rate* can be achieved in the host computer. The problem is to move the robotic hand from an initial kinematic attitude $\mathbf{H}(0)$ to a destination kinematic attitude $\mathbf{H}(t)$ at time t. It is natural to describe the motion in much more detail than just specifying the beginning and the destination in order to avoid collisions with objects present in the workspace. Thus, one defines *via points* at which the kinematic attitude of the robotic hand must be specified. For more detailed trajectories the generalized velocity attitude as well as the generalized acceleration attitude must be specified. Obviously, to create a time-dependent kinematic attitude for the robotic hand, $\mathbf{H}(t)$, one has to resort to a set of time-dependent joint angles or a time-dependent vector of the joint angles $\boldsymbol{\theta}(t)$, such that

$$\boldsymbol{\theta}(t) = (\boldsymbol{\theta}_1 \quad \boldsymbol{\theta}_2 \quad \cdots \quad \boldsymbol{\theta}_n)^T, \tag{4.7.1}$$

where $\boldsymbol{\theta}(t)$ is nothing but the time-dependent solution of the inverse kinematics problem in connection with the initial $\mathbf{H}(0)$ and final $\mathbf{H}(t)$ kinematic attitudes of the robotic hand.

Next, we describe a number of techniques available for the design and generation of the desired joint vectors of robotic joint trajectories $\boldsymbol{\theta}(t)$.

4.7.2 Cubic Joint Trajectories

The simplest and most common way of specifying a joint trajectory $\theta_i(t)$ is to specify the initial and final values of $\theta_i(t)$ and $\dot{\theta}_i(t)$. These are normally stated as

$$\theta_i(0) = \theta_{io}, \tag{4.7.2}$$

$$\theta_i(t_f) = \theta_{if}, \tag{4.7.3}$$

$$\dot{\theta}_i(0) = 0, \tag{4.7.4}$$

and

$$\dot{\theta}(t_f) = 0, \tag{4.7.5}$$

where t_f is the final time and the robotic hand is required to be at rest initially at time $t = 0$ and to come to rest at time $t = t_f$.

These constraints, (4.7.2)–(4.7.5), can be satisfied by third-degree polynomials in time; that is,

$$\theta_i(t) = \theta_{io} + a_{1i}t + a_{2i}t^2 + a_{3i}t^3, \tag{4.7.6}$$

such that

$$\theta_{if} = \theta_{io} + a_{1i}t_f + a_{2i}t_f^2 + a_{3i}t_f^3, \tag{4.7.7}$$

$$0 = a_{1i}, \tag{4.7.8}$$

and

$$0 = 2a_{2i}t_f + 3a_{3i}t_f^2, \tag{4.7.9}$$

where a_{2i} and a_{3i} are found to be

$$a_{2i} = 3(\theta_{if} - \theta_{io})t_f^{-2} \tag{4.7.10}$$

and

$$a_{3i} = 2(\theta_{io} - \theta_{if})t_f^{-3}. \tag{4.7.11}$$

If the initial and final velocities are not zero, such as a robot in a conveyer tracking situation, then the coefficients of the polynomial are obtained by requiring that

$$\dot{\theta}_i(0) = \dot{\theta}_{io} \tag{4.7.12}$$

and

$$\dot{\theta}_i(t_f) = \theta_{if}. \tag{4.7.13}$$

The coefficients of Equation (4.7.6) are then obtained easily:

$$a_{ix} = \dot{\theta}_{io}, \tag{4.7.14}$$

$$a_{2i} = 3(\theta_{if} - \theta_{io})t_f^{-2} - 2\dot{\theta}_{io}t_f^{-1} - \dot{\theta}_{if}t_f^{-1}, \tag{4.7.15}$$

and

$$a_{3i} = 2(\theta_0 - \theta_{if})t_f^{-3} + (\dot{\theta}_{io} + \dot{\theta}_{if})t_f^{-2}. \tag{4.7.16}$$

Note that Equations (4.7.14)–(4.7.16) are general enough to be applicable to any *via point* between the initial and the final destination point for the robot. However, because of the requirement of continuity of position, velocity, and acceleration, the solutions of multiple cubic trajectory equations for the coefficients become more complex. The following example will clarify the situation.

EXAMPLE 4.7.1 ■

Find the coefficients of two cubic trajectories, assuming the same duration for both segments of the trajectory.

Solution. Note that in this case, for the first and second segments of the motion trajectory,

$$\theta_1(t) = a_{10} + a_{11}t + a_{12}t^2 + a_{13}t^3 \tag{4.7.17}$$

and

$$\theta_2(t) = a_{20} + a_{21}t + a_{22}t^2 + a_{23}t^3. \tag{4.7.18}$$

The constraints are

$$\theta_1(0) = \theta_{10}, \qquad \theta_2(0) = \theta_1(t_{f1}); \tag{4.7.19}$$

$$\dot{\theta}_1(t) = \dot{\theta}_{10}, \qquad \dot{\theta}_2(0) = \dot{\theta}_1(t_{f1}); \tag{4.7.20}$$

$$\theta_2(t_{f2}) = \theta_{2f}, \qquad \dot{\theta}_2(t_{f2}) = \dot{\theta}_{2f}; \tag{4.7.21}$$

and

$$\ddot{\theta}_1(0) = \ddot{\theta}_{10}, \qquad \ddot{\theta}_2(0) = \ddot{\theta}_1(t_{f1}). \tag{4.7.22}$$

From the above eight equations the eight coefficients a_{10}, a_{11}, a_{12}, a_{13}, a_{20}, a_{21}, a_{22}, and a_{23} can be found (see Homework Problem 6). ■ ■

EXAMPLE 4.7.2 ■

Plot the position, velocity, and acceleration of a two-segment continuous-acceleration trajectory such that $\theta_0 = 6°$, $\theta_2(0) = 16° = \theta_{20}$, and $\theta_{2f} = 39°$, each segment lasts about a second, and the velocity at a via point is $\dot{\theta}_{20} = 18°/\text{s}$.

Solution. Note that

$$\theta(t) = a_0 + a_1 t + a_2 t^2 + a_3 t^3,$$
$$\dot{\theta}(t) = a_1 + 2a_2 t + 3a_3 t^2,$$

and

$$\ddot{\theta}(t) = 2a_2 + 6a_3 t.$$

For segment 1, $a_{01} = \theta_0$, $a_{11} = 0$, $a_{21} = 3t_f^{-2}(\theta_{20} - \theta_0) - \dot{\theta}_{20} t_{f1}^{-1}$, and $a_{31} = 2t_{f1}^{-3}(\theta_0 - \theta_{20}) + \dot{\theta}_{20} t_{f1}^{-2}$, where $\dot{\theta}_{20} = \dot{\theta}_1(t_{f1})$. For segment 2, $a_{02} = \theta_{20}$, $a_{22} = 3t_{f2}^{-2}(\theta_{2f} - \theta_{20}) - 2\theta_{20} t_{f2}^{-2}$, $a_{32} = 2t_{f2}^{-3}(\theta_{20} - \theta_{2f}) + \dot{\theta}_{20} t_{f2}^{-2}$, and $a_{12} = \dot{\theta}_{20}$.

Based on the above values, graphs for the position, velocity, and acceleration can be plotted: see Figures 4.7.2–4.7.5. ■ ■

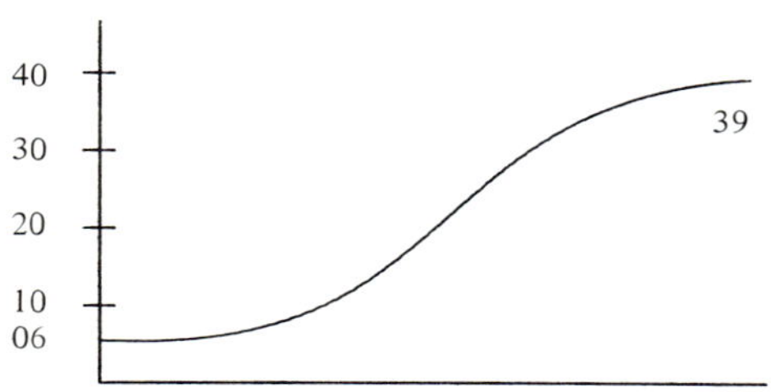

Figure 4.7.2 Position trajectory.

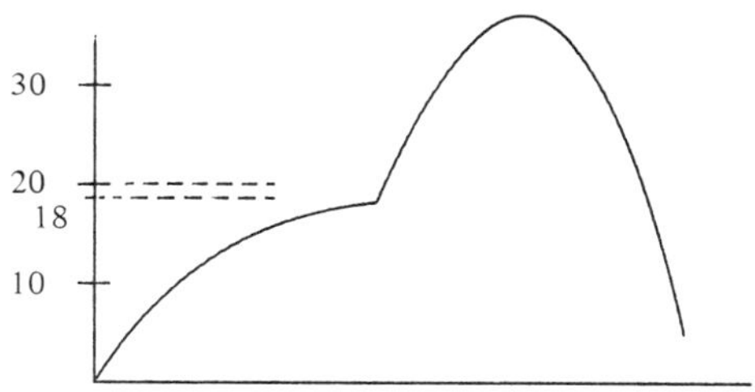

Figure 4.7.3 Velccity trajectory.

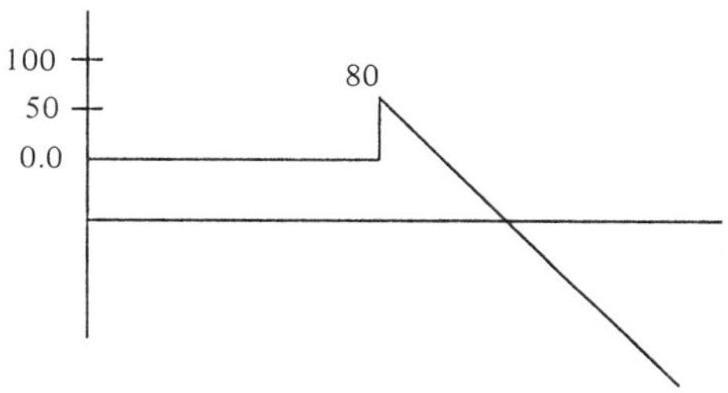

Figure 4.7.4 Acceleration trajectory.

Figure 4.7.5 Complex robotic trajectory for sealant application. (Courtesy of Cincinnati Milacron.)

4.8 GENERAL DESIGN CONSIDERATIONS ON TRAJECTORIES

Generally speaking, to design a robotic trajectory, the following points should be taken into account:

1. As the gripper is picking up an object its motion should be away from the object to avoid collision with the surface supporting the object.
2. The lift-off point should be along the normal to the surface and the origin of the gripper coordinate frame should pass through the lift-off point. Thus, one will have an admissible motion. The speed at which the object is to be picked off also can be controlled by keeping track of the time required to reach this position.
3. As a rule of thumb, the position of the lift-off point from the supporting surface is recommended to be at least 25% of the length of the last link from the surface ($0.25d_6$ + tool length) (Fig. 4.8.1).
4. The same set of requirements—1, 2, and 3—are applicable to the set-down point of the final position motion; that is, the gripper should move in a direction normal to the surface and slow down to the final position to let go of the object.
5. From the above considerations, one has four distinct positions for each robotic trajectory—initial, lift-off, set-down, and final positions (Fig. 4.8.2).

Thus, we may impose the following eight constraints on designing a motion trajectory:

1. Initial position is fixed.
2. Initial velocity is normally zero.
3. Initial acceleration is normally zero.
4. Final position is fixed.
5. Final velocity is normally zero.
6. Final acceleration is normally zero.
7. Lift-off position should be $0.25d_6$ + tool length.
8. Set-down position should be $0.25d_6$ + tool length.

We may also want to add the velocity and acceleration limitations on the intermediate positions, that is, lift-off and set-down, to the above constraints. However, by relaxing these constraints, we would have eight constraints to which we could fit a minimum polynomial, that is, a seventh-order polynomial with unknown coefficients. Thus, for any joint angle θ_i, we may write

$$\theta_i(t) = \sum_{n=0}^{7} a_{ni} t^n, \tag{4.8.1}$$

where a_{ni} is a $8 \times i$ matrix.

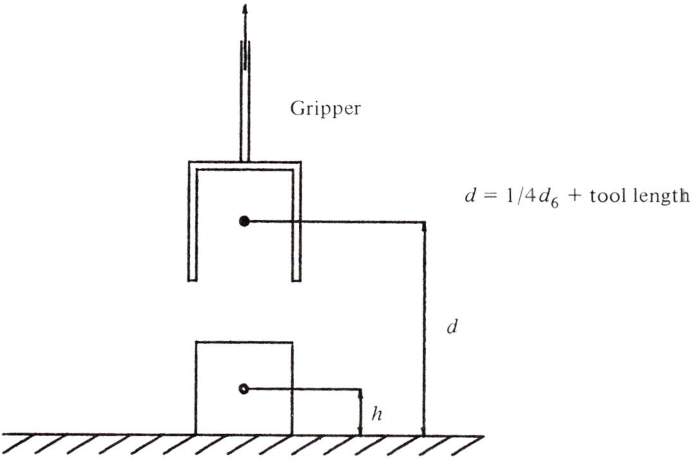

Figure 4.8.1 Recommended lift-on position.

Since this is a polynomial of the seventh order, it is conceivable that it may have a significant number of extrema (minima and maxima) which would be undesirable for robotic motion trajectories. Also, it may take a long time to compute all unknown coefficients, for example, $8 \times i$ of them (if $i = 6$, there are 48 coefficients). One must also evaluate extremum values of the robotic motion trajectory to make sure that such extrema do not penetrate beyond the work

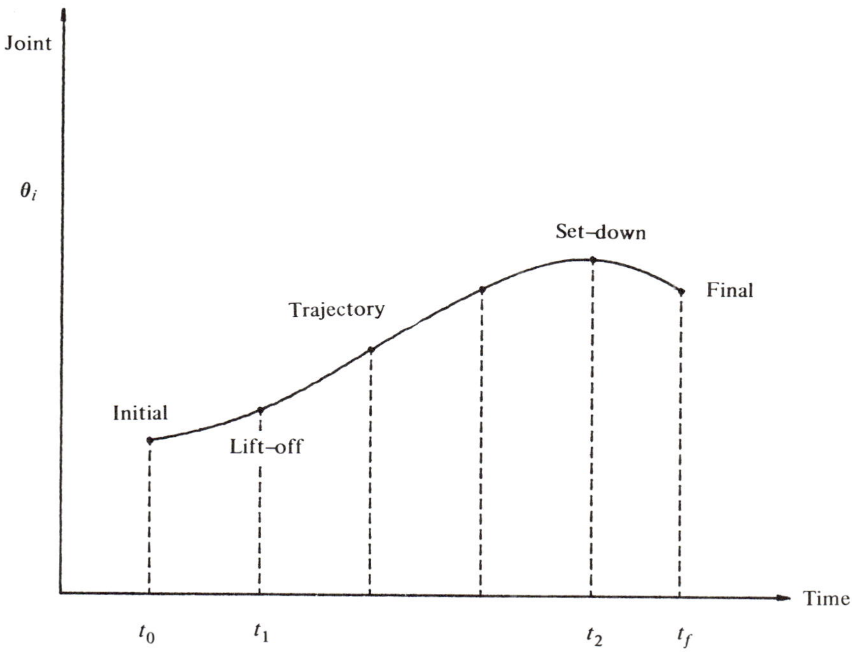

Figure 4.8.2 Typical *i*th joint motion trajectory.

TABLE 4.8.1 JOINT TRAJECTORY CONSTRAINTS

Number	Constraint description	Constraint equation
1	Initial position of gripper	$\theta_i(t_0) = \theta^*_{i0}$
2	Initial velocity of robot	$\dot{\theta}_i(t_0) = \dot{\theta}^*_{i0}$
3	Initial acceleration of robot	$\ddot{\theta}_i(t_0) = \ddot{\theta}^*_{i0}$
4	Lift-off position of robot	$\theta_i(t_1) = \theta^*_{i1}$
5	Position continuity at t_1	$\theta_i(t_1^-) = \theta_i(t_1^+)$
6	Velocity continuity at t_1	$\dot{\theta}_i(t_1^-) = \dot{\theta}_i(t_1^+)$
7	Acceleration continuity at t_1	$\ddot{\theta}_i(t_i^-) = \ddot{\theta}_i(t_1^+)$
8	Set-down position of robot	$\theta_i(t_2) = \theta^*_{i2}$
9	Position continuity at t_2	$\theta_i(t_2^-) = \theta_i(t_2^+)$
10	Velocity continuity at t_2	$\dot{\theta}_i(t_2^-) = \dot{\theta}_i(t_2^+)$
11	Acceleration continuity at t_2	$\ddot{\theta}_i(t_2^-) = \ddot{\theta}_i(t_2^+)$
12	Final position of robot	$\theta_i(t_3) = \theta^*_{i3}$
13	Final velocity of robot	$\dot{\theta}_i(t_3) = \dot{\theta}^*_{i3}$
14	Final acceleration of robot	$\ddot{\theta}_i(t_3) = \ddot{\theta}^*_{i3}$

surface. Thus, it is desirable to resort to lower-order polynomials by dividing the trajectory into segments that can be computed rather easily. The advantage of such lower-order polynomials is that it is easier to compute the unknown coefficients and to evaluate the roots of their derivatives when computing the extrema.

All such segmented trajectories must be continuous in position, velocity, and acceleration to achieve smooth robotic motion. To ensure this, such values must be matched at segment intersections.

Having taken all such constraints into account, one ends up with 14 constraints for calculating the unknown coefficients for a three-segment trajectory, that is, a 4–3–4 and a 3–5–3 trajectory. The three segments are initial, intermediate, and final positions. Table 4.8.1 displays the 14 constraints pertaining to robotic motion trajectories that are segmented. (See Fig. 4.8.3.)

In order for three lower-order polynomials to be determined completely by the above 14 conditions, the collective (sum) degrees of the powers of t must be $14 - 3$, because there are three free constants for these three different polynomials. In other words, there must be 14 unknowns that are the coefficients of the polynomials. Three of these coefficients are free because they are not multiplied by the t's. Thus, because any mth degree polynomial in t must have $m + 1$ coefficients, the sum of the degrees of the three segment polynomials must be at least 11. There are a number of ways to satisfy this condition, namely, a $4 + 3 + 4$, $3 + 5 + 3$, or $5 - 2 - 4$. We now elaborate on some of these trajectories.

4.8.1 The 4–3–4 Trajectory

For the three segments of a motion trajectory, namely, lift-off, intermediate, and set-down, there exist polynomials of fourth, third, and fourth degree that could fairly well approximate them. To solve for each segment it is convenient to

Figure 4.8.3 Robotic application in laser trimming trajectories. (Courtesy of Cincinnati Milacron.)

introduce a dimensionless variable for time, τ, such that

$$\tau = \frac{t - t_{m-1}}{t_m - t_{m-1}} \quad \text{and} \quad \tau_m = t_m - t_{m-1}, \tag{4.8.2}$$

where the subscripts m and $m - 1$ refer to the mth and $(m - 1)$th segment of the robotic motion trajectory. Thus, during motion along the mth trajectory segment, the nondimensional time τ ranges from 0 to 1 as the real time changes from t_{m-1} to t_m. The introduction of equations (4.8.2) will simplify the equations considerably.

Segment 1: The Fourth-Order Polynomial Consider $\rho_{mi}(\tau)$ to be the polynomial representation of the mth segment of motion trajectory and that of the ith joint. Thus,

$$\rho_{1i}(t) = \sum_{j=0}^{4} C_{1j}\tau^j = C_{10} + C_{11}\tau + C_{12}\tau^2 + C_{13}\tau^3 + C_{14}\tau^4, \tag{4.8.3}$$

$$\tau_1\dot{\rho}_{1i}(t) = \sum_{j=0}^{4} iC_{1j}\tau^{j-1} = C_{11} + 2C_{12} + 3C_{13}\tau^2 + 4C_{14}\tau^3, \tag{4.8.4}$$

and

$$\tau_1^2\ddot{\rho}_{1i}(t) = \sum_{j=0}^{4} j(j-1)C_{1j}\tau^{j-2} = 2C_{12} + 6C_{13}\tau + 12C_{14}\tau^2. \tag{4.8.5}$$

Note that these polynomials must satisfy the conditions shown in Table 4.8.1. Thus,

$$\rho_{1i}(0) = \theta^*_{i0}, \qquad \dot{\rho}_1(0) = \dot{\theta}^*_{i0}, \qquad \ddot{\rho}_1(0) = \ddot{\theta}^*_{i0}. \tag{4.8.6}$$

Applying the above conditions to Equations (4.8.3), (4.8.4), and (4.8.5) yields

$$\rho_{1i}(\tau) = C_{14}\tau^4 + C_{13}\tau^3 + \tfrac{1}{2}\ddot{\theta}^*_{i0}\tau_1^2\tau^2 + \dot{\theta}^*_{i0}\tau_1\tau + \theta^*_{i0}. \tag{4.8.7}$$

EXAMPLE 4.8.1 ■

Determine the motion trajectory of the gripper during the lift-off position.

Solution. The trajectory pertains to θ_6 and thus one has to set $i = 6$ in Equation (4.8.7). Thus,

$$\rho_{16}(\tau) = C_{14}\tau^4 + C_{13}\tau^3 + \tfrac{1}{2}\ddot{\theta}^*_{60}\tau_1^2\tau^2 + \dot{\theta}^*_{60}\tau_1 + \theta^*_{60}. \qquad ■■ \tag{4.8.8}$$

Note that in Equation (4.8.7) there are two unknowns to be found—C_{14} and C_{13}—and these unknowns are found by using the continuity conditions:

$$\rho_{1i}(1) = \rho_{2i}(0), \qquad \dot{\rho}_{1i}(1) = \dot{\rho}_{2i}(0), \qquad \ddot{\rho}_{1i}(1) = \ddot{\rho}_{2i}(0). \tag{4.8.9}$$

Thus, we need to calculate $\rho_{2i}(\tau)$ first

Segment 2: The Third-Order Polynomial Similar to the treatment given for segment 1, one would have

$$\rho_{2i}(\tau) = \sum_{j=0}^{3} C_{2j}\tau^j = C_{20} + C_{21}\tau + C_{22}\tau^2 + C_{23}\tau^3, \tag{4.8.10}$$

$$\tau_2\dot{\rho}_{2i}(\tau) = \sum_{j=0}^{3} iC_{2j}\tau^{j-1} = C_{21} + 2C_{22}\tau + 3C_{23}\tau^2, \tag{4.8.11}$$

and

$$\tau_2^2\ddot{\rho}_{2i}(\tau) = \sum_{j=0}^{3} j(j-1)C_{2j}\tau^{j-2} = 2C_{22} + 6C_{23}\tau. \tag{4.8.12}$$

Again, these equations have to satisfy the following conditions:

$$\rho_{2i}(0) = \theta^*_{i1}, \qquad \dot{\rho}_{2i}(0) = \dot{\theta}^*_{i1}, \qquad \ddot{\rho}_{2i}(0) = \theta^*_{i1}. \tag{4.8.13}$$

Equations (4.8.2) and (4.8.9) determine some of the unknown coefficients. Namely, one obtains

$$C_{20} = \theta^*_{i1}, \qquad C_{21} = \dot{\theta}^*_{i1}\tau_2, \qquad C_{22} = \tfrac{1}{2}\ddot{\theta}^*_{i1}\tau_2^2, \tag{4.8.14}$$

$$C_{13} + C_{14} = \theta^*_{i1} - (\theta^*_{i0} + \dot{\theta}^*_{i0}\tau_1 + \tfrac{1}{2}\ddot{\theta}^*_{i0}\tau_1^2), \tag{4.8.15}$$

$$4\tau_1^{-1}C_{14} + 3\tau_1^{-1}C_{13} - \tau_2^{-1}C_{21} = -(\dot{\theta}^*_{i0} + \tau_1\ddot{\theta}^*_{i0}), \tag{4.8.16}$$

and

$$6\tau_1^{-2}C_{13} + 12\tau^{-2}C_{14} - 2\tau_2^{-2}C_{22} = -\ddot{\theta}^*_{i0}. \tag{4.8.17}$$

At the end of segment 2 we must still satisfy the continuity equations, that is,

$$\rho_{2i}(1) = \rho_{3i}(0), \qquad \dot{\rho}_{2i}(1) = \dot{\rho}_{3i}(0), \qquad \ddot{\rho}_{2i}(1) = \ddot{\rho}_{3i}(0). \tag{4.8.18}$$

Thus, one has to calculate the third segment polynomial to calculate the additional coefficients.

Segment 3: The Fourth-Order Polynomial Similar to the two previous treatments given for segments 1 and 2, one would have

$$\rho_{3i}(\tau) = \sum_{j=0}^{4} C_{3j}\tau^j = C_{30} + C_{31}\tau + C_{32}\tau^2 + C_{33}\tau^3 + C_{34}\tau^4, \tag{4.8.19}$$

$$\tau_3\dot{\rho}_{3i}(\tau) = \sum_{j=0}^{4} iC_{3j}\tau^{j-1} = C_{31} + 2C_{32}\tau + 3C_{33}\tau^2 + 4C_{34}\tau^3, \tag{4.8.20}$$

and

$$\tau_3^2\rho_{3i}(\tau) = \sum_{j=0}^{4} j(j-1)C_{3j}\tau^{j-2} = 2C_{32} + 6C_{33}\tau + 12C_{34}\tau^2. \tag{4.8.21}$$

Again we have the initial conditions:

$$\rho_{3i}(0) = \theta_{i2}^*, \qquad \dot{\rho}_3(0) = \dot{\theta}_{i2}, \qquad \ddot{\rho}_3(0) = \ddot{\theta}_{i2}^*, \tag{4.8.22}$$

which in conjunction with Equations (4.8.8), (4.8.9), and (4.8.10) yield

$$C_{30} = \theta_{i2}^*, \qquad C_{31} = \dot{\theta}_{i2}^*\tau_3, \qquad C_{32} = \tfrac{1}{20}\ddot{\theta}_{i2}^*\tau_3^{-2}. \tag{4.8.23}$$

The continuity equations (4.8.18) now yield

$$C_{21} + C_{22} + C_{23} = \theta_{i2}^* - \theta_{i1}^*, \tag{4.8.24}$$

$$\tau_2^{-1}C_{21} + 2\tau_2^{-1}C_{22} + 3\tau_2^{-1}C_{23} - \tau_3^{-1}C_{31} = 0, \tag{4.8.25}$$

and

$$2\tau_2^{-2}C_{22} + 6\tau_2^{-2}C_{23} - 2\tau_3^{-2}C_{32} = 0. \tag{4.8.26}$$

Finally, one has the end conditions:

$$\rho_{3i}(1) = \theta_{if}^*, \qquad \dot{\rho}_{3i}(1) = \dot{\theta}_{if}^*, \qquad \ddot{\rho}_{3i}(1) = \ddot{\theta}_{if}^*. \tag{4.8.27}$$

Employing Equations (4.8.27) in (4.8.19)–(4.8.21), one obtains

$$C_{31} + C_{32} + C_{33} + C_{34} = \theta_{if}^* - \theta_{i2}^*, \tag{4.8.28}$$

$$C_{31} + 2C_{32} + 2C_{33} + 4C_{34} = \tau_3\dot{\theta}_{if}^*, \tag{4.8.29}$$

and

$$2C_{32} + 6C_{33} + 12C_{34} = \tau_3^2\ddot{\theta}_{if}^*. \tag{4.8.30}$$

Note that since, generally, $\dot{\theta}_{i1}^*$, $\ddot{\theta}_{i1}^*$, $\dot{\theta}_{i2}^*$, and $\ddot{\theta}_{i2}^*$ are not known, we have calculated explicitly only five coefficients out of the 14 unknown coefficients, which are $C_{10}, C_{11}, C_{12}, C_{13}, C_{14}, C_{20}, C_{21}, C_{22}, C_{23}, C_{30}, C_{31}, C_{32}, C_{33}$, and C_{34}. The explicitly known coefficients are

$$(C_{10}, C_{11}, C_{12}, C_{20}, C_{30}) \equiv (\theta_{io}^*, \tau_1\dot{\theta}_{io}^*, \tfrac{1}{2}\tau_1^2\ddot{\theta}_{io}^*, \theta_{i1}^*, \theta_{i2}^*) \tag{4.8.31}$$

The remaining nine equations are (4.8.15)–(4.8.17), (4.8.24)–(4.8.26), and (4.8.28)–(4.8.30), which can be put into matrix form (9 × 9) such that

$$\begin{bmatrix} 1 & 1 & 0 & 0 & 0 & 0 & 0 & 0 & 0 \\ 3\tau_1^{-1} & 4\tau_1^{-1} & -\tau_2^{-1} & 0 & 0 & 0 & 0 & 0 & 0 \\ 6\tau_1^{-2} & 12\tau_1^{-2} & 0 & -2\tau_2^{-2} & 0 & 0 & 0 & 0 & 0 \\ 0 & 0 & 1 & 1 & 1 & 0 & 0 & 0 & 0 \\ 0 & 0 & \tau_2^{-1} & 2\tau_2^{-1} & 3\tau_2^{-2} & -\tau_3^{-1} & 0 & 0 & 0 \\ 0 & 0 & 0 & 2\tau_2^{-2} & 6\tau_2^{-2} & 0 & -2\tau_3^{-2} & 0 & 0 \\ 0 & 0 & 0 & 0 & 0 & 1 & 1 & 1 & 1 \\ 0 & 0 & 0 & 0 & 0 & 1 & 2 & 3 & 4 \\ 0 & 0 & 0 & 0 & 0 & 0 & 2 & 6 & 12 \end{bmatrix} \begin{bmatrix} C_{13} \\ C_{14} \\ C_{21} \\ C_{22} \\ C_{23} \\ C_{31} \\ C_{32} \\ C_{33} \\ C_{34} \end{bmatrix}$$

$$= \begin{bmatrix} \theta_{i1}^* - (\theta_{io}^* + \tau_1\dot{\theta}_{io}^* + \frac{1}{2}\tau_1^2\ddot{\theta}_{io}^* \\ -(\dot{\theta}_{io}^* + \ddot{\theta}_{io}^*\tau_1) \\ -\ddot{\theta}_{io}^* \\ \theta_{i2}^* - \theta_{i1}^* \\ 0 \\ 0 \\ \theta_{if}^* - \theta_{i2}^* \\ \tau_3\dot{\theta}_{if}^* \\ \tau_3^2\ddot{\theta}_{if}^* \end{bmatrix}. \tag{4.8.32}$$

Simplifying Equation (4.8.32) into a compact form, one has

$$\mathbf{BC} = \boldsymbol{\delta}, \tag{4.8.33}$$

where **B** is the 9 × 9 matrix of the coefficients **C** and $\boldsymbol{\delta}$ is some vectorial array pertaining to differential angular changes in the joint angles and their time derivatives as depicted in (4.8.32). The complete solution to the array **C** (the nine remaining coefficients to be found) is

$$\mathbf{C} = \mathbf{B}^{-1}\boldsymbol{\delta}. \tag{4.8.34}$$

EXAMPLE 4.8.2 ■

Find the complete explicit solutions for the three polynomials $\rho_1(\tau)$, $\rho_2(\tau)$, and $\rho_3(\tau)$.

Solution. It can be shown that the solutions for the coefficients C_{13}, C_{14}, C_{21}, C_{22}, C_{23}, C_{31}, C_{32}, C_{33}, and C_{34} are (see Appendix A)

$$\begin{aligned} C_{13} &= (\tau_2^{-1}\tau_3 + 2\tau_1^{-1}\tau_3 + 2 + 3\tau_1^{-1}\tau_2)^{-1} \\ &\quad \times \{2\delta_1^*(4 + 2\tau_2^{-1}\tau_3 + 2\tau_1^{-1}\tau_3 + 3\tau_1^{-1}\tau_2) - \delta_2^*\tau_2^{-1}\tau_1(3 + \tau_2^{-1}\tau_3) \\ &\quad + 2\delta_3^*\tau_3^{-1}\tau_1 - \dot\theta_{io}^*\tau_1(6 + 6\tau_1^{-1}\tau_2 + 4\tau_1^{-1}\tau_3 + 3\tau_2^{-1}\tau_3) \\ &\quad - \dot\theta_{if}^*\tau_1 - \ddot\theta_{io}^*\tau_1\tau_3[\tfrac{5}{3} + 2\tau_3^{-1}\tau_1 + \tau_1\tau_2^{-1} + \tfrac{5}{2}\tau_2\tau_3^{-1}] + \ddot\theta_{if}^*\tau_1\tau_3\} \\ &= C_{13}(t), \end{aligned} \tag{4.8.35}$$

$$C_{14} = [\delta_1^* - \dot\theta_{io}^*\tau_1 - \tfrac{1}{2}\ddot\theta_{io}^*\tau_1^2 - C_{13}(t)], \tag{4.8.36}$$

$$C_{21} = \dot\theta_{i1}^*\tau_2, \qquad C_{22} = \tfrac{1}{2}\ddot\theta_{i1}^*\tau_2^2, \qquad C_{23} = (\delta_2^* - \dot\theta_{i1}^*\tau_2 - \tfrac{1}{2}\ddot\theta_{i1}^*\tau_2^2, \tag{4.8.37}$$

$$C_{31} = \dot\theta_{i2}^*\tau_2, \qquad C_{22} = \tfrac{1}{2}\ddot\theta_{i2}^*\tau_3^2, \tag{4.8.38}$$

$$C_{33} = (-8\delta_3^* + 5\dot\theta_{if}^*\tau_3 - \tfrac{1}{2}\ddot\theta_{if}^*\tau_3^2 + 3\ddot\theta_{i2}^*\tau_3), \tag{4.8.39}$$

$$\delta_1^* = \theta_{i1}^* - \theta_{io}^*, \qquad \delta_2^* = \theta_{i2}^* - \theta_{i1}^*, \qquad \delta_3^* = \theta_{if}^* - \theta_{i2}^*. \tag{4.8.40}$$

Thus, the complete solutions for the three segments of the motion trajectory are

$$\begin{aligned} \theta_{i1} = \rho_{1i}(\tau) &= \theta_{io}^* + (\theta_{io}^*\tau_1)\tau + (\tfrac{1}{2}\ddot\theta_{io}^*\tau_1^2)\tau^2 + C_{13}\tau^3 \\ &\quad + (\delta_1 - \dot\theta_{io}^*\tau_1 - \tfrac{1}{2}\ddot\theta_{io}^*\tau_1^2 - C_{13})\tau^4, \end{aligned} \tag{4.8.41}$$

$$\dot\theta_{i1}^* = \dot\rho_{1i}(1)\tau_1^{-1} = 4\tau_1^{-1} - 3\dot\theta_{io}^* - \ddot\theta_{io}^*\tau_1 - \tau_1^{-1}C_{13}, \tag{4.8.42}$$

and

$$\ddot\theta_{i1}^* = \ddot\rho_{1i}(1)\tau_1^{-2} = 12\tau_1^{-2}\delta_1 - 12\tau_1^{-1}\dot\theta_{io}^* - 5\ddot\theta_{io}^* - 6\tau_1^{-2}C_{13} \tag{4.8.43}$$

for the first segment and

$$\begin{aligned} \theta_{i2} = \rho_{2i}(\tau) &= \theta_{i1}^* + (\dot\theta_{i1}^*\tau_2)\tau + (\tfrac{1}{2}\ddot\theta_{i1}^*\tau_2^2)\tau^2 \\ &\quad + (\delta_2 - \dot\theta_{i1}^*\tau_2 - \tfrac{1}{2}\ddot\theta_{i1}^*\tau_2^2)\tau^3, \end{aligned} \tag{4.8.44}$$

$$\dot\theta_{i2}^* = \dot\rho_{2i}(1)\tau_2^{-1} = 3\tau_2^{-1}\delta_2 - 2\dot\theta_{i1}^* - \tfrac{1}{2}\ddot\theta_{i1}^*\tau_2, \tag{4.8.45}$$

and

$$\ddot\theta_{i2}^* = \ddot\rho_{2i}(1)\tau_1^{-2} = 6\tau_2^{-2}\delta_2 - 6\tau_2 - 1\dot\theta_{i1}^* - 2\ddot\theta_{i1}^*\tau_2 \tag{4.8.46}$$

for the second segment, and

$$\begin{aligned} \theta_{if} = \rho_{3i}(\tau) &= \theta_{i2}^* + (\dot\theta_{i2}^*\tau_3)\tau + \tfrac{1}{2}(\ddot\theta_{2i}^*\tau_3^2)\tau^2 \\ &\quad + (5\dot\theta_{if}^*\tau_3 - 8\delta_3 - \tfrac{1}{2}\ddot\theta_{if}^*\tau_3^2 + 3\dot\theta_{i2}^*\tau_3)\tau^3 \\ &\quad + (9\delta_3 - 4\dot\theta_{i2}^*\tau_3 - \tfrac{1}{2}\ddot\theta_{i2}^*\tau_3^2 - 5\dot\theta_{if}^*\tau_3 + \tfrac{1}{2}\ddot\theta_{if}^*\tau_3^2)\tau^4. \end{aligned} \tag{4.8.47}$$

for the third segment. ■■

4.8.2 The 3–5–3 Trajectory

Following the exact procedure for deriving Equations (4.8.41)–(4.8.47), one can easily show that for the three segments of the trajectory one may derive the following expressions for joint motion trajectories:

$$\theta_{i1} = \rho_{1i}(\tau) = \theta^*_{io} + (\dot{\theta}^*_{io}\tau_1)\tau + \tfrac{1}{2}(\ddot{\theta}^*_{io}\tau_1^2)\tau^2 + (\delta_1 - \dot{\theta}^*_{io}\tau_1 - \tfrac{1}{2}\ddot{\theta}^*_{io}\tau_1^2)\tau^3, \tag{4.8.48}$$

$$\dot{\theta}^*_{i1} = \dot{\rho}_{i1}(1)\tau_1^{-1} = 3\delta_1\tau_1^{-1} - 2\dot{\theta}^*_{io} - \tfrac{1}{2}\ddot{\theta}^*_{io}\tau_1, \tag{4.8.49}$$

and

$$\ddot{\theta}^*_{i1} = \ddot{\rho}_{i1}(1)\tau_1^{-2} = 6\delta_1\tau_1^{-2} - 6\tau_1^{-1}\dot{\theta}^*_{io} - 2\ddot{\theta}^*_{io} \tag{4.8.50}$$

for the first segment,

$$\begin{aligned}\theta_{i2} = \rho_{2i}(\tau) = \theta^*_{i1} &+ (\dot{\theta}^*_{i1}\tau_2)\tau + \tfrac{1}{2}(\ddot{\theta}^*_{i1}\tau_2^2)\tau^2 \\ &+ (10\delta_2 - 6\dot{\theta}^*_{i1}\tau_2 - 4\dot{\theta}^*_{i2}\tau_2 - \tfrac{3}{2}\ddot{\theta}^*_{i1}\tau_2^2 + \tfrac{1}{2}\ddot{\theta}^*_{i2}\tau_2^2)\tau^3 \\ &+ (8\dot{\theta}^*_{i1}\tau_2 - 15\delta_2 + 7\dot{\theta}^*_{i2}\tau_2 + \tfrac{3}{2}\ddot{\theta}^*_{i1}\tau_2^2 - \ddot{\theta}^*_{i2}\tau_2^2)\tau^4 \\ &+ (-\tfrac{1}{2}\ddot{\theta}^*_{i1}\tau_2^2 + 6\delta_2 - 3\dot{\theta}^*_{i1}\tau_2 - 3\theta^*_{i2}\tau_2 - \tfrac{1}{2}\ddot{\theta}^*_{i2}\tau_2^2)\tau^5,\end{aligned} \tag{4.8.51}$$

$$\dot{\theta}^*_{i2} = \dot{\rho}_{2i}(1)\tau_2^{-1} = 3\tau_3^{-1}\delta_3 - 2\dot{\theta}^*_{if} + \tfrac{1}{2}\ddot{\theta}^*_{if}\tau_3, \tag{4.8.52}$$

and

$$\ddot{\theta}^*_{i2} = \ddot{\rho}_{2i}(1)\tau_2^{-2} = -6\tau_3^{-2}\delta_3 + 6\tau_3^{-1}\dot{\theta}^*_{if} - 2\ddot{\theta}^*_{if}, \tag{4.8.53}$$

for the second segment, and

$$\begin{aligned}\theta_{if} = \rho_{3i}(\tau) = \dot{\theta}^*_{i2} &+ (3\delta_3 - 2\dot{\theta}^*_{if}\tau_3 + \tfrac{1}{2}\ddot{\theta}^*_{if}\tau_3^2)\tau \\ &+ (3\dot{\theta}^*_{if}\tau_3 - 3\delta_3 - \ddot{\theta}^*_{if}\tau_3)^2\tau^2 + (\delta_3 - \dot{\theta}^*_{if}\tau_3 + \tfrac{1}{2}\ddot{\theta}^*_{if}\tau_3^2)\tau^3\end{aligned} \tag{4.8.54}$$

for the third segment, where

$$\delta_1 = \theta^*_{i1} - \theta^*_{io}, \qquad \delta_2 = \theta_{i2} - \theta^*_{i1}, \qquad \delta_3 = \theta^*_{if} - \theta^*_{i2}. \tag{4.8.55}$$

4.9 ADMISSIBLE MOTION TRAJECTORIES

The 4–3–4 and 3–5–3 motion trajectories that were derived in the previous two sections for the segment polynomials ρ_{1i}, ρ_{2i}, and ρ_{3i} are indeed subject to geometrical and other types of constraints, such as not being allowed to go through any inaccessible points in the joint space. If the robot in performing a motion trajectory collides with an obstacle, it may stall and draw heavy current, which may blow fuses, transistors, and other electronic components. Sustained and prolonged robotic stalls may cause serious damage to the motors and their driving circuits and controller. Needless to say, it also may seriously damage and permanently deform the linkages.

Let us consider the 4–3–4 trajectory and check for inadmissible points. All we have to do is check for the extrema of each polynomial and make sure that such extrema do not extend beyond the admissible workspace. Thus, we have to check the values of $\rho_{im}(B_1)$, $\rho_{im}(B_2)$, and $\rho_{im}(B_k)$, where B_k, $k = 1, 2, \ldots, n$, are the n real roots of the derivative of ρ_{im} with respect to time τ. Since for a 4–3–4 trajectory the highest degree polynomial is a fourth degree, it follows that there are at most three local extrema. Here we are dealing with polynomials of the form

$$\tau^3 + \alpha_1\tau^2 + \alpha_2\tau + \alpha_3 = 0, \tag{4.9.1}$$

for which we have to find the roots. If the extrema of ρ_{i1} or ρ_{i3} lie outside the admissible region, the easiest solution is to move the lift-off or set-down points until the entire segment lies in the interior of the workspace. However, it is the ρ_{i2} segment that presents problems and tends to pierce through the work envelope. In this case, we must split the mid-trajectory into two or more segments by introducing some additional admissible points. For example, having computed $\rho_{i1} - \rho_{i2} - \rho_{i3}$, we then compute the $\rho'_{i2} - \rho''_{i2} - \rho_{i3}$ trajectory using the constraints at the end of ρ_{i1}.

APPENDIX A: Inversion Procedure

Consider Equation (4.8.32), which is in the form

$$\mathbf{MC} = \boldsymbol{\delta}. \tag{A1}$$

To obtain a set of unique solutions to (A1) we intend to solve Equation (4.8.32) or (A1) by an inversion procedure such that

$$\mathbf{C} = \mathbf{M}^{-1}\boldsymbol{\delta}, \tag{A2}$$

where $\mathbf{M}^{-1}$ is the inverse of the 9×9 matrix of the coefficients. Let the matrices **A**, **B**, **C**, **D**, **E**, **F**, and **G** be defined such that

$$\mathbf{A} = \begin{bmatrix} 1 & 1 & 0 \\ 3\tau_1^{-1} & 4\tau_1^{-1} & -\tau_2^{-1} \\ 6\tau_1^{-2} & 12\tau_1^{-2} & 0 \end{bmatrix}, \qquad \mathbf{B} = \begin{bmatrix} 1 & 1 & 0 \\ 2\tau_2^{-1} & 3\tau_2^{-2} & -\tau_3^{-1} \\ 2\tau_2^{-2} & 6\tau_2^{-2} & 0 \end{bmatrix}, \tag{A3}$$

$$\mathbf{C} = \begin{bmatrix} 1 & 1 & 1 \\ 2 & 3 & 4 \\ 2 & 6 & 12 \end{bmatrix}, \qquad \mathbf{D} = \begin{bmatrix} 0 & 0 & 1 \\ 0 & 0 & \tau_2^{-1} \\ 0 & 0 & 0 \end{bmatrix}, \tag{A4}$$

$$\mathbf{E} = \begin{bmatrix} 0 & 0 & 0 \\ 0 & 0 & 0 \\ -2\tau_2^{-2} & 0 & 0 \end{bmatrix}, \qquad \mathbf{F} = \begin{bmatrix} 0 & 0 & 0 \\ 0 & 0 & 0 \\ -2\tau_3^{-2} & 0 & 0 \end{bmatrix}, \tag{A5}$$

and

$$\mathbf{G} = \begin{bmatrix} 0 & 0 & 1 \\ 0 & 0 & 1 \\ 0 & 0 & 0 \end{bmatrix}. \tag{A6}$$

Note that the matrix of the coefficients, **M**, has the form

$$\mathbf{M} = \begin{bmatrix} A & E & 0 \\ D & B & F \\ 0 & G & C \end{bmatrix}. \tag{A7}$$

The inverse of the above matrix, after a lengthy computation (Appendix B), is found to be

$$\mathbf{M}^{-1} = \begin{bmatrix} A^* & B^* & C^* \\ D^* & E^* & F^* \\ G^* & H & I^* \end{bmatrix}, \tag{A8}$$

where

$$A^* = [A^{-1} + A^{-1}EH^{-1}DA^{-1} + A^{-1}EH^{-1}FI^{-1}GH^{-1}DA^{-1}], \tag{A9}$$

$$B^* = -[A^{-1}EH^{-1} + A^{-1}EH^{-1}FI^{-1}GH^{-1}], \tag{A10}$$

$$C^* = A^{-1}EH^{-1}FI^{-1}, \qquad F^* = -H^{-1}FI^{-1}, \qquad I^* = I^{-1}, \tag{A11}$$

$$D^* = -[H^{-1}DA^{-1} + H^{-1}FI^{-1}GH^{-1}DA^{-1}], \qquad G^* = [I^{-1}GH^{-1}DA^{-1}], \tag{A12}$$

$$E^* = [H^{-1} + H^{-1}FI^{-1}GH^{-1}], \qquad H^* = [I^{-1}GH^{-1}], \tag{A13}$$

and

$$H = [-DA^{-1}E + B], \qquad I = [-GH^{-1}F + C]. \tag{A14}$$

After a lengthy computation and direct substitution of the given matrices and their inverses, an explicit expression for the matrix $\mathbf{M}^{-1}$ can be obtained as depicted below. Let

$$a = \frac{3\tau_1 + 9\tau_2 - \tau_1\tau_2 - 6\tau_2^2}{2(\tau_1 + 2\tau_2)}, \tag{A15}$$

$$b = \frac{\tau_3(\tau_1 + 6\tau_2 - 3)}{\tau_2(\tau_1 + 2\tau_2)}, \tag{A16}$$

$$e = \frac{\tau_2^2}{\tau_3(\tau_1 + 2\tau_2)(2a + \tau_3)}, \tag{A17}$$

$$l = \frac{\tau_1}{\tau_1 + 2\tau_2} + \tau_1 eb, \tag{A18}$$

$$n = \frac{1 + eb(\tau_1 + 2\tau_2)}{\tau_1 + 2\tau_2}, \tag{A19}$$

$$h = n\tau_2 - e\tau_3, \tag{A20}$$

$$m = \frac{\tau_2}{\tau_1 + 2\tau_2} + eb(\tau_1 + 3\tau_2), \tag{A21}$$

$$k = \frac{\tau_3(b\tau_2 - \tau_3)}{2a + \tau_3}, \tag{A22}$$

$$f = m\tau_2 - \tau_3 e(\tau_1 + 3\tau_2), \tag{A23}$$

and

$$z = l - \left(\frac{\tau_1\tau_3}{\tau_2}\right)e. \tag{A24}$$

Matrix $\mathbf{M}^{-1}$ is then given by the matrix equation (A25).

$$\mathbf{M}^{-1} = \begin{bmatrix}
2(1+Z) & -\tau_1 Z & -\frac{1}{6}\tau_1^2(1-Z) & -\tau_1\tau_2^{-1}l & \tau_1^2\tau_2^{-1}\tau_3 e & \frac{1}{6}\tau_1^2\tau_2^{-1}\tau_3^2 e & 2\tau_1^2\tau_2^{-1}e & -\tau_1^2\tau_2^{-1}e & \frac{1}{6}\tau_1^2\tau_2^{-1}e \\
-(1+2Z) & \tau_1 Z & \frac{1}{6}\tau_1^2(1-Z) & \tau_1\tau_2^{-1}l & -\tau_1^2\tau_2^{-1}\tau_3 e & -\frac{1}{6}\tau_1^2\tau_2^{-1}\tau_3^2 e & -2\tau_1^2\tau_2^{-1}e & \tau_1^2\tau_2^{-1}e & -\frac{1}{6}\tau_1^2\tau_2^{-1}e \\
2\tau_2\tau_1^{-1}(1-Z) & -\tau_2(1-Z) & \frac{1}{6}\tau_2\tau_1(1-Z) & l & -\tau_1\tau_3 e & -\frac{1}{6}\tau_1\tau_3^2 e & -2\tau_1 e & \tau_1 e & -\frac{1}{6}\tau_1 e \\
-6\tau_1^{-1}\tau_2 h & 3\tau_2 h & -\frac{1}{2}\tau_1\tau_2 h & 3\tau_2 n & -3\tau_2\tau_3 e & -\frac{1}{2}\tau_2\tau_3^2 e & -6\tau_2 e & 3\tau_2 e & -\frac{1}{2}\tau_2 e \\
2\tau_1^{-1}f & -f & \frac{1}{6}\tau_1 f & -m & e\tau_3(\tau_1+3\tau_2) & \frac{1}{6}e\tau_3^2(\tau_1+3\tau_2) & 2e(\tau_1+3\tau_2) & -e(\tau_1+3\tau_2) & \frac{1}{6}e(\tau_1+3\tau_2) \\
-2\tau_1^{-1}k & k & -\frac{1}{6}\tau_1 k & b\tau_3(2a+\tau_3)^{-1} & -\tau_3^2(2a+\tau_3)^{-1} & \frac{1}{3}a\tau_3^2(2a+\tau_3)^{-1} & 4a(2a+\tau_3)^{-1} & -2a(2a+\tau_3)^{-1} & \frac{1}{3}a(2a+\tau_3)^{-1} \\
6\tau_1^{-1} & -3k & \frac{1}{2}\tau_1 k & -3\tau_3 b(2a+\tau_3)^{-1} & 3\tau_3^2(2a+\tau_3)^{-1} & -a\tau_3^2(2a+\tau_3)^{-1} & 6\tau_3(2a+\tau_3)^{-1} & -3\tau_3(2a+\tau_3)^{-1} & \frac{1}{2}\tau_3(2a+\tau_3)^{-1} \\
-6\tau_1^{-1}k & 3k & -\frac{1}{2}\tau_1 k & 3\tau_3 b(2a+\tau_3)^{-1} & -3\tau_3^2(2a+\tau_3)^{-1} & a\tau_3^2(2a+\tau_3)^{-1} & -4(a+2\tau_3)(2a+\tau_3)^{-1} & \left(\dfrac{4a+5\tau_3}{2a+\tau_3}\right)^{-1} & -(a+\tau_3)(2a+\tau_3)^{-1} \\
2\tau_1^{-1}k & -k & \frac{1}{6}\tau_1 k & -\tau_3 b(2a+\tau_3)^{-1} & \tau_3^2(2a+\tau_3)^{-1} & -\dfrac{a\tau_3^2}{3(2a+\tau_3)} & \dfrac{2a+3\tau_3}{2a+\tau_3} & -\left(\dfrac{2a+2\tau_3}{2a+\tau_3}\right) & \dfrac{4a+3\tau_3}{6(2a+\tau_3)}
\end{bmatrix}. \tag{A25}$$

Now we can compute the various coefficients for each segment of a trajectory. For any given set of segment time intervals—τ_1, τ_2, τ_3—and any set of data—θ_{i0}^*, θ_{i1}^*, θ_{i2}^*, θ_{i3}^*, $\dot{\theta}_{i0}^*$, $\dot{\theta}_{i3}^*$, $\ddot{\theta}_{i0}^*$, $\ddot{\theta}_{i3}^*$—a complete set of segment trajectories and the associated coefficients can be calculated by using the matrix equation (A2) and a simple computer program (see Appendix C). In the next section we present a set of tables pertaining to various sets of data of interest.

Numerical Results

For clarity we have displayed only the numerical values associated with a six-axis robot and a given segment trajectory. Let $P(i, m)$ denote the mth segment of the trajectory polynomial of the ith joint. For example, $P(1, 1)$ as given in Table A2 implies that the trajectory polynomial pertaining to the first motion segment of the first joint is represented by

$$\theta_{11}(t) = 20 + 40\tau + 600\tau^2 - 1100.356\tau^3 + 470.36\tau^4,$$

provided the time segments τ_1, τ_2, and τ_3 and the initial, via, and final data, θ_{i0}^*, θ_{i1}^*, θ_{i2}^*, θ_{i3}^*, $\dot{\theta}_{i0}^*$, and $\dot{\theta}_{i3}^*$, are as given in Table A1. For additional numerical data refer to Tables A3–A30.

TABLE A1 TIME SEGMENTS AND MOTION DATA FOR $\tau_1 = 20$, $\tau_2 = 30$, AND $\tau_3 = 10$

	Joint					
Data	1	2	3	4	5	6
θ^*_{i0}	20	30	10	40	3	20
θ^*_{i1}	30	4	2	13	40	20
θ^*_{i2}	20	30	50	29	10	20
θ^*_{i3}	60	70	80	90	10	20
$\dot{\theta}^*_{i0}$	2	4	1	2	3	3
$\dot{\theta}^*_{i3}$	3	4	2	1	3	2
$\ddot{\theta}^*_{i0}$	3	2	1	5	4	3
$\ddot{\theta}^*_{i3}$	3	2	1	1	2	3

TABLE A2 JOINT SEGMENT TRAJECTORY COEFFICIENTS FOR THE DATA OF TABLE A1

Trajectory of joint 1
$P(1,1) = 20, 40, 600, -1100.356, 470,3559$
$P(1,2) = 30, -269.4661, 272.4025, -12.93622$
$P(1,3) = 20, 91.34838, 25.95483, -195.9548, 118.6516$

Trajectory of joint 2
$P(2,1) = 30, 80, 400, -914.3353, 408.3355$
$P(2,2) = 4, -344.4969, 240.7639, 129.7331$
$P(2,3) = 30, 50.00136, 69.9959, -169.9959, 89.99862$

Trajectory of joint 3
$P(3,1) = 10, 20, 200, -408.7557, 180.7557$
$P(3,2) = 2, -124.8666, 131.1006, 41.76594$
$P(3,3) = 50, 47.17043, 28.48872, -98.48872, 52.82957$

Trajectory of joint 4
$P(4,1) = 40, 40, 1000, -1861.636, 794.636$
$P(4,2) = 13, -549.5463, 411.5421, 154.0041$
$P(4,3) = 29, 96.3128, 97.06154, -249.0615, 116.6872$

Trajectory of joint 5
$P(5,1) = 3, 60, 800, -1398.862, 575.8625$
$P(5,2) = 40, -349.7065, 131.8214, 187.8851$
$P(5,3) = 10, -22.42505, 77.2751, -117.2751, 62.42505$

Trajectory of joint 6
$P(6,1) = 20, 60, 600, -1145.544, 485.5442$
$P(6,2) = 20, -351.6838, 172.423, 179.2608$
$P(6,3) = 20, 3.696091, 78.91169, -188.9117, 106.3039$

TABLE A3 TIME SEGMENTS AND MOTION DATA FOR $\tau_1 = 20$, $\tau_2 = 30$, AND $\tau_3 = 10$ WITH INCREASED FINAL ACCELERATION

	Joint					
Data	1	2	3	4	5	6
θ^*_{i0}	20	30	10	40	3	20
θ^*_{i1}	30	4	2	13	40	20
θ^*_{i2}	20	30	50	29	10	20
θ^*_{i3}	60	70	80	90	10	20
$\dot{\theta}^*_{i0}$	2	4	1	2	3	3
$\dot{\theta}^*_{i3}$	3	4	2	1	3	2
$\ddot{\theta}^*_{i0}$	3	2	1	5	4	3
$\ddot{\theta}^*_{i3}$	30	20	10	10	20	30

TABLE A4 JOINT SEGMENT TRAJECTORY COEFFICIENTS FOR THE DATA OF TABLE A3

Trajectory of joint 1
$P(1,1) = 20, 40, 600, -1211.239, 581.2389$
$P(1,2) = 30, -103.1416, 1020.862, -927.7206$
$P(1,3) = 20, 615.2703, -195.8111, -1324.189, 944.7296$

Trajectory of joint 2
$P(2,1) = 30, 80, 400, -988.2572, 482.2574$
$P(2,2) = 4, -233.614, 739.7371, -480.1232$
$P(2,3) = 30, 399.2827, -77.84804, -922.152, 640.7173$

Trajectory of joint 3
$P(3,1) = 10, 20, 200, -445.7166, 217.7167$
$P(3,2) = 2, -69.4251, 380.5872, -263.1622$
$P(3,3) = 50, 221.8111, -45.43325, -474.5668, 328.1889$

Trajectory of joint 4
$P(4,1) = 40, 40, 1000, -1898.597, 831.597$
$P(4,2) = 13, -494.1049, 661.0287, -150.924$
$P(4,3) = 29, 270.9534, 23.13958, -625.1395, 392.0465$

Trajectory of joint 5
$P(5,1) = 3, 60, 800, -1472.784, 649.7844$
$P(5,2) = 40, -238.8236, 630.7946, -421.9712$
$P(5,3) = 10, 326.8563, -70.56883, -869.4311, 613.1437$

Trajectory of joint 6
$P(6,1) = 20, 60, 600, -1256.427, 596.4271$
$P(6,2) = 20, -185.3594, 920.8829, -735.5236$
$P(6,3) = 20, 527.6181, -142.8542, -1317.146, 932.3819$

TABLE A5 TIME SEGMENTS AND MOTION DATA FOR $\tau_1 = 20$, $\tau_2 = 30$, AND $\tau_3 = 10$ WITH INCREASED INITIAL ACCELERATION

	Joint					
Data	1	2	3	4	5	6
θ^*_{i0}	20	30	10	40	3	20
θ^*_{i1}	30	4	2	13	40	20
θ^*_{i2}	20	30	50	29	10	20
θ^*_{i3}	60	70	80	90	10	20
$\dot{\theta}^*_{i0}$	2	4	1	2	3	3
$\dot{\theta}^*_{i3}$	3	4	2	1	3	2
$\ddot{\theta}^*_{i0}$	30	20	10	50	40	30
$\ddot{\theta}^*_{i3}$	3	2	1	1	2	3

TABLE A6 JOINT SEGMENT TRAJECTORY COEFFICIENTS FOR THE DATA OF TABLE A5

Trajectory of joint 1
$P(1,1) = 20, 40, 6000, -10226.02, 4196.024$
$P(1,2) = 30, -2780.966, 1120.657, 1650.309$
$P(1,3) = 20, -124.8734, 674.6199, -844.6199, 334.8734$

Trajectory of joint 2
$P(2,1) = 30, 80, 4000, -6998.113, 2892.114$
$P(2,2) = 4, -2018.83, 806.267, 1238.563$
$P(2,3) = 30, -94.14647, 502.4393, -602.4392, 234.1465$

Trajectory of joint 3
$P(3,1) = 10, 20, 2000, -3450.645, 1422.645$
$P(3,2) = 2, -962.0332, 413.8522, 596.181$
$P(3,3) = 50, -24.9035, 244.7104, -314.7104, 124.9035$

Trajectory of joint 4
$P(4,1) = 40, 40, 10000, -17071.08, 7004.082$
$P(4,2) = 13, -4735.378, 1825.3, 2926.077$
$P(4,3) = 29, -264.0568, 1178.17, -1330.17, 477.0568$

Trajectory of joint 5
$P(5,1) = 3, 60, 8000, -13566.42, 5543.42$
$P(5,2) = 40, -3698.372, 1262.828, 2405.545$
$P(5,3) = 10, -310.7207, 942.162, -982.162, 350.7207$

Trajectory of joint 6
$P(6,1) = 20, 60, 6000, -10271.21, 4211.212$
$P(6,2) = 20, -2863.183, 1020.678, 1842.506$
$P(6,3) = 20, -212.5257, 727.5769, -837.5769, 322.5257$

TABLE A7 TIME SEGMENTS AND MOTION DATA FOR $\tau_1 = 20$, $\tau_2 = 30$, AND $\tau_3 = 10$ WITH INCREASED FINAL VELOCITY

	Joint					
Data	1	2	3	4	5	6
θ_{i0}^*	20	30	10	40	3	20
θ_{i1}^*	30	4	2	13	40	20
θ_{i2}^*	20	30	50	29	10	20
θ_{i3}^*	60	70	80	90	10	20
$\dot{\theta}_{i0}^*$	2	4	1	2	3	3
$\dot{\theta}_{i3}^*$	30	40	20	10	30	20
$\ddot{\theta}_{i0}^*$	3	2	1	5	4	3
$\ddot{\theta}_{i3}^*$	3	2	1	1	2	3

TABLE A8 JOINT SEGMENT TRAJECTORY COEFFICIENTS FOR THE DATA OF TABLE A7

Trajectory of joint 1
$P(1,1) = 20, 40, 600, -1033.826, 403.8262$
$P(1,2) = 30, -369.2608, -176.6735, 535.9343$
$P(1,3) = 20, -223.0048, 159.0144, 210.9856, -106.9952$

Trajectory of joint 2
$P(2,1) = 30, 80, 400, -825.629, 319.6291$
$P(2,2) = 4, -477.5565, -358.0041, 861.5605$
$P(2,3) = 30, -369.1362, 247.4087, 372.5914, -210.8638$

Trajectory of joint 3
$P(3,1) = 10, 20, 200, -364.4025, 136.4025$
$P(3,2) = 2, -191.3964, -168.2834, 407.6797$
$P(3,3) = 50, -162.3984, 117.1951, 172.8049, -97.60163$

Trajectory of joint 4
$P(4,1) = 40, 40, 1000, -1839.459, 772.4594$
$P(4,2) = 13, -582.8112, 261.8501, 336.961$
$P(4,3) = 29, -8.471598, 141.4147, -113.4147, 41.4716$

Trajectory of joint 5
$P(5,1) = 3, 60, 800, -1332.333, 509.3327$
$P(5,2) = 40, -449.5012, -317.2546, 736.7557$
$P(5,3) = 10, -336.7782, 210.3347, 289.6654, -163.2218$

Trajectory of joint 6
$P(6,1) = 20, 60, 600, -1101.191, 441.191$
$P(6,2) = 20, -418.2136, -126.961, 545.1746$
$P(6,3) = 20, -205.8727, 167.6181, 82.38196, -44.1273$

TABLE A9 TIME SEGMENTS AND MOTION DATA FOR $\tau_1 = 20$, $\tau_2 = 30$, AND $\tau_3 = 10$ WITH INCREASED INITIAL VELOCITY

	Joint					
Data	1	2	3	4	5	6
θ_{i0}^*	20	30	10	40	3	20
θ_{i1}^*	30	4	2	13	40	20
θ_{i2}^*	20	30	50	29	10	20
θ_{i3}^*	60	70	80	90	10	20
$\dot{\theta}_{i0}^*$	20	40	10	20	30	30
$\dot{\theta}_{i3}^*$	3	4	2	1	3	2
$\ddot{\theta}_{i0}^*$	3	2	1	5	4	3
$\ddot{\theta}_{i3}^*$	3	2	1	1	2	3

TABLE A10 JOINT SEGMENT TRAJECTORY COEFFICIENTS FOR THE DATA OF TABLE A9

Trajectory of joint 1
$P(1,1) = 20, 400, 600, -1845.489, 855.4893$
$P(1,2) = 30, -771.766, 442.0534, 319.7126$
$P(1,3) = 20, 48.10403, 155.6879, -325.6879, 161.896$

Trajectory of joint 2
$P(2,1) = 30, 800, 400, -2404.602, 1178.602$
$P(2,2) = 4, -1349.096, 580.0658, 795.0309$
$P(2,3) = 30, -36.48736, 329.462, -429.4621, 176.4874$

Trajectory of joint 3
$P(3,1) = 10, 200, 200, -781.3223, 373.3224$
$P(3,2) = 2, -376.0165. 215.9261, 208.0904$
$P(3,3) = 50, 25.54825, 93.35523, -163.3552, 74.45175$

Trajectory of joint 4
$P(4,1) = 40, 400, 1000, -2606.769, 1179.769$
$P(4,2) = 13, -1051.846, 581.193, 486.6532$
$P(4,3) = 29, 53.06843, 226.7946, -378.7946, 159.9316$

Trajectory of joint 5
$P(5,1) = 3, 600, 800, -2516.563, 1153.563$
$P(5,2) = 40, -1103.156, 386.2978, 686.8585$
$P(5,3) = 10, -87.29159, 271.8747, -311.8747, 127.2916$

Trajectory of joint 6
$P(6,1) = 20, 600, 600, -2263.246, 1063.244$
$P(6,2) = 20, -1105.134, 426.8994, 678.2341$
$P(6,3) = 20, -61.17042, 273.5113, -383.5113, 171.1704$

TABLE A11 TIME SEGMENTS AND MOTION DATA FOR $\tau_1 = 200$, $\tau_2 = 300$, AND $\tau_3 = 100$ WITH INCREASED TIME

	Joint					
Data	1	2	3	4	5	6
θ_{i0}^*	20	30	10	40	3	20
θ_{i1}^*	30	4	2	13	40	20
θ_{i2}^*	20	30	50	29	10	20
θ_{i3}^*	60	70	80	90	10	20
$\dot{\theta}_{i0}^*$	2	4	1	2	3	3
$\dot{\theta}_{i3}^*$	3	4	2	1	3	2
$\ddot{\theta}_{i0}^*$	3	2	1	5	4	3
$\ddot{\theta}_{i3}^*$	3	2	1	1	2	3

TABLE A12 JOINT SEGMENT TRAJECTORY COEFFICIENTS FOR THE DATA OF TABLE A11

Trajectory of joint 1
$P(1,1) = 20, 400, 60000, -103444.4, 43054.43$
$P(1,2) = 30, -26573.35, 17984.88, 8578.484$
$P(1,3) = 20, 3160.729, 4857.812, -19337.81, 11359.27$

Trajectory of joint 2
$P(2,1) = 30, 800, 40000, -70099.48, 29273.48$
$P(2,2) = 4, -18606.78, 12020.47, 6612.318$
$P(2,3) = 30, 1833.43, 3539.711, -12819.71, 7486.571$

Trajectory of joint 3
$P(3,1) = 10, 200, 20000, -34627.43, 14419.44$
$P(3,2) = 2, -9006.838, 5927.22, 3127.619$
$P(3,3) = 50, 959.6271, 1701.119, -6361.119, 3730.373$

Trajectory of joint 4
$P(4,1) = 40, 400, 100000, -170488.8, 70061.82$
$P(4,2) = 13, -46228.77, 20034.98, 26209.81$
$P(4,3) = 29, -1965.569, 10962.71, -15884.71, 6948.57$

Trajectory of joint 5
$P(5,1) = 3, 600, 80000, -137238.1, 56675.13$
$P(5,2) = 40, -36620.8, 18756.86, 17833.96$
$P(5,3) = 100, 357.085, 8028.743, -17428.74, 9042.916$

Trajectory of joint 6
$P(6,1) = 200, 600, 60000, -103885.6, 43285.58$
$P(6,2) = 20, -26871.63, 18127.67, 8743.953$
$P(6,3) = 20, 3157.054, 4928.836, -19528.84, 11442.95$

TABLE A13 TIME SEGMENTS AND MOTION DATA FOR $\tau_1 = 200$, $\tau_2 = 300$, AND $\tau_3 = 100$ WITH INCREASED FINAL ACCELERATION

	Joint					
Data	1	2	3	4	5	6
θ^*_{i0}	20	30	10	40	3	20
θ^*_{i1}	30	4	2	13	40	20
θ^*_{i2}	20	30	50	29	10	20
θ^*_{i3}	60	70	80	90	10	20
$\dot{\theta}^*_{i0}$	2	4	1	2	3	3
$\dot{\theta}^*_{i3}$	3	4	2	1	3	2
$\ddot{\theta}^*_{i0}$	3	2	1	5	4	3
$\ddot{\theta}^*_{i3}$	30	20	10	10	20	30

TABLE A14 JOINT SEGMENT TRAJECTORY COEFFICIENTS FOR THE DATA OF TABLE A13

Trajectory of joint 1
$P(1,1) = 20, 400, 60000, -113895.4, 53505.37$
$P(1,2) = 30, -10896.95, 88528.71, -77641.77$
$P(1,3) = 20, 55128.03, -16044.07, -133435.9, 94391.98$

Trajectory of joint 2
$P(2,1) = 30, 800, 40000, -77066.78, 36240.77$
$P(2,2) = 4, -8155.843, 59049.7, -50867.85$
$P(2,3) = 30, 36478.29, -10394.88, -88885.12, 62841.71$

Trajectory of joint 3
$P(3,1) = 10, 200, 20000, -38111.08, 17903.09$
$P(3,2) = 2, -3781.369, 29441.83, -25612.46$
$P(3,3) = 50, 18282.06, -5266.174, -44393.83, 31407.94$

Trajectory of joint 4
$P(4,1) = 40, 400, 100000, -173972.4, 73545.47$
$P(4,2) = 13, -41003.3, 43549.59, -2530.274$
$P(4,3) = 29, 15356.86, 3995.416, -53917.42, 34626.14$

Trajectory of joint 5
$P(5,1) = 3, 600, 80000, -144205.4, 63642.42$
$P(5,2) = 40, -26169.86, 65786.08, -39646.2$
$P(5,3) = 10, 35001.95, -5905.843, -93494.16, 64398.06$

Trajectory of joint 6
$P(6,1) = 20, 600, 60000, -114336.5, 53736.52$
$P(6,2) = 20, -11195.22, 88671.51, -77476.3$
$P(6,3) = 20, 55124.35, -15973.04, -133627, 94475.65$

TABLE A15 TIME SEGMENTS AND MOTION DATA FOR $\tau_1 = 200$, $\tau_2 = 300$, AND $\tau_3 = 100$ WITH INCREASED INITIAL ACCELERATION

	Joint					
Data	1	2	3	4	5	6
θ^*_{i0}	20	30	10	40	3	20
θ^*_{i1}	30	4	2	13	40	20
θ^*_{i2}	20	30	50	29	10	20
θ^*_{i3}	60	70	80	90	10	20
$\dot{\theta}^*_{i0}$	2	4	1	2	3	3
$\dot{\theta}^*_{i3}$	3	4	2	1	3	2
$\ddot{\theta}^*_{i0}$	30	20	10	50	40	30
$\ddot{\theta}^*_{i3}$	3	2	1	1	2	3

TABLE A16 JOINT SEGMENT TRAJECTORY COEFFICIENTS FOR THE DATA OF TABLE A15

Trajectory of joint 1
$P(1,1) = 20, 400, 600000, -1017170, 416780$
$P(1,2) = 30, -275985, 110632.4, 165342.6$
$P(1,3) = 20, -17688.89, 67406.68, -81886.69, 32208.89$

Trajectory of joint 2
$P(2,1) = 30, 800, 400000, -679249.9, 278423.9$
$P(2,2) = 4, -184881.1, 73785.51, 111121.8$
$P(2,3) = 30, -12066.32, 45238.95, -54518.95, 21386.33$

Trajectory of joint 3
$P(3,1) = 10, 200, 200000, -339202.6, 138994.6$
$P(3,2) = 2, -92144, 36809.73, 55382.32$
$P(3,3) = 50, -5990.247, 22550.73, -27210.73, 10680.25$

Trajectory of joint 4
$P(4,1) = 40, 400, 1000000, -1693365, 692937.8$
$P(4,2) = 13, -461914.8, 174447.5, 287483.3$
$P(4,3) = 29, -36714.95, 115210.9, -120132.9, 41697.94$

Trajectory of joint 5
$P(5,1) = 3, 600, 800000, -1355539, 554975.9$
$P(5,2) = 40, -369196.5, 142286.9, 226852.8$
$P(5,3) = 10, -27442.42, 91427.19, -100827.2, 36842.42$

Trajectory of joint 6
$P(6,1) = 20, 600, 600000, -1017611, 417011.2$
$P(6,2) = 20, -276283.1, 110775.2, 165508.1$
$P(6,3) = 20, -17692.57, 67477.71, -82077.71, 32292.57$

TABLE A17 TIME SEGMENTS AND MOTION DATA FOR $\tau_1 = 200$, $\tau_2 = 300$, AND $\tau_3 = 100$ WITH INCREASED FINAL VELOCITY

	Joint					
Data	1	2	3	4	5	6
θ^*_{i0}	20	30	10	40	3	20
θ^*_{i1}	30	4	2	13	40	20
θ^*_{i2}	20	30	50	29	10	20
θ^*_{i3}	60	70	80	90	10	20
$\dot{\theta}^*_{i0}$	2	4	1	2	3	3
$\dot{\theta}^*_{i3}$	30	40	20	10	30	20
$\ddot{\theta}^*_{i0}$	3	2	1	5	4	3
$\ddot{\theta}^*_{i3}$	3	2	1	1	2	3

TABLE A18 JOINT SEGMENT TRAJECTORY COEFFICIENTS FOR THE DATA OF TABLE A17

Trajectory of joint 1
$P(1,1) = 20, 400, 60000, -102817.4, 42427.37$
$P(1,2) = 30, -27513.94, 13752.24, 13751.7$
$P(1,3) = 20, 42.69092, 6111.925, -15191.93, 9077.309$

Trajectory of joint 2
$P(2,1) = 30, 800, 40000, -69263.41, 28437.41$
$P(2,2) = 4, -19860.9, 6376.966, 13509.94$
$P(2,3) = 30, -2323.954, 5211.861, -7291.861, 4443.954$

Trajectory of joint 3
$P(3,1) = 10, 200, 20000, -34209.4\ 14001.4$
$P(3,2) = 2, -9633.894, 3105.466, 6576.428$
$P(3,3) = 50, -1119.065, 2537.194, -3597.194, 2209.065$

Trajectory of joint 4
$P(4,1) = 40, 400, 100000, -170279.8, 69852.81$
$P(4,2) = 13, -46542.3, 18624.1, 27934.22$
$P(4,3) = 29, -3004.915, 11380.75, -14502.75, 6187.916$

Trajectory of joint 5
$P(5,1) = 3, 600, 80000, -136611.1, 56048.07$
$P(5,2) = 40, -37561.38, 14524.23, 23007.18$
$P(5,3) = 10, -2760.953, 9282{,}856, -13282.86, 6760.953$

Trajectory of joint 6
$P(6,1) = 20, 600, 60000, -103467.6, 42867.55$
$P(6,2) = 20, -27498.68, 15305.92, 12192.76$
$P(6,3) = 20, 1078.363, 5764.911, -5764.91, 9921.638$

TABLE A19 TIME SEGMENTS AND MOTION DATA FOR $\tau_1 = 200$, $\tau_2 = 300$, AND $\tau_3 = 100$ WITH INCREASED INITIAL VELOCITY

	Joint					
Data	1	2	3	4	5	6
θ^*_{i0}	20	30	10	40	3	20
θ^*_{i1}	30	4	2	13	40	20
θ^*_{i2}	20	30	50	29	10	20
θ^*_{i3}	60	70	80	90	10	20
$\dot{\theta}^*_{i0}$	20	40	10	20	30	30
$\dot{\theta}^*_{i3}$	3	4	2	1	3	2
$\ddot{\theta}^*_{i0}$	3	2	1	5	4	3
$\ddot{\theta}^*_{i3}$	3	2	1	1	2	3

TABLE A20 JOINT SEGMENT TRAJECTORY COEFFICIENTS FOR THE DATA OF TABLE A19

Trajectory of joint 1
$P(1,1) = 20, 4000, 60000, -110918.9, 46928.94$
$P(1,2) = 30, -31561.59, 19837.83, 11713.77$
$P(1,3) = 20, 2743.736, 6108.791, -20588.79, 11776.26$

Trajectory of joint 2
$P(2,1) = 30, 8000, 40000, -85048.49, 37022.5$
$P(2,2) = 4, -28583.24, 15726.37, 12882.87$
$P(2,3) = 30, 999.4451, 6041.666, -15321.67, 8320.556$

Trajectory of joint 3
$P(3,1) = 10, 2000, 20000, -38364.69, 16356.7$
$P(3,2) = 2, -11500.96, 6853.695, 4695.26$
$P(3,3) = 50, 751.1305, 2326.609, -6986.609, 3938.87$

Trajectory of joint 4
$P(4,1) = 40, 4000, 100000, -177963.3, 73936.34$
$P(4,2) = 13, -51217.02, 21887.93, 29345.09$
$P(4,3) = 29, -2382.564, 12213.68, -17135.68, 7365.564$

Trajectory of joint 5
$P(5,1) = 3, 6000, 80000, -148449.9, 62486.9$
$P(5,2) = 40, -44103.14, 21536.29, 22536.88$
$P(5,3) = 10, -268.4033, 9905.212, -19305.21, 9668.404$

Trajectory of joint 6
$P(6,1) = 20, 6000, 60000, -115097.4, 49097.35$
$P(6,2) = 20, -34353.97, 20907.1, 13446.88$
$P(6,3) = 20, 2531.566, 6805.301, -21405.3, 12068.43$

TABLE A21 TIME SEGMENTS AND MOTION DATA FOR $\tau_1 = 2$, $\tau_2 = 3$, AND $\tau_3 = 1$ WITH REDUCED TIME

	Joint					
Data	1	2	3	4	5	6
θ^*_{i0}	20	30	10	40	3	20
θ^*_{i1}	30	4	2	13	40	20
θ^*_{i2}	20	30	50	29	10	20
θ^*_{i3}	60	70	80	90	10	20
$\dot{\theta}^*_{i0}$	2	4	1	2	3	3
$\dot{\theta}^*_{i3}$	3	4	2	1	3	2
$\ddot{\theta}^*_{i0}$	3	2	1	5	4	3
$\ddot{\theta}^*_{i3}$	3	2	1	1	2	3

TABLE A22 JOINT SEGMENT TRAJECTORY COEFFICIENTS FOR THE DATA OF TABLE A21

Trajectory of joint 1
$P(1,1) = 20, 4, 6, -51.22807, 51.22807$
$P(1,2) = 30, 100.8421, 359.2895, -470.1316$
$P(1,3) = 20, 116.4298, -116.7895, 41.28948, -.9298246$

Trajectory of joint 2
$P(2,1) = 30, 8, 4, -109.1228, 71.1228$
$P(2,2) = 4, -40.31579, 232.5789, -166.2632$
$P(2,3) = 30, 86.19298, -29.57894, -43.42106, 26.80702$

Trajectory of joint 3
$P(3,1) = 10, 2, 2, -44.63158, 32.63158$
$P(3,2) = 2, 3.947374, 143.7632, -99.71054$
$P(3,3) = 50, 63.92105, -17.26316, -39.23684, 22.57895$

Trajectory of joint 4
$P(4,1) = 40, 4, 10, -144.2456, 103.2456$
$P(4,2) = 13, 6.368439, 442.6579, -433.0263$
$P(4,3) = 29, 152.886, -95.15788, -25.34211, 28.61404$

Trajectory of joint 5
$P(5,1) = 3, 6, 8, 35.05264, -12.05263$
$P(5,2) = 40, 118.421, 91.89473, -240.3158$
$P(5,3) = 10, 20.63158, -69.89474, 74.89474, -25.63158$

Trajectory of joint 6
$P(6,1) = 20, 6, 6, -20.21053, 8.210526$
$P(6,2) = 20, -14.68421, -12.07895, 26.76316$
$P(6,3) = 20, -4.026316, 7.578947, -5.078947, 1.526316$

TABLE A23 TIME SEGMENTS AND MOTION DATA FOR $\tau_1 = 2$, $\tau_2 = 3$, AND $\tau_3 = 1$ WITH INCREASED FINAL ACCELERATION

	Joint					
Data	1	2	3	4	5	6
θ^*_{i0}	20	30	10	40	3	20
θ^*_{i1}	30	4	2	13	40	20
θ^*_{i2}	20	30	50	29	10	20
θ^*_{i3}	60	70	80	90	10	20
$\dot{\theta}^*_{i0}$	2	4	1	2	3	3
$\dot{\theta}^*_{i3}$	3	4	2	1	3	2
$\ddot{\theta}^*_{i0}$	3	2	1	5	4	3
$\ddot{\theta}^*_{i3}$	30	20	10	10	20	30

TABLE A24 JOINT SEGMENT TRAJECTORY COEFFICIENTS FOR THE DATA OF TABLE A23

Trajectory of joint 1
$P(1,1) = 20, 4, 6, -54.07018, 54.07018$
$P(1,2) = 30, 105.1053, 378.4737, -493.579$
$P(1,3) = 20, 122.8246, -122.4737, 33.47368, 6.175439$

Trajectory of joint 2
$P(2,1) = 30, 8, 4, -111.0175, 73.01754$
$P(2,2) = 4, -37.47368, 245.3684, -181.8948$
$P(2,3) = 30, 90.45613, -33.36841, -48.63159, 31.54386$

Trajectory of joint 3
$P(3,1) = 10, 2, 2, -45.57895, 33.57895$
$P(3,2) = 2, 5.368427, 150.1579, -107.5263$
$P(3,3) = 50, 66.05263, -19.1579, 41.8421, 24.94737$

Trajectory of joint 4
$P(4,1) = 40, 4, 10, -145.193, 104.193$
$P(4,2) = 13, 7.789492, 449.0526, -440.8421$
$P(4,3) = 29, 155.0175, -97.05262, -27.94737, 30.98246$

Trajectory of joint 5
$P(5,1) = 3, 6, 8, 33.1579, -10.1579$
$P(5,2) = 40, 121.2632, 104.6842, -255.9474$
$P(5,3) = 10, 24.89474, -73.68421, 69.68421, -20.89474$

Trajectory of joint 6
$P(6,1) = 20, 6, 6, -23.05263, 11.05263$
$P(6,2) = 20, -10.42105, 7.105265, 3.31579$
$P(6,3) = 20, 2.368421, 1.894736, -12.89474, 8.631579$

TABLE A25 TIME SEGMENTS AND MOTION DATA FOR $\tau_1 = 2$, $\tau_2 = 3$, AND $\tau_3 = 1$ WITH INCREASED INITIAL ACCELERATION

	Joint					
Data	1	2	3	4	5	6
θ^*_{i0}	20	30	10	40	3	20
θ^*_{i1}	30	4	2	13	40	20
θ^*_{i2}	20	30	50	29	10	20
θ^*_{i3}	60	70	80	90	10	20
$\dot{\theta}^*_{i0}$	2	4	1	2	3	3
$\dot{\theta}^*_{i3}$	3	4	2	1	3	2
$\ddot{\theta}^*_{i0}$	30	20	10	50	40	30
$\ddot{\theta}^*_{i3}$	3	2	1	1	2	3

TABLE A26 JOINT SEGMENT TRAJECTORY COEFFICIENTS FOR THE DATA OF TABLE A25

Trajectory of joint 1
$P(1,1) = 20, 4, 60, -139.3333, 85.33334$
$P(1,2) = 30, 71.00001, 346.5, -427.5$
$P(1,3) = 20, 112.1667, -104, 28.5, 3.333332$

Trajectory of joint 2
$P(2,1) = 30, 8, 40, -167.8596, 93.85965$
$P(2,2) = 4, -60.21052, 224.0526, -137.8421$
$P(2,3) = 30, 83.35088, -21.05263, -51.94737, 29.64912$

Trajectory of joint 3
$P(3,1) = 10, 2, 20, -74, 44$
$P(3,2) = 2, -5.999995, 139.5, -85.50001$
$P(3,3) = 50, 62.49999, -13, -43.5, 24$

Trajectory of joint 4
$P(4,1) = 40, 4, 100, -291.0877, 160.0877$
$P(4,2) = 13, -43.3684, 421.3421, -361.9737$
$P(4,3) = 29, 145.7807, -73.8421, -46.6579, 35.7193$

Trajectory of joint 5
$P(5,1) = 3, 6, 80, -82.42105, 33.42105$
$P(5,2) = 40, 78.63158, 74.8421, -183.4737$
$P(5,3) = 10, 14.94737, -52.8421, 57.8421, -19.94737$

Trajectory of joint 6
$P(6,1) = 20, 6, 60, -108.3158, 42.31578$
$P(6,2) = 20, -44.52632, -24.86841, 69.39475$
$P(6,3) = 20, -8.289474, 20.36842, -17.86842, 5.789474$

TABLE A27 TIME SEGMENTS AND MOTION DATA FOR $\tau_1 = 2$, $\tau_2 = 3$, AND $\tau_3 = 1$ WITH INCREASED FINAL VELOCITY

	Joint					
Data	1	2	3	4	5	6
θ^*_{i0}	20	30	10	40	3	20
θ^*_{i1}	30	4	2	13	40	20
θ^*_{i2}	20	30	50	29	10	20
θ^*_{i3}	60	70	80	90	10	20
$\dot{\theta}^*_{i0}$	2	4	1	2	3	3
$\dot{\theta}^*_{i3}$	30	40	20	10	30	20
$\ddot{\theta}^*_{i0}$	3	2	1	5	4	3
$\ddot{\theta}^*_{i3}$	3	2	1	1	2	3

TABLE A28 JOINT SEGMENT TRAJECTORY COEFFICIENTS FOR THE DATA OF TABLE A27

Trajectory of joint 1
$P(1,1) = 20, 4, 6, -34.17544, 34.17544$
$P(1,2) = 30, 75.26316, 244.1842, -329.4474$
$P(1,3) = 20, 78.0614, -82.68421, 61.18421, -16.5614$

Trajectory of joint 2
$P(2,1) = 30, 8, 4, -86.38596, 48.38597$
$P(2,2) = 4, -74.42105, 79.10526, 21.31579$
$P(2,3) = 30, 35.03509, 15.89474, -16.89474, 5.964911$

Trajectory of joint 3
$P(3,1) = 10, 2, 2, -33.26316, 21.26316$
$P(3,2) = 2, -13.10526, 67.02632, -5.921064$
$P(3,3) = 50, 38.34211, 5.473682, -25.97368, 12.15789$

Trajectory of joint 4
$P(4,1) = 40, 4, 10, -138.5614, 97.56141$
$P(4,2) = 13, -2.157875, 404.2895, -386.1316$
$P(4,3) = 29, 140.0965, -83.78947, -18.71053, 23.40351$

Trajectory of joint 5
$P(5,1) = 3, 6, 8, 52.10527, -29.10526$
$P(5,2) = 40, 92.84209, -23.21053, -99.63158$
$P(5,3) = 10, -17.73684, -35.78948, 94.78948, -41.26316$

Trajectory of joint 6
$P(6,1) = 20, 6, 6, -8.842108, -3.157896$
$P(6,2) = 20, -31.73684, -88.81578, 120.5526$
$P(6,3) = 20, -29.60526, 30.31579, 8.184212, -8.894736$

TABLE A29 TIME SEGMENTS AND MOTION DATA FOR $\tau_1 = 2$, $\tau_2 = 3$, AND $\tau_3 = 1$ WITH INCREASED INITIAL VELOCITY

	Joint					
Data	1	2	3	4	5	6
θ^*_{i0}	20	30	10	40	3	20
θ^*_{i1}	30	4	2	13	40	20
θ^*_{i2}	20	30	50	29	10	20
θ^*_{i3}	60	70	80	90	10	20
$\dot{\theta}^*_{i0}$	20	40	10	20	30	30
$\dot{\theta}^*_{i3}$	3	4	2	1	3	2
$\ddot{\theta}^*_{i0}$	3	2	1	5	4	3
$\ddot{\theta}^*_{i3}$	3	2	1	1	2	3

TABLE A30 JOINT SEGMENT TRAJECTORY COEFFICIENTS FOR THE DATA OF TABLE A29

Trajectory of joint 1
$P(1,1) = 20, 40, 6, -119.4386, 83.4386$
$P(1,2) = 30, 41.1579, 333.7105, -384.8684$
$P(1,3) = 20, 107.9035, -91.21052, 15.71052, 7.59649$

Trajectory of joint 2
$P(2,1) = 30, 80, 4, -245.5439, 135.5439$
$P(2,2) = 4, -159.6842, 181.4211, 4.263162$
$P(2,3) = 30, 69.14035, 21.57895, -94.57895, 43.85965$

Trajectory of joint 3
$P(3,1) = 10, 20, 2, -78.73684, 48.73684$
$P(3,2) = 2, -25.89474, 130.9737, -57.07895$
$P(3,3) = 50, 59.65789, -4.473686, -52.02632, 26.84211$

Trajectory of joint 4
$P(4,1) = 40, 40, 10, -212.4562, 135.4562$
$P(4,2) = 13, -53.31578, 417.079, -347.7631$
$P(4,3) = 29, 144.3596, -69.57893, -50.92106, 37.14036$

Trajectory of joint 5
$P(5,1) = 3, 60, 8, -67.26316, 36.26315$
$P(5,2) = 40, 28.89474, 53.52632, -112.4211$
$P(5,3) = 10, 7.842105, -31.52632, 36.52632, -12.84211$

Trajectory of joint 6
$P(6,1) = 20, 60, 6, -122.5263, 56.52632$
$P(6,2) = 20, -104.2105, -50.44736, 154.6579$
$P(6,3) = 20, -16.81579, 45.94736, -43.44736, 14.31579$

APPENDIX B: Detail of Matrix Inversion

Let

$$\mathbf{G} = \begin{bmatrix} A & E & \mathbf{0} \\ D & B & F \\ \mathbf{0} & g & C \end{bmatrix}. \tag{B1}$$

A^{-1} already exists. Then

$$\mathbf{G} = \begin{bmatrix} A & \mathbf{0} & \mathbf{0} \\ D & I & \mathbf{0} \\ \mathbf{0} & \mathbf{0} & \mathbf{0} \end{bmatrix} \begin{bmatrix} I & A^{-1}E & \mathbf{0} \\ \mathbf{0} & [-DA^{-1}E + B] & F \\ \mathbf{0} & g & C \end{bmatrix}. \tag{B2}$$

If $[-DA^{-1}E + B]^{-1}$ exists, then

$$\begin{bmatrix} I & A^{-1}E \\ \mathbf{0} & [-DA^{-1}E + B] \end{bmatrix}^{-1} \tag{B3}$$

exists. Thus,

$$\begin{bmatrix} I & A^{-1}E & \mathbf{0} \\ \mathbf{0} & [-DA^{-1}E + B] & F \\ \mathbf{0} & g & C \end{bmatrix} = \begin{bmatrix} I & A^{-1}E & \mathbf{0} \\ \mathbf{0} & [-DA^{-1}E + B] & \mathbf{0} \\ \mathbf{0} & g & I \end{bmatrix} \begin{bmatrix} I & \mathbf{0} & -A^{-1}E[\,]^{-1}F \\ \mathbf{0} & I & [\,]^{-1}F \\ \mathbf{0} & \mathbf{0} & [[-g[\,]^{-1}F + C]] \end{bmatrix}, \tag{B4}$$

where

$$[\,] = [-DA^{-1}E + B]. \tag{B5}$$

Therefore,

$$\mathbf{G} = \begin{bmatrix} A & E & \mathbf{0} \\ D & B & F \\ \mathbf{0} & g & C \end{bmatrix} = \begin{bmatrix} A & \mathbf{0} & \mathbf{0} \\ D & I & \mathbf{0} \\ \mathbf{0} & \mathbf{0} & I \end{bmatrix} \begin{bmatrix} I & A^{-1}\mathrm{E} & \mathbf{0} \\ \mathbf{0} & [\,] & \mathbf{0} \\ \mathbf{0} & g & I \end{bmatrix} \begin{bmatrix} I & \mathbf{0} & -A^{-1}E[\,]^{-1}F \\ \mathbf{0} & I & [\,]^{-1}F \\ \mathbf{0} & \mathbf{0} & [[\,]] \end{bmatrix}, \tag{B6}$$

where $[\,[\,]\,] = [-g[\,]^{-1}F + C]$. Then, if $[\,[\,]\,]^{-1}$ exists, we have

$$\mathbf{G}^{-1} = \begin{bmatrix} A & E & \mathbf{0} \\ D & B & F \\ \mathbf{0} & g & C \end{bmatrix}^{-1}$$

$$= \begin{bmatrix} I & \mathbf{0} & -A^{-1}E[\,]^{-1}F \\ \mathbf{0} & I & [\,]^{-1}F \\ \mathbf{0} & \mathbf{0} & [\,[\,]\,] \end{bmatrix}^{-1} \begin{bmatrix} I & A^{-1}E & \mathbf{0} \\ \mathbf{0} & [\,] & \mathbf{0} \\ \mathbf{0} & g & I \end{bmatrix}^{-1} \begin{bmatrix} A & \mathbf{0} & \mathbf{0} \\ D & I & \mathbf{0} \\ \mathbf{0} & \mathbf{0} & I \end{bmatrix}^{-1}. \tag{B7}$$

Since

$$\begin{bmatrix} A & \mathbf{0} & \mathbf{0} \\ D & I & \mathbf{0} \\ \mathbf{0} & \mathbf{0} & I \end{bmatrix}^{-1} = \begin{bmatrix} A^{-1} & \mathbf{0} & \mathbf{0} \\ -DA^{-1} & I & \mathbf{0} \\ \mathbf{0} & \mathbf{0} & I \end{bmatrix}, \tag{B8}$$

$$\begin{bmatrix} I & A^{-1}E & \mathbf{0} \\ \mathbf{0} & [\,] & \mathbf{0} \\ \mathbf{0} & g & I \end{bmatrix}^{-1} = \begin{bmatrix} I & -A^{-1}E[\,]^{-1} & \mathbf{0} \\ \mathbf{0} & [\,]^{-1} & \mathbf{0} \\ \mathbf{0} & -g[\,]^{-1} & I \end{bmatrix}^{-1}, \tag{B9}$$

and

$$\begin{bmatrix} I & \mathbf{0} & -A^{-1}E[\,]^{-1}F \\ \mathbf{0} & I & [\,]^{-1}F \\ \mathbf{0} & \mathbf{0} & [\,[\,]\,] \end{bmatrix}^{-1} = \begin{bmatrix} I & \mathbf{0} & A^{-1}E[\,]^{-1}F[\,[\,]\,]^{-1} \\ \mathbf{0} & I & -[\,]^{-1}F[\,[\,]\,]^{-1} \\ \mathbf{0} & \mathbf{0} & [\,[\,]\,]^{-1} \end{bmatrix}, \tag{B10}$$

then

$$\mathbf{G}^{-1} = \begin{bmatrix} A & E & \mathbf{0} \\ D & B & \mathbf{0} \\ \mathbf{0} & g & C \end{bmatrix}^{-1}$$

$$= \begin{bmatrix} I & \mathbf{0} & A^{-1}E[\,]^{-1}F[\,[\,]\,]^{-1} \\ \mathbf{0} & I & -[\,]^{-1}F[\,[\,]\,]^{-1} \\ \mathbf{0} & \mathbf{0} & [\,[\,]\,]^{-1} \end{bmatrix} \begin{bmatrix} I & A^{-1}E[\,]^{-1} & \mathbf{0} \\ \mathbf{0} & [\,]^{-1} & \mathbf{0} \\ \mathbf{0} & -g[\,]^{-1} & I \end{bmatrix} \begin{bmatrix} A^{-1} & \mathbf{0} & \mathbf{0} \\ -DA^{-1} & I & \mathbf{0} \\ \mathbf{0} & \mathbf{0} & I \end{bmatrix}$$

$$= \begin{bmatrix} I & \mathbf{0} & A^{-1}E[\,]^{-1}F[\,[\,]\,]^{-1} \\ \mathbf{0} & I & -[\,]^{-1}F[\,[\,]\,]^{-1} \\ \mathbf{0} & \mathbf{0} & [\,[\,]\,]^{-1} \end{bmatrix} \begin{bmatrix} A^{-1} + A^{-1}E[\,]^{-1}DA^{-1} & -A^{-1}E[\,]^{-1} \\ -[\,]^{-1}DA^{-1} & [\,]^{-1} \\ g[\,]^{-1}DA^{-1} & -g[\,]^{-1} \end{bmatrix} \tag{B11}$$

Finally,

$$\mathbf{G}^{-1} = \begin{bmatrix} [A^{-1} + A^{-1}E[\ ]^{-1}DA^{-1} + A^{-1}E[\ ]^{-1}F[\ [\ ]\ ]^{-1}g[\ ]^{-1}DA^{-1}] & -[A^{-1}E[\ ]^{-1} + A^{-1}E[\ ]^{-1}F[\ [\ ]\ ]^{-1}g[\ ]^{-1}] & A^{-1}E[\ ]^{-1}F[\ [\ ]\ ]^{-1} \\ -[[\ ]^{-1}DA^{-1} + [\ ]^{-1}F[[\ ]]^{-1}g[\ ]^{-1}DA^{-1}] & [[\ ]^{-1} + [\ ]^{-1}F[[\ ]]^{-1}g[\ ]^{-1}] & -[\ ]^{-1}F[\ [\ ]\ ]^{-1} \\ [\ [\ [\ ]\ ]^{-1}g[\ ]^{-1}DA^{-1}] & -[\ [\ [\ ]\ ]^{-1}g[\ ]^{-1}] & [\ [\ ]\ ]^{-1} \end{bmatrix}. \tag{B12}$$

APPENDIX C: Computer Algorithm

```
 10 LPRINT "4-3-4 ROBOTIC TRAJECTORY, SIX JOINTS"
 20 DIM AE(6,8)
 30 DATA 200,300,100
 40 READ T1, T2, T3
 50 LPRINT "DATA"
 60 LPRINT   "T1="T1    "     T2="T2   "      T3="T3
 70 LPRINT "AI0, AI1,AI2,AI3,AVI0,AVI3,AAI0,AAI3 "
 80 DATA 20,30,20,60,2,3,3,3
 90 DATA 30,4,30,70,4,4,2,2
100 DATA 10,2,50,80,1,2,1,1
110 DATA 40,13,29,90,2,1,5,1
120 DATA 3,40,10,10,3,3,4,2
130 DATA 20,20,20,20,3,2,3,3
140 FOR I=1 TO 6
150 FOR J=1 TO 8
160 READ AE(I,J)
170 NEXT J
180 LPRINT AE(I,1),AE(I,2),AE(I,3),AE(I,4),AE(I,5),AE(I,6),
    AE(I,7),AE(I,8)
190 NEXT I
200 LPRINT "-----------"
210 A=(3*T1+9*T2-T1*T2-6*(T2↑2))/(2*(T1+2*T2))
220 B=(T3*(T1+6*T2-3))/(T2*(T1+2*T2))
230 E=(T2↑2)/(T3*(T1+2*T2)*(2*A+T3))
240 AN=(1+E*B*(T1+2*T2))/(T1+2*T2)
250 AM=((T2)/(T1+2*T2))+(E*B*(T1+3*T2))
260 AF=(AM*T2)-(T3*E*(T1+3*T2))
270 H=(AN*T2)-(T3*E)
280 AK=(T3*(B*T2-T3))/(2*A+T3)
290 AL=(T1/(T1+2*T2))+(T1*E*B)
300 AZ=AL-((T1*T3*E)/T2)
310 DIM G(9,9)
320 G(1,1)=2*(1+AZ)
330 G(1,2)=-(T1*AZ)
340 G(1,3)=-((T1↑2)*(1-AZ))/6
350 G(1,4)=-(T1*AL)/(T2)
360 G(1,5)=(E*T3*(T1↑2))/(T2)
370 G(1,6)=(E*(T1↑2)*(T3↑2))/(6*T2)
380 G(1,7)+ (2*E(T1↑2))/(T2)
390 G(1,8)=-(E*(T1↑2))/(T2)
400 G(1,9)=(E*(T1↑2))/(6*T2)
410 G(2,1)=-(1+2*AZ)
420 G(2,2)=(T1*AZ)
430 G(2,3)=((T1↑2)*(1-AZ))/6
440 G(2,4)=(T1*AL)/(T2)
450 G(2,5)=-(E*T3*(T1↑2))/(T2)
460 G(2,6)=-(E*(T1↑2)*(T3↑2))/(6*T2)
470 G(2,7)=-(2*E*(T1↑2))/(T2)
```

```
480 G(2,8)=(E*(T1↑2))/(T2)
490 G(2,9)=-(E*(T1↑2))/(6*T2)
500 G(3,1)=(2*T2*(1-AZ))/(T1)
510 G(3,2)=-(1-AZ)*T2
520 G(3,3)=(T1*T2*(1-AZ))/6
530 G(3,4)=AL
540 G(3,5)=-(T1*T3*E)
550 G(3,6)=-(T1*E*(T3 2))/6
560 G(3,7)=-(2*E*T1)
570 G(3,8)=(T1*E)
580 G(3,9)=-(T1*E)/(6)
590 G(4,1)=-(6*H*T2)/(T1)
600 G(4,2)=(3*T2*H)
610 G(4,3)=-(T1*T2*H)/2
620 G(4,4)=(3*T2*AN)
630 G(4,5)=-(3*T2*T3*E)
640 G(4,6)=-(T2*E*(T3↑2))/2
650 G(4,7)=-(6*T2*E)
660 G(4,8)=(3*T2*E)
670 G(4,9)=-(E*T2)/2
680 G(5,1)=(2*AF)/(T1)
690 G(5,2)=-AF
700 G(5,3)=(T1*AF)/6
710 G(5,4)=-AM
720 G(5,5)=E*T3*(T1+3*T2)
730 G(5,6)=(E*(T3↑2)*(T1+3*T2))/6
740 G(5,7)=2*E*(T1+3*T2)
750 G(5,8)=-E*(T1+3*T2)
760 G(5,9)=(E*(T1+3*T2))/6
770 G(6,1)=-(2*AK)/(T1)
780 G(6,2)=AK
790 G(6,3)=-(T1*AK)/6
800 G(6,4)=(B*T3)/(2*A+T3)
810 G(6,5)=-(T3↑2)/(2*A+T3)
820 G(6,6)=(A*(T3↑2))/(3*(2*A+T3))
830 G(6,7)=(4*A)/(2*A+T3)
840 G(6,8)=-(2*A)/(2*A+T3)
850 G(6,9)=(A)/(3*(2*A+T3))
860 G(7,1)=(6*AK)/(T1)
870 G(7,2)=-(3*AK)
880 G(7,3)=(T1*AK)/2
890 G(7,4)=-(3*T3*B)/(2*A+T3)
900 G(7,5)=(3*(T3↑2))/(2*A+T3)
910 G(7,6)=-(A*(T3↑2))/(2*A+T3)
920 G(7,7)=(6*T3)/(2*A+T3)
930 G(7,8)=-(3*T3)/(2*A+T3)
940 G(7,9)=(T3)/(2*(2*A+T3))
950 G(8,1)=-(6*AK)/(T1)
960 G(8,2)=3*AK
970 G(8,3)=-(T1*AK)/2
```

```
 980 G(8,4)=(3*T3*B)/(2*A+T3)
 990 G(8,5)=-(3*(T3↑2))/(2*A+T3)
1000 G(8,6)=(A*(T3↑2))/(2*A+T3)
1010 G(8,7)=-(4*A+8*T3)/(2*A+T3)
1020 G(8,8)=(4*A+5*T3)/(2*A+T3)
1030 G(8,9)=-(A+T3)/(2*A+T3)
1040 G(9,1)=(2*AK)/(T1)
1050 G(9,2)=-AK
1060 G(9,3)=(T1*AK)/6
1070 G(9,4)=-(T3*B)/(2*A+T3)
1080 G(9,5)=(T3↑2)/(2*A+T3)
1090 G(9,6)=-(A*(T3↑2))/(6*A+3*T3)
1100 G(9,7)=(2*A+3*T3)/(2*A+T3)
1110 G(9,8)=-(2*A+2*T3)/(2*A+T3)
1120 G(9,9)=(4*A+3*T3)/(12*A+6*T3)
1130 DIM D(9,6)
1140 DIM AS(6,1)
1150 DIM AA(6,1)
1160 FOR M=1 TO 6
1170 D(1,M)=AE(M,2)-AE(M,1)-(AE(M,5)*T1)-((AE(M,7)*(T1↑2))/2)
1180 D(2,M)=-AE(M,5)-(AE(M,7)*T1)
1190 D(3,M)=-AE(M,7)
1200 D(4,M)=AE(M,3)-AE(M,2)
1210 D(5,M)=0!
1220 D(6,M)=0!
1230 D(7,M)=AE(M,4)-AE(M,3)
1240 D(8,M)=T3*AE(M,6)
1250 D(9,M)=(T3↑2)*AE(M,8)
1260 AS(M,1)=AE(M,5)*T1
1270 AA(M,1)=((T1↑2)*AE(M,7))/2
1280 NEXT M
1290 FOR I=1 TO 9
1300 FOR J=1 TO 6
1310 C(I,J)=0
1320 FOR K=1 TO 9
1330 C(I,J)=C(I,J)+G(I,K)*D(K,J)
1340 NEXT K
1350 NEXT J
1360 NEXT I
1370 LPRINT "COEFFICIENTS"
1380 FOR N=1 TO 6
1390 LPRINT "TRAJECTORY OF JOINT ("N")"
1400 LPRINT "P("N"1)=" AE(N,1)","AS(N,1)","AA(N,1)","C(1,N)",
     "C(2,N)
1410 LPRINT "P("N"2)=" AE(N,2)","C(3,N)","C(4,N)","C(5,N)
1420 LPRINT "P("N"3)=" AE(N,3)","C(6,N)","C(7,N)","C(8,N)",
     "C(9,N)
1430 NEXT N
1440 END
```

BIBLIOGRAPHY

Gupta, K. C., and Roth, B. (1982). Design Considerations for Manipulator Workspace, *Trans. ASME J. Mech. Design*, **104**(4), 705–711.

Kumar, A., and Waldron, K. J. (1981). The Workspace of a Mechanical Manipulator, *Trans. ASME J. Mech. Design*, **103**(3), 665–672.

Lee, T. W., and Yang, D. C. H. (1983). On the Evaluation of Manipulator Workspace, *Trans. ASME J. Mechanism, Transm. Autom. Design*, **105**(5), 70–77.

Lin, C. S., Chang, P. R., and Lun, J. Y. S. (1983). Formulation and Optimization of Cubic Polynomial Joint Trajectories for Industrial Robots, *IEEE Trans. Autom. Control*, **AC-28** (12), 1066–1071.

Sugimoto, K., and Duffy, J. (1981). Determination of Extreme Distances of a Robot Hand—Part I. A General Theory, *Trans. ASME J. Mech. Design*, **103**(3), 631–636.

Tsai, Y. C., and Soni, A. H. (1983). An Algorithm for the Workspace of a General $n - R$ Robot, *Trans. ASME J. Mechanisms, Transm. Autom. Design*, **105**(1), 55–57.

Vukobratovic, M., and Kircanski, M. (1982). A Method for Optimal Synthesis of Manipulator Robot Trajectories, *Trans. ASME J. Dyn. Syst. Meas. Control*, **104**, 188–193.

Yang, D. C. H., and Lee, T. W. (1983). On the Workspace of Mechanical Manipulators, *Trans ASME, J. Mechanisms, Transm. Autom. Design*, **105**(5), 63–69.

Yang, D. C. H., and Lee, T. W. (1984). Heuristic Combinatorial Optimization in the Design of Manipulator Workspace, *IEEE Trans. Syst. Man Cybern.*, **SMC-14**(4), 571–580.

HOMEWORK PROBLEMS

1. If P is an end point on the gripper off set from the nth axis by a distance d and a length a, show that the equation for the torus$(n - 1)$ in the $(n - 1)$th frame is

$$\{x_{n-1}^2 + y_{n-1}^2 + (z_{n-1} - d_{i-1})^2 - [a_{n-1}^2 + a^2 + (d + d_n)^2]\}^2$$
$$= 4a_{n-1}^2\left[a^2 - \left(\frac{(Z_{n-1} - d_{n-1}) - (d + d_n)\cos\alpha_{n-1}}{\sin\alpha_{n-1}}\right)^2\right]$$

2. For the cases presented in Figure P4.2 determine the manipulator workspaces $W_i(H)$ as well as $W_i(P)$ for $i = 1, 2$, and 3 (also refer to Fig. 4.2.3).

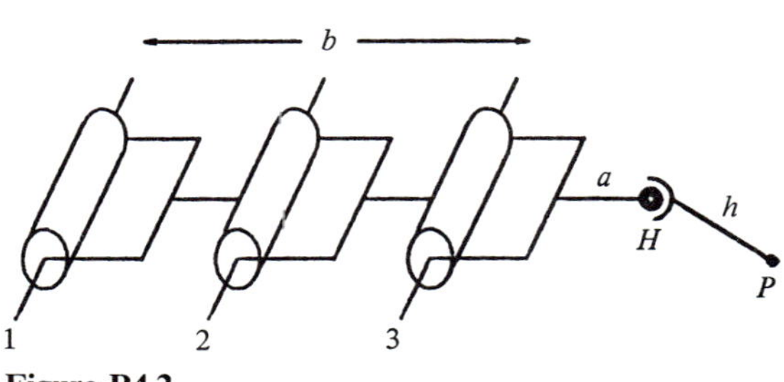

Figure P4.2

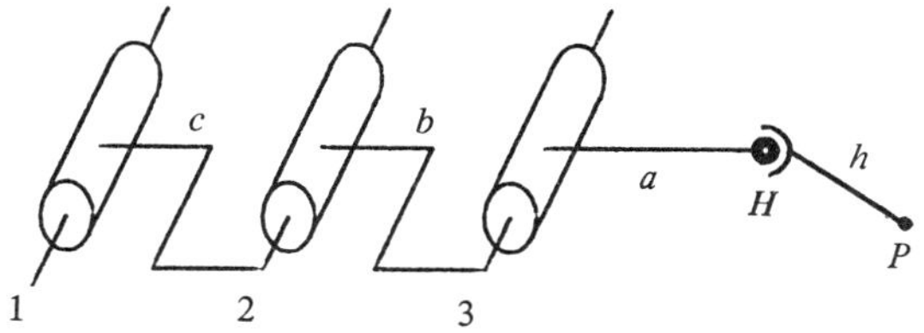

Figure P4.2 *(continued)*

3. Find the extreme distances of a general $3R$ robot shown in Figure P4.3.

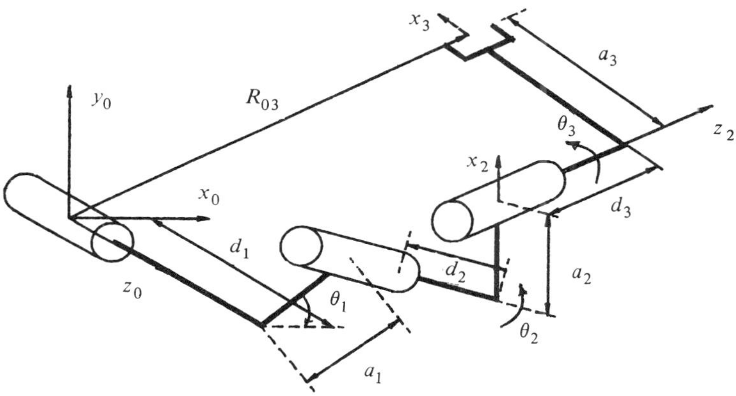

Figure P4.3 General $3R$ robotic manipulator.

4. Completely describe the robotic task of inserting a peg in a hole, ignoring the compliance of mating parts. Assume general 4×4 homogeneous transformations for the location and orientation of the robotic hand and the mating parts.
5. Consider a single-axis robot that is at rest at $\theta = 25°$. It is to move an angular distance of $\theta = 100°$ in 5 s. Find the coefficients of a cubic joint polynomial that achieves this motion and brings the arm to rest at the proper destination. Plot the position, velocity, and acceleration of the joint as a function of time.
6. Referring to Example 4.7.1 and Equations (4.7.17)–(4.7.22), find the values of the eight coefficients if

$$\dot{\theta}_1(0) = 0 \quad \text{and} \quad \dot{\theta}_2(t_{f2}) = 0.$$

7. Expand the equations that result from the matrix equation (4.8.32). Having done the expansion, derive an expression for C_{13} or other coefficients in the vector $\mathbf{C}$. Note that

$$\mathbf{C} \equiv [C_{13}\, C_{14}\, C_{21}\, C_{22}\, C_{23}\, C_{31}\, C_{32}\, C_{33}\, C_{34}]^T.$$

Chapter 5

Dynamics of Robotic Manipulators

5.1 INTRODUCTION

So far we have analyzed the kinematics, the kinematic equations, direct and inverse kinematic solutions for the vector of the joint angles, and some peripheral subjects pertaining to differential robotic transformations and trajectory planning. We are now in a position to discuss the governing equations pertaining to the dynamics of robotic manipulators. In discussing such equations we shall use some of the results that were obtained in the analyses of the aforementioned subjects. Essentially, dynamic equations are needed for the dynamic control of robotic manipulators. The purpose of robotic dynamic control is to maintain the proper dynamic response of a computer-based robotic manipulator such that the response is in accord with some prescribed set of objectives and goals. These objectives and goals can be translated in terms of impact and collisional forces and inertias experienced by either the gripper or the object. In general, the control problem consists of obtaining the governing equations for the robotics dynamics in the form of a dynamic model of the physical robotic manipulator and then specifying the required control laws to achieve the desired robotic response. In this chapter, we concentrate on obtaining the physical model for the robotic dynamics while in the next chapter we discuss the control theories in connection with robotic manipulators.

As discussed before, a computer-based robotic manipulator can be modeled as an open-loop kinematic and dynamic chain with several rigid bodies (links) connected in series by either revolute or prismatic joints. Since we already know the solution for the inverse kinematics problem, we are looking for a set of joint angles θ_i that would place the gripper at a location and orientation specified by T_i^0 with respect to the base coordinate. In robotic statics and dynamics we are dealing with a set of joint forces $\mathbf{f}_i$ and joint torques $\mathbf{t}_i$ that achieve a required set of force $\mathbf{f}$ and torque $\mathbf{t}$ for the gripper. Thus, we are concerned with the inverse problem for manipulator dynamics—the problem of computing the joint torques required to produce given joint positions, velocities, and accelerations. There are essentially three approaches available to arrive at a set of governing coupled severely nonlinear differential equations describing the dynamics of a robotic manipulator:

1. Bond Graph Dynamic Modeling.
2. Newton–Euler Dynamic Modeling.
3. Lagrange–Euler Dynamic Modeling.

In addition to the above techniques for the dynamic modeling of robotic manipulators, two recursive approaches have been developed. One is called the Newton–Euler recursive formulation and another Lagrangian recursive formulation. These alternate approaches are capable of reducing the number of computations drastically. The efficiency of these techniques lies in a set of recurrence relationships for the velocities, accelerations, and generalized forces. The number of additions and multiplications in these methods varies directly with the number of joints (n) as opposed to past formulations with dependence on higher powers of n.

We shall formulate the dynamic equations for robotic manipulators by the above three approaches and present briefly the recursive formulations. The governing equations are generally nonlinear coupled second-order differential equations. Each differential equation contains a large number of torque or force terms classified into four groups:

1. Inertial forces or torques due to the weights of the links.
2. Reaction forces or torques generated by acceleration at other joints.
3. Centripetal and Coriolis type forces and torques between joints.
4. Gravitational or loading type forces and torques on links.

In the next section we discuss the bond graph modeling of robotic manipulators.

5.2 BOND GRAPH MODELING OF ROBOTIC MANIPULATORS

To make the presentation self-contained, we introduce the basics of bond graph modeling originally conceived by Hank (Henry) Paynter in 1950. This is a technique that is used widely in the analysis and design of dynamic systems. The system is defined loosely to include mechanical, electrical, hydraulic, and other more obvious systems, as well as a mixture of these. The social, economic, and management aspects of the systems could also be considered.

Here we intend to apply bond graph dynamic modeling to complex robotic systems (not necessarily open-loop kinematic chains) such as a multifingered hand or multihanded robot. For single open-loop manipulators, however, we shall use the other techniques.

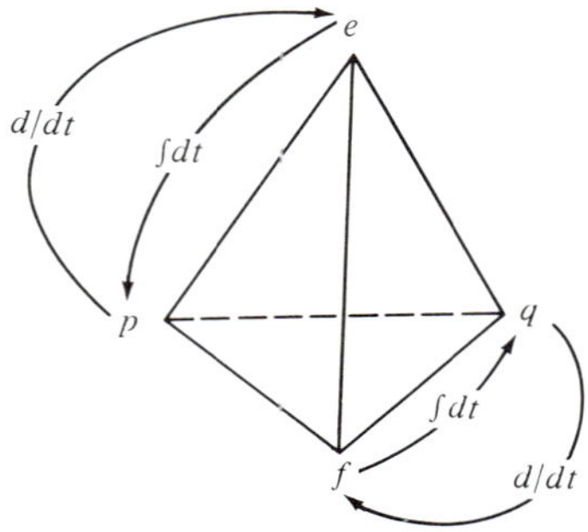

Figure 5.2.1 Tetrahedron of state in the bond graph theory.

Essentially, bond graph modeling considers the flow of energy and information from one system port to another and thus appears to be especially suitable for robotic manipulators. Four generalized system variables—effort e, flow f, displacement q, and momentum p—are defined. These variables are related through what is called the *tetrahedron of states*, as shown in Figure 5.2.1. The p-q link is called a memristor but is hard to define in practical terms. The e-q link is a capacitor element, e-f is a resistive element, and p-f is an inertial element. These elements will be discussed in detail later.

Bond graphs are used to relate the above elements with others to be defined shortly and consequently to model the system under consideration. Essentially, a bond graph is a drawing. Half arrows ($\rightharpoonup$) are used to indicate the flow of energy and information, and full arrows ($\rightarrow$) are used as active bonds (control). Each bond has an effort and a flow associated with it.

To understand the bond graph modeling it is essential to understand its elements, which we now define.

One-Port Resistor Elements (Energy Dissipating) These elements are represented symbolically as

$$\xrightharpoonup[f]{e} \quad R,$$

where R is resistance. The constitutive equations for a generalized resistor have the following form:

$$e = \phi_R(f) \tag{5.2.1}$$

and

$$f = \phi_R^{-1}(e), \tag{5.2.2}$$

where ϕ_R is generally a nonlinear function. Some examples of linear resistors include the mechanical dashpot, the rotational damper such as in robotic joints, hydraulic resistance in pipe flow, and the electrical resistor.

One-Port Capacitor Elements (Energy Storing) These elements are represented symbolically as

$$\xrightarrow[f=\dot{q}]{e} \quad C,$$

where C is capacitance. The constitutive equations have the following form:

$$e = \phi_C^{-1}(q) \tag{5.2.3}$$

and

$$q = \phi_C(e), \tag{5.2.4}$$

where ϕ_C is generally a nonlinear function. Note that $f = \dot{q}$. Some examples of linear capacitors include a linear spring or linear joint stiffness (robotics), a rod in torsion (infinitesimal elasticity), a tank filled with fluid, and an electrical capacitor.

One-Port Inertia Elements (Energy Storing) These are represented symbolically as

$$\xrightarrow[f]{e=\dot{p}} \quad I,$$

where I is an inertial element. The constitutive equations have the following form:

$$p = \phi_I(f) \tag{5.2.5}$$

and

$$f = \phi_I^{-1}(p), \tag{5.2.6}$$

where again ϕ_I is generally nonlinear. Note that $e = \dot{p}$. Some examples of linear systems are a moving mass, a rotating flywheel, and an electrical coil.

In addition to one-port elements, there are multiport elements that relate the basic elements. Two-port elements consist of transformers and gyrators.

Transformers These are represented symbolically as

$$\xrightarrow[f_1]{e_1} \overset{m}{\dot{\text{T}}\dot{\text{F}}} \xrightarrow[f_2]{e_2},$$

where m is the transformation modulus and TF is the transformer. The constitutive equations have the following form:

$$e_1 = me_2 \tag{5.2.7}$$

and

$$f_2 = mf_1. \tag{5.2.8}$$

Note that $e_1 f_1 = m e_2 f_1 = m e_2 m^{-1} f_2 = e_2 f_2$. Some examples of typical transformers include a simple lever, two gears, a piston (positive displacement pump), an electrical transformer, and hydraulic jacks.

It is possible to have a modulated transformer, represented symbolically as

$$\xrightarrow[f_1]{e_1} \quad \overset{m}{\dot{\text{M}}\dot{\text{T}}\text{F}} \quad \xrightarrow[f_2]{e_2},$$

where m can vary in time but is not necessarily an explicit function of time.

Gyrators These are represented symbolically as

$$\xrightarrow[f_1]{e_1} \quad \overset{r}{\dot{\text{G}}\dot{\text{Y}}} \quad \xrightarrow[f_2]{e_2},$$

where r is the gyration modulus and GY is the gyrator. The constitutive equations have the following form:

$$e_1 = r f_2 \tag{5.2.9}$$

and

$$e_2 = r f_1. \tag{5.2.10}$$

Note that $e_1 f_1 = r f_2 f_1 = r f_2 r^{-1} e_2 = f_2 e_2$. Some examples of typical gyrators include gyroscopes, centrifugal pumps, and electrical voice coils.

Again, it is possible to have a modulated gyrator. The following items should be noted:

1. Two cascaded gyrators form a transformer.
2. Two cascaded transformers form a transformer.
3. A gyrator and an I element form a C element.
4. A gyrator and a C element form an I element.

These relationships are shown easily by the constitutive equations. Finally, multiport junction elements are defined in two classes: *common effort junctions* and *common flow junctions.* In both cases it is assumed that power is conserved.

Common Effort Junctions These are represented symbolically as 0-junctions such that

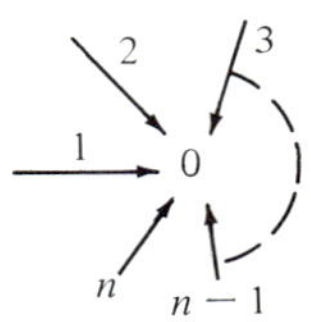

0–Junction

The constitutive equations have the following form:

$$e_1 = e_2 = e_3 = \cdots = e_n \quad \text{(common efforts)} \tag{5.2.11}$$

and

$$f_1 + f_2 + f_3 + \cdots + f_n = 0 \quad \text{(Kirchhoff's law).} \tag{5.2.12}$$

Power conservation (power = effort × flow) is represented by

$$e_1 f_1 + e_2 f_2 + \cdots + e_n f_n = 0. \tag{5.2.13}$$

This implies that the input power of the junction equals the power output of the junction.

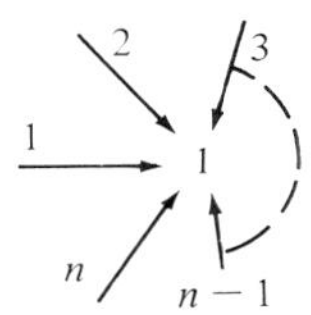

1-Junction

Common Flow Junctions These are represented symbolically as 1-junctions such that

The constitutive equations have the following form:

$$f_1 = f_2 = \cdots = f_n \tag{5.2.14}$$

and

$$e_1 + e_2 + \cdots + e_n = 0. \tag{5.2.15}$$

Again, power is conserved, that is,

$$e_1 f_1 + e_2 f_2 + \cdots + e_n f_n = 0. \tag{5.2.16}$$

One more concept is the key to understanding bond graphs. This is the idea of causality. Causality is defined in terms of what causes the flow as a result of an effort. Symbolically, it is represented by a vertical line on the end of the arrow that is adjacent to the element causing the flow:

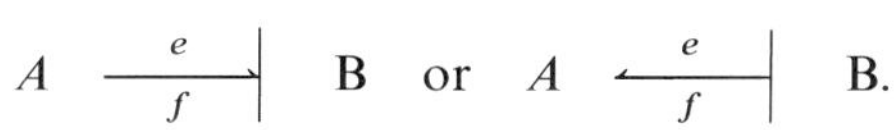

Both imply that B causes the flow and A is the effort exerted. Similarly,

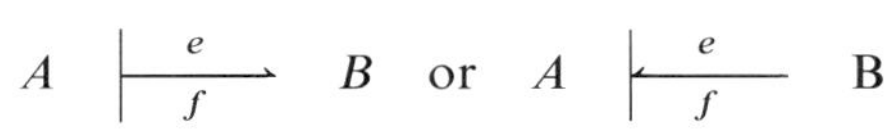

both show that A causes the flow and B is the effort exerted.

Multiport causality is defined as follows:

$$S_f \equiv \text{source of flow} \quad \text{and} \quad S_e \equiv \text{source of effort.}$$

Transformers have their own causality assignments:

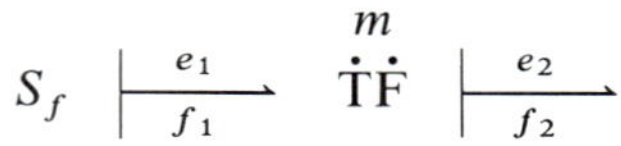

and

Gyrators similarly have their own special causalities:

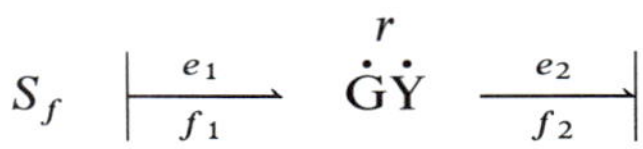

and

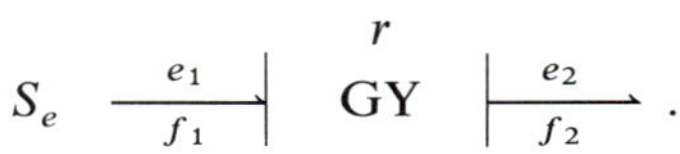

0-*Junctions* take the following causality assignments: The source of *flow* for one bond only is represented by

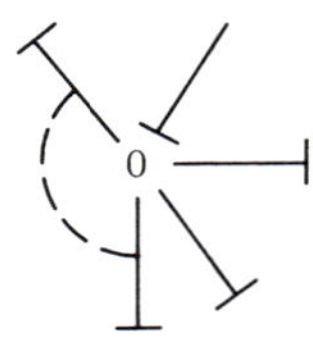

1-*Junctions* take the following causality assignments: The source of *effort* for one bond only is represented by

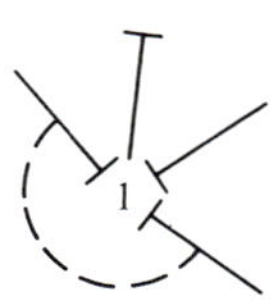

One-port causalities are best shown as block diagrams as follows:

R Elements

C Elements

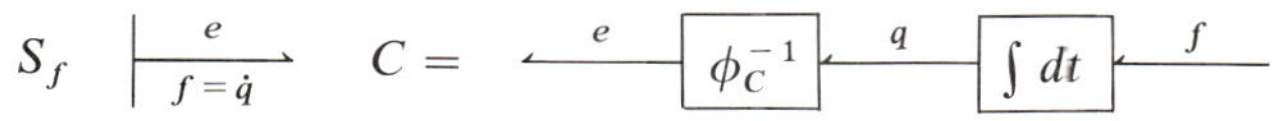

(Integral causality)

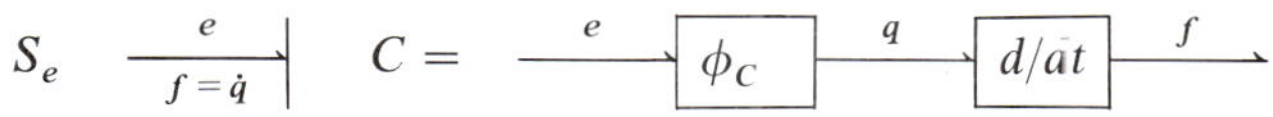

(Differential causality)

I Elements

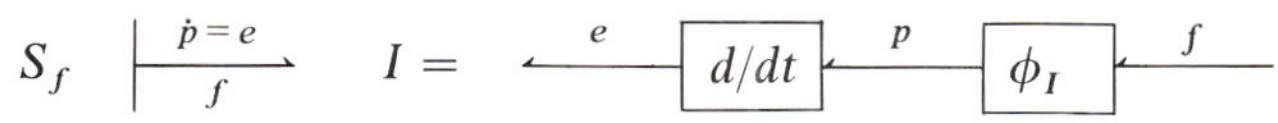

(Differential causality)

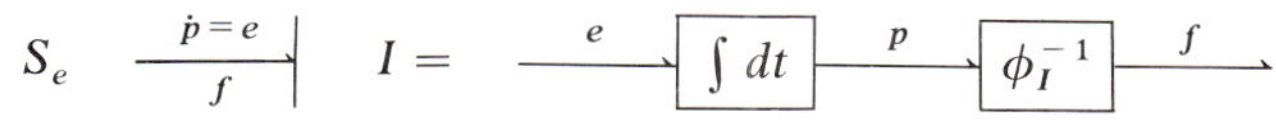

(Integral causality)

Whenever possible, it is preferable to use integral causality for C and I elements, that is,

It should be noted that while causality is important to a complete understanding of the system, it is not essential to the determination of the state-space equations.

Some bond graphs also can be described using fields. Initially, we described one-port R, C, and I elements and two-port TF and GY elements; however, in some cases it is possible to use multiport fields. Fields are used to combine many similar elements on a bond graph into one element whose constitutive equation is a mathematical combination of the individual equations. There are energy storing C and I fields, energy dissipating R fields, and power conserving junction structures consisting of TF, GY, 0, and I elements. While it is very difficult to describe the field concept in detail without an example, a brief explanation of the various elements should offer some insight.*

* For a complete description refer to *System Dynamics—A Unified Approach*, by D. Karnop and R. Rosenberg, Wiley, New York, 1975.

First consider the C field which is represented symbolically as

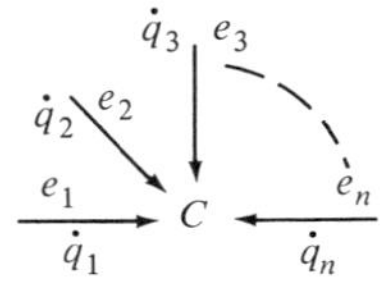

There are as many bonds as there are ports in the field. The energy E can be expressed as

$$E = \int_{t_0}^{t} \sum_{i=1}^{n} (e_i f_i) dt = \int_{t_0}^{t} \sum_{i=1}^{n} e_i \dot{q}_i \, dt \tag{5.2.17}$$

As an example of a C field, consider a supported beam that is acted on by a number of forces. In the study of such a beam, one must form a stiffness matrix, relating deflections to these forces. This matrix, in a sense, is the set of constitutive equations for the elastic behavior of the beam. As is well known, the deflection of a beam can store potential energy, just like a spring. Therefore, a beam can be seen as a C field.

The second energy storing field is the I field. I fields are analogous to C fields, with the difference being that efforts are studied as opposed to flows. Symbolically, I fields are represented as

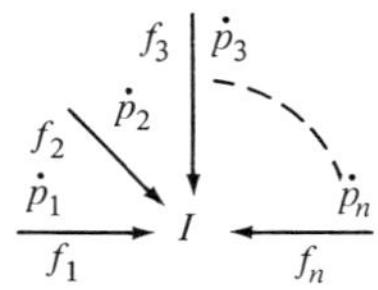

and the energy stored is

$$E = \int_{t_0}^{t} \sum_{i=1}^{n} (f_i e_i) dt = \int_{t_0}^{t} \sum_{i=1}^{n} f_i \dot{p}_i \, dt. \tag{5.2.18}$$

An example of an I field is a rigid body that is undergoing translational as well as rotational motion. In this case, there are two types of inertia: the inertia of the translational motion and the inertia of the object about its center of gravity. Therefore, such a rigid body can be described as an I field.

The energy dissipative field is the R field, which is represented by

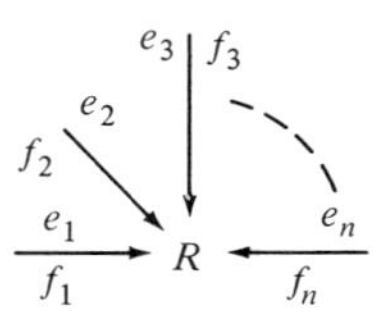

and its constitutive equations are

$$e_i = \phi_{Ri}(f_1, f_2, \ldots, f_n) \tag{5.2.19}$$

and

$$f_i = \phi_{Ri}^{-1}(e_1, e_2, \ldots, e_n), \tag{5.2.20}$$

where $i = 1, 2, \ldots, n$. As an example of an R field, consider a shaft running through several bearings. Each bearing has a resistance associated with it, and all of them together form an R field.

One final concept, which is extremely important to our discussion, is that of multiport modeling in robotics. Multiport modeling allows us to study systems of masses, rigid bodies, and force fields and to determine the set of differential equations that describe the behavior of such systems.

There are several required steps in this modeling technique that lead to the systems bond graph. Initially, one must define some key vectors as follows:

$\mathbf{q}_k \equiv$ generalized coordinate vector that defines the configuration of the system at any instant, (5.2.21)

$\mathbf{q}_C \equiv$ the set of displacements defining the potential energy, (5.2.22)

and

$\mathbf{v}_I = \dot{\mathbf{q}}_I \equiv$ the set of inertial velocities defining the kinetic co-energy. (5.2.23)

The next step is to obtain the transformations relating $\mathbf{q}_C$ to $\mathbf{q}_k$ and $\mathbf{q}_I$ to $\mathbf{q}_k$ and to differentiate them with respect to time to obtain the transformations relating $\dot{\mathbf{q}}_C$ to $\dot{\mathbf{q}}_k$ and $\dot{\mathbf{q}}_I = \mathbf{v}_I$ to $\dot{\mathbf{q}}_k$. These results make up the MTF elements in the bond graph. It should be noted that it is very important to define $\mathbf{q}_I$ properly, such that $\dot{\mathbf{q}}_I = \mathbf{v}_I$; otherwise, $\dot{\mathbf{q}}_k$ must be related directly to $\mathbf{v}_I$. Finally, by creating columns of 1-junctions of $\dot{\mathbf{q}}_k$, $\dot{\mathbf{q}}_C$, and $\mathbf{v}_I$, the bond graph can be drawn with C and I elements at their appropriate $\dot{\mathbf{q}}_C$ and $\mathbf{v}_I$ junctions.

It is this technique that will be used to develop the bond graph for the mechanical hand described below. However, this bond graph is arrived at gradually. First, let us consider the problem of Fig. 5.2.2, which has some similarity to a single-jointed finger manipulator.

EXAMPLE 5.2.1 ■

Derive the governing dynamic equations for a single-jointed manipulator (Fig. 5.2.2) using bond graph modeling.

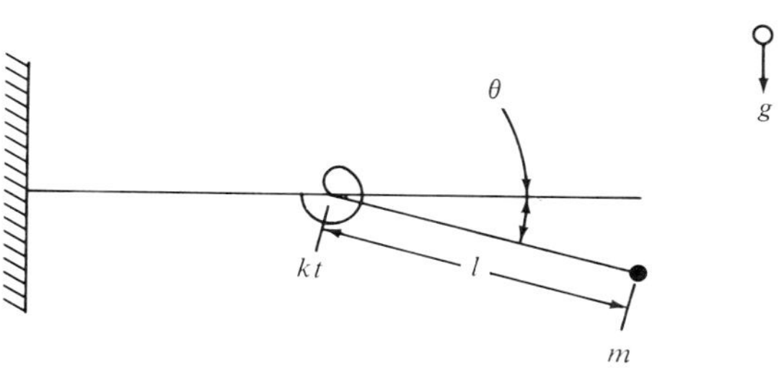

Figure 5.2.2 Cantilever pendulum mechanism with elastic joint modeling of a simple mechanical hand.

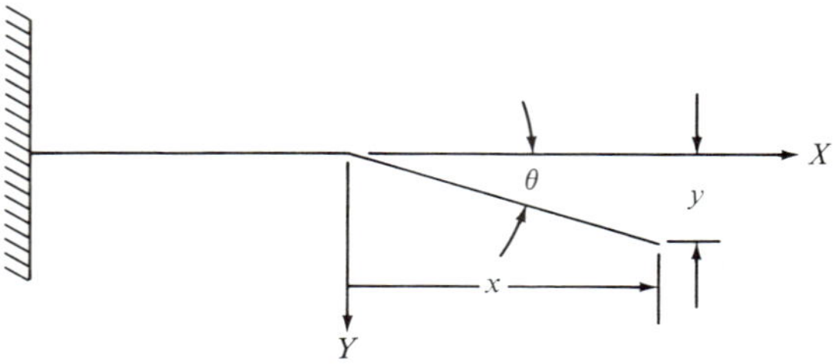

Figure 5.2.3 Geometry of Figure 5.2.2.

Solution. First, we define the geometric constraints in Figure 5.2.3. The key vectors may be defined as

$$\mathbf{q}_k = [\theta], \quad \mathbf{q}_C = [\theta], \quad \mathbf{v}_I = \begin{bmatrix} \dot{x} \\ \dot{y} \end{bmatrix}, \quad \text{or} \quad \mathbf{q}_I = \begin{bmatrix} x \\ y \end{bmatrix}.$$

we then relate $\mathbf{q}_C$ to $\mathbf{q}_k$:

$$\theta = \theta \quad \text{or} \quad \dot{\theta} = \dot{\theta}.$$

Therefore $\mathbf{T}_{Ck} = [1]$. We then relate $\mathbf{q}_I$ to $\mathbf{q}_k$:

$$x = l \cos \theta \tag{5.2.24}$$

and

$$y = l \sin \theta \tag{5.2.25}$$

or

$$\dot{x} = (-l \sin \theta)\dot{\theta} \tag{5.2.26}$$

and

$$\dot{y} = (l \cos \theta)\dot{\theta}. \tag{5.2.27}$$

Therefore

$$\mathbf{T}_{Ik} = \begin{bmatrix} -l \sin \theta \\ l \cos \theta \end{bmatrix}. \tag{5.2.28}$$

Now establish 1-junctions for $\dot{\mathbf{q}}_I$, $\dot{\mathbf{q}}_k$, and $\mathbf{v}_I$ and insert the MTFs. The bond graph will appear as shown in Figure 5.2.4. By using Kirchhoff's law around the 1-junctions, we obtain the state equations:

$$\dot{p}_1 = \frac{-1}{l \sin \theta} e_7, \qquad e_7 = -e_3 - e_8; \tag{5.2.29}$$

$$e_3 = k_t q_3, \qquad e_8 = (l \cos \theta) e_6; \tag{5.2.30}$$

$$e_6 = \dot{p}_2 - mg. \tag{5.2.31}$$

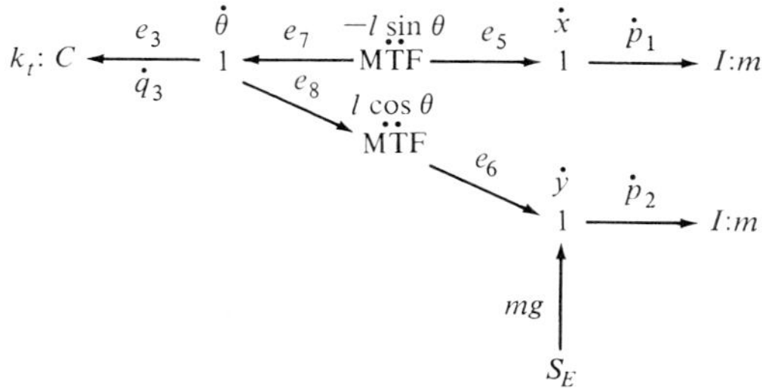

Figure 5.2.4 Bond graph for a single-jointed manipulator.

Thus,

$$\dot{p}_1 = \frac{-1}{l \sin \theta}[-k_t q_3 - (\dot{p}_2 - mg)l \cos \theta]. \tag{5.2.32}$$

Similarly,

$$\dot{p}_2 = \frac{1}{l \cos \theta}(-k_t q_3 + \dot{p}_1 l \sin \theta) + mg \tag{5.2.33}$$

and

$$\dot{q}_3 = \dot{\theta} = \frac{-p_1}{ml \sin \theta}. \tag{5.2.34}$$

To point out the significance of these equations, we consider the following analysis:

$$p_1 = m\dot{x} \quad \text{and} \quad p_2 = m\dot{y}\} \quad I \text{ field relations} \tag{5.2.35}$$

so that

$$\dot{p}_1 = m\ddot{x} \quad \text{and} \quad \dot{p}_2 = m\ddot{y}, \tag{5.2.36}$$

where

$$\ddot{x} = (-l \sin \theta)\ddot{\theta} + (-l \cos \theta)\dot{\theta}^2 \quad \text{and} \quad \ddot{y} = (l \cos \theta)\ddot{\theta} - (l \sin \theta)\dot{\theta}^2. \tag{5.2.37}$$

These may be inserted into the equation for either $\dot{p}_1$ or $\dot{p}_2$, since both are the same. We choose to use the equation for $\dot{p}_2$:

$$(ml \cos \theta)\ddot{\theta} - (ml \sin \theta)\dot{\theta}^2$$
$$= \frac{-k_t(q_3)}{l \cos \theta} - (ml \sin \theta \tan \theta)\ddot{\theta} - (ml \sin \theta)\dot{\theta}^2 + mg. \tag{5.2.38}$$

Now, $(-ml \sin \theta)\dot{\theta}^2$ may be canceled from both sides and the terms multiplied through by $l \cos \theta$:

$$(ml^2 \cos^2 \theta)\ddot{\theta} = -k_t(q_3) - (ml^2 \sin^2 \theta)\ddot{\theta} + mgl \cos \theta. \tag{5.2.39}$$

These may be rearranged to obtain

$$ml^2(\sin^2 \theta + \cos^2 \theta)\ddot{\theta} = -k_t(q_3) + mgl \cos \theta. \tag{5.2.40}$$

Noting that $\sin^2 \theta + \cos^2 \theta = 1$ and $q_3 = \theta$, we see that

$$ml^2\ddot{\theta} = -k_t\theta + mgl \cos \theta. \tag{5.2.41}$$

Now we analyze the same problem using Lagrange's equation. The kinetic energy T is

$$T = \tfrac{1}{2}ml^2\dot{\theta}^2 \tag{5.2.42}$$

and the potential energy V is

$$V = \tfrac{1}{2}k_t\theta^2 - mgl \sin \theta \quad \text{(assume } y = 0 \text{ as datum).} \tag{5.2.43}$$

Recalling Lagrange's equation,

$$\frac{d}{dt}\left(\frac{\partial T}{\partial \dot{q}}\right) - \frac{\partial T}{\partial q} + \frac{\partial V}{\partial q} = 0, \tag{5.2.44}$$

where $q = \theta$ in this case, it is seen that

$$\frac{\partial T}{\partial \dot{\theta}} = ml^2\dot{\theta}, \qquad \frac{d}{dt}(ml^2\dot{\theta}) = ml^2\ddot{\theta}; \tag{5.2.45}$$

and

$$\frac{\partial T}{\partial \theta} = 0, \qquad \frac{\partial V}{\partial \theta} = k_t\theta - mgl \cos \theta. \tag{5.2.46}$$

Therefore,

$$ml^2\ddot{\theta} + k_t\theta - mgl \cos \theta = 0 \tag{5.2.47}$$

or

$$ml^2\ddot{\theta} = -k_t\theta + mgl \cos \theta. \tag{5.2.48}$$

It is seen that (5.2.41) and (5.2.48) are identical. While Lagrange's equations are easier to use in this case, and possibly even in the next problem to be considered, there will be problems that are easier to examine using bond graph theory.

To arrive at the standard forms we must relate $\dot{p}_1$ to $\dot{p}_2$. Using the I field relations we see that

$$\frac{p_1}{\dot{x}} = \frac{p_2}{\dot{y}} \tag{5.2.49}$$

or

$$p_1 = \frac{\dot{x}}{\dot{y}} p_2, \tag{5.2.50}$$

where

$$\frac{\dot{x}}{\dot{y}} = \tan\theta. \tag{5.2.51}$$

Therefore,

$$p_1 = (\tan\theta)p_2, \tag{5.2.52}$$

which yields

$$\dot{p}_1 = (\sec^2\theta)\dot{\theta}p_2 + (\tan\theta)\dot{p}_2. \tag{5.2.53}$$

Note that

$$\dot{\theta}_1 = \frac{-p_1}{ml\sin\theta} \tag{5.2.54}$$

or, finally, that

$$\dot{p}_1 = \frac{-(\sec^2\theta)p_1p_2}{ml\sin\theta} + (\tan\theta)\dot{p}_2. \tag{5.2.55}$$

Substitute (5.2.55) into the equation for $\dot{p}_2$ to obtain

$$\dot{p}_2 = \frac{1}{l\cos\theta}\left[-k_tq_3 + \left(\frac{(-\sec^2\theta)p_1p_2}{ml\sin\theta} + (\tan\theta)\dot{p}_2\right)l\sin\theta\right] + mg \tag{5.2.56}$$

or

$$\dot{p}_2 = \left(\frac{-k_tq_3}{l\cos\theta} - \frac{(\sec^2\theta)p_1p_2}{ml\cos\theta} + mg\right)(1 - l\tan^2\theta)^{-1}. \tag{5.2.57}$$

Therefore, the state-space equations are

$$\dot{p}_1 = \frac{(-\sec^2\theta)p_1p_2}{ml\sin\theta} + (\tan\theta)\left[\left(\frac{-k_tq_3}{l\cos\theta} - \frac{(\sec^2\theta)p_1p_2}{ml\cos\theta} + mg\right)(1 - l\tan^2\theta)^{-1}\right], \tag{5.2.58}$$

$$\dot{p}_2 = \left(\frac{k_tq_3}{l\cos\theta} - \frac{(\sec^2\theta)p_1p_2}{ml\cos\theta} + mg\right)(1 - l\tan^2\theta), \tag{5.2.59}$$

and

$$\dot{q}_3 = \dot{\theta} = \frac{-p_1}{ml\sin\theta}. \quad ■■ \tag{5.2.60}$$

Now let us consider the more difficult problem of Figure 5.2.5, which has similarities to a two-jointed robotic manipulator.

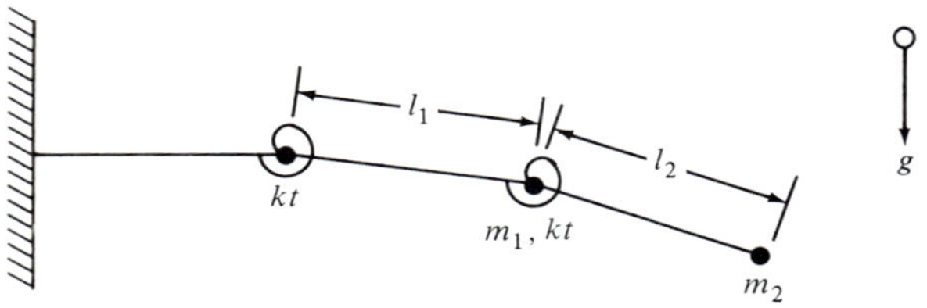

Figure 5.2.5 Cantilever-double pendulum mechanism with elastic joints.

EXAMPLE 5.2.2 ■

For a two-jointed manipulator (Fig. 5.2.5), derive the dynamic equations using bond graph modeling.

Solution. First, we define the geometric constraints in Figure 5.2.6. The key vectors may be defined as

$$\mathbf{q}_k = \begin{bmatrix} \theta_1 \\ \theta_2 \end{bmatrix}, \quad \mathbf{q}_C = \begin{bmatrix} \theta_1 \\ \delta \end{bmatrix}, \quad \mathbf{v}_I = \begin{bmatrix} \dot{x}_1 \\ \dot{y}_1 \\ \dot{x}_2 \\ \dot{y}_2 \end{bmatrix}, \quad \text{or} \quad \mathbf{q}_I = \begin{bmatrix} x_1 \\ y_1 \\ x_2 \\ y_2 \end{bmatrix}.$$

We then relate $\mathbf{q}_C$ to $\mathbf{q}_k$:

$$\theta_1 = \theta_1 \quad \text{and} \quad \delta = -\theta_1 + \theta_2,$$

or

$$\dot{\theta}_1 = \dot{\theta}_1 \quad \text{and} \quad \dot{\delta} = -\dot{\theta}_1 + \dot{\theta}_2.$$

Therefore,

$$\mathbf{T}_{Ck} = \begin{bmatrix} 1 & 0 \\ -1 & 1 \end{bmatrix}.$$

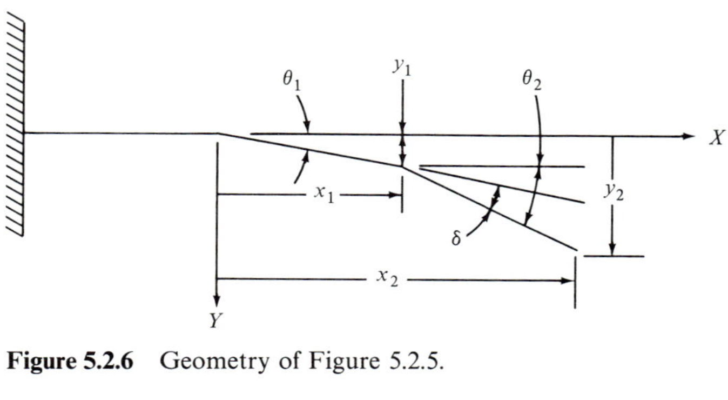

Figure 5.2.6 Geometry of Figure 5.2.5.

We then relate $\mathbf{q}_I$ to $\mathbf{q}_k$:

$$x_1 = l_1 \cos \theta_1,$$
$$y_1 = l_1 \sin \theta_1,$$
$$x_2 = l_1 \cos \theta_1 + l_2 \cos \theta_2,$$

and

$$y_2 = l_1 \sin \theta_1 + l_2 \sin \theta_2,$$

or

$$\dot{x}_1 = (-l_1 \sin \theta_1)\dot{\theta}_1,$$
$$\dot{y}_1 = (l_1 \cos \theta_1)\dot{\theta}_1,$$
$$\dot{x}_2 = (-l_1 \sin \theta_1)\dot{\theta}_1 + (-l_2 \sin \theta_2)\dot{\theta}_2,$$

and

$$\dot{y}_2 = (l_1 \cos \theta_1)\dot{\theta}_1 + (l_2 \cos \theta_2)\dot{\theta}_2.$$

Therefore,

$$\mathbf{T}_{Ik} = \begin{bmatrix} -l_1 \sin \theta_1 & 0 \\ l_1 \cos \theta_1 & 0 \\ -l_1 \sin \theta_1 & -l_2 \sin \theta_2 \\ l_1 \cos \theta_1 & l_2 \cos \theta_2 \end{bmatrix}.$$

By establishing 1-junctions for $\dot{\mathbf{q}}_C$, $\dot{\mathbf{q}}_k$, and $\mathbf{v}_I$ and inserting the proper MTFs, the bond graph will appear as shown in Figure 5.2.7. Note that MTFs of zero modulus have been eliminated, and MTFs of unit modulus have been made into direct links.

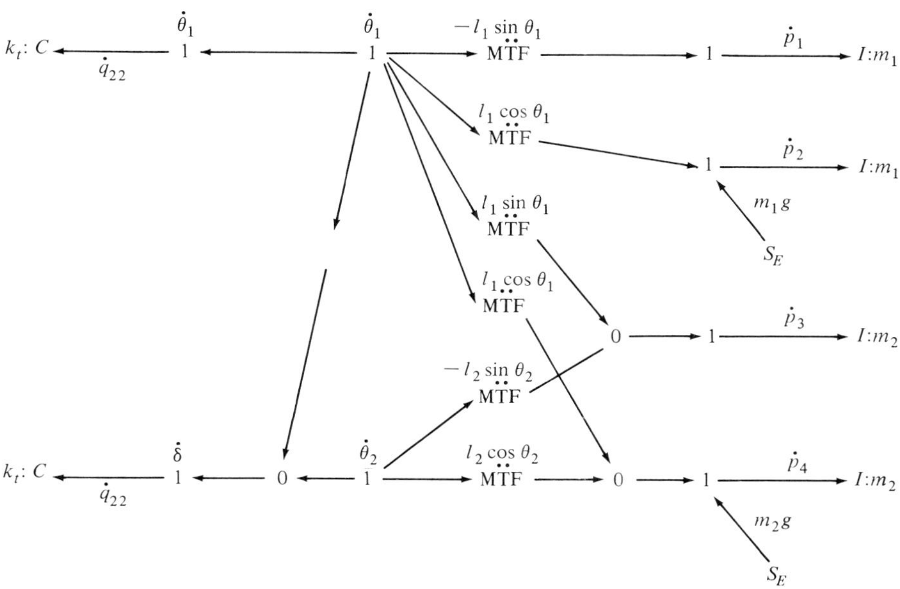

Figure 5.2.7 Bond graph for a two-axis elastic manipulator.

From this bond graph we can derive the initial set of state-space equations:

$$\dot{p}_1 = (\dot{p}_2 + \dot{p}_4 - m_1 g - m_2 g) \cot \theta_1 - \dot{p}_3 + \frac{k_t}{l_1 \sin \theta_1} (\theta_1 - \delta), \tag{5.2.61}$$

$$\dot{p}_2 = (\dot{p}_1 + \dot{p}_3) \tan \theta_1 - \dot{p}_4 + m_1 g + m_2 g + \frac{k_t}{l_1 \cos \theta_1} (\delta - \theta_1), \tag{5.2.62}$$

$$\dot{p}_3 = (\dot{p}_4 - m_2 g) \cot \theta_2 + \frac{k_t(\delta)}{l_2 \sin \theta_2}, \tag{5.2.63}$$

$$\dot{p}_4 = (\dot{p}_3) \tan \theta_2 + m_2 g - \frac{k_t(\delta)}{l_2 \cos \theta_2}, \tag{5.2.64}$$

$$\dot{q}_{22} = \dot{\theta}_1 = \frac{-p_1}{m_1 l_1 \sin \theta_1}, \tag{5.2.65}$$

and

$$\dot{q}_{27} = \dot{\delta} = \dot{\theta}_2 - \dot{\theta}_1 = \frac{1}{l_2 \sin \theta_2} \left(\frac{p_1}{m_1} - \frac{p_3}{m_2} \right) + \frac{p_1}{m_1 l_1 \sin \theta_1}. \tag{5.2.66}$$

By applying the same type of analysis used in the previous example to this example, we can determine the equations of motion. At this point it should be mentioned that Equations (5.2.61) and (5.2.62) are the same, and so are Equations (5.2.63) and (5.2.64).

In this example the I field relations are

$$p_1 = m_1 \dot{x}, \quad p_2 = m_1 \dot{y}_1, \quad p_3 = m_2 \dot{x}_2, \quad \text{and} \quad p_4 = m_2 \dot{y}_2. \tag{5.2.67}$$

Therefore,

$$\dot{p}_1 = m_1 \ddot{x}_1, \quad \dot{p}_2 = m_1 \ddot{y}_1, \quad \dot{p}_3 = m_2 \ddot{x}_2, \quad \text{and} \quad \dot{p}_4 = m_2 \ddot{y}_2. \tag{5.2.68}$$

Taking the appropriate derivatives yields

$$\dot{p}_1 = (-m_1 l_1 \sin \theta_1)\ddot{\theta}_1 + (-m_1 l_1 \cos \theta_1)\dot{\theta}_1^2, \tag{5.2.69a}$$

$$\dot{p}_2 = (m_1 l_1 \cos \theta_1)\ddot{\theta}_1 + (-m_1 l_1 \sin \theta_1)\dot{\theta}_1^2, \tag{5.2.69b}$$

$$\begin{aligned} \dot{p}_3 = {} & (-m_2 l_1 \sin \theta_1)\ddot{\theta}_1 + (-m_2 l_1 \cos \theta_1)\dot{\theta}_1^2 \\ & + (-m_2 l_2 \sin \theta_2)\ddot{\theta}_2 + (-m_2 l_2 \cos \theta_2)\dot{\theta}_2^2, \end{aligned} \tag{5.2.69c}$$

and

$$\begin{aligned} \dot{p}_4 = {} & (m_2 l_1 \cos \theta_1)\ddot{\theta}_1 + (-m_2 l_1 \sin \theta_1)\dot{\theta}_1^2 \\ & + (m_2 l_2 \cos \theta_2)\ddot{\theta}_2 + (-m_2 l_2 \sin \theta_2)\dot{\theta}_2^2. \end{aligned} \tag{5.2.69d}$$

By inserting Equations (5.2.69a)–(5.2.69d) into either (5.2.61) or (5.2.62) and manipulating as in the previous example, we obtain

$$\begin{aligned} m_1 l_1^2 \ddot{\theta}_1 + m_2 l_1^2 \ddot{\theta}_1 + m_2 l_1 l_2 [\cos(\theta_1 - \theta_2)] \ddot{\theta}_2 \\ = -m_2 l_1 l_2 [\sin(\theta_1 - \theta_2)] \dot{\theta}_2^2 + m_1 l_1 g \cos\theta_1 \\ + m_2 l_1 g \cos\theta_1 + k_t(\theta_2 - 2\theta_1). \end{aligned} \tag{5.2.70a}$$

Then, by inserting Equations (5.2.69c) and (5.2.69d) into either (5.2.58) or (5.2.64), we obtain

$$\begin{aligned} m_2 l_1 l_2 [\cos(\theta_1 - \theta_2)] \ddot{\theta}_1 + m_2 l_2^2 \ddot{\theta}_2 \\ = m_2 l_1 l_2 [\sin(\theta_1 - \theta_2)] \dot{\theta}_1^2 + m_2 l_2 g \cos\theta_2 - k_t(\theta_2 - \theta_1). \end{aligned} \tag{5.2.70b}$$

As in the previous example we also use Lagrange's equation to determine the equations of motion. The kinetic energy T is

$$T = \tfrac{1}{2} m_1 l_1^2 \dot{\theta}_1^2 + \tfrac{1}{2} m_2 l_1^2 \dot{\theta}_1^2 + \tfrac{1}{2} m_2 l_2 \dot{\theta}_2^2 + m_2 l_1 l_2 \dot{\theta}_1 \dot{\theta}_2 [\cos(\theta_1 - \theta_2)]$$

and if $y = 0$ is a datum, the potential energy V is

$$\begin{aligned} V = \tfrac{1}{2} k_t \theta_1^2 + \tfrac{1}{2} k_t (\theta_2^2 + \theta_1^2 - 2\theta_1\theta_2) - m_1 g (l_1 \sin\theta_1) \\ - m_2 g l_1 \sin\theta_1 - m_2 g l_2 \sin\theta_2. \end{aligned}$$

Using Lagrange's equation for θ_1 yields

$$\begin{aligned} m_1 l_1^2 \ddot{\theta}_1 + m_2 l_1^2 \ddot{\theta}_1 + m_2 l_1 l_2 [\cos(\theta_1 - \theta_2)] \ddot{\theta}_2 \\ = -m_2 l_1 l_2 [\sin(\theta_1 - \theta_2)] \dot{\theta}_2^2 + m_1 l_1 g \cos\theta_1 + m_2 l_1 g \cos\theta_1 \\ + k_t(\theta_2 - 2\theta_1), \end{aligned} \tag{5.2.71a}$$

and using them for θ_2 yields

$$\begin{aligned} m_2 l_1 l_2 [\cos(\theta_1 - \theta_2)] \ddot{\theta}_1 + m_2 l_2^2 \ddot{\theta}_2 \\ = m_2 l_1 l_2 [\sin(\theta_1 - \theta_2)] \dot{\theta}_1^2 + m_2 l_2 g \cos\theta_2 - k_t(\theta_2 - \theta_1). \end{aligned} \tag{5.2.71b}$$

Now note that Equation (5.2.70a) is identical to (5.2.71a) and that Equation (5.2.70b) is identical to (5.2.71b).

Now we must put the initial equations into standard form, which is more difficult in this problem than in the last problem. If we solve Equations (5.2.61) or (5.2.62) and (5.2.63) or (5.2.64) for the gravity forces, we obtain

$$\begin{aligned} m_1 g = \dot{p}_2 - \dot{p}_1 \tan\theta_1 - \dot{p}_3 \tan\theta_1 + \dot{p}_3 \tan\theta_2 \\ + \frac{k_t(\theta_1 - \delta)}{l_1 \cos\theta_1} - \frac{k_t(\delta)}{l_2 \cos\theta_2} \end{aligned} \tag{5.2.72}$$

and

$$m_2 g = \dot{p}_4 - \dot{p}_3 \tan\theta_2 + \frac{k_t(\delta)}{l_2 \cos\theta_2}. \tag{5.2.73}$$

From Equation (5.2.67) we know that

$$p_1 = \frac{\dot{x}_1}{\dot{y}_1} p_2 \tag{5.2.74a}$$

and

$$p_3 = \frac{\dot{x}_2}{\dot{y}_2} p_4 \tag{5.2.74b}$$

and from the inertia relationships we know that

$$\dot{x}_1 = (-l_1 \sin \theta_1)\dot{\theta}_1, \tag{5.2.75a}$$

$$\dot{y}_1 = (l_1 \cos \theta_1)\dot{\theta}_1, \tag{5.2.75b}$$

$$\dot{x}_2 = (-l_1 \sin \theta_1)\dot{\theta}_1 + (-l_2 \sin \theta_2)\dot{\theta}_2, \tag{5.2.75c}$$

and

$$\dot{y}_2 = (l_1 \cos \theta_1)\dot{\theta}_1 + (l_2 \cos \theta_2)\dot{\theta}_2. \tag{5.2.75d}$$

At this point it is necessary to assume that $\dot{\theta}_2 = f_c(t)\dot{\theta}_1$, where $f_c(t)$ is defined as a control function and is a function of the masses, rod lengths, k_t, and time.

From Equations (5.2.74a), (5.2.75a), and (5.2.75b), we see that

$$p_1 = (-\tan \theta_1)p_2.$$

Therefore,

$$\dot{p}_1 = (-\tan \theta_1)\dot{p}_2 + (-\sec^2 \theta_1)\dot{\theta}_1 p_2,$$

but since

$$\dot{\theta}_1 = \frac{-\dot{x}_1}{l_1 \sin \theta_1} = \frac{-p_1}{m_1 l_1 \sin \theta_1},$$

this yields

$$\dot{p}_1 = (-\tan \theta_1)\dot{p}_2 + (-\sec^3 \theta_1)\frac{-p_1}{m_1 l_1} p_2.$$

Here we define

$$A = (\sec^3 \theta_1)\frac{p_1}{m_1 l_1} \tag{5.2.76}$$

or, finally,

$$\dot{p}_1 = (-\tan \theta_1)\dot{p}_2 + A p_2. \tag{5.2.77}$$

From Equations (5.2.74b), (5.2.75c), and (5.2.75d) and by using the assumption that $\dot{\theta}_2 = f_c(t)\dot{\theta}_1$, we arrive at

$$p_3 = \frac{(-l_1 \sin\theta_1)\dot{\theta}_1 + f_c(t)(-l_2 \sin\theta_2)\dot{\theta}_1}{(l_1 \cos\theta_1)\dot{\theta}_1 + f_c(t)(l_2 \cos\theta_2)\dot{\theta}_1} p_4.$$

Now we define

$$B = \frac{(-l_1 \sin\theta_1)\dot{\theta}_1 + f_c(t)(-l_2 \sin\theta_2)\dot{\theta}}{(l_1 \cos\theta_1)\dot{\theta}_1 + f_c(t)(l_2 \cos\theta_2)\dot{\theta}_1}. \tag{5.2.78}$$

Therefore,

$$\dot{p}_3 = B\dot{p}_4 + \dot{B}p_4. \tag{5.2.79}$$

With the appropriate substitution for $\dot{\theta}_1$ and $\dot{\theta}_2$, we obtain

$$\begin{aligned}\dot{B} = \Bigg(\Bigg\{&[l_1 \cos\theta_1 + f_c(t) l_2 \cos\theta_2] \\ &\times\left[\frac{p_1}{m_1 l_1}\cot\theta_1 + f_c(t)^2\left(\frac{p_1 l_2 \cos\theta_2}{m_1 l_1 \sin\theta_1}\right) - \dot{f}_c(t) l_2 \sin\theta_2\right]\Bigg\} \\ &-\Bigg\{[-l_1 \sin\theta_1 - f_c(t)\sin\theta_2] \\ &\times\left[\frac{p_1}{m_1} + f_c(t)^2\left(\frac{p_1 l_2 \sin\theta_2}{m_1 l_1 \sin\theta_1}\right) + \dot{f}_c(t) l_2 \cos\theta_2\right]\Bigg\}\Bigg) \\ &\times [l_1 \cos\theta_1 + f_c(t) l_2 \cos\theta_2]^{-2}.\end{aligned}$$

By substituting Equations (5.2.77) and (5.2.79) into (5.2.72) and (5.2.73), we obtain

$$\begin{aligned}m_1 g = \dot{p}_2 &- [(-\tan\theta_1)\dot{p}_2 + Ap_2]\tan\theta_1 - (B\dot{p}_4 + \dot{B}p_4)\tan\theta_1 \\ &+ (B\dot{p}_4 + \dot{B}p_4)\tan\theta_2 + \frac{k_t(\theta_1 - \delta)}{l_1 \cos\theta_1} - \frac{k_t(\delta)}{l_2 \cos\theta_2}\end{aligned} \tag{5.2.80}$$

and

$$m_2 g = \dot{p}_4 - (B\dot{p}_4 + \dot{B}p_4)\tan\theta_2 + \frac{k_t(\delta)}{l_2 \cos\theta_2}. \tag{5.2.81}$$

Solving Equation (5.2.80) for $\dot{p}_4$ yields

$$\dot{p}_4 = \left(m_2 g + \dot{B}p_4 \tan\theta_2 - \frac{k_t(\delta)}{l_2 \cos\theta_2}\right)(1 - B\tan\theta_2)^{-1}. \tag{5.2.82}$$

Using Equation (5.2.82) and solving (5.2.80) for $\dot{p}_2$ yields

$$\begin{aligned}\dot{p}_2 = \Bigg(& m_1 g + A p_2 \tan\theta_1 \\ & + \left\{ \left[B\left(m_2 g + \dot{B} p_4 \tan\theta_2 - \frac{k_t(\delta)}{l_2 \cos\theta_2} \right) \right.\right. \\ & \left.\left. \times (1 - B\tan\theta_2)^{-1} \right] + \dot{B} p_4 \right\} \tan\theta_1 \\ & - \left\{ \left[B\left(m_2 g + \dot{B} p_4 \tan\theta_2 - \frac{k_t(\delta)}{l_2 \cos\theta_2} \right) \right.\right. \\ & \left.\left. \times (1 - B\tan\theta_2)^{-1} \right] + \dot{B} p_4 \right\} \tan\theta_2 \\ & - \frac{k_t(\theta_1 - \delta)}{l_1 \cos\theta_1} + \frac{k_t(\delta)}{l_2 \cos\theta_2} \Bigg) (1 + \tan^2\theta_1)^{-1}. \end{aligned} \tag{5.2.83}$$

Recall that

$$\dot{p}_1 = (-\tan\theta_1)\dot{p}_2 + A p_2 \tag{5.2.84}$$

and that

$$\dot{p}_3 = B\dot{p}_4 + \dot{B} p_4. \tag{5.2.85}$$

In addition,

$$\dot{q}_{22} = \dot{\theta}_1 = \frac{-p_1}{m_1 l_1 \sin\theta_1} \tag{5.2.86}$$

and

$$\dot{q}_{27} = \dot{\delta} = \dot{\theta}_2 - \dot{\theta}_1 = \frac{1}{l_2 \sin\theta_2}\left(\frac{p_1}{m_1} - \frac{p_3}{m_2}\right) + \frac{p_1}{m_1 l_1 \sin\theta_1}. \tag{5.2.87}$$

These equations—(5.2.87), (5.2.84), (5.2.85), (5.2.82), and (5.2.83)—are the state-space equations for this problem. ■ ■

5.3 MORE COMPLEX MANIPULATORS

Hopefully, these two examples have offered some insight into the bond graph technique and its abilities. Next, we consider a detailed analysis of one robotic finger in the XZ plane (i.e., no gravity effect), which is also under the influence of a reference velocity. A schematic diagram of the problem is shown in Figure 5.3.1.

EXAMPLE 5.3.1 ■

For a three-axis robot (Fig. 5.3.1) with a reference velocity, find the governing dynamics equations using bond graph modeling.

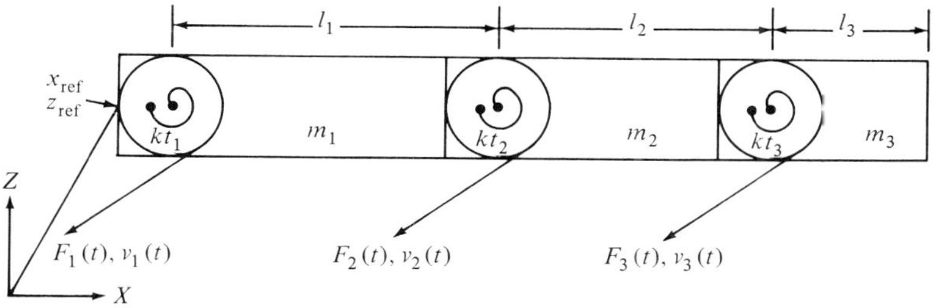

Figure 5.3.1 Schematic for a robotic finger.

Solution. In analyzing this problem the following assumptions are made:

1. The mass of each link is concentrated at a point in the center of the link (i.e., $l_n/2$)
2. Because of assumption 1, the rotational inertia can be neglected.

Based on these assumptions, the geometric variables are defined as shown in Figure 5.3.2.

The key vectors may be defined as

$$\mathbf{q}_k = \begin{bmatrix} x_{ref} \\ z_{ref} \\ \theta_1 \\ \theta_2 \\ \theta_3 \end{bmatrix}, \quad \mathbf{q}_C = \begin{bmatrix} \theta_1 \\ \delta_1 \\ \delta_2 \end{bmatrix}, \quad \mathbf{v}_I = \begin{bmatrix} \dot{x}_{ref} \\ \dot{z}_{ref} \\ \dot{x}_1 \\ \dot{z}_1 \\ \dot{x}_2 \\ \dot{z}_2 \\ \dot{x}_3 \\ \dot{z}_3 \end{bmatrix}, \quad \text{or} \quad \mathbf{q}_I = \begin{bmatrix} x_{ref} \\ z_{ref} \\ x_1 \\ z_1 \\ x_2 \\ z_2 \\ x_3 \\ z_3 \end{bmatrix}. \tag{5.3.1}$$

We then relate $\mathbf{q}_C$ to $\mathbf{q}_k$:

$$\begin{aligned} \theta_1 &= \theta_1, \\ \delta_1 &= -\theta_1 + \theta_2, \\ \delta_2 &= -\theta_2 + \theta_3; \end{aligned} \tag{5.3.2a}$$

or

$$\begin{aligned} \dot{\theta}_1 &= \dot{\theta}_1, \\ \dot{\delta}_1 &= -\dot{\theta}_1 + \dot{\theta}_2, \\ \dot{\delta}_2 &= -\dot{\theta}_2 + \dot{\theta}_3. \end{aligned} \tag{5.3.2b}$$

Therefore,

$$\mathbf{T}_{Ck} = \begin{bmatrix} 0 & 0 & 1 & 0 & 0 \\ 0 & 0 & -1 & 1 & 0 \\ 0 & 0 & 0 & -1 & 1 \end{bmatrix}. \tag{5.3.3}$$

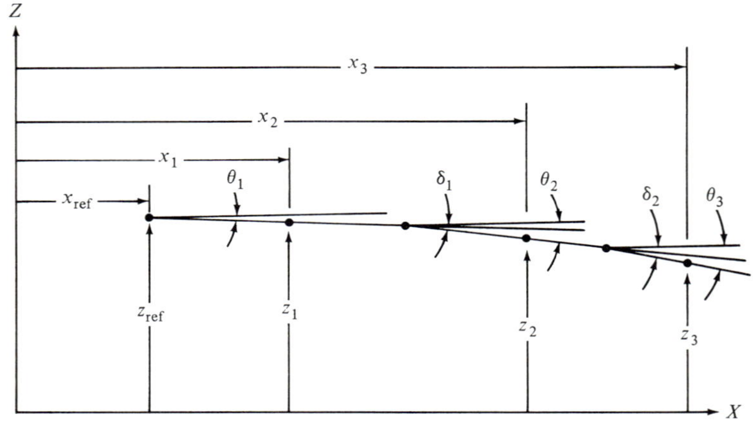

Figure 5.3.2 Geometric variables for a robotic finger.

We then relate $\mathbf{q}_I$ to $\mathbf{q}_k$:

$$x_{\text{ref}} = x_{\text{ref}}, \tag{5.3.4}$$

$$z_{\text{ref}} = z_{\text{ref}}, \tag{5.3.5}$$

$$x_1 = x_{\text{ref}} + \frac{l_1}{2}\cos\theta_1, \tag{5.3.6}$$

$$z_1 = z_{\text{ref}} + \left(-\frac{l_1}{2}\sin\theta_1\right), \tag{5.3.7}$$

$$x_2 = x_{\text{ref}} + l_1\cos\theta_1 + \frac{l_2}{2}\cos\theta_2, \tag{5.3.8}$$

$$z_2 = z_{\text{ref}} + (-l_1\sin\theta_1) + \left(-\frac{l_2}{2}\sin\theta_2\right), \tag{5.3.9}$$

$$x_3 = x_{\text{ref}} + l_1\cos\theta_1 + l_2\cos\theta_2 + \frac{l_3}{2}\cos\theta_3, \tag{5.3.10}$$

$$z_3 = z_{\text{ref}} + (-l_1\sin\theta_1) + (-l_2\sin\theta_2) + \left(-\frac{l_3}{2}\sin\theta_3\right); \tag{5.3.11}$$

or

$$\dot{x}_{\text{ref}} = \dot{x}_{\text{ref}}, \tag{5.3.12}$$

$$\dot{z}_{\text{ref}} = \dot{z}_{\text{ref}}, \tag{5.3.13}$$

$$\dot{x}_1 = \dot{x}_{\text{ref}} + \left(-\frac{l_1}{2}\sin\theta_1\right)\dot{\theta}_1, \tag{5.3.14}$$

$$\dot{z}_1 = \dot{z}_{ref} + \left(-\frac{l_1}{2}\cos\theta_1\right)\dot{\theta}_1, \tag{5.3.15}$$

$$\dot{x}_2 = \dot{x}_{ref} + (-l_1 \sin\theta_1)\dot{\theta}_1 + \left(-\frac{l_2}{2}\sin\theta_2\right)\dot{\theta}_2, \tag{5.3.16}$$

$$\dot{z}_2 = \dot{z}_{ref} + (-l_1 \cos\theta_1)\dot{\theta}_1 + \left(-\frac{l_2}{2}\cos\theta_2\right)\dot{\theta}_2, \tag{5.3.17}$$

$$\dot{x}_3 = \dot{x}_{ref} + (-l_1 \sin\theta_1)\dot{\theta}_1 + (-l_2 \sin\theta_2)\dot{\theta}_2 + \left(-\frac{l_3}{2}\sin\theta_3\right)\dot{\theta}_3, \tag{5.3.18}$$

$$\dot{z}_3 = \dot{z}_{ref} + (-l_1 \cos\theta_1)\dot{\theta}_1 + (-l_2 \cos\theta_2)\dot{\theta}_2 + \left(-\frac{l_3}{2}\cos\theta_3\right)\dot{\theta}_3. \tag{5.3.19}$$

Therefore,

$$\mathbf{T}_{Ik} = \begin{bmatrix} 1 & 0 & 0 & 0 & 0 \\ 0 & 1 & 0 & 0 & 0 \\ 1 & 0 & -\frac{l_1}{2}\sin\theta_1 & 0 & 0 \\ 0 & 1 & -\frac{l_1}{2}\cos\theta_1 & 0 & 0 \\ 1 & 0 & -l_1\sin\theta_1 & -\frac{l_2}{2}\sin\theta_2 & 0 \\ 0 & 1 & -l_1\cos\theta_1 & -\frac{l_2}{2}\cos\theta_2 & 0 \\ 1 & 0 & -l_1\sin\theta_1 & -l_2\sin\theta_2 & -\frac{l_3}{2}\sin\theta_3 \\ 0 & 1 & -l_1\cos\theta_1 & -l_2\cos\theta_2 & -\frac{l_3}{2}\cos\theta_3 \end{bmatrix}. \tag{5.3.20}$$

The bond graph corresponding to this finger is shown in Figure 5.3.3.

The initial state-space equations are

$$\dot{p}_6 = F_4(t) - \dot{p}_8 - \dot{p}_{10} - \dot{p}_{12}, \tag{5.3.21}$$

$$\dot{p}_7 = F_5(t) - \dot{p}_9 - \dot{p}_{11} - \dot{p}_{13}, \tag{5.3.22}$$

$$\dot{p}_8 = \frac{-2F_1(t)}{r_1 l_1 \sin\theta_1} - (\dot{p}_9 + 2\dot{p}_{11} + 2\dot{p}_{13})\cot\theta_1 - 2\dot{p}_{10} - 2\dot{p}_{12} - \frac{2k_{t2}(\delta_1)}{l_1\sin\theta_1} + \frac{2k_{t1}(\theta_1)}{l_1\sin\theta_1}, \tag{5.3.23}$$

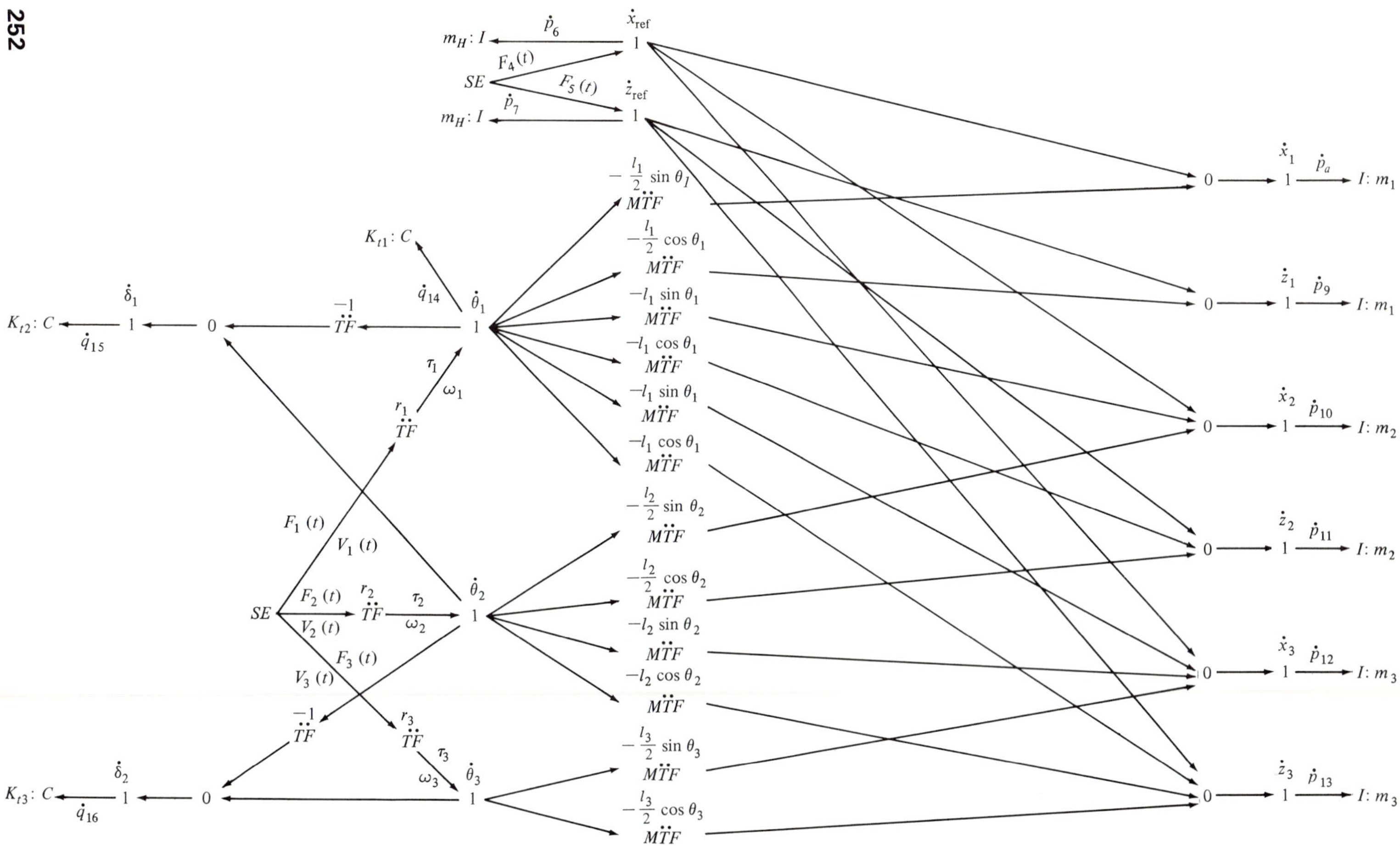

Figure 5.3.3 Bond graph for a three-axis robotic manipulator.

$$\dot{p}_9 = \frac{-2F_1(t)}{r_1 l_1 \cos\theta_1} - (\dot{p}_8 + 2\dot{p}_{10} + 2\dot{p}_{12})\tan\theta_1 - 2\dot{p}_{11} - 2\dot{p}_{13}$$

$$- \frac{2k_{t2}(\delta_1)}{l_1\cos\theta_1} + \frac{2k_{t1}(\theta_1)}{l_1\cos\theta_1}, \tag{5.3.24}$$

$$\dot{p}_{10} = \frac{-2F_2(t)}{r_2 l_2 \sin\theta_2} - (\dot{p}_{11} + 2\dot{p}_{13})\cot\theta_2 - 2\dot{p}_{12}$$

$$- \frac{2k_{t3}(\delta_2)}{l_2\sin\theta_2} + \frac{2k_{t2}(\delta_1)}{l_2\sin\theta_2}, \tag{5.3.25}$$

$$\dot{p}_{11} = \frac{-2F_2(t)}{r_2 l_2 \cos\theta_2} - (\dot{p}_{10} + 2\dot{p}_{12})\tan\theta_2 - 2\dot{p}_{13}$$

$$- \frac{2k_{t3}(\delta_2)}{l_2\cos\theta_2} + \frac{2k_{t2}(\delta_1)}{l_2\cos\theta_2}, \tag{5.3.26}$$

$$\dot{p}_{12} = \frac{-2F_3(t)}{r_3 l_3 \sin\theta_3} - \dot{p}_{13}\cot\theta_3 + \frac{2k_{t3}(\delta_2)}{l_3\sin\theta_3}, \tag{5.3.27}$$

$$\dot{p}_{13} = \frac{-2F_3(t)}{r_3 l_3 \cos\theta_3} - \dot{p}_{12}\tan\theta_3 + \frac{2k_{t3}(\delta_2)}{l_3\cos\theta_3}, \tag{5.3.28}$$

$$\dot{q}_{14} = \dot{\theta}_1 = r_1^{-1} v_1(t), \tag{5.3.29}$$

$$\dot{q}_{15} = \dot{\delta}_1 = \dot{\theta}_2 - \dot{\theta}_1 = r_2^{-1} v_2(t) - r_1^{-1} v_1(t), \tag{5.3.30}$$

and

$$\dot{q}_{16} = \dot{\delta}_2 = \dot{\theta}_3 - \dot{\theta}_2 = r_3^{-1} v_3(t) - r_2^{-1} v_2(t). \tag{5.3.31}$$

The I field relations are

$$p_6 = m_H \dot{x}_{\text{ref}}, \tag{5.3.32}$$

$$p_7 = m_H \dot{z}_{\text{ref}}, \tag{5.3.33}$$

$$p_8 = m_1 \dot{x}_1, \tag{5.3.34}$$

$$p_9 = m_1 \dot{z}_1, \tag{5.3.35}$$

$$p_{10} = m_2 \dot{x}_2, \tag{5.3.36}$$

$$p_{11} = m_2 \dot{z}_2, \tag{5.3.37}$$

$$p_{12} = m_3 \dot{x}_3, \tag{5.3.38}$$

and

$$p_{13} = m_3 \dot{z}_3. \tag{5.3.39}$$

To determine the equations of motion we use Equations (5.3.24), (5.3.26), and (5.3.28). We have neglected Equations (5.3.21) and (5.3.22) because $\dot{p}_6$ and $\dot{p}_7$ do not appear in the other equations and because they are the result of an arbitrary reference velocity. Note that we could have used Equations (5.3.23), (5.3.25), and (5.3.27) in place of (5.3.24), (5.3.26), and (5.3.28). By an analysis similar to the previous examples, we arrive at the following equations of motion:

$$\begin{aligned}\left(\frac{m_1}{2} + 2m_2 + 2m_3\right)l_1^2\ddot{\theta}_1 &+ (m_2 + 2m_3)l_1 l_2[\cos(\theta_1 - \theta_2)]\ddot{\theta}_2 \\ &+ m_3 l_1 l_3[\cos(\theta_1 - \theta_3)]\ddot{\theta}_3 + (m_2 + 2m_3)l_1 l_2[\sin(\theta_1 - \theta_2)]\dot{\theta}_2^2 \\ &+ m_3 l_1 l_3[\sin(\theta_1 - \theta_3)]\dot{\theta}_3^2 \\ &= \frac{2F_1(t)}{r_1} + l_1(m_1 + 2m_2 + 2m_3) \\ &\quad \times (\ddot{z}_{\text{ref}} \cos\theta_1 + \ddot{x}_{\text{ref}} \sin\theta_1) + 2k_{t2}(\delta_1) - 2k_{t1}(\theta_1),\end{aligned} \tag{5.3.40}$$

$$\begin{aligned}(m_2 + m_3)l_1 l_2[\cos(\theta_1 - \theta_2)]\ddot{\theta}_1 &+ \left(\frac{m_2}{2} + 2m_3\right)l_2^2\ddot{\theta}_2 \\ &+ m_3 l_2 l_3[\cos(\theta_2 - \theta_3)]\ddot{\theta}_3 - (m_2 + 2m_3)l_1 l_2[\sin(\theta_1 - \theta_2)]\dot{\theta}_1^2 \\ &+ m_3 l_2 l_3[\sin(\theta_2 - \theta_3)]\dot{\theta}_3^2 \\ &= \frac{2F_2(t)}{r_2} + l_2(m_2 + 2m_3) \\ &\quad \times (\ddot{z}_{\text{ref}} \cos\theta_2 + \ddot{x}_{\text{ref}} \sin\theta_2) + 2k_{t3}(\delta_2) - 2k_{t2}(\delta_1),\end{aligned} \tag{5.3.41}$$

and

$$\begin{aligned}m_3 l_1 l_3[\cos(\theta_1 - \theta_3)]\ddot{\theta}_1 &+ m_3 l_2 l_3[\cos(\theta_2 - \theta_3)]\ddot{\theta}_2 + \frac{m_3}{2} l_3^2\ddot{\theta}_3 \\ &- m_3 l_1 l_3[\sin(\theta_1 - \theta_3)]\dot{\theta}_1^2 - m_2 l_2 l_3[\sin(\theta_2 - \theta_3)]\dot{\theta}_2^2 \\ &= \frac{2F_3(t)}{r_3} + m_3 l_3(\ddot{z}_{\text{ref}} \cos\theta_3 + \ddot{x}_{\text{ref}} \sin\theta_3) - 2k_{t3}(\delta_2).\end{aligned} \tag{5.3.42}$$

The standard form of the equations are

$$\dot{p}_{13} = \left(\frac{-2F_3(t)}{r_3 l_3 \cos\theta_3} - \dot{E}p_{13} \tan\theta_3 + \frac{2k_{t3}(\delta_2)}{l_3 \cos\theta_3}\right)(1 + E\tan\theta_3)^{-1}, \tag{5.3.43}$$

$$\begin{aligned}\dot{p}_{11} = \Bigg(\frac{-2F_2(t)}{r_2 l_2 \cos\theta_2} &- \dot{D}p_{11} \tan\theta_2 - 2E\dot{p}_{13} \tan\theta_2 - 2\dot{E}p_{13} \tan\theta_2 - 2\dot{p}_{13} \\ &- \frac{2k_{t3}(\delta_2)}{l_2 \cos\theta_2} + \frac{2k_{t2}(\delta_1)}{l_2 \cos\theta_2}\Bigg)(1 - D\tan\theta_2)^{-1},\end{aligned} \tag{5.3.44}$$

$$\begin{aligned}\dot{p}_9 = \Bigg(&\frac{-2F_1(t)}{r_1 l_1 \cos\theta_1} - \dot{C}p_9 \tan\theta_1 - 2D\dot{p}_{11}\tan\theta_1 \\ &- 2\dot{D}p_{11}\tan\theta_1 - 2E\dot{p}_{13}\tan\theta_1 \\ &- 2\dot{E}p_{13}\tan\theta_1 - 2\dot{p}_{13} - 2\dot{p}_{11} - \frac{2k_{t2}(\delta_1)}{l_1\cos\theta_1} + \frac{2k_{t1}(\theta_1)}{l_1\cos\theta_1}\Bigg) \\ &\times (1 + C\tan\theta_1)^{-1}, \end{aligned} \tag{5.3.45}$$

$$\dot{p}_8 = C\dot{p}_9 + \dot{C}p_9, \tag{5.3.46}$$

$$\dot{p}_{10} = D\dot{p}_{11} + \dot{D}p_{11}, \tag{5.3.47}$$

and

$$\dot{p}_{12} = E\dot{p}_{13} + \dot{E}p_{13}, \tag{5.3.48}$$

where

$$C = \frac{\dot{x}_{\text{ref}} + (-\frac{1}{2}l_1 \sin\theta_1)\dot{\theta}_1}{\dot{z}_{\text{ref}} + (-\frac{1}{2}l_1 \cos\theta_1)\dot{\theta}_1}, \tag{5.3.49}$$

$$D = \frac{\dot{x}_{\text{ref}} + (-l_1 \sin\theta_1)\dot{\theta}_1 + (-\frac{1}{2}l_2 \sin\theta_2)\dot{\theta}_2}{\dot{z}_{\text{ref}} + (-l_1 \cos\theta_1)\dot{\theta}_1 + (-\frac{1}{2}l_1 \cos\theta_2)\dot{\theta}_2}, \tag{5.3.50}$$

and

$$E = \frac{\dot{x}_{\text{ref}} + (-l_1 \sin\theta_1)\dot{\theta}_1 + (-l_2 \sin\theta_2)\dot{\theta}_2 + (-\frac{1}{2}l_3 \sin\theta_3)\dot{\theta}_3}{\dot{z}_{\text{ref}} + (-l_1 \cos\theta_1)\dot{\theta}_1 + (-l_2 \cos\theta_2)\dot{\theta}_2 + (-\frac{1}{2}l_3 \cos\theta_3)\dot{\theta}_3}. \tag{5.3.51}$$

5.4 SOME EXAMPLES OF BOND GRAPH DYNAMIC MODELING OF ROBOTIC MANIPULATORS

5.4.1 Summary

Common efforts and common flow junctions can exist in multifaceted systems. The bond graph shows the direction of energy and information as well as causality, for example,

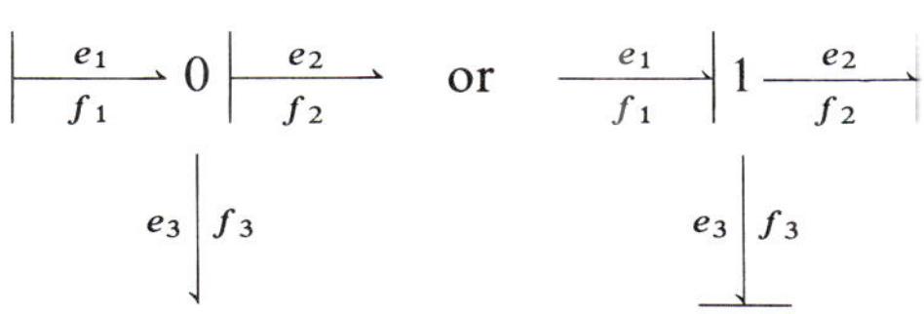

For the 0-junctions

$$\sum_{i=1}^{n} f_i = 0, \qquad e_1 = e_2 = \cdots = e_n, \tag{5.4.1}$$

and for the 1-junctions

$$\sum_{i=1}^{n} e_i = 0, \qquad f_1 = f_2 = \cdots = f_n. \tag{5.4.2}$$

These are Kirchhoff's laws for 0-junctions and 1-junctions, and they are especially suitable for joint actions.

5.4.2 Examples of Bond Graph Modeling

In what follows we illustrate the power of bond graph modeling as applied to robotic manipulators.

EXAMPLE 5.4.1 ■

Construct a bond graph for a single upright manipulator (Fig. 5.4.1) and derive the governing equations by means of the associated Kirchhoff laws at the common junctions.

Solution. To construct a bond graph for this system, we first must determine the common flow relationships between $\dot{x}$, $\dot{y}$, and the rotary speed $\dot{\theta}$. These are obtained easily as shown in Figure 5.4.2.

$$x = l \sin \theta, \tag{5.4.3}$$

$$\dot{x} = (l \cos \theta)\dot{\theta}, \tag{5.4.4}$$

$$y = l \cos \theta, \tag{5.4.5}$$

$$\dot{y} = -(l \sin \theta)\dot{\theta}, \tag{5.4.6}$$

$$p_1 \equiv m\dot{x}, \tag{5.4.7}$$

$$p_2 = m\dot{y}, \tag{5.4.8}$$

$$e_7 = (l \cos \theta)e_5 = m\ddot{x}l \cos \theta, \tag{5.4.9}$$

$$e_6 + mg = m\ddot{y}, \tag{5.4.10}$$

$$e_8 = (-l \sin \theta)e_6 = (-l \sin \theta)(m\ddot{y} - mg), \tag{5.4.11}$$

$$\begin{aligned} \tau_1 &= e_7 + e_8 + k_\tau(\theta - \theta_0) \\ &= m\ddot{x}l \cos \theta - (l \sin \theta)(m\ddot{y} - mg) + k_t(\theta - \theta_0), \end{aligned} \tag{5.4.12}$$

$$\ddot{x} = (l \cos \theta)\ddot{\theta} - (l \sin \theta)\dot{\theta}^2, \tag{5.4.13}$$

and

$$\ddot{y} = (-l \sin \theta)\ddot{\theta} - (l \cos \theta)\dot{\theta}^2. \quad ■■ \tag{5.4.14}$$

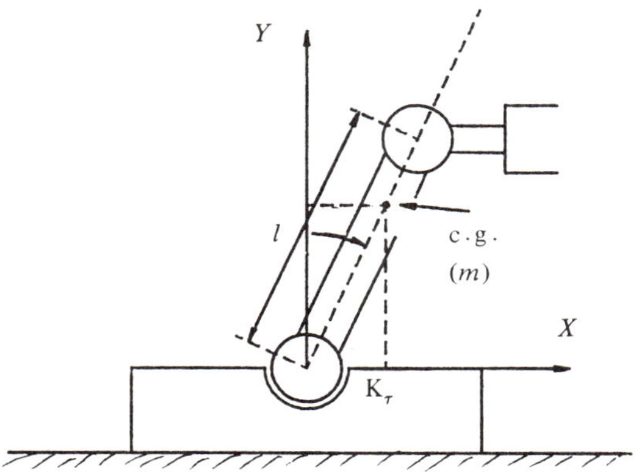

Figure 5.4.1 Geometry of a simple manipulator.

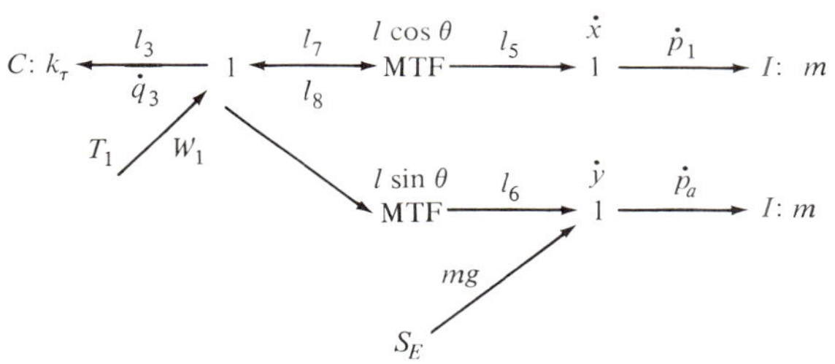

Figure 5.4.2 Bond graph for a single upright manipulator.

EXAMPLE 5.4.2 ■

Referring to the Figure 5.4.3, construct the relevant bond graph and the governing dynamic equations.

Solution. Again we first relate the $\dot{x}$, $\dot{y}$, and $\dot{z}$ flows to $\dot{\theta}$, $\dot{\phi}$, and $\dot{l}$ to identify the common flow junctions. Thus,

$$x = l \sin\phi \cos\theta, \tag{5.4.15}$$

$$y = l \sin\phi \sin\theta, \tag{5.4.16}$$

$$z = l \cos\phi, \tag{5.4.17}$$

$$\dot{x} = (l \cos\theta \cos\phi)\dot{\phi} - (l \sin\phi \sin\theta)\dot{\theta} + \dot{l} \cos\theta \sin\phi, \tag{5.4.18}$$

$$\dot{y} = (l \sin\theta \cos\phi)\dot{\phi} + (l \sin\phi \cos\theta)\dot{\theta} + \dot{l} \sin\phi \sin\theta, \tag{5.4.19}$$

and

$$\dot{z} = \dot{l} \cos\phi - (l \sin\phi)\dot{\phi}. \tag{5.4.20}$$

The relevant bond graph and the governing dynamic equations for this spherical robot are shown in Figure 5.4.4.

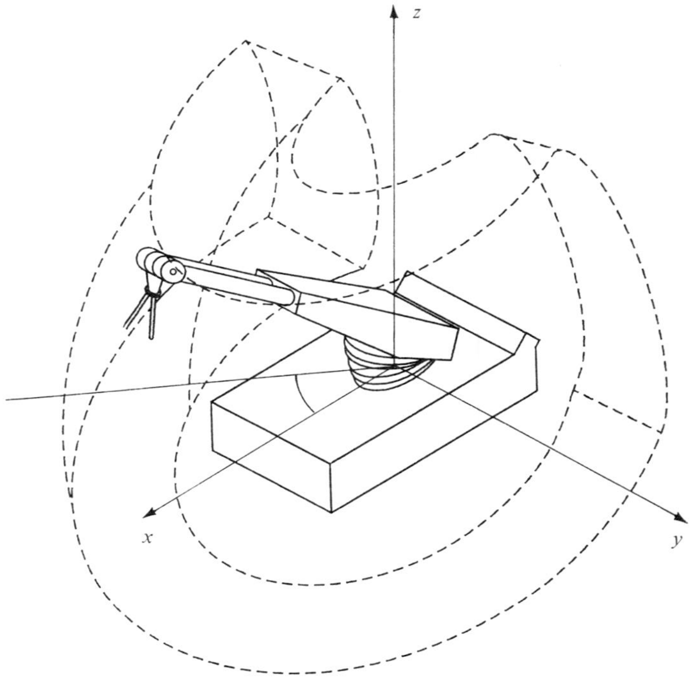

Figure 5.4.3 Geometry of a spherical robot.

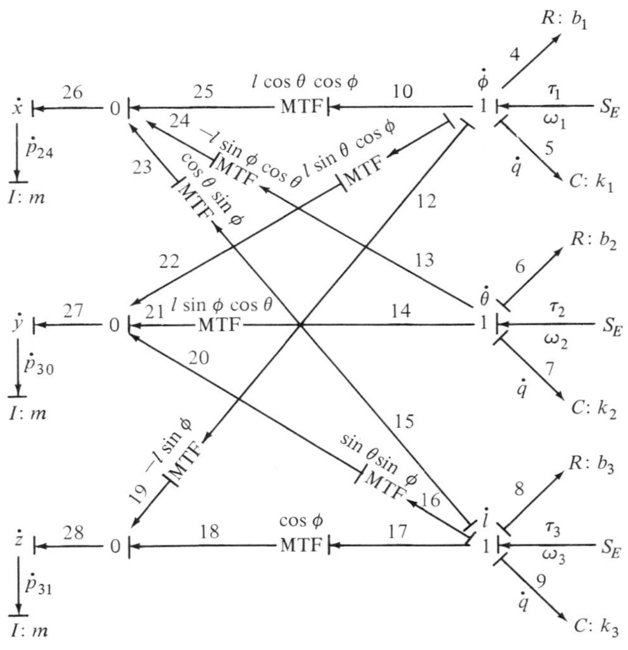

Figure 5.4.4 Bond graph corresponding to a spherical robot.

The governing equations (Kirchhoff's) are

$$T_1 = b_1\dot{\phi} + k_1\phi - m\ddot{z}l\sin\phi + m\ddot{y}l\sin\theta\cos\phi + m\ddot{x}l\cos\theta\cos\phi, \tag{5.4.21}$$

$$T_2 = b_2\dot{\theta} + k_2\theta + m\ddot{y}l\sin\phi\cos\theta - m\ddot{x}l\sin\phi\cos\theta, \tag{5.4.22}$$

and

$$T_3 = b_3\dot{l} + k_3 l + m\ddot{z}\cos\phi + m\ddot{y}\sin\theta\sin\phi + m\ddot{x}\cos\theta\sin\phi. \tag{5.4.23}$$

Substitute for $\ddot{x}$, $\ddot{y}$, and $\ddot{z}$ from the equations presented before. ■ ■

EXAMPLE 5.4.3 ■

Referring to Figure 5.4.5, which corresponds to a cylindrical polar robotic manipulator, construct the relevant bond graph and the governing dynamic equations.

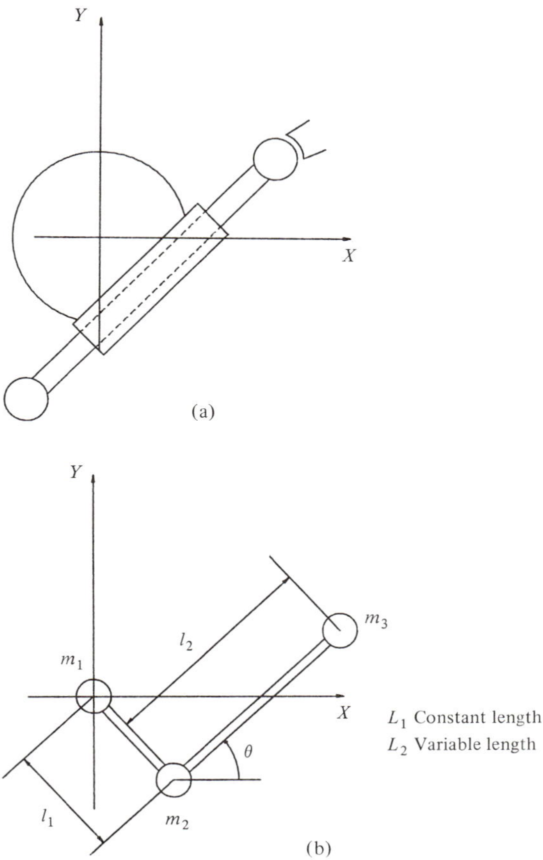

Figure 5.4.5 (a) Cylindrical polar manipulator. (b) Geometry of a polar manipulator.

Solution. Similar to the previous examples, we first match the common flows $\dot{x}$ and $\dot{y}$ to the common flows $\dot{\theta}$ and $\dot{L}_2$. Thus, corresponding to masses m_2 and m_3, we have

$$x_2 = L_1 \sin \theta, \tag{5.4.24}$$

$$y_2 = -L_1 \cos \theta, \tag{5.4.25}$$

$$x_3 = L_1 \sin \theta + L_2 \cos \theta, \tag{5.4.26}$$

and

$$y_3 = -L_1 \cos \theta + L_2 \sin \theta. \tag{5.4.27}$$

Let S_{e2}, $S_{e\theta}$, and S_{eL}, respectively, be the three sources of effort into common junctions $\dot{z}$ in the direction perpendicular to both x and y, $\dot{\theta}$, and $\dot{L}_2$. Now the governing equations for the construction of common flows are

$$\dot{x}_2 = L_1\dot{\theta} \cos \theta, \tag{5.4.28}$$

$$\dot{y}_2 = L_1\dot{\theta} \sin \theta, \tag{5.4.29}$$

$$\dot{x}_3 = L_1\dot{\theta} \cos \theta - L_2\dot{\theta} \sin \theta + \dot{L}_2 \cos \theta, \tag{5.4.30}$$

and

$$y_3 = L_1\dot{\theta} \sin \theta - L_2\dot{\theta} \cos \theta + \dot{L}_2 \sin \theta. \tag{5.4.31}$$

Thus, the relevant bond graph can be constructed, keeping in mind that the $\dot{z}$ coordinate is decoupled from $\dot{x}$ and $\dot{y}$, as shown in Figure 5.4.6.

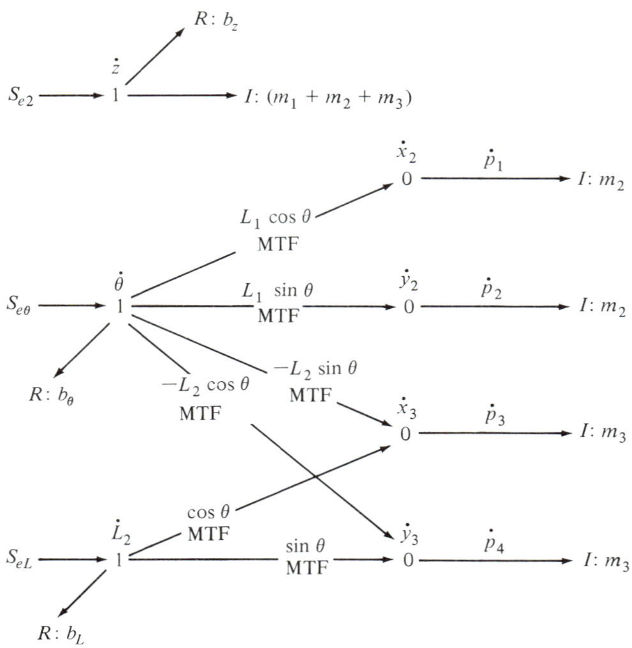

Figure 5.4.6 Bond graph corresponding to a polar cylindrical robot.

The governing equations are

$$S_{e2} = \dot{\theta} b_\theta + \dot{p}_1 L_1 \cos\theta + \dot{p}_2 L_1 \sin\theta - \dot{p}_3 L_2 \sin\theta - \dot{p}_4 L_2 \cos\theta, \tag{5.4.32}$$

$$S_{eL} = \dot{L}_2 b_L + \dot{p}_3 \cos\theta + \dot{p}_4 \sin\theta, \tag{5.4.33}$$

$$\dot{x}_2 = (L_1 \cos\theta)\dot{\theta}, \tag{5.4.34}$$

$$\dot{y}_2 = (L_1 \sin\theta)\dot{\theta}, \tag{5.4.35}$$

$$\dot{x}_3 = (L_1 \cos\theta)\dot{\theta} - (L_2 \sin\theta)\dot{\theta} + \dot{L}_2 \cos\theta, \tag{5.4.36}$$

$$\dot{y}_3 = (L_1 \sin\theta)\dot{\theta} - (L_2 \cos\theta)\dot{\theta} + \dot{L}_2 \sin\theta, \tag{5.4.37}$$

$$\ddot{x}_2 = (-L_1 \sin\theta)\dot{\theta}^2 + (L_1 \cos\theta)\ddot{\theta}, \tag{5.4.38}$$

$$\ddot{y}_2 = (L_1 \cos\theta)\dot{\theta}^2 + (L_1 \sin\theta)\ddot{\theta}, \tag{5.4.39}$$

$$\ddot{x}_3 = (-L_1 \sin\theta)\dot{\theta}^2 + (L_1 \cos\theta)\ddot{\theta} - (L_2 \cos\theta)\dot{\theta}^2 - (L_2 \sin\theta)\ddot{\theta} + \ddot{L}_2 \cos\theta - (\dot{L}_2 \sin\theta)\dot{\theta}, \tag{5.4.40}$$

$$\ddot{y}_3 = (L_1 \cos\theta)\dot{\theta}^2 + (L_1 \sin\theta)\ddot{\theta} + (L_2 \sin\theta)\dot{\theta}^2 - (L_2 \cos\theta)\ddot{\theta} + \ddot{L}_2 \sin\theta + (\dot{L}_2 \cos\theta)\dot{\theta}, \tag{5.4.41}$$

$$\begin{aligned} S_{e\theta} = {} & \dot{\theta} b_\theta + m_2 L_1^2[(\cos\theta)\ddot{\theta} - (\sin\theta)\dot{\theta}^2] \cos\theta \\ & + m_2 L_1^2[(\cos\theta)\dot{\theta}^2 + (\sin\theta)\ddot{\theta}] \sin\theta \\ & + m_3[(L_1 \sin\theta)\dot{\theta}^2 \\ & - (L_1 \cos\theta)\ddot{\theta} + (L_2 \cos\theta)\dot{\theta}^2 + (L_2 \sin\theta)\ddot{\theta} \\ & - \ddot{L}_2 \cos\theta + (\dot{L}_2 \sin\theta)\dot{\theta}]L_2 \sin\theta \\ & - m_3(L_1 \cos\theta)\dot{\theta}^2 + (L_1 \sin\theta)\ddot{\theta} \\ & + (L_2 \sin\theta)\dot{\theta}^2 - (L_2 \cos\theta)\ddot{\theta} \\ & + \ddot{L}_2 \sin\theta + (\dot{L}_2 \cos\theta)\dot{\theta}]L_2 \cos\theta, \end{aligned} \tag{5.4.42}$$

$$\begin{aligned} S_{eL} = {} & \dot{L}_2 b_L + m_3[(-L_1 \sin\theta)\dot{\theta}^2 + (L_1 \cos\theta)\ddot{\theta} \\ & - (L_2 \cos\theta)\dot{\theta}^2 - (L_2 \sin\theta)\ddot{\theta} \\ & + \ddot{L}_2 \cos\theta - (\dot{L}_2 \sin\theta)\dot{\theta}] \cos\theta + m_2[(L_1 \cos\theta)\dot{\theta}^2 + (L_1 \sin\theta)\ddot{\theta} \\ & + (\ddot{L}_2 \sin\theta)\dot{\theta}^2 - (L_2 \cos\theta)\ddot{\theta} + L_2 \sin\theta + (\dot{L}_2 \cos\theta)\dot{\theta}] \sin\theta, \end{aligned} \tag{5.4.43}$$

and

$$S_{e2} = (m_1 + m_2 + m_3)g + \dot{z} b_z + \ddot{z}(m_1 + m_2 + m_3). \quad ■■ \tag{5.4.44}$$

EXAMPLE 5.4.4 ■

For the three-axis robotic manipulator shown in Figure 5.4.7 derive the governing dynamic equations using bond graph modeling.

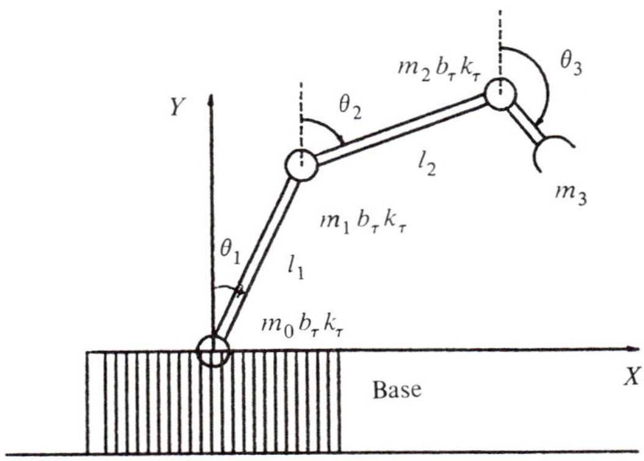

Figure 5.4.7 Simplified geometry of a five-axis robotic manipulator for bond graph modeling.

Solution. Again we first relate the linear and angular velocities:

$$x_1 = l_1 \sin \theta_1, \tag{5.4.45}$$

$$\dot{x}_1 = (l_1 \cos \theta_1)\dot{\theta}_1, \tag{5.4.46}$$

$$\ddot{x}_1 = (l_1 \cos \theta_1)\ddot{\theta}_1 - (l_1 \sin \theta_1)\dot{\theta}_1^2, \tag{5.4.47}$$

$$y_1 = l_1 \cos \theta_1, \tag{5.4.48}$$

$$\dot{y}_1 = -(l_1 \sin \theta_1)\dot{\theta}_1, \tag{5.4.49}$$

$$\ddot{y}_1 = -(l_1 \sin \theta_1)\theta_1 - (l_1 \cos \theta_1)\dot{\theta}_1^2, \tag{5.4.50}$$

$$x_2 = l_1 \sin \theta_1 + l_2 \sin \theta_2, \tag{5.4.51}$$

$$\dot{x}_2 = (l_1 \cos \theta_1)\dot{\theta}_1 + (l_2 \cos \theta_2)\dot{\theta}_2, \tag{5.4.52}$$

$$\ddot{x}_2 = (l_1 \cos \theta_1)\ddot{\theta}_1 - (l_1 \sin \theta_1)\dot{\theta}_1^2 + (l_2 \cos \theta_2)\ddot{\theta}_2 - (l_2 \sin \theta_2)\dot{\theta}_2^2, \tag{5.4.53}$$

$$y_2 = l_1 \cos \theta_1 + l_2 \cos \theta_2, \tag{5.4.54}$$

$$\dot{y}_2 = -(l_1 \sin \theta_1)\dot{\theta}_1 - (l_2 \sin \theta_2)\dot{\theta}_2, \tag{5.4.55}$$

$$\ddot{y}_2 = -(l_1 \sin \theta_1)\ddot{\theta}_1 - (l_1 \cos \theta_1)\dot{\theta}_1^2 - (l_2 \sin \theta_2)\ddot{\theta}_2 - (l_2 \cos \theta_2)\dot{\theta}_2^2, \tag{5.4.56}$$

$$x_3 = l_1 \sin \theta_1 + l_2 \sin \theta_2 + l_3 \sin \theta_3, \tag{5.4.57}$$

$$\dot{x}_3 = (l_1 \cos \theta_1)\dot{\theta}_1 + (l_2 \cos \theta_2)\dot{\theta}_2 + (l_3 \cos \theta_3)\dot{\theta}_3, \tag{5.4.58}$$

$$\begin{aligned}\ddot{x}_3 = {} & (l_1 \cos \theta_1)\ddot{\theta}_1 - (l_1 \sin \theta_1)\dot{\theta}_1^2 + (l_2 \cos \theta_2)\ddot{\theta}_2 \\ & + (l_2 \sin \theta_2)\dot{\theta}_2^2 + (l_3 \cos \theta_3)\ddot{\theta}_3 - (l_3 \sin \theta_3)\dot{\theta}_3^2,\end{aligned} \tag{5.4.59}$$

$$y_3 = l_1 \cos \theta_1 + l_2 \cos \theta_2 + l_3 \cos \theta_3, \tag{5.4.60}$$

$$\dot{y}_3 = -(l_1 \sin \theta_1)\dot{\theta}_1 - (l_2 \sin \theta_2)\dot{\theta}_2 - (l_3 \sin \theta_3)\dot{\theta}_3, \tag{5.4.61}$$

and

$$\begin{aligned}\ddot{y}_3 = &-(l_1 \sin \theta_1)\ddot{\theta}_1 - (l_1 \cos \theta_1)\dot{\theta}_1^2 - (l_2 \sin \theta_2)\ddot{\theta}_2 \\ &- (l_2 \cos \theta_2)\dot{\theta}_2^2 - (l_3 \sin \theta_3)\ddot{\theta}_3 - (l_3 \cos \theta_3)\dot{\theta}_3^2. \qquad (5.4.62)\end{aligned}$$

Note that $\dot{p}_7 = m_1\ddot{x}_1$, $\dot{p}_8 = m_1\ddot{y}_1$, $\dot{p}_9 = m_2\ddot{x}_2$, $\dot{p}_{10} = m_2\ddot{y}_2$, $\dot{p}_{11} = m_3\ddot{x}_3$, and $\dot{p}_{12} = m_3\ddot{y}_3$.

For $\dot{\theta}_1$,

$$\tau_1(t) = \sum_{i=1}^{9} e_i^{(1)}(t)$$

and

$$\begin{aligned}\tau_1(t) = &\tau_2(t) - b_2(\dot{\theta}_2 - \dot{\theta}_1) - k_{\tau_2}(\theta_2 - \theta_1) + k_{\tau_1}\theta_1 + b_1\dot{\theta}_1 \\ &+ m_1(\alpha_1 l_1 \cos \theta_1)(l_1\ddot{\theta}_1 \cos \theta_1 - l_1\dot{\theta}_1^2 \sin \theta_1) \\ &+ m_1(\alpha_1 l_1 \sin \theta_1)(l_1\ddot{\theta}_1 \sin \theta_1 + l_1\dot{\theta}_1^2 \cos \theta_1) \\ &+ m_2(l_1 \cos \theta_1)(l_1\ddot{\theta}_1 \cos \theta_1 - l_1\dot{\theta}_1^2 \sin \theta_1 \\ &+ l_2\ddot{\theta}_2 \cos \theta_2 - l_2\dot{\theta}_2^2 \sin \theta_2) \\ &+ m_2(l_1 \sin \theta_1)(l_1\ddot{\theta}_1 \sin \theta_1 \\ &+ l_1\dot{\theta}_1^2 \cos \theta_1 + l_2\ddot{\theta}_2 \sin \theta_2 + l_2\dot{\theta}_2^2 \cos \theta_2) \\ &+ m_3(l_1 \cos \theta_1)(l_1\ddot{\theta}_1 \cos \theta_1 - l_1\dot{\theta}_1^2 \sin \theta_1 \\ &+ l_2\ddot{\theta}_2 \cos \theta_2 - l_2\dot{\theta}_2^2 \sin \theta_2 + l_3\ddot{\theta}_3 \cos \theta_3 - l_3\dot{\theta}_3^2 \sin \theta_3) \\ &+ m_3(l_1 \sin \theta_1)(l_1\ddot{\theta}_1 \sin \theta_1 + l_1\dot{\theta}_1^2 \cos \theta_1 + l_2\ddot{\theta}_2 \sin \theta_2 \\ &+ l_2\dot{\theta}_2^2 \cos \theta_2 + l_3\ddot{\theta}_3 \sin \theta_3 + l_3\dot{\theta}_3^2 \cos \theta_3), \qquad (5.4.63)\end{aligned}$$

where $\tau_1(t)$ and $\tau_2(t)$ are the time-dependent actuating torques at joints 1 and 2, b_1 and b_2 are the corresponding linear damping coefficients, and k_{τ_1} and k_{τ_2} are the torsional elasticities, respectively.

For $\dot{\theta}_2$,

$$\sum_{i=1}^{6} e_i^{(2)}(t) = 0$$

and

$$\begin{aligned}&k_{\tau_2}(\theta_2 - \theta_1) - b_2(\dot{\theta}_2 - \dot{\theta}_1) - \tau_2(t) \\ &+ (m_2\alpha_2 l_2 \cos \theta_2)(l_1\ddot{\theta}_1 \cos \theta_1 - l_1\dot{\theta}_1^2 \sin \theta_1 \\ &+ l_2\ddot{\theta}_2 \cos \theta_2 - l_2\dot{\theta}_2^2 \sin \theta_2) \\ &+ (m_2\alpha_2 l_2 \sin \theta_2) \\ &\times (l_1\ddot{\theta}_1 \sin \theta_1 + l_1\dot{\theta}_1^2 \cos \theta_1 + l_2\ddot{\theta}_2 \sin \theta_2 + l_2\dot{\theta}_2^2 \cos \theta_2) \\ &+ m_3 l_2 \cos \theta_2(l_1\ddot{\theta}_1 \cos \theta_1 - l_1\dot{\theta}_1^2 \sin \theta_1 + l_2\ddot{\theta}_2 \cos \theta_2 \\ &- l_2\dot{\theta}_2^2 \sin \theta_2 + l_3\ddot{\theta}_3 \cos \theta_3 - l_3\dot{\theta}_3^2 \sin \theta_3) \\ &+ m_3 l_2 \sin \theta_2(l_1\ddot{\theta}_1 \sin \theta_1 + l_1\dot{\theta}_1^2 \cos \theta_1 + l_2\ddot{\theta}_2 \sin \theta_2 \\ &+ l_2\dot{\theta}_2^2 \cos \theta_2 + l_3\ddot{\theta}_3 \sin \theta_3 + l_3\dot{\theta}_3^2 \cos \theta_3) \\ &+ \tau_3(t) - k_{\tau_3}(\theta_3 - \theta_2) - b_3(\dot{\theta}_3 - \dot{\theta}_2) = 0. \qquad (5.4.64)\end{aligned}$$

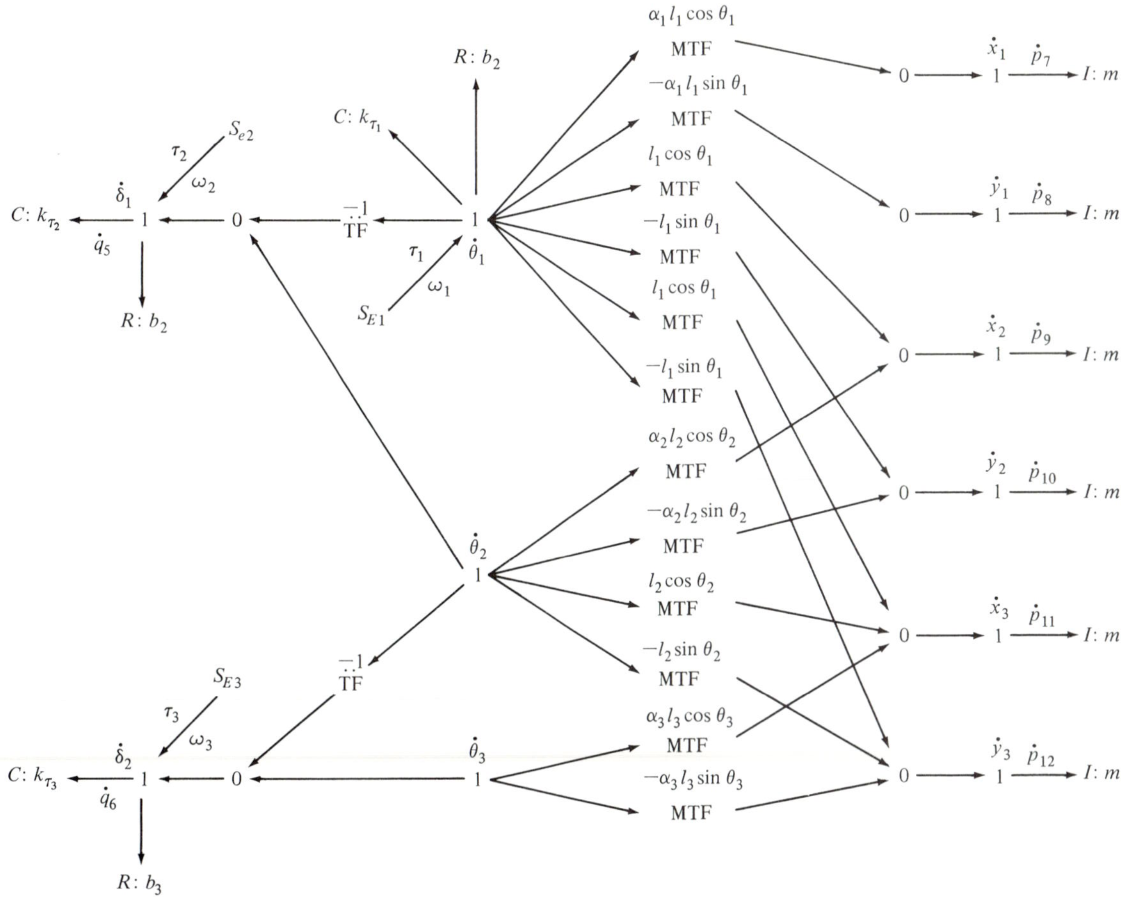

Figure 5.4.8 Bond graph corresponding to the robotic manipuator in Figure 5.4.7.

For $\dot{\theta}_3$,

$$\sum_{i=1}^{3} e_i^{(3)}(t) = 0$$

and

$$\begin{aligned} k_{\tau_3}(\theta_3 - \theta_2) &+ b_3(\dot{\theta}_3 - \dot{\theta}_2) - \tau_3(t) + (m_3\alpha_3 l_3 \cos\theta_3) \\ &\times (l_1\ddot{\theta}_1 \cos\theta_1 - l_1\dot{\theta}_1^2 \sin\theta_1 + l_2\ddot{\theta}_2 \cos\theta_2 \\ &- l_2\dot{\theta}_2^2 \sin\theta_2 + l_3\ddot{\theta}_3 \cos\theta_3 - l_3\dot{\theta}_3^2 \sin\theta_3) \\ &+ m_3\alpha_3 l_3 \sin\theta_3(l_1\ddot{\theta}_1 \sin\theta_1 + l_1\dot{\theta}_1^2 \cos\theta_1 \\ &+ l_2\ddot{\theta}_3 \sin\theta_2 + l_2\dot{\theta}_2^2 \cos\theta_2 \\ &+ l_3\ddot{\theta}_3 \sin\theta_3 + l_3\dot{\theta}_3^2 \cos\theta_3) = 0. \end{aligned} \tag{5.4.65}$$

Simplifying Equations (5.4.63), (5.4.64), and (5.4.65) yields

$$\mathbf{I}(\theta_1, \theta_2, \theta_3)\ddot{\boldsymbol{\theta}} + \mathbf{J}(\theta_1, \theta_2, \theta_3)\dot{\boldsymbol{\theta}}^2 + \mathbf{b}\dot{\boldsymbol{\theta}} + \mathbf{k}\boldsymbol{\theta} = \mathbf{T}(t), \tag{5.4.66}$$

where

$$\mathbf{I} = \begin{bmatrix} l_1^2(\alpha_1 m_1 + m_2 + m_3) & l_1 l_2(m_2 + m_3)\cos(\theta_1 - \theta_2) & m_3 l_1 l_3 \cos(\theta_1 - \theta_3) \\ l_1 l_2(\alpha_2 m_2 + m_3)\cos(\theta_1 - \theta_2) & (\alpha_2 m_2 + m_3) l_2^2 & m_3 l_2 l_3 \cos(\theta_2 - \theta_3) \\ \alpha_3 l_3 l_1 m_3 \cos(\theta_1 - \theta_3) & \alpha_3 l_2 l_3 \cos(\theta_2 - \theta_3) & \alpha_3 l_3^2 m_3 \end{bmatrix} \tag{5.4.67}$$

and

$$\mathbf{J} = \begin{bmatrix} 0 & l_1 l_2(m_2 + m_3)\sin(\theta_1 - \theta_2) & m_3 l_1 l_3 \sin(\theta_1 - \theta_3) \\ -l_1 l_2(\alpha_2 m_2 + m_3)\sin(\theta_1 - \theta_2) & 0 & m_3 l_2 l_3 \sin(\theta_3 - \theta_2) \\ -\alpha_3 m_3 l_1 l_3 \sin(\theta_1 - \theta_3) & -m_3\alpha_3 l_2 l_3 \sin(\theta_2 - \theta_3) & 0 \end{bmatrix}. \tag{5.4.68}$$

See Figure 5.4.8 for the bond graph corresponding to Figure 5.4.7.

■ ■

EXAMPLE 5.4.5 ■

For the two-handed robot shown in Figure 5.4.9, derive the governing dynamic equations.

Solution. As discussed before we first relate the linear and angular velocities to construct the common flow junctions:

$$z_2 = z_0 + \alpha l \sin\theta_2, \tag{5.4.69}$$

$$z_3 = z_0 + \alpha l \sin\theta_3, \tag{5.4.70}$$

$$x_2 = -x_0 \cos\theta_1 - \alpha l \cos\theta_2 \sin\theta_2, \tag{5.4.71}$$

$$x_3 = x_0 \cos\theta_1 - \alpha l \cos\theta_3 \sin\theta_1. \tag{5.4.72}$$

$$y_2 = -x_0 \sin\theta_1 + \alpha l \cos\theta_2 \cos\theta_1, \tag{5.4.73}$$

$$y_3 = x_0 \sin\theta_1 + \alpha l \cos\theta_3 \cos\theta_1, \tag{5.4.74}$$

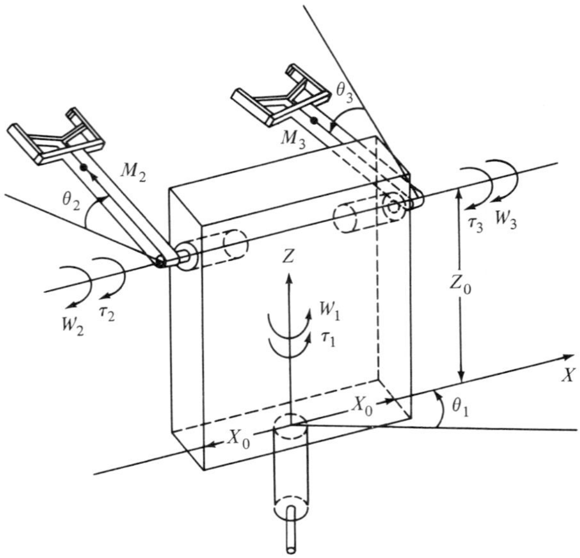

Figure 5.4.9 Geometry of a two-handed robotic manipulator.

$$\dot{z}_2 = \alpha l\dot{\theta}_2 \cos\theta_2, \tag{5.4.75}$$

$$\dot{z}_3 = \alpha l\dot{\theta}_3 \cos\theta_3, \tag{5.4.76}$$

$$\dot{x}_2 = x_0\dot{\theta}_1 \sin\theta_1 + \alpha l\dot{\theta}_2 \sin\theta_2 \sin\theta_1 - \alpha l\dot{\theta}_1 \cos\theta_2 \cos\theta_1, \tag{5.4.77}$$

$$\dot{x}_3 = -x_0\dot{\theta}_1 \sin\theta_1 + \alpha l\dot{\theta}_3 \sin\theta_3 \sin\theta_1 - \alpha l\dot{\theta}_1 \cos\theta_3 \cos\theta_1, \tag{5.4.78}$$

$$\dot{y}_2 = -x_0\dot{\theta}_1 \cos\theta_1 - \alpha l\dot{\theta}_2 \sin\theta_2 \cos\theta_1 - \alpha l\dot{\theta}_1 \cos\theta_2 \sin\theta_1, \tag{5.4.79}$$

$$\dot{y}_3 = x_0\dot{\theta}_1 \cos\theta_1 - \alpha l\dot{\theta}_3 \sin\theta_3 \cos\theta_3 \cos\theta_1 - \alpha l\dot{\theta}_1 \cos\theta_3 \sin\theta_1, \tag{5.4.80}$$

and

$$\begin{aligned} \dot{p}_4 &= m_2\ddot{x}_2, & \dot{p}_7 &= m_3\ddot{x}_3, & \dot{p}_5 &= m_2\ddot{z}_2, \\ \dot{p}_8 &= m_3\ddot{z}_3, & \dot{p}_6 &= m_2\ddot{y}_2, & \dot{p}_9 &= m_3\ddot{y}_2. \end{aligned} \tag{5.4.81}$$

See Figure 5.4.10 for the bond graph corresponding to this problem.

$$\begin{aligned} \dot{\theta}_1 \Rightarrow T_1 - b_1\dot{\theta}_1 \frac{d}{dt}(J_1\dot{\theta}_1) - k_{T_1}\theta_1 \\ &= (x_0 \sin\theta_1 - \alpha l \cos\theta_1 \cos\theta_2)m_2\ddot{x}_2 \\ &\quad - (x_0 \cos\theta_1 - \alpha l \cos\theta_2 \sin\theta_1)m_2\ddot{y}_2 \\ &\quad - (x_0 \sin\theta_1 + \alpha l \cos\theta_1 \cos\theta_3)m_3\ddot{x}_3 \\ &\quad + (x_0 \cos\theta_1 - \alpha l \cos\theta_3 \sin\theta_1)m_3\ddot{y}_3, \end{aligned} \tag{5.4.82}$$

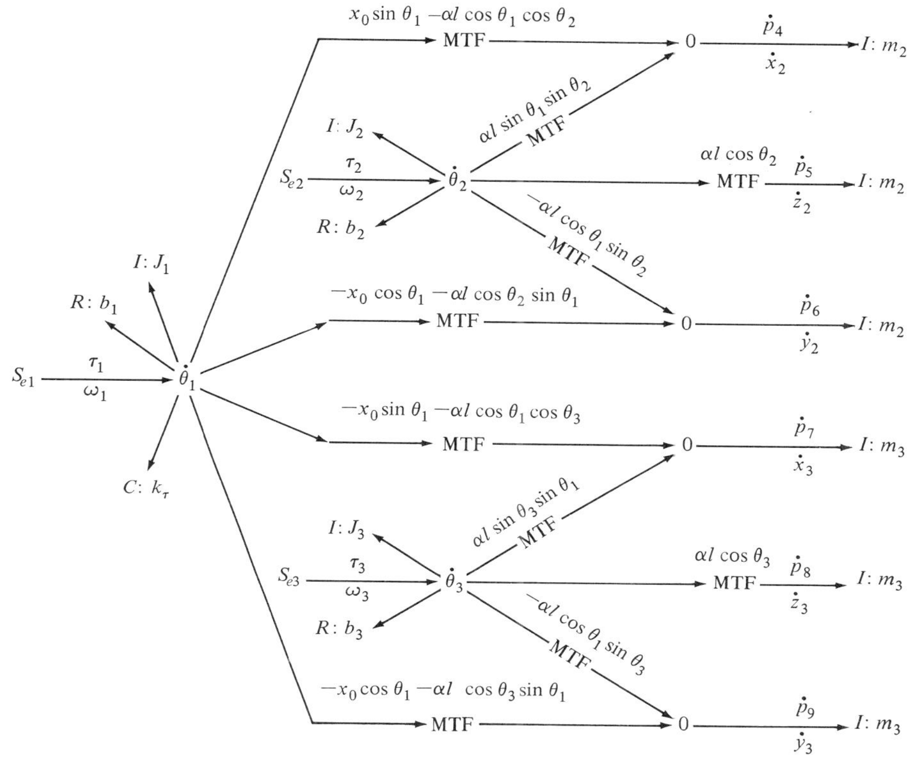

Figure 5.4.10 Bond graph corresponding to a two-handed robot.

$$
\begin{aligned}
\dot{\theta}_2 \Rightarrow T_2 - b_2\dot{\theta}_2 - J_2\ddot{\theta}_2 - k_{T_2}\theta_2 \\
= (\alpha l \sin\theta_1 \sin\theta_2)m_2\ddot{x}_2 + (\alpha l \cos\theta_2)m_2\ddot{z}_2 - (\alpha l \cos\theta_1 \sin\theta_2)m_2\ddot{y}_2,
\end{aligned}
\tag{5.4.83}
$$

and

$$
\begin{aligned}
\theta_3 \Rightarrow T_3 - b_3\dot{\theta}_3 - J_3\ddot{\theta}_3 - k_{T_3}\theta_3 \\
= (\alpha l \sin\theta_3 \sin\theta_1)m_3\ddot{x}_3 + (\alpha l \cos\theta_3)m_3\ddot{z}_3 - (\alpha l \cos\theta_1 \sin\theta_3)m_3\ddot{y}_3.
\end{aligned}
\tag{5.4.84}
$$

Substitute for x_2, y_2, z_2, x_3, y_3, and z_3 as functions of θ_i, $\dot{\theta}_i$, and $\ddot{\theta}_i$.

■ ■

5.5 LAGRANGE–EULER DYNAMIC MODELING OF ROBOTIC MANIPULATORS IN BRIEF

Here we apply the Lagrangian formulation for nonconservative systems. Thus, we employ the Lagrangian technique in conjunction with the Denavit–Hartenberg representation of coordinate frames in open-loop kinematic chains.

The general formulation of the Lagrange–Euler approach is based on the following equation:

$$\frac{d}{dt}\left(\frac{\partial L}{\partial \dot{q}_i}\right) - \left(\frac{\partial L}{\partial q_i}\right) = \tau_i, \tag{5.5.1}$$

where $L \equiv$ Lagrangian function $\equiv K - V$
$T \equiv$ kinetic energy of the whole system
$V \equiv$ potential energy of the whole system
$q_i \equiv$ generalized coordinate, that is, θ_i
$\tau_i \equiv$ generalized forces or torques

In what follows we derive a set of expressions for T, V, and L so as to derive the governing equations of robotic dynamics.

5.5.1 Preliminary Definitions

We consider the robotic arm to be composed of rigid links that are subjected to rigid body dynamics formulation. Realistically, all robotic manipulators are made up of flexible structures. However, such treatments are beyond the scope of this book because of the complexity. We now review a number of fundamental definitions, such as the generalized robotic coordinates, general dynamic constraints, robotic velocity and acceleration in robotic generalized coordinates, dynamic transformation laws between generalized robotic coordinates, robotic velocity and acceleration in moving frames, the inertia tensor, Newton's equation of motion, Euler's equations of motion, the Lagrangian and Lagrange's equations, and Hamilton's principle of least action, which leads to the Lagrange–Euler governing equations for robot dynamics.

5.5.2 Generalized Robotic Coordinates

Consider a robotic arm as a series of rigid links connected by a series of N prismatic or revolute joints. The independent freedom of motion at each joint can be considered as a degree of freedom associated with the independent actuation of a displacement field q_m, $m = 1, 2, \ldots, N$, in a robotic manipulator. On the other hand, the robotic attitude change at the end effector can be described by changes in a six-dimensional coordinate system $\mathbf{x}_i = [p_x \; p_y \; p_z \; \theta \; \phi \; \psi]$ of the end effector frame T_0^n, such that $\mathbf{p} = [p_x \, p_y \, p_z]^T$ is the position vector of the origin of the T_0^n frame and θ, ϕ, and ψ are either the Eulerian angles or the RPY angles of the T_0^n with respect to the base frame (x_0, y_0, z_0).

From the material covered in Chapter 4, it is clear that

$$x_i = x_i(q_1, q_2, \ldots, q_N, t) \tag{5.5.2}$$

or conversely

$$q_m = q_m(x_1, x_2, x_3, x_4, x_5, x_6, t) \tag{5.5.3}$$

where t is the time. Given Equation (5.5.2), a necessary condition for solving for q_m is that the Jacobian of transformation must be nonsingular; that is, its determinant must be different from zero:

$$|\mathbf{J}(x_i, q_m)| = \left|\frac{\partial x_i}{\partial q_m}\right| \neq 0. \tag{5.5.4}$$

This means that we can obtain a unique solution only when $m = 6$ and the robot is a six-axis robotic manipulator.

5.5.3 Dynamic Constraints

Dynamic constraints may be present for a robotic manipulator in motion. Generally, these constraints are expressed in the form

$$\phi_c(x_1, x_2, x_3, x_4, x_5, x_6, t) = 0 \tag{5.5.5}$$

or

$$\phi_c^*(q_1, q_2, \ldots, q_N, t) = 0. \tag{5.5.6}$$

5.5.4 Robotic Velocity and Acceleration

For an N-jointed robotic manipulator with no constraints, one may assign a Cartesian rectangular coordinate frame at each joint and define a generalized Cartesian joint coordinate $x_i(j)$ pertaining to the jth frame, that is, the $(j + 1)$th joint. Thus, one has the following relationships:

$$x_{i(j)} = \hat{x}_{i(j)}(q_1, q_2, \ldots, q_N, t), \qquad i, j = 1, 2, \ldots, N. \tag{5.5.7}$$

Then the velocity is defined as

$$\dot{x}_{i(j)} = \sum_{m=1}^{N} \frac{\partial x_{i(j)}}{\partial q_m} \dot{q}_m + \frac{\partial x_{i(j)}}{\partial t}, \tag{5.5.8}$$

and the acceleration is defined as

$$\begin{aligned} \ddot{x}_{i(j)} = {} & \sum_{m=1}^{N} \left[\frac{\partial x_{i(j)}}{\partial q_m}\right] \ddot{q}_m + \sum_{m=1}^{N} \sum_{n=1}^{N} \left[\frac{\partial x_{i(j)}^2}{\partial q_m \partial q_n}\right] \dot{q}_m \dot{q}_n \\ & + 2 \sum_{m=1}^{N} \left[\frac{\partial x_{i(j)}^2}{\partial q_m \partial t}\right] \dot{q}_m + \frac{\partial x_{i(j)}^2}{\partial t^2}. \end{aligned} \tag{5.5.9}$$

EXAMPLE 5.5.1 ■

Determine the components of the velocity and acceleration vectors at the origin of the gripper frame of a cylindrical polar robotic manipulator.

Solution. Referring to Figure 5.4.5, we see that at the origin of the gripper frame

$$x = r \cos \theta, \qquad y = r \sin \theta, \qquad z = z, \tag{5.5.10}$$

$$r = (x^2 + y^2)^{1/2}, \qquad \theta = \tan^{-1}(y/x). \tag{5.5.11}$$

The components of the velocity vector are

$$\dot{x} = \dot{r} \cos \theta - r\dot{\theta} \sin \theta, \tag{5.5.12}$$

$$\dot{y} = \dot{r} \sin \theta + r\dot{\theta} \cos \theta, \tag{5.5.13}$$

$$\dot{r} = (\dot{x}x + \dot{y}y)(x^2 + y^2)^{-1/2}, \tag{5.5.14}$$

and

$$\dot{\theta} = (x\dot{y} - y\dot{x})(x^2 + y^2)^{-1/2}. \tag{5.5.15}$$

The components of the acceleration vector are

$$\ddot{x} = (\ddot{r} - r\dot{\theta}^2) \cos \theta - (r\ddot{\theta} + 2\dot{r}\dot{\theta}) \sin \theta \tag{5.5.16}$$

and

$$\ddot{y} = (\ddot{r} - r\dot{\theta}^2) \sin \theta - (r\ddot{\theta} + 2\dot{r}\dot{\theta}) \cos \theta. \tag{5.5.17}$$

The radial and tangential components of the velocity $\mathbf{v}$ and the acceleration $\mathbf{a}$, respectively, are given by

$$\mathbf{v} = \dot{r}\mathbf{e}_r + r\dot{\theta}\mathbf{e}_\theta + \dot{z}\mathbf{e}_z \tag{5.5.18}$$

and

$$\mathbf{a} = (\ddot{r} - r\dot{\theta}^2)\mathbf{e}_r + (r\ddot{\theta} + 2\dot{r}\dot{\theta})\mathbf{e}_\theta + \ddot{z}\mathbf{e}_z, \tag{5.5.19}$$

where $\mathbf{e}_r$, $\mathbf{e}_\theta$, and $\mathbf{e}_z$ are the unit vectors in the radial, tangential, and axial directions, respectively. ■ ■

EXAMPLE 5.5.2 ■

Determine the components of the velocity and acceleration vectors at the origin of the gripper frame of a spherical robotic manipulator (see Fig. 5.4.3).

Solution. The origin of the gripper frame is at

$$r = (x^2 + y^2 + z^2)^{1/2}, \tag{5.5.20}$$

$$\theta = \cot^{-1}\left(\frac{z}{(x^2 + y^2)^{1/2}}\right), \qquad \phi = \tan^{-1}(y/x), \tag{5.5.21}$$

$$x = r \sin \theta \cos \phi, \qquad y = r \sin \theta \sin \phi, \qquad z = r \cos \theta. \tag{5.5.22}$$

The Cartesian and curvilinear components of the velocity vector are, respectively,

$$\dot{x} = \dot{r} \sin \theta \cos \phi + r\dot{\theta} \cos \theta \cos \phi - r\dot{\phi} \sin \theta \sin \phi, \tag{5.5.23}$$

$$\dot{y} = \dot{r} \sin \theta \sin \phi + r\dot{\theta} \cos \theta \sin \phi - r\dot{\phi} \sin \theta \cos \phi, \tag{5.5.24}$$

$$\dot{z} = \dot{r} \cos \theta - r\dot{\theta} \sin \theta, \tag{5.5.25}$$

$$\dot{r} = \frac{x\dot{x} + y\dot{y} + z\dot{z}}{(x^2 + y^2 + z^2)^{1/2}}, \tag{5.5.26}$$

and

$$\dot{\theta} = \frac{(x\dot{x} + y\dot{y})z - \dot{z}(x^2 + y^2)}{(x^2 + y^2)^{1/2}(x^2 + y^2 + z^2)}. \tag{5.5.27}$$

The Cartesian and curvilinear components of the acceleration vector are, respectively,

$$\ddot{x} = (\ddot{r} - r\dot{\theta}^2 - r\dot{\phi}^2) \sin\theta \cos\phi + (2\dot{r}\dot{\theta} + r\ddot{\theta}) \cos\theta \cos\phi \\ - (r\ddot{\phi} + 2\dot{r}\dot{\phi}) \sin\theta \sin\phi - 2r\dot{\theta}\dot{\phi} \cos\theta \sin\phi, \tag{5.5.28}$$

$$\ddot{y} = (\ddot{r} - r\dot{\theta}^2 - r\dot{\phi}^2) \sin\theta \sin\phi + (2\dot{r}\dot{\theta} + r\ddot{\theta}) \cos\theta \sin\phi \\ + (r\ddot{\phi} + 2\dot{r}\dot{\phi}) \sin\theta \cos\phi + 2r\dot{\theta}\dot{\phi} \cos\theta \cos\phi, \tag{5.5.29}$$

$$\ddot{z} = (\ddot{r} - r\dot{\theta}^2) \cos\theta - (r\ddot{\theta} + 2\dot{r}\dot{\theta}) \sin\theta, \tag{5.5.30}$$

$$\ddot{r} = (x\ddot{x} + \dot{x}^2 + y\ddot{y} + \dot{y}^2 + z\ddot{z} + \dot{z}^2)(x^2 + y^2 + z^2)^{-1/2} \\ - (x\dot{x} + y\dot{y} + z\dot{z})^2(x^2 + y^2 + z^2)^{-3/2}, \tag{5.5.31}$$

$$\ddot{\theta} = [(x\ddot{x} + \dot{x}^2 + y\ddot{y} + \dot{y}^2)z - \dot{z}(x\dot{x} + y\dot{y}) - \ddot{z}(x^2 + y^2)] \\ \times (x^2 + y^2)^{1/2}(x^2 + y^2 + z^2)^{-1} + [(x\dot{x} + y\dot{y})z - \dot{z}(x^2 + y^2)] \\ \times [-(x\dot{x} + y\dot{y})(x^2 + y^2)^{-3/2}(x^2 + y^2 + z^2)] \\ - 2(x^2 + y^2)^{-1/2}(x\dot{x} + y\dot{y} + z\dot{z}), \tag{5.5.32}$$

and

$$\ddot{\phi} = (x\ddot{y} - \ddot{x}y)(x^2 + y^2)^{-1} - 2(x\dot{x} + y\dot{y})(x^2 + y^2)^{-3/2}(x\dot{y} - y\dot{x}). \tag{5.5.33}$$

The radial and the two tangential components of the velocity and acceleration vectors are, respectively,

$$\mathbf{v} = \dot{r}\mathbf{e}_r + r\dot{\theta}\mathbf{e}_\theta + (r \sin\theta)\dot{\phi}\mathbf{e}_\phi \tag{5.5.34}$$

and

$$\mathbf{a} = (\ddot{r} - r\dot{\theta}^2 - r\dot{\phi}^2 \sin^2\theta)\mathbf{e}_r + (r\ddot{\theta} + 2\dot{r}\dot{\theta}^2 - r\dot{\phi}^2 \sin\theta \cos\theta)\mathbf{e}_\theta \\ + [(r \sin\phi)\ddot{\phi} + (2\dot{r} \sin\theta)\dot{\phi} + (2r \cos\theta)\dot{\theta}\dot{\phi})]\mathbf{e}_\phi. \quad ■■ \tag{5.5.35}$$

5.5.5 Velocity and Acceleration in Moving Frames

Since a typical robotic manipulator is a natural setting for moving and rotating frames where origins are attached to other moving and rotating frames, it is useful to express the velocity and acceleration vectors in such frames with respect to each other. Furthermore, we can develop relationships for the angular moment and kinetic energies referred to either a fixed or inertial (Newtonian)

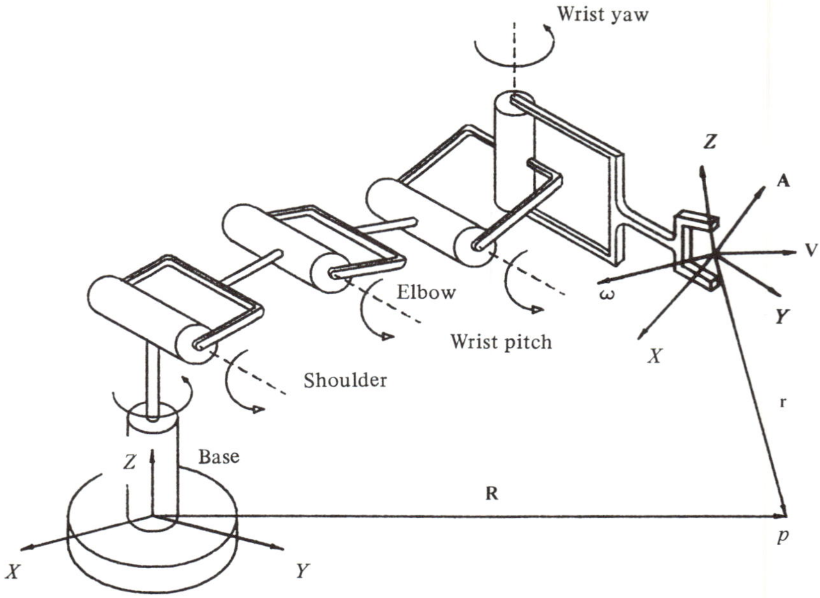

Figure 5.5.1 Moving robotic frames for velocity and acceleration.

frame of reference, or a noninertial moving frame of reference. Referring to Figure 5.5.1, let $OXYZ$ be a Newtonian frame of reference and $oxyz$ be a moving and rotating coordinate frame attached to a point in a robotic manipulator. Therefore, $\mathbf{r}(xyz)$ and $\mathbf{r}(XYZ)$ are the position vectors of a material point p relative to $oxyz$ and $OXYZ$, respectively, $\mathbf{v}$ is the velocity of the moving origin o, $\mathbf{a}$ is the acceleration of the moving origin o, and $\boldsymbol{\omega}$ is the angular velocity of the $oxyz$ frame, all with respect to the base frame $OXYZ$. Let the subscripts (xyz) and (XYZ) mean relative to $oxyz$ and $OXYZ$, respectively. Then the velocity $\mathbf{v}_{(XYZ)}$ is given by

$$\left(\frac{d^2\mathbf{r}}{dt^2}\right)_{(XYZ)} = \mathbf{a}_{(XYZ)} = \mathbf{A} + \left(\frac{d^2\mathbf{r}}{dt^2}\right)_{(xyz)} + \left(\frac{d\boldsymbol{\omega}}{dt}\right)_{(xyz)} \times \mathbf{r}_{(xyz)} + 2\boldsymbol{\omega} \times \mathbf{v}_{(xyz)} + \boldsymbol{\omega} \times (\boldsymbol{\omega} \times \mathbf{r}_{(xyz)}). \tag{5.5.36}$$

Note that

$$\boldsymbol{\omega} \times (\boldsymbol{\omega} \times \mathbf{r}_{(xyz)}) = \boldsymbol{\omega}(\boldsymbol{\omega} \cdot \mathbf{r}_{(xyz)}) - \mathbf{r}_{(xyz)}\omega^2, \tag{5.5.37}$$

and $(d\boldsymbol{\omega}/dt)_{(XYZ)} = (d\boldsymbol{\omega}/dt)_{(xyz)}$ because $\boldsymbol{\omega} \times \boldsymbol{\omega} = 0$. Now let $\mathbf{h}_{(xyz)}$ and $\mathbf{h}_{(XYZ)}$ be the angular momentum vectors of a link in $oxyz$ and $OXYZ$ frames, respectively, $\mathbf{R}$ the position vector of the origin of the $oxyz$ frame with respect to the $OXYZ$ frame, $\mathbf{r}_c$ the position vector of the center of mass of the robotic link with

Figure 5.5.2 Dynamics of an auto assembly line.

respect to the *oxyz* frame, and m the mass of the link. Then it is easy to establish that

$$\mathbf{h}_{(XYZ)} = \mathbf{h}_{(xyz)} + \mathbf{R} \times m\dot{\mathbf{R}} + \mathbf{R} \times m\dot{\mathbf{r}}_c + \mathbf{r}_c \times m\dot{\mathbf{R}} \tag{5.5.38}$$

and

$$\dot{\mathbf{h}}_{(XYZ)} = \dot{\mathbf{h}}_{(xyz)} + \mathbf{R} \times m\ddot{\mathbf{R}} + \mathbf{R} \times m\ddot{\mathbf{r}}_c + \mathbf{r}_c \times m\ddot{\mathbf{R}}. \tag{5.5.39}$$

Similarly, the kinetic energies $T_{(xyz)}$ and $T_{(XYZ)}$ are related by the following relationship:

$$T_{(XYZ)} = T_{(xyz)} + \tfrac{1}{2}m(\dot{\mathbf{R}} \cdot \dot{\mathbf{R}} + \dot{\mathbf{R}} \cdot \dot{\mathbf{r}}_c). \tag{5.5.40}$$

5.5.6 Robotic Mass Distribution and the Inertia Tensor

For robotic manipulators that are free to move in three-dimensional space, there are a finite number of possible rotation axes. To express the governing dynamic equations about a rotation axis, the mass distribution of the robot about that axis is required. To describe such rotational mass distributions, the moment of inertia tensor is employed.

The inertia tensor ${}^{A}\mathbf{I}$ with respect to a robotic frame A is a 3×3 matrix such that

$${}^{A}\mathbf{I} = \begin{bmatrix} I_{xx} & I_{xy} & I_{xz} \\ I_{xy} & I_{yy} & I_{yz} \\ I_{xz} & I_{yz} & I_{zz} \end{bmatrix}, \tag{5.5.41}$$

where for a discrete robotic system

$$I_{xx} = \sum_{i=1}^{N} m_i(z_i^2 + y_i^2), \tag{5.5.42}$$

$$I_{yy} = \sum_{i=1}^{N} m_i(x_i^2 + z_i^2), \tag{5.5.43}$$

$$I_{zz} = \sum_{i=1}^{N} m_i(x_i^2 + y_i^2), \tag{5.5.44}$$

$$I_{xy} = -\sum_{i=1}^{N} m_i x_i y_i, \tag{5.5.45}$$

$$I_{xz} = -\sum_{i=1}^{N} m_i x_i z_i, \tag{5.5.46}$$

and

$$I_{yz} = -\sum_{i=1}^{N} m_i y_i z_i, \tag{5.5.47}$$

and for a continuous robotic system

$$I_{xx} = \iiint \rho(z^2 + y^2)\, dx\, dy\, dz, \tag{5.5.48}$$

$$I_{yy} = \iiint \rho(x^2 + z^2)\, dx\, dy\, dz, \tag{5.5.49}$$

$$I_{zz} = \iiint \rho(x^2 + y^2)\, dx\, dy\, dz. \tag{5.5.50}$$

$$I_{xy} = -\iiint \rho xy\, dx\, dy\, dz, \tag{5.5.51}$$

$$I_{xz} = -\iiint \rho xz\, dx\, dy\, dz, \tag{5.5.52}$$

and

$$I_{yz} = -\iiint \rho yzy\, dx\, dy\, dz, \tag{5.5.53}$$

where ρ is the mass density of the robot, and the integration is taken over the volume and with respect to a frame whose origin either coincides with the center of mass of the robotic link or is attached to the link.

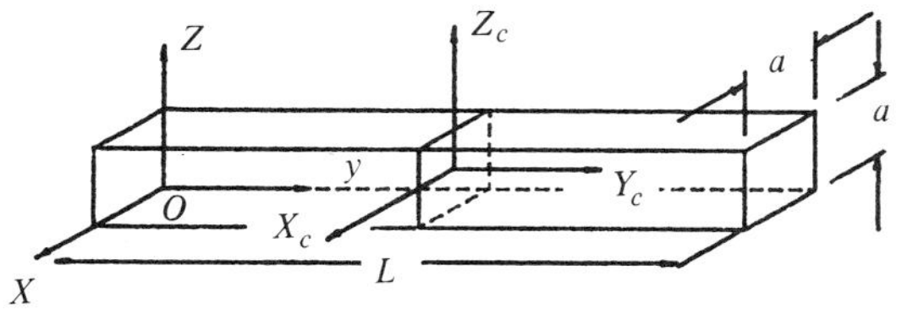

Figure 5.5.3 Robotic link with a square cross section.

EXAMPLE 5.5.3 ■

Find the moment of inertia of a robotic link with a square cross section both with respect to an end coordinate frame $oxyz$ (Fig. 5.5.3) and the center of mass coordinate frame $o_c x_c y_c z_c$.

Solution. Expressions (5.5.48)–(5.5.53) for this case simplify to the following integrals, first with respect to the coordinate frame xyz and then with respect to the coordinate frame $x_c y_c z_c$:

$$I_{xx} = \int_0^a \int_0^L \int_0^a (y^2 + z^2)\rho \, dx \, dy \, dz, \tag{5.5.54}$$

$$= \int_0^a \int_0^L \rho(y^2 + z^2)a \, dy \, dz, \tag{5.5.55}$$

$$= \int_0^a (\tfrac{1}{3}L^3 + z^2 L)a\rho \, dz, \tag{5.5.56}$$

or, finally,

$$I_{xx} = \rho\left(\frac{a^2L^3}{3} + \frac{a^4L}{3}\right) = \frac{m}{3}(L^2 + a^2). \tag{5.5.57}$$

Similarly,

$$I_{zz} = \frac{m}{3}(L^2 + a^2) \tag{5.5.58}$$

and

$$I_{yy} = \frac{m}{3}(a^2 + a^2) = \frac{2ma^2}{3}. \tag{5.5.59}$$

Now

$$I_{xy} = -\int_0^a \int_0^L \int_0^a xy\rho \, dx \, dy \, dz, \tag{5.5.60}$$

$$= \int_0^a \int_0^L \frac{a^2}{3} y\rho \, dy \, dz = -\int_0^a \frac{a^2L^2}{4} \rho \, dz, \tag{5.5.61}$$

or, finally,

$$I_{xy} = -\frac{m}{4} aL. \tag{5.5.62}$$

Similarly,

$$-I_{xz} = \frac{m}{4}a^2 \quad \text{and} \quad -I_{yz} = \frac{m}{4}aL. \tag{5.5.63}$$

Thus, the inertia tensor with respect to the xyz frame becomes

$$^{A}\mathbf{I} = \begin{bmatrix} \frac{m}{2}(L^2 + a^2) & -\frac{m}{4}aL & -\frac{m}{4}a^2 \\ -\frac{m}{4}aL & \frac{2m}{3}a^2 & -\frac{m}{4}aL \\ -\frac{m}{4}a^2 & -\frac{m}{4}aL & \frac{m}{3}(L^2 + a^2) \end{bmatrix}. \tag{5.5.64}$$

Now to find the inertia tensor with respect to the center of mass coordinate frame, $o_c x_c y_c z_c$, all one has to do is to repeat the previous integrals (5.5.54)–(5.5.63) with different limits; that is,

$$^{C}I_{xx} = \int_{-a/2}^{a/2} \int_{-L/2}^{L/2} \int_{-a/2}^{a/2} (y^2 + z^2)\rho\, dx\, dy\, dz, \tag{5.5.65}$$

$$= \int_{-a/2}^{a/2} \int_{-L/2}^{L/2} \rho(y^2 + z^2)a\, dy\, dz, \tag{5.5.66}$$

$$= \int_{-a/2}^{a/2} \left(\frac{L^3}{12} + z^2 L\right)a\rho\, dz, \tag{5.5.67}$$

or, finally,

$$^{C}I_{xx} = \phi\left(\frac{a^2 L^3}{12} + \frac{a^4 L}{12}\right) = \frac{m}{12}(L^2 + a^2). \tag{5.5.68}$$

Similarly,

$$^{C}I_{zz} = \frac{m}{12}(L^2 + a^2) \quad \text{and} \quad ^{C}I_{yy} = \frac{ma^2}{6}. \tag{5.5.69}$$

Thus,

$$^{C}I_{xy} = 0, \qquad ^{C}I_{xz} = 0, \qquad ^{C}I_{yz} = 0, \tag{5.5.70}$$

and

$$^{C}\mathbf{I} = \begin{bmatrix} \frac{m}{12}(a^2 + L^2) & 0 & 0 \\ 0 & \frac{ma^2}{6} & 0 \\ 0 & 0 & \frac{m}{12}(L^2 + a^2) \end{bmatrix}. \quad ■■ \tag{5.5.71}$$

Parallel Axis Theorem Comparing matrix equations (5.5.64) and (5.5.71) clearly indicates that

$$^{A}I_{zz} = {}^{C}I_{zz} + m(x_c^2 + y_c^2), \tag{5.5.72}$$

$$^{A}I_{xy} = {}^{C}I_{xy} + mx_c y_c, \tag{5.5.73}$$

$$^{A}I_{yy} = {}^{C}I_{yy} + m(x_c^2 + z_c^2), \tag{5.5.74}$$

$$^{A}I_{xx} = {}^{C}I_{xx} + m(y_c^2 + z_c^2), \tag{5.5.75}$$

$$^{A}I_{xz} = {}^{C}I_{xz} + mx_c z_c, \tag{5.5.76}$$

and

$$^{A}I_{yz} = {}^{C}I_{yz} + my_c z_c, \tag{5.5.77}$$

which are the familar expressions of the parallel axis theorem for the moment of inertia tensor ${}^{A}\mathbf{I}$ or ${}^{C}\mathbf{I}$.

5.5.7 Newton's Equation

If we consider the robotic arm as a series of rigid links connected by joints, then obviously for each link member a center of mass can be defined. Furthermore, an inertia tensor $\mathbf{I}_{ij}$ also can be defined; we elaborate on this tensor later. To cause motion for the links one must apply forces and couples. Now Newton's equation of motion states that the force $\mathbf{F}$ and the acceleration $\dot{\mathbf{v}}$ of a body are related to each other by the total mass of the body such that

$$\mathbf{F} = M\dot{\mathbf{v}}, \tag{5.5.78}$$

where M is the total mass of the body. Normally, in complex discrete structures, the mass M is replaced by a matrix $\mathbf{M}$, which is called the mass matrix of the body.

5.5.8 Euler's Equations

In the same manner, one may want to describe the rotational motion and acceleration of rigid bodies. Under such circumstances it can be shown that if $\boldsymbol{\tau}$ is the torque and $\boldsymbol{\omega}$ is the angular velocity vector, then

$$\boldsymbol{\tau} = \mathbf{I}\dot{\boldsymbol{\omega}} + \boldsymbol{\omega} \times \mathbf{I}\boldsymbol{\omega}, \tag{5.5.79}$$

where $\dot{\boldsymbol{\omega}}$ is the angular acceleration and $\mathbf{I}$ is the inertia tensor of the body with respect to its center of mass. This means that $\mathbf{I}$ is derived in a body frame whose origin is coincident with the center of mass of the body under consideration.

EXAMPLE 5.5.4 ■

Expand the Eulerian equations of motion of a robotic body if its rotation is the same as the rotation of a coordinate frame $oxyz$ embedded in it.

Solution. Consider Equation (5.5.79) and note that

$$\mathbf{I}\boldsymbol{\omega} = \begin{bmatrix} I_{xx} & I_{xy} & I_{xz} \\ I_{xy} & I_{yy} & I_{yz} \\ I_{xz} & I_{yz} & I_{zz} \end{bmatrix} \begin{bmatrix} \boldsymbol{\omega}_x \\ \boldsymbol{\omega}_y \\ \boldsymbol{\omega}_z \end{bmatrix} = \begin{bmatrix} I_{xx}\omega_x + I_{xy}\omega_y + I_{xz}\omega_z \\ I_{xy}\omega_x + I_{yy}\omega_y + I_{yz}\omega_z \\ I_{xz}\omega_x + I_{yz}\omega_y + I_{zz}\omega_z \end{bmatrix} = \begin{bmatrix} a_x \\ a_y \\ a_z \end{bmatrix}. \tag{5.5.80}$$

Now

$$\boldsymbol{\omega} \times (\mathbf{I}\boldsymbol{\omega}) = \begin{bmatrix} \mathbf{e}_x & \mathbf{e}_y & \mathbf{e}_z \\ \omega_x & \omega_y & \omega_z \\ a_x & a_y & a_z \end{bmatrix} \tag{5.5.81}$$

$$= \mathbf{e}_x(\omega_y a_z - \omega_z a_y) - \mathbf{e}_y(\omega_x a_z - \omega_z a_x) + \mathbf{e}_x(\omega_x a_y + \omega_y a_x) \tag{5.5.82}$$

Thus, the governing equations of motion (5.5.79) expand to the following:

$$\tau_x = I_{xx}\dot{\omega}_x + \omega_y\omega_z(I_{zz} - I_{yy}) + I_{xy}(\omega_z\omega_x - \dot{\omega}_y) - I_{xz}(\dot{\omega}_z + \omega_y\omega_x) - I_{yz}(\omega_y^2 - \omega_z^2), \tag{5.5.83}$$

$$\tau_y = I_{yy}\dot{\omega}_y + \omega_z\omega_x(I_{xx} - I_{zz}) + I_{yz}(\omega_x\omega_y - \dot{\omega}_z) - I_{xy}(\dot{\omega}_x + \omega_z\omega_y) - I_{xz}(\omega_z^2 - \omega_x^2), \tag{5.5.84}$$

$$\tau_z = I_{zz}\dot{\omega}_z + \omega_x\omega_y(I_{yy} - I_{xx}) + I_{xz}(\omega_y\omega_z - \dot{\omega}_x) - I_{yz}(\dot{\omega}_y + \omega_x\omega_z) - I_{xy}(\omega_x^2 - \omega_y^2), \tag{5.5.85}$$

where $\boldsymbol{\omega} = [\omega_x\ \omega_y\ \omega_z]^T$ is the vector of the rotation of the *oxyz* frame and is related to the Euler angles ϕ, θ, and ψ by the following equations:

$$\omega_x = \dot{\theta}\cos\psi + \dot{\phi}\sin\theta\sin\psi, \tag{5.5.86}$$

$$\omega_y = -\dot{\theta}\sin\psi + \dot{\phi}\sin\theta\cos\psi, \tag{5.5.87}$$

and

$$\omega_z = \dot{\psi} + \dot{\phi}\cos\theta. \quad ■■ \tag{5.5.88}$$

5.5.9 The Lagrangian and Lagrange's Equations

The Lagrangian L is defined as the difference beween the kinetic and potential energies of a dynamic system, such that

$$L = T - V, \tag{5.5.89}$$

where T is the kinetic energy and V is the potential energy. If the generalized coordinate is denoted by q_i, it is then clear that

$$L = L(q_i, \dot{q}_i, t). \tag{5.5.90}$$

Hamilton's Principle of Least Action For a dynamic system the variation of the action integral defined by

$$I = \int_{t_i}^{t_2} L(q_i, \dot{q}_i, t)\, dt \tag{5.5.91}$$

is equal to the work done by the external forces. Furthermore, for a conservative system the motion of the system from time t_1 to time t_2 is such that I is an extremum for the path of the motion.

Lagrange's Equations Based on Hamilton's principle, the governing equations of motion for a dynamic system can be shown to be

$$\frac{d}{dt}\left(\frac{\partial L}{\partial \dot{q}_i}\right) - \frac{\partial L}{\partial q_i} = F_i, \tag{5.5.92}$$

where F_i is the vector of nonconservative generalized forces; that is, it is not derivable from a potential field.

The generalized momentum of a dynamic system is defined as

$$p_i = \frac{\partial L}{\partial \dot{q}_i}, \tag{5.5.93}$$

which from Equation (5.5.92) implies that

$$p_i = \frac{\partial L}{\partial q_i} + F_i. \tag{5.5.94}$$

The Hamiltonian H is defined further by

$$H(q_i, p_i, t) = \sum_{i=1}^{N} p_i \dot{q}_i - L. \tag{5.5.95}$$

The Hamilton canonical equations are defined as

$$\dot{q}_i = \frac{\partial H}{\partial p_i}, \tag{5.5.96}$$

$$\dot{p}_i = -\frac{\partial H}{\partial q_i}, \tag{5.5.97}$$

and

$$\frac{\partial H}{\partial t} = -\frac{\partial L}{\partial t}. \tag{5.5.98}$$

5.6 LAGRANGE–EULER DYNAMIC MODELING OF ROBOTIC MANIPULATORS

In Section 5.5, we briefly discussed the nature of the Lagrange–Euler (LE) dynamic equations. Here we apply these equations to robotic manipulator dynamics, considering them to be nonconservative systems. In deriving the pertinent equations, we consistently use the Denavit–Hartenberg transformations of robotic coordinate frames.

Let us recall the Lagrange–Euler equations

$$\frac{d}{dt}\left(\frac{\partial L}{\partial \dot{q}_i}\right) - \left(\frac{\partial L}{\partial q_i}\right) = \tau_i, \tag{5.6.1}$$

where

$$L = L(q_i, \dot{q}_i, t) \equiv T - V \equiv \text{Lagrangian}, \tag{5.6.2}$$

$$T \equiv T(q_i, \dot{q}_i, t) \equiv \text{kinetic energy of the whole system}, \tag{5.6.3}$$

$$V \equiv V(q_i, t) \equiv \text{potential energy of the whole system}, \tag{5.6.4}$$

$$q_i \equiv \text{generalized coordinate, that is, } \theta_i \tag{5.6.5}$$

and

$$\tau_i \equiv \text{generalized force or torque.} \tag{5.6.6}$$

To familiarize ourselves with these equations, let us solve a typical example.

EXAMPLE 5.6.1 ■

Derive the governing dynamic equations for an n-tuple pendulum shown in Figure 5.6.1 and simplify your results for a simple plane double pendulum.

Solution. Since there are no nonconservative forces and the gravitational forces are derivable from potential fields, one has from the Lagrange–Euler equations (5.6.1)

$$\frac{d}{dt}\left(\frac{\partial K}{\partial \dot{q}_i}\right) - \left(\frac{\partial T}{\partial q_i}\right) + \left(\frac{\partial V}{\partial q_i}\right) = 0, \qquad i = 1, 2, \ldots, n. \tag{5.6.7}$$

Let $q_i = \theta_i$ be the generalized coordinates. Therefore, the position of the ith bob is given by

$$x_i = \sum_{\alpha=1}^{i} l_\alpha \sin\theta_\alpha \tag{5.6.8}$$

and

$$y_i = \sum_{\alpha=1}^{i} l_\alpha \cos\theta_\alpha. \tag{5.6.9}$$

Then, by definition,

$$\dot{x}_i = \sum_{\alpha=1}^{i} l_\alpha \dot{\theta}_\alpha \cos\theta_\alpha \tag{5.6.10}$$

and

$$\dot{y}_i = \sum_{\alpha=1}^{i} - l_\alpha \dot{\theta}_\alpha \sin\theta_\alpha, \tag{5.6.11}$$

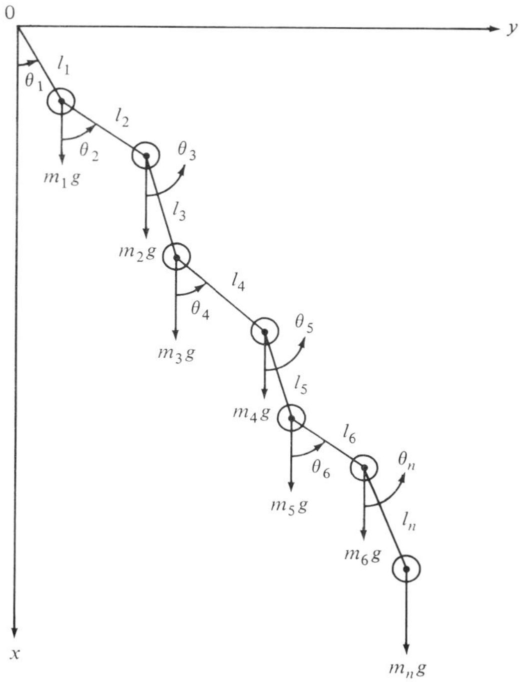

Figure 5.6.1 An n-tuple pendulum.

and the squares of $\dot{x}_i$ and $\dot{y}_i$ are found to be

$$\dot{x}_i^2 = \sum_{\alpha=1}^{i} \sum_{\beta=1}^{i} l_\alpha l_\beta \dot{\theta}_\alpha \dot{\theta}_\beta \cos\theta_\alpha \cos\theta_\beta \tag{5.6.12}$$

and

$$\dot{y}_i^2 = \sum_{\alpha=1}^{i} \sum_{\beta=1}^{i} l_\alpha l_\beta \dot{\theta}_\alpha \dot{\theta}_\beta \sin\theta_\alpha \sin\theta_\beta. \tag{5.6.13}$$

Thus,

$$T = \frac{1}{2} \sum_{i=1}^{n} m_i(\dot{x}_i^2 + \dot{y}_i^2), \tag{5.6.14}$$

which upon substitution of $\dot{x}_i^2$ and $\dot{y}_i^2$ from Equations (5.6.12) and (5.6.13) yields

$$T = \frac{1}{2} \sum_{i=1}^{n} m_i \left(\sum_{\alpha=1}^{i} l_\alpha^2 \dot{\theta}_\alpha^2 + \sum_{\alpha=1}^{i} \sum_{\beta=1}^{i} l_\alpha l_\beta \dot{\theta}_\alpha \dot{\theta}_\beta \cos(\theta_\alpha - \theta_\beta) \right). \tag{5.6.15}$$

For convenience we have separated all terms for which $\alpha = \beta$. Note that since the double summation argument is a completely symmetric function in θ_α and θ_β, every term will be repeated when summed. This means that

$$\sum_{\alpha=1}^{i} \sum_{\substack{\beta=1 \\ \beta \neq \alpha}}^{i} l_\alpha l_\beta \dot{\theta}_\alpha \dot{\theta}_\beta \cos(\theta_\alpha - \theta_\beta) = 2 \sum_{\alpha=1}^{i-1} \sum_{\beta=\alpha+1}^{i} l_\alpha l_\beta \dot{\theta}_\alpha \dot{\theta}_\beta \cos(\theta_\alpha - \theta_\beta), \tag{5.6.16}$$

and thus T reduces to

$$T = \frac{1}{2}\left(\sum_{i=1}^{n} \sum_{\alpha=1}^{i} m_i l_\alpha^2 \dot{\theta}_\alpha^2 + 2 \sum_{i=1}^{n} \sum_{\alpha=1}^{i-1} \sum_{\beta=\alpha+1}^{i} m_i l_\alpha l_\beta \dot{\theta}_\alpha \dot{\theta}_\beta \cos(\theta_\alpha - \theta_\beta) \right). \tag{5.6.17}$$

Now to compute the potential energy, the equilibrium position of the pendulum is considered as the reference datum ($P = 0$), which means that if the first pendulum has a deflection of $\theta_1 \neq 0$, then all other pendulums remain vertical. Thus, the system potential energy is

$$V = \sum_{\alpha=1}^{n} \sum_{i=\alpha}^{n} m_i g l_\alpha (1 - \cos \theta_\alpha). \tag{5.6.18}$$

Now Equation (5.6.7) can be expanded by noting that

$$\begin{aligned} \frac{d}{dt}\left(\frac{\partial T}{\partial \dot{\theta}_i}\right) = \frac{1}{2}\Bigg(& \sum_{i=1}^{n} \sum_{\alpha=1}^{i} 2 m_i l_\alpha^2 \dot{\theta}_\alpha \ddot{\theta}_\alpha \delta_{\alpha i} \\ & + 2 \sum_{i=1}^{n} \sum_{\alpha=1}^{i-1} \sum_{\beta=\alpha+1}^{i} [m_i l_\alpha l_\beta (\ddot{\theta}_\alpha \dot{\theta}_\beta \delta_{\alpha i} - \dot{\theta}_\alpha \ddot{\theta}_\beta \delta_{\beta i}) \cos(\theta_\alpha - \theta_\beta) \\ & - m_i l_\alpha l_\beta \dot{\theta}_\alpha \dot{\theta}_\beta (\dot{\theta}_\alpha \delta_{\alpha i} - \dot{\theta}_\beta \delta_{\beta i}) \sin(\theta_\alpha - \theta_\beta)] \Bigg), \end{aligned} \tag{5.6.19}$$

where $\delta_{\alpha i}$ and $\delta_{\beta i}$ are the Kronecker deltas; that is, $\delta_{\alpha i} = 1$ if $\alpha = i$ and $\delta_{\alpha i} = 0$ if $\alpha \neq i$. Furthermore,

$$\frac{\partial T}{\partial \theta_i} = - \sum_{i=1}^{n} \sum_{\alpha=1}^{i-1} \sum_{\beta=\alpha+1}^{i} m_i l_\alpha l_\beta \dot{\theta}_\alpha \dot{\theta}_\beta (\delta_{\alpha i} - \delta_{\beta i}) \sin(\theta_\alpha - \theta_\beta) \tag{5.6.20}$$

and

$$\frac{\partial V}{\partial \theta_i} = \sum_{\alpha=1}^{n} \sum_{i=\alpha}^{n} m_i g l_\alpha \delta_{\alpha i} \sin \theta_\alpha. \tag{5.6.21}$$

Now let us simplify the equations for a double pendulum. For $n = 2$ Equation (5.6.19) reduces to

$$\frac{d}{dt}\left(\frac{\partial T}{\partial \dot{\theta}_1}\right) = (m_1 + m_2)l_1^2\ddot{\theta}_1 + m_2 l_1 l_2 \cos(\theta_1 - \theta_2)\ddot{\theta}_2 - m_1 l_1 l_2 \dot{\theta}_2(\dot{\theta}_1 - \dot{\theta}_2)\sin(\theta_1 - \theta_2) \tag{5.6.22}$$

and

$$\frac{d}{dt}\left(\frac{\partial T}{\partial \dot{\theta}_2}\right) = m_2 l_2^2\ddot{\theta}_2 + m_2 l_1 l_2 \cos(\theta_1 - \theta_2)\ddot{\theta}_1 - m_2 l_1 l_2 \dot{\theta}_1(\dot{\theta}_1 - \dot{\theta}_2)\sin(\theta_1 - \theta_2). \tag{5.6.23}$$

Equations (5.6.20) and (5.6.21) now reduce to

$$\frac{\partial T}{\partial \dot{\theta}_1} = -m_2 l_1 l_2 \dot{\theta}_1\dot{\theta}_2 \sin(\theta_1 - \theta_2), \tag{5.6.24}$$

$$\frac{\partial T}{\partial \dot{\theta}_2} = -m_2 l_1 l_2 \dot{\theta}_1\dot{\theta}_2 \sin(\theta_1 - \theta_2), \tag{5.6.25}$$

$$\frac{\partial V}{\partial \theta_2} = m_2 g l_2 \sin\theta_2, \tag{5.6.26}$$

and

$$\frac{\partial P}{\partial \theta_1} = (m_1 + m_2)gl \sin\theta_1. \tag{5.6.27}$$

The final forms of the Lagrange–Euler equations for the double pendulum reduce to

$$(m_1 + m_2)l_1^2\ddot{\theta}_1 + m_2 l_1 l_2 \cos(\theta_1 - \theta_2)\ddot{\theta}_2 + m_2 l_1 l_2 \dot{\theta}_2^2 \sin(\theta_1 - \theta_2) + (m_1 + m_2)gl_1 \sin\theta_1 = 0 \tag{5.6.28}$$

and

$$m_2 l_2^2\ddot{\theta}_2 + m_2 l_1 l_2 \cos(\theta_1 - \theta_2)\ddot{\theta}_1 + m_2 l_1 l_2 \dot{\theta}_1^2 \sin(\theta_1 - \theta_2) + m_2 g l_2 \sin\theta_2 = 0. \tag{5.6.29}$$

In the following section we derive a set of expressions for the total kinetic energy T, the total potential energy V, and the Lagrangian L for a general n-axis robotic manipulator.

5.6.1 Velocity of the Joints

Consider $\mathbf{r}_0^i$ to be the position vector of a point P fixed with respect to the coordinate frame i (moving with it) and expressed with respect to the base coordinate (x_0, y_0, z_0). Referring to Figure 5.6.2, the point $\mathbf{r}_0^i$ has a fixed

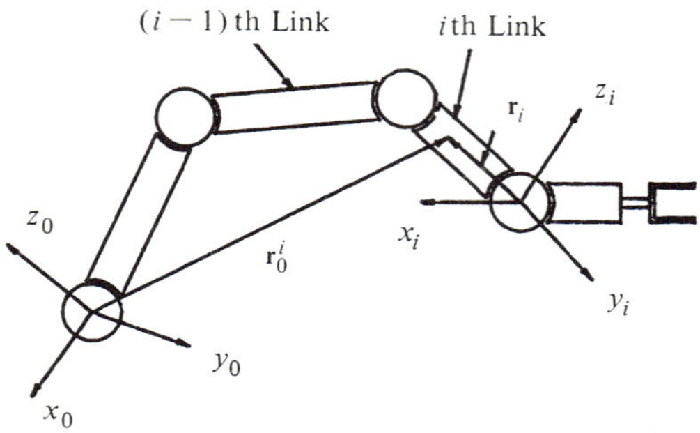

Figure 5.6.2 Geometry of a planar manipulator.

coordinate with respect to the ith frame, such that the point P as well as other points fixed in link i and expressed in coordinate frame i will have zero velocity with respect to the coordinate frame i. Note that the coordinate frame i is not an inertial frame. Clearly, the velocity of P with respect to the fixed or inertial frame or the base coordinate (see Fig. 5.6.3) can be given by

$$\mathbf{v}_i = \frac{d}{dt}(\mathbf{r}_0^i) \tag{5.6.30}$$

Recall that if $\mathbf{A}_{i-1}^i$ is the D-H transformation, then

$$\mathbf{A}_0^i = \mathbf{A}_0^1 \mathbf{A}_1^2 \cdots \mathbf{A}_{i-1}^i \tag{5.6.31}$$

and

$$\mathbf{r}_0^i = \mathbf{A}_0^i \mathbf{r}_i \quad \text{and} \quad \mathbf{r}_{i-1}^i = \mathbf{A}_{i-1}^i \mathbf{r}_i. \tag{5.6.32}$$

Thus,

$$\mathbf{v}_0^i = \mathbf{v}_i = \frac{d}{dt}(\mathbf{A}_0^i \mathbf{r}_i) = \sum_{j=1}^{i} \left(\frac{\partial A_0^i}{\partial q_i} \frac{dq_j}{dt} \right) \mathbf{r}_i, \tag{5.6.33}$$

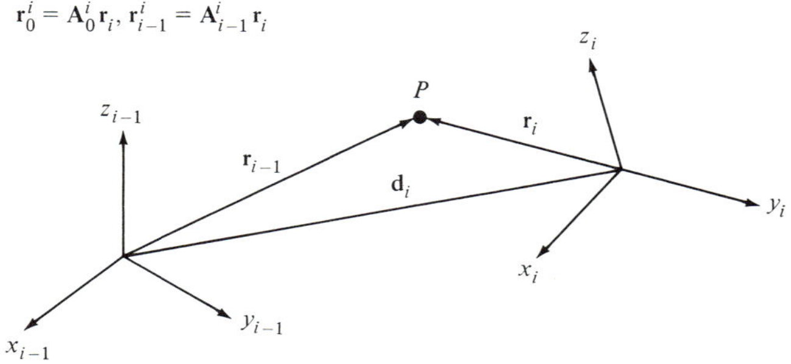

Figure 5.6.3 Vectorial representation of joint coordinates.

where the q_j are the independent variables, that is, θ_i $(i = 1, 2, \ldots, n)$ is. Therefore,

$$\mathbf{v}_0^i = \sum_{j=1}^{i} \left(\frac{\partial A_0^i}{\partial q_i} \dot{q}_j \right) \mathbf{r}_i, \qquad i = 1, 2, \ldots, n. \tag{5.6.34}$$

It can be shown that

$$\frac{\partial A_{i-1}^i}{\partial q_j} = Q_i A_{i-1}^i, \tag{5.6.35}$$

where Q_i is a constant matrix. For example, for A_{i-1}^i we have

$$\frac{\partial A_{i-1}^i}{\partial \theta_i} - \begin{bmatrix} -\sin\theta_i & -\cos\alpha_i \cos\theta_i & \sin\alpha_i \cos\theta_i & -a_i \sin\theta_i \\ \cos\theta_i & -\cos\alpha_i \sin\theta_i & \sin\alpha_i \sin\theta_i & a_i \cos\theta_i \\ 0 & 0 & 0 & 0 \\ 0 & 0 & 0 & 0 \end{bmatrix}$$

$$= \begin{bmatrix} 0 & -1 & 0 & 0 \\ 1 & 0 & 0 & 0 \\ 0 & 0 & 0 & 0 \\ 0 & 0 & 0 & 0 \end{bmatrix} \begin{bmatrix} \cos\theta_i & -\cos\theta_i \sin\theta_i & \sin\alpha_i \sin\theta_i & a_i \cos\theta_i \\ \sin\theta_i & \cos\alpha_i \cos\theta_i & -\sin\alpha_i \cos\theta_i & a_i \sin\theta_i \\ 0 & \sin\alpha_i & \cos\alpha_i & d_i \\ 0 & 0 & 0 & 0 \end{bmatrix}. \tag{5.6.36}$$

Thus,

$$\frac{\partial A_{i-1}^i}{\partial \theta_i} = Q_i A_{i-1}^i, \tag{5.6.37}$$

where

$$\mathbf{Q}_i = \begin{bmatrix} 0 & -1 & 0 & 0 \\ 1 & 0 & 0 & 0 \\ 0 & 0 & 0 & 0 \\ 0 & 0 & 0 & 0 \end{bmatrix}. \tag{5.6.38}$$

Similarly, it can be shown that for a prismatic joint

$$\mathbf{Q}_j = \begin{bmatrix} 0 & 0 & 0 & 0 \\ 0 & 0 & 0 & 0 \\ 0 & 0 & 0 & 1 \\ 0 & 0 & 0 & 0 \end{bmatrix}. \tag{5.6.39}$$

Generally,

$$\frac{\partial A_0^i}{\partial q_j}$$

$$= \begin{cases} A_0^1 A_1^2 \cdots A_{j-2}^{j-1} Q_j A_{j-1}^j A_j^{j+1} \cdots A_{i-1}^i, & \text{for } j \le i, \quad (5.6.40) \\ 0, & \text{for } j > i,\ i = 1, 2, 3, \ldots, n. \quad (5.6.41) \end{cases}$$

■ ■

EXAMPLE 5.6.2 ■

Prove that expressions (5.6.40) and (5.6.41) are true.

Solution. Note that $A_0^i = A_0^1 A_1^2 \cdots A_{j-1}^j \cdots A_{i-1}^i$. The only matrix that depends on q_j is A_{j-1}^j and thus if $j \le i$ we have

$$\frac{\partial A_0^i}{\partial q_j} = A_0^1 A_1^2 \cdots A_{j-2}^{j-1} \frac{\partial A_{j-1}^j}{\partial q_j} A_j^{j+1} \cdots A_{i-1}^i. \tag{5.6.42}$$

Now we know that $\partial A_{j-1}^j / \partial q_j = Q_j A_{j-1}^j$, and therefore the proof follows. Clearly, for $j > i$ no dependence on q_j exists and $\partial A_0^i / \partial q_j = 0$ for $j > i$. Q.E.D. ■ ■

For convenience we define a quantity called D_{ij} such that

$$D_{ij} = \frac{\partial A_0^i}{\partial q_j} = \begin{cases} A_0^{j-1} Q_j A_{j-1}^i, & j \le i, \quad (5.6.43) \\ 0, & j > i. \quad (5.6.44) \end{cases}$$

In a similar manner the higher-order derivatives of A_0^i are defined; that is,

$$\frac{\partial D_{ij}}{\partial q_k} = D_{ijk} = \begin{cases} A_0^{j-1} Q_j A_{j-1}^{k-1} Q_k A_{k-1}^i, & i \ge k \ge j, \quad (5.6.45) \\ A_0^{k-1} Q_k A_{k-1}^{j-1} Q_j A_{j-1}^i, & i \ge j \ge k, \quad (5.6.46) \\ 0, & i < j \text{ or } i < k. \quad (5.6.47) \end{cases}$$

To derive the governing dynamic equations for robotic manipulators, we employ the Lagrange–Euler formulation; namely, for a generally nonconservative system, Equation (5.6.1) holds. Thus, we need to calculate the Lagrangian $L = T - V$, where T is the kinetic energy and V is the potential energy of the manipulator. For a proof of Equation (5.6.1) refer to any text on Lagrangian dynamics.

5.6.2 Kinetic Energy T of the Robotic Arm

The incremental kinetic energy dT associated with a differential mass dm of the manipulator is given by

$$dT = \sum_{i=1}^{n} dT_i, \tag{5.6.48}$$

where dT_i is the incremental kinetic energy associated with a differential mass dm_i of the ith link of the manipulator and is given by

$$dT_i = \tfrac{1}{2}\,\mathrm{Tr}(\mathbf{v}_i \mathbf{v}_i^T)\, dm_i. \tag{5.6.49}$$

$$\mathbf{v}_i \mathbf{v}_i^T = \begin{bmatrix} v_{1i} \\ v_{2i} \\ v_{3i} \\ 0 \end{bmatrix} [v_{1i} \quad v_{2i} \quad v_{3i} \quad 0] = \begin{bmatrix} v_{1i}^2 & v_{1i}v_{2i} & v_{1i}v_{3i} & 0 \\ v_{2i}v_{1i} & v_{2i}^2 & v_{2i}v_{3i} & 0 \\ v_{3i}v_{1i} & v_{3i}v_{2i} & v_{3i}^2 & 0 \\ 0 & 0 & 0 & 0 \end{bmatrix} \tag{5.6.50}$$

and

$$\operatorname{Tr} \mathbf{A} = \sum_{i=1}^{n} A_{ii}. \tag{5.6.51}$$

Thus, based on Equations (5.6.35) and (5.6.43),

$$dT_i = \tfrac{1}{2}\operatorname{Tr}\left[\sum_{p=1}^{i} D_{ip}\dot{q}_p \mathbf{r}_i \left(\sum_{r=1}^{i} D_{ir}\dot{q}_r \mathbf{r}_i\right)^T\right] dm_i \tag{5.6.52}$$

or, finally,

$$dT_i = \tfrac{1}{2}\operatorname{Tr}\left(\sum_{p=1}^{i}\sum_{r=1}^{i} D_{ip}(\mathbf{r}_i\, dm_i \mathbf{r}_i^T) D_{ir}^T \dot{q}_p \dot{q}_r\right). \tag{5.6.53}$$

Integrating, we obtain

$$T_i = \int dT_i = \tfrac{1}{2}\operatorname{Tr}\left[\sum_{p=1}^{i}\sum_{r=1}^{j} D_{ip}\left(\int \mathbf{r}_i \mathbf{r}_i^T\, dm_i\right) D_{ir}^T \dot{q}_p \dot{q}_r\right]. \tag{5.6.54}$$

The integral term is nothing more than the moment of inertia matrix of the link i about the origin of the joint coordinate i, that is, the moment of inertia tensor $\mathbf{J}_i$ of the ith link.

$$\mathbf{J}_i = \int \mathbf{r}_i \mathbf{r}_i^T\, dm_i = \begin{bmatrix} \int x_i^2\, dm_i & \int x_i y_i\, dm_i & \int x_i z_i\, dm_i & \int x_i\, dm_i \\ \int x_i y_i\, dm_i & \int y_i^2\, dm_i & \int y_i z_i\, dm_i & \int y_i\, dm_i \\ \int z_i x_i\, dm_i & \int y_i z_i\, dm_i & \int z_i^2\, dm_i & \int z_i\, dm_i \\ \int x_i\, dm_i & \int y_i\, dm_i & \int z_i\, dm_i & \int dm_i \end{bmatrix}. \tag{5.6.55}$$

Expressing $\mathbf{J}_i$ in terms of the inertia tensor $\mathbf{I}$,

$$I_{ij} = \mathbf{I} = \begin{cases} \displaystyle\int\left[\delta_{ij}\left(\sum_k x_k^2\right) - x_i x_j\right] dm, & i = j \\ \displaystyle\int x_i x_j\, dm, & i \neq j, \end{cases}$$

we have

$$\mathbf{J}_i = \begin{bmatrix} (-I_{xx} + I_{yy} + I_{zz})/2 & I_{xy} & I_{xz} & m_i \bar{x}_i \\ I_{xy} & (I_{xx} - I_{yy} + I_{zz})/2 & I_{yz} & m_i \bar{y}_i \\ I_{xz} & I_{yz} & (I_{xx} + I_{yy} - I_{zz})/2 & m_i \bar{z}_i \\ m_i \bar{x}_i & m_i \bar{y}_i & m_i \bar{z}_i & m_i \end{bmatrix}, \tag{5.6.56}$$

where $\bar{x}_i$, $\bar{y}_i$, and $\bar{z}_i$ are the coordinates of the center of mass of link i with respect to the origin of the ith frame. Thus, the total kinetic energy of the robotic arm is

$$T = \sum_{i=1}^{n} T_i = \frac{1}{2} \sum_{i=1}^{n} \sum_{p=1}^{i} \sum_{r=1}^{i} [\mathrm{Tr}(D_{ip} J_i D_{ir}^T) \dot{q}_p \dot{q}_r]. \tag{5.6.57}$$

5.6.3 Potential Energy V of the Robotic Arm

The total potential energy owing to the weight of the robotic manipulator again is given as a sum of all potential energies of the individual links. Thus,

$$V = \sum_{i=1}^{n} V_i = \sum_{i=1}^{n} [-m_i \mathbf{g} \cdot (\mathbf{A}_0^i \bar{\mathbf{r}}_i)], \tag{5.6.58}$$

where $\bar{\mathbf{r}}_i$ is the location of the center of mass of link i expressed in the ith frame and $\mathbf{g}$ is the gravitational vector.

5.6.4 The Lagrangian $L = T - V$

From Equations (5.6.57) and (5.6.58) we obtain

$$L = \frac{1}{2} \sum_{i=1}^{n} \sum_{j=1}^{i} \sum_{p=1}^{i} [\mathrm{Tr}(D_{ij} J_i D_{ip}^T) \dot{q}_p \dot{q}_r] + \sum_{i=1}^{n} m_i \mathbf{g} \cdot (\mathbf{A}_0^i \bar{\mathbf{r}}_i). \tag{5.6.59}$$

Remember that we are trying to derive the dynamics equations by means of the Lagrange–Euler equations; that is,

$$\frac{d}{dt}\left(\frac{\partial L}{\partial \dot{q}_i}\right) - \frac{\partial L}{\partial q_i} = \tau_i, \qquad i = 1, 2, \ldots, n. \tag{5.6.60}$$

We note that

$$\begin{aligned} \frac{\partial L}{\partial \dot{q}_k} &= \frac{1}{2} \sum_{i=1}^{n} \sum_{j=1}^{i} \sum_{p=1}^{i} [\mathrm{Tr}(D_{ij} J_i D_{ip}^T)(\dot{q}_p \delta_{jk} + \dot{q}_j \delta_{kp}) \Big) \\ &= \sum_{i=k}^{n} \sum_{p=1}^{i} \mathrm{Tr}[D_{ip} J_i D_{ik}^T] \dot{q}_p \end{aligned} \tag{5.6.61}$$

and

$$\begin{aligned} \frac{d}{dt}\left(\frac{\partial L}{\partial \dot{q}_k}\right) &= \sum_{i=k}^{n} \sum_{p=1}^{i} \mathrm{Tr}(D_{ip} J_i D_{ik}^T) \ddot{q}_p \\ &\quad + \sum_{i=k}^{n} \sum_{p=1}^{i} \sum_{m=1}^{i} \mathrm{Tr}(D_{ipm} J_i D_{ik}^T) \dot{q}_p \dot{q}_m \\ &\quad + \sum_{i=1}^{n} \sum_{p=1}^{i} \sum_{m=1}^{i} \mathrm{Tr}(D_{ikm} J_i D_{ip}^T) \dot{q}_p \dot{q}_m. \end{aligned} \tag{5.6.62}$$

Similarly,

$$\frac{\partial L}{\partial q_k} = \frac{1}{2}\sum_{i=k}^{n}\sum_{j=1}^{i}\sum_{p=1}^{i} \mathrm{Tr}(D_{ijk}J_iD_{ip}^T)\dot{q}_j\dot{q}_p + \frac{1}{2}\sum_{i=k}^{n}\sum_{j=1}^{i}\sum_{p=1}^{i} \mathrm{Tr}(D_{ipk}J_iJ_iD_{ij}^T)\dot{q}_j\dot{q}_p + \sum_{i=p}^{n} m_i\mathbf{g}\cdot D_{ik}\bar{\mathbf{r}}_i. \tag{5.6.63}$$

Thus, Equation (5.6.60) expands to

$$\tau_k = \sum_{i=k}^{n}\sum_{p=1}^{i} \mathrm{Tr}(D_{ip}J_iD_{ik}^T)\ddot{q}_k + \sum_{i=k}^{n}\sum_{p=1}^{i}\sum_{m=1}^{i} \mathrm{Tr}(D_{ipm}J_iD_{ik}^T)\dot{q}_p\dot{q}_m + \sum_{i=p}^{n}\sum_{p=1}^{i}\sum_{m=1}^{i} \mathrm{Tr}(D_{ikm}J_iD_{ip}^T)\dot{q}_p\dot{q}_m - \frac{1}{2}\sum_{i=k}^{n}\sum_{j=1}^{i}\sum_{p=1}^{i} \mathrm{Tr}(D_{ijk}J_iD_{ip}^T)\dot{q}_j\dot{q}_p - \frac{1}{2}\sum_{i=k}^{n}\sum_{j=1}^{i}\sum_{p=1}^{i} \mathrm{Tr}(D_{ipk}J_iD_{ij}^T)\dot{q}_j\dot{q}_p - \sum_{i=k}^{n} m_i\mathbf{g}\cdot D_{ik}\bar{\mathbf{r}}_i. \tag{5.6.64}$$

Note that we may simplify the above equation further by the following definitions: Let

$$U_{ijk} = \mathrm{Tr}(D_{jk}J_jD_{ji}^T) = \mathrm{Tr}(D_{kj}J_kD_{ki}^T] = U_{ikj}, \tag{5.6.65}$$

so that Equation (5.6.64) simplifies to

$$\tau_i = \sum_{k=1}^{n} M_{ik}\ddot{q}_k + \sum_{k=1}^{n}\sum_{m=1}^{n} M_{ikm}\dot{q}_k\dot{q}_m + M_i, \tag{5.6.66}$$

where M_i, M_{ik}, and M_{ikm} are generalized mass matrices such that

$$M_i = \sum_{j=i}^{n}(-m_j\mathbf{g}\cdot \mathrm{D}_{ji}\bar{\mathbf{r}}_j), \tag{5.6.67}$$

$$M_{ik} = \sum_{j=\max(i,k)}^{n} \mathrm{Tr}(D_{jk}J_jD_{ji}^T), \tag{5.6.68}$$

and

$$M_{ikm} = \sum_{j=\max(i,k,m)}^{n} \mathrm{Tr}(D_{jkm}J_jD_{ji}^T), \tag{5.6.69}$$

where

$$\mathbf{g} \equiv [g_x \quad g_y \quad g_z \quad 0]. \tag{5.6.70}$$

Equation (5.6.65) has also been written in the robotics literature in a more compact vectorial form:

$$\boldsymbol{\tau} = \mathbf{D}(\boldsymbol{\theta})\ddot{\boldsymbol{\theta}} + \mathbf{H}(\boldsymbol{\theta}, \dot{\boldsymbol{\theta}}) + \mathbf{G}(\boldsymbol{\theta}). \tag{5.6.71}$$

The numerical computations involved with these equations are extremely large and can involve approximately 66,394 multiplications and 51,456 additions for a six-axis robot for which $n = 6$.

It must be noted that the Newton–Euler equations are much more efficient than the Lagrange–Euler equations when numerical calculations are involved. For example, for an n-axis robot the Newton–Euler (NE) equations require

$$126n - 99 \text{ multiplications} \quad \text{and} \quad 106n - 92 \text{ additions},$$

while the Lagrange–Euler (LE) equations require

$$32n^4 + 86n^3 + 171n^2 + 53n - 128 \text{ multiplications}$$

and

$$25n^4 + 66n^3 + 129n^2 + 42n - 96 \text{ additions}.$$

For a typical robot of $n = 6$, the computation time for the latter (LE) is two orders of magnitude more than the former (NE).

5.6.5 Two-Link Robotic Dynamics with Distributed Mass

Consider the following two-link robotic arm, where the masses are concentrated at the middle of each link as shown in Figure 5.6.4.

Note from the governing equations (5.6.66) that we have

$$\begin{bmatrix}\tau_1\\ \tau_2\end{bmatrix} = \begin{bmatrix}M_{11} & M_{12}\\ M_{21} & M_{22}\end{bmatrix}\begin{bmatrix}\ddot{\theta}_1\\ \ddot{\theta}_2\end{bmatrix} + \begin{bmatrix}\sum_{k=1}^{n}\sum_{m=1}^{n} M_{1km}\dot{\theta}_k\dot{\theta}_m\\ \sum_{k=1}^{n}\sum_{m=1}^{n} M_{2km}\dot{\theta}_k\dot{\theta}_m\end{bmatrix} + \begin{bmatrix}M_1\\ M_2\end{bmatrix}. \tag{5.6.72}$$

In expanded form we have

$$M_{11} = \mathrm{Tr}(\mathbf{D}_{11}\mathbf{J}_1\mathbf{D}_{11}^T) + \mathrm{Tr}(\mathbf{D}_{21}\mathbf{J}_2\mathbf{D}_{21}^T), \tag{5.6.73}$$

$$M_{12} = M_{21} = \mathrm{Tr}(\mathbf{D}_{22}\mathbf{J}_2\mathbf{D}_{21}^T), \tag{5.6.74}$$

$$M_{22} = \mathrm{Tr}(\mathbf{D}_{22}\mathbf{J}_2\mathbf{D}_{22})^{\mathrm{T}} \tag{5.6.75}$$

$$M_1 = m_1\mathbf{g}\cdot\mathbf{D}_{11}\bar{\mathbf{r}}_1 + m_1\mathbf{g}\cdot\mathbf{D}_{21}\bar{\mathbf{r}}_2, \tag{5.6.76}$$

and

$$M_1 = m_2\mathbf{g}\cdot\mathbf{D}_{22}\bar{\mathbf{r}}_2. \tag{5.6.77}$$

Note that for this two-link system $\alpha_1 = \alpha_2 = 0$, $d_1 = d_2 = 0$, and $a_2 = a_1 = 1$. Thus the D-H matrices reduce to

$$\mathbf{A}_0^1 = \begin{bmatrix}C_1 & -S_1 & 0 & lC_1\\ S_1 & C_1 & 0 & lS_1\\ 0 & 0 & 1 & 0\\ 0 & 0 & 0 & 1\end{bmatrix} \quad \text{and} \quad \mathbf{A}_1^2 = \begin{bmatrix}C_2 & -S_2 & 0 & lC_2\\ S_2 & C_2 & 0 & lS_2\\ 0 & 0 & 1 & 0\\ 0 & 0 & 0 & 1\end{bmatrix}, \tag{5.6.78}$$

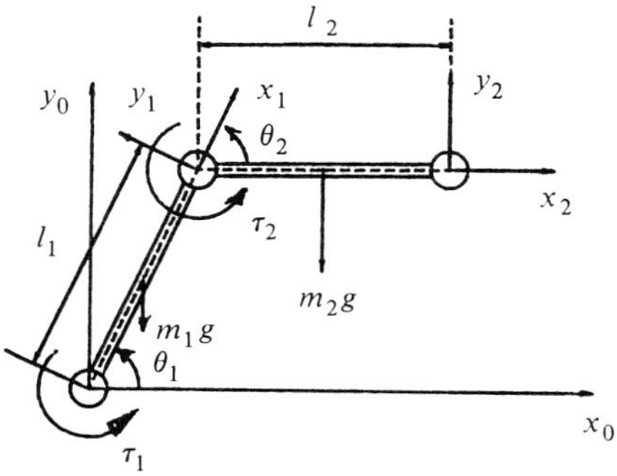

Figure 5.6.4 Geometry of a two-axis robot with distributed mass.

so that

$$\mathbf{A}_0^2 = \begin{bmatrix} C_{12} & -S_{12} & 0 & l(C_{12}+C_1) \\ S_{12} & C_{12} & 0 & l(S_{12}+S_1) \\ 0 & 0 & 1 & 0 \\ 0 & 0 & 0 & 1 \end{bmatrix}. \tag{5.6.79}$$

Since the joints are both revolute, from Equation (5.6.43) we have

$$\mathbf{D}_{11} = \mathbf{A}_0^0\mathbf{Q}_1\mathbf{A}_0^1 = \mathbf{Q}_1\mathbf{A}_0^1 = \begin{bmatrix} 0 & -1 & 0 & 0 \\ 1 & 0 & 0 & 0 \\ 0 & 0 & 0 & 0 \\ 0 & 0 & 0 & 0 \end{bmatrix} \begin{bmatrix} C_1 & -S_1 & 0 & lC_1 \\ S_1 & C_1 & 0 & lS_1 \\ 0 & 0 & 1 & 0 \\ 0 & 0 & 0 & 1 \end{bmatrix} \tag{5.6.80}$$

or

$$\mathbf{D}_{11} = \begin{bmatrix} -S_1 & -C_1 & 0 & lS_1 \\ C_1 & -S_1 & 0 & lC_1 \\ 0 & 0 & 1 & 0 \\ 0 & 0 & 0 & 1 \end{bmatrix}. \tag{5.6.81}$$

Furthermore, $\mathbf{D}_{21} = \mathbf{A}_0^0\mathbf{Q}_1\mathbf{A}_0^2 = \mathbf{Q}_1\mathbf{A}_0^2$ such that

$$\mathbf{D}_{21} = \begin{bmatrix} 0 & -1 & 0 & 0 \\ 1 & 0 & 0 & 0 \\ 0 & 0 & 0 & 0 \\ 0 & 0 & 0 & 0 \end{bmatrix} \begin{bmatrix} C_{12} & -S_{12} & 0 & l(C_{12}+C_1) \\ S_{12} & C_{12} & 0 & l(S_{12}+S_1) \\ 0 & 0 & 1 & 0 \\ 0 & 0 & 0 & 1 \end{bmatrix} \tag{5.6.82}$$

or

$$\mathbf{D}_{21} = \begin{bmatrix} -S_{12} & -C_{12} & 0 & -l(S_{12}+S_1) \\ C_{12} & -S_{12} & 0 & l(C_{12}+C_1) \\ 0 & 0 & 1 & 0 \\ 0 & 0 & 0 & 1 \end{bmatrix}. \tag{5.6.83}$$

Now $\mathbf{D}_{22} = \mathbf{A}_0^1 \mathbf{Q}_2 \mathbf{A}_1^2$ such that

$$\mathbf{D}_{22} = \begin{bmatrix} C_1 & -S_1 & 0 & lC_1 \\ S_1 & C_1 & 0 & lS_1 \\ 0 & 0 & 1 & 0 \\ 0 & 0 & 0 & 1 \end{bmatrix} \begin{bmatrix} 0 & -1 & 0 & 0 \\ 1 & 0 & 0 & 0 \\ 0 & 0 & 0 & 0 \\ 0 & 0 & 0 & 0 \end{bmatrix} \begin{bmatrix} C_1 & -S_1 & 0 & lC_1 \\ S_1 & C_1 & 0 & lS_1 \\ 0 & 0 & 1 & 0 \\ 0 & 0 & 0 & 1 \end{bmatrix}. \tag{5.6.84}$$

Finally,

$$\mathbf{D}_{22} = \begin{bmatrix} -S_{12} & -C_{12} & 0 & -lS_{12} \\ C_{12} & -S_{12} & 0 & lC_{12} \\ 0 & 0 & 1 & 0 \\ 0 & 0 & 0 & 1 \end{bmatrix}. \tag{5.6.85}$$

From Equation (5.6.56) we obtain

$$\mathbf{J}_1 = \begin{bmatrix} \frac{1}{3}m_1 l_1^2 & 0 & 0 & -\frac{1}{2}m_1 l_1 \\ 0 & 0 & 0 & 0 \\ 0 & 0 & 0 & 0 \\ -\frac{1}{2}m_1 l_1 & 0 & 0 & m_1 \end{bmatrix} \begin{bmatrix} \frac{1}{3}m_2 l_2^2 & 0 & 0 & -\frac{1}{2}m_2 l_2 \\ 0 & 0 & 0 & 0 \\ 0 & 0 & 0 & 0 \\ -\frac{1}{2}m_2 l_2 & 0 & 0 & m_2 \end{bmatrix} = \mathbf{J}_2. \tag{5.6.86}$$

Now

$$\begin{aligned} \mathbf{D}_{11}\mathbf{J}_1\mathbf{D}_{11}^T &= \begin{bmatrix} -S_1 & -C_1 & 0 & -lS_1 \\ C_1 & -S_1 & 0 & lC_1 \\ 0 & 0 & 0 & 0 \\ 0 & 0 & 0 & 0 \end{bmatrix} \begin{bmatrix} \frac{1}{3}m_1 l_1^2 & 0 & 0 & -\frac{1}{2}m_1 l_1 \\ 0 & 0 & 0 & 0 \\ 0 & 0 & 0 & 0 \\ -\frac{1}{2}m_1 l_1 & 0 & 0 & m_1 \end{bmatrix} \mathbf{D}_{11}^T \\ &= \begin{bmatrix} \frac{1}{6}m_1 S_1 l_1^2 & 0 & 0 & -\frac{1}{2}m_1 S_1 l_1 \\ -\frac{1}{6}m_1 C_1 l_1^2 & 0 & 0 & \frac{1}{2}m_1 C_1 l_1 \\ 0 & 0 & 0 & 0 \\ 0 & 0 & 0 & 0 \end{bmatrix} \begin{bmatrix} -S_1 & -C_1 & 0 & -lS_1 \\ C_1 & -S_1 & 0 & lC_1 \\ 0 & 0 & 0 & 0 \\ 0 & 0 & 0 & 0 \end{bmatrix}, \end{aligned} \tag{5.6.87}$$

so that

$$\mathbf{D}_{11}\mathbf{J}_1\mathbf{D}_{11}^T = \begin{bmatrix} -\frac{1}{6}m_1 S_1^2 l_1^2 + \frac{1}{2}m_1 S_1^2 l_1^2 & \frac{1}{6}m_1 S_1 C_1 l_1 - \frac{1}{2}m_1 S_1 C_1 l_1^2 & 0 & 0 \\ \frac{1}{6}m_1 C_1 S_1 l_1^2 - \frac{1}{2}m_1 C_1 S_1 l_1^2 & -\frac{1}{6}m_1 C_1^2 l_1^2 + \frac{1}{2}m_1 C_1^2 l_1^2 & 0 & 0 \\ 0 & 0 & 0 & 0 \\ 0 & 0 & 0 & 0 \end{bmatrix}. \tag{5.6.88}$$

Thus,

$$\mathrm{Tr}(D_{11}J_1D_{11})^T = \frac{1}{3}m_1 l_1^2 S_1^2 + \frac{1}{3}m_1 l_1^2 C_1^2 = \frac{1}{3}m_1 l_1^2. \tag{5.6.89}$$

Now

$$
\begin{aligned}
(\mathbf{D}_{21}\mathbf{J}_2\mathbf{D}_{21})^T
&= \begin{bmatrix} -S_{12} & -C_{12} & 0 & -l(S_{12}+S_1) \\ C_{12} & -S_{12} & 0 & l(C_{12}+C_1) \\ 0 & 0 & 1 & 0 \\ 0 & 0 & 0 & 1 \end{bmatrix}
\begin{bmatrix} \frac{1}{3}m_2 l_2^2 & 0 & 0 & -\frac{1}{2}m_2 l_2 \\ 0 & 0 & 0 & 0 \\ 0 & 0 & 0 & 0 \\ \frac{1}{2}m_2 l_2 & 0 & 0 & m_2 \end{bmatrix} \mathbf{D}_{21}^T \\
&= \begin{bmatrix} \frac{1}{6}m_2 l^2 S_{12} + \frac{1}{2}m_2 l^2 S_1 & 0 & 0 & -\frac{1}{2}m_1 S_{12} l - m_2 S_1 l \\ -\frac{1}{6}m_2 l^2 C_{12} - \frac{1}{2}m_2 l^2 C_1 & 0 & 0 & \frac{1}{2}m_2 C_{12} l + m_2 C_1 l \\ 0 & 0 & 0 & 0 \\ 0 & 0 & 0 & 0 \end{bmatrix} \mathbf{D}_{21}^T
\end{aligned} \tag{5.6.90}
$$

$$
(\mathbf{D}_{21}\mathbf{J}_2\mathbf{D}_{21})^T
= \begin{bmatrix} -\frac{1}{6}m_2 l^2 S_{12}^2 - \frac{1}{2}m_2 l^2 S_1 & S_{12} + \frac{1}{2}m_2 S_{12} l^2 (S_{12}+S_1) + m_2 l^2 S_1 (S_{12}+S_1) & 0 & 0 \\ -\frac{1}{6}m_2 l^2 C_{12}^2 & -\frac{1}{2}m_2 l^2 C_1 C_{12} + \frac{1}{2}m_2 l^2 C_{12}(C_{12}+C_1) + m_2 C_1 l^2 (C_{12}+C_1) & 0 & 0 \\ 0 & 0 & 0 & 0 \\ 0 & 0 & 0 & 0 \end{bmatrix}, \tag{5.6.91}
$$

so that

$$
\begin{aligned}
\mathrm{Tr}(D_{21}J_2D_{21}^T) &= -m_2 l^2(\tfrac{1}{6}S_{12} + \tfrac{1}{2}S_1)S_{12} + m_2 l^2(\tfrac{1}{2}S_{12} + S_1)(S_{12}+S_1) \\
&\quad - m_2 l^2(\tfrac{1}{6}C_{12} + \tfrac{1}{2}C_1)C_{12} + m_2 l^2(\tfrac{1}{2}C_{12} + C_1)(C_{12}+C_1) \\
&= m_2 l^2(\tfrac{4}{3} + C_2).
\end{aligned} \tag{5.6.92}
$$

Therefore, from Equation (5.6.73),

$$
M_{11} = \tfrac{1}{3}m_1 l^2 + \tfrac{4}{3}m_2 l^2 + m_2 C_2 l^2. \tag{5.6.93}
$$

Now we intend to find $M_{12} = M_{21} = \mathrm{Tr}(D_{22}J_2D_{21}^T)$ explicitly. Note that in this case

$$
\mathbf{D}_{22}\mathbf{J}_2\mathbf{D}_{21}^T = \begin{bmatrix} -S_{12} & -C_{12} & 0 & -lS_{12} \\ C_{12} & -S_{12} & 0 & lC_{12} \\ 0 & 0 & 0 & 0 \\ 0 & 0 & 0 & 0 \end{bmatrix}
\begin{bmatrix} \frac{1}{3}m_2 l^2 & 0 & 0 & -\frac{1}{2}m_2 l \\ 0 & 0 & 0 & 0 \\ 0 & 0 & 0 & 0 \\ -\frac{1}{2}m_2 l & 0 & 0 & m_2 \end{bmatrix} \mathbf{D}_{21}^T \tag{5.6.94}
$$

$$
= \begin{bmatrix} \frac{1}{6}m_2 l^2 S_{12} & 0 & 0 & -\frac{1}{2}m_2 l S_{12} \\ -\frac{1}{6}m_2 l^2 C_{12} & 0 & 0 & \frac{1}{2}m_2 l C_{12} \\ 0 & 0 & 0 & 0 \\ 0 & 0 & 0 & 0 \end{bmatrix}
\begin{bmatrix} -S_{12} & C_{12} & 0 & 0 \\ -C_{12} & -S_{12} & 0 & 0 \\ 0 & 0 & 0 & 0 \\ -l(S_{12}+S_1) & l(C_{12}+C_1) & 0 & 0 \end{bmatrix} \tag{5.6.95}
$$

$$
= \begin{bmatrix} -\frac{1}{6}m_2 l^2 S_{12}^2 + \frac{1}{2}m_2 l^2 S_{12}(S_{12}+S_1) & \frac{1}{6}m_2 l S_{12}^2 C_{12} - \frac{1}{2}m_2 l S_{12}^2 (C_{12}+C_1) & 0 & 0 \\ -\frac{1}{6}m_2 l^2 C_{12}^1 S_{12} - \frac{1}{2}m_2 l^2 C_{12}(S_{12}+S_1) & -\frac{1}{6}m_2 l^2 C_{12}^2 + \frac{1}{2}m_2 l^2 C_{12}(C_{12}+C_1) & 0 & 0 \\ 0 & 0 & 0 & 0 \\ 0 & 0 & 0 & 0 \end{bmatrix}. \tag{5.6.96}
$$

Therefore $M_{12} = M_{21} = \text{Tr}(D_{22}J_2D_{21}^T)$ reduces to

$$M_{12} = M_{21} = \tfrac{1}{3}m_2l^2 + \tfrac{1}{2}m_2l^2C_2. \tag{5.6.97}$$

To find $M_{22} = \text{Tr}(D_{22}J_2D_{22}^T)$ we note that

$$\mathbf{D}_{22}\mathbf{J}_2\mathbf{D}_{22}^T = \begin{bmatrix} -S_{12} & -C_{12} & 0 & -lS_{12} \\ C_{12} & -S_{12} & 0 & lC_{12} \\ 0 & 0 & 1 & 0 \\ 0 & 0 & 0 & 1 \end{bmatrix} \begin{bmatrix} \frac{1}{3}m_2l^2 & 0 & 0 & -\frac{1}{2}m_2l \\ 0 & 0 & 0 & 0 \\ 0 & 0 & 0 & 0 \\ -\frac{1}{2}m_2l & 0 & 0 & m_2 \end{bmatrix} \mathbf{D}_{22}^T \tag{5.6.98}$$

$$= \begin{bmatrix} \frac{1}{6}m_2l^2S_{12} & 0 & 0 & -\frac{1}{2}m_2lS_{12} \\ -\frac{1}{6}m_2l^2C_{12} & 0 & 0 & \frac{1}{2}m_2lC_{12} \\ 0 & 0 & 0 & 0 \\ 0 & 0 & 0 & 0 \end{bmatrix} \begin{bmatrix} -S_{12} & C_{12} & 0 & 0 \\ -C_{12} & -S_{12} & 0 & 0 \\ 0 & 0 & 0 & 0 \\ -lS_{12} & lC_{12} & 0 & 0 \end{bmatrix} \tag{5.6.99}$$

$$= \begin{bmatrix} -\frac{1}{6}m_2l^2S_{12}^2 + \frac{1}{2}m_2l^2S_{12}^2 & \frac{1}{6}m_2l^2S_{12}C_{12} - \frac{1}{2}m_2l^2S_{12}C_{12} & 0 & 0 \\ \frac{1}{6}m_2l^2S_{12}C_{12} - \frac{1}{2}m_2l^2C_{12}S_{12} & -\frac{1}{6}m_2l^2C_{12}^2 + \frac{1}{2}m_2l^2C_{12}^2 & 0 & 0 \\ 0 & 0 & 0 & 0 \\ 0 & 0 & 0 & 0 \end{bmatrix}. \tag{5.6.100}$$

Therefore,

$$M_{22} = \text{Tr}(D_{22}J_2D_{22}^T) = \tfrac{1}{3}m_2l^2S_{12}^2 + \tfrac{1}{3}m_2l^2C_{12}^2 = \tfrac{1}{3}m_2l^2. \tag{5.6.101}$$

To find the terms pertaining to Coriolis and centrifugal accelerations, we note from Equations (5.6.45), (5.6.46), (5.6.69), and (5.6.72) that we should calculate terms like M_{111}, M_{112}, M_{211}, M_{212}, M_{221}, and M_{222}. Note that

$$M_{ikm} = \sum_{j=\max(i,k,m)}^{n} \text{Tr}(D_{jkm}J_jD_{ji}^T). \tag{5.6.102}$$

Thus,

$$M_{111} = \text{Tr}(D_{111}J_1D_{11}^T) + \text{Tr}(D_{211}J_2D_{21}^T) = 0, \tag{5.6.103}$$

$$M_{112} = \text{Tr}(D_{212}J_2D_{21}^T) = \tfrac{1}{2}m_2S_2l_1^2 = M_{121} = (D_{121}J_2D_{21}^T), \tag{5.6.104}$$

$$M_{211} = \text{Tr}(D_{211}J_2D_{22}^T) = \tfrac{1}{2}m_2S_2l^2, \tag{5.6.105}$$

$$M_{212} = \text{Tr}(D_{212}J_2D_{22}^T) = 0, \tag{5.6.106}$$

$$M_{221} = \text{Tr}(D_{231}J_2D_{22}^T) = 0, \tag{5.6.107}$$

$$M_{222} = \text{Tr}(D_{222}J_2D_{22}^T) = 0, \tag{5.6.108}$$

and

$$M_{122} = \tfrac{1}{2}m_2S_2l^2. \tag{5.6.109}$$

Now we must evaluate the gravitational term M_i from Equation (5.5.67). Note that

$$M_1 = -m_1\mathbf{g}\cdot D_{11}\bar{\mathbf{r}}_1 - m_2\mathbf{g}\cdot D_{21}\bar{\mathbf{r}}_2 \tag{5.6.110}$$

$$= -m_1(0, -g, 0, 0)\begin{bmatrix} -S_1 & -C_1 & 0 & -lS_1 \\ C_1 & -S_1 & 0 & lC_1 \\ 0 & 0 & 0 & 0 \\ 0 & 0 & 0 & 0 \end{bmatrix}\begin{bmatrix} -l/2 \\ 0 \\ 0 \\ 1 \end{bmatrix}$$

$$- m_2(0, -g, 0, 0)\begin{bmatrix} -S_{12} & -C_{12} & 0 & -l(S_{12}+S_1) \\ C_{12} & -S_{12} & 0 & l(C_{12}+C_1) \\ 0 & 0 & 0 & 0 \\ 0 & 0 & 0 & 0 \end{bmatrix}\begin{bmatrix} -l/2 \\ 0 \\ 0 \\ 1 \end{bmatrix}; \tag{5.6.111}$$

$$M_1 = -\tfrac{1}{2}m_1glC_1 - \tfrac{1}{2}m_2glC_{12} - m_2glC_1. \tag{5.6.112}$$

Similarly,

$$M_2 = -m_2\mathbf{g}\cdot D_{22}\bar{\mathbf{r}}_2 = -\tfrac{1}{2}m_2glC_{12}. \tag{5.6.113}$$

Finally, the governing equations reduce to

$$\begin{aligned}\tau_1 = {} & [\tfrac{1}{3}m_1l^2 + \tfrac{4}{3}m_2l^2 + m_2C_2l^2]\ddot{\theta}_1 + [\tfrac{1}{3}m_2l^2 + \tfrac{1}{2}m_2l^2C_2]\ddot{\theta}_2 - \tfrac{1}{2}m_2S_2l^2\dot{\theta}_2^2 \\ & - m_2S_2l^2\dot{\theta}_1\dot{\theta}_2 - \tfrac{1}{2}m_1glC_1 - \tfrac{1}{2}m_2glC_{12} - m_2glC_1\end{aligned} \tag{5.6.114}$$

and

$$\tau_2 = [\tfrac{1}{3}m_2l^2 + \tfrac{1}{2}m_2l^2C_2]\ddot{\theta}_2 + \tfrac{1}{3}m_2l^2\ddot{\theta}_2 + \tfrac{1}{2}m_2S_2l^2\dot{\theta}_1^2 - \tfrac{1}{2}m_2glC_{12}. \tag{5.6.115}$$

5.6.6 Dynamic Equations of Motion for a General Six-Axis Robotic Manipulator

Equation (5.6.64) can be expressed in matrix form by Equation (5.6.66) such that

$$\boldsymbol{\tau} = \mathbf{M}(\boldsymbol{\theta})\ddot{\boldsymbol{\theta}} + \mathbf{H}(\boldsymbol{\theta}, \dot{\boldsymbol{\theta}}) + \mathbf{G}(\boldsymbol{\theta}). \tag{5.6.116}$$

Here $\boldsymbol{\tau}$ is a 6×1 vector, $\mathbf{M}(\boldsymbol{\theta})$ is a 6×6 matrix, $\mathbf{H}$ is a 6×1 vector involving all Coriolis and centripetal acceleration terms, and $\mathbf{G}$ is a 6×1 gravitational vector. (See Fig. 5.6.5.) For a six-axis robot the above equations expand to

$$\begin{aligned}\tau_1 = {} & M_{11}\ddot{\theta}_1 + M_{12}\ddot{\theta}_2 + M_{13}\ddot{\theta}_3 + M_{14}\ddot{\theta}_4 + M_{15}\ddot{\theta}_5 + M_{16}\ddot{\theta}_6 + M_{111}\dot{\theta}_1^2 \\ & + M_{122}\dot{\theta}_2^2 + M_{133}\dot{\theta}_3^2 + M_{144}\dot{\theta}_4^2 + M_{155}\dot{\theta}_5^2 + M_{166}\dot{\theta}_6^2 + M_{112}\dot{\theta}_1\dot{\theta}_2 \\ & + M_{113}\dot{\theta}_1\dot{\theta}_3 + M_{114}\dot{\theta}_1\dot{\theta}_4 + M_{115}\dot{\theta}_1\dot{\theta}_5 + M_{116}\dot{\theta}_1\dot{\theta}_6 + M_{123}\dot{\theta}_2\dot{\theta}_3 \\ & + M_{124}\dot{\theta}_2\dot{\theta}_4 + M_{125}\dot{\theta}_2\dot{\theta}_5 + M_{126}\dot{\theta}_2\dot{\theta}_6 + M_{134}\dot{\theta}_3\dot{\theta}_4 + M_{135}\dot{\theta}_3\dot{\theta}_5 \\ & + M_{136}\dot{\theta}_3\dot{\theta}_6 + M_{145}\dot{\theta}_4\dot{\theta}_5 + M_{146}\dot{\theta}_4\dot{\theta}_6 + M_{156}\dot{\theta}_5\dot{\theta}_6 + G_1,\end{aligned} \tag{5.6.117}$$

Figure 5.6.5 Dynamics of a crowded work cell.

$$\begin{aligned}\tau_2 = {} & M_{12}\ddot{\theta}_1 + M_{22}\ddot{\theta}_2 + M_{23}\ddot{\theta}_3 + M_{24}\ddot{\theta}_4 + M_{25}\ddot{\theta}_5 + M_{26}\ddot{\theta}_6 + M_{211}\dot{\theta}_1^2 \\ & + M_{222}\dot{\theta}_2^2 + M_{233}\dot{\theta}_3^2 + M_{244}\dot{\theta}_4^2 + M_{255}\dot{\theta}_5^2 + M_{266}\dot{\theta}_6^2 + M_{212}\dot{\theta}_1\dot{\theta}_2 \\ & + M_{213}\dot{\theta}_1\dot{\theta}_3 + M_{214}\dot{\theta}_1\dot{\theta}_4 + M_{215}\dot{\theta}_1\dot{\theta}_5 + M_{216}\dot{\theta}_1\dot{\theta}_6 + M_{223}\dot{\theta}_2\dot{\theta}_3 \\ & + M_{224}\dot{\theta}_2\dot{\theta}_4 + M_{225}\dot{\theta}_2\dot{\theta}_5 + M_{226}\dot{\theta}_2\dot{\theta}_6 + M_{234}\dot{\theta}_3\dot{\theta}_4 + M_{235}\dot{\theta}_3\dot{\theta}_5 \\ & + M_{236}\dot{\theta}_3\dot{\theta}_6 + M_{245}\dot{\theta}_4\dot{\theta}_5 + M_{246}\dot{\theta}_4\dot{\theta}_6 + M_{256}\dot{\theta}_5\dot{\theta}_6 + G_2,\end{aligned} \tag{5.6.118}$$

$$\begin{aligned}\tau_3 = {} & M_{13}\ddot{\theta}_1 + M_{23}\ddot{\theta}_2 + M_{33}\ddot{\theta}_3 + M_{34}\ddot{\theta}_4 + M_{35}\ddot{\theta}_5 + M_{36}\ddot{\theta}_6 + M_{322}\dot{\theta}_2^2 \\ & + M_{333}\dot{\theta}_3^2 + M_{344}\dot{\theta}_4^2 + M_{355}\dot{\theta}_2^2 + M_{366}\dot{\theta}_6^2 + M_{312}\dot{\theta}_1\dot{\theta}_2 \\ & + M_{313}\dot{\theta}_1\dot{\theta}_3 + M_{314}\dot{\theta}_1\dot{\theta}_4 + M_{315}\dot{\theta}_1\dot{\theta}_5 + M_{316}\dot{\theta}_1\dot{\theta}_6 + M_{323}\dot{\theta}_2\dot{\theta}_3 \\ & + M_{324}\dot{\theta}_2\dot{\theta}_4 + M_{325}\dot{\theta}_2\dot{\theta}_5 + M_{326}\dot{\theta}_2\dot{\theta}_6 + M_{345}\dot{\theta}_4\dot{\theta}_5 + M_{346}\dot{\theta}_4\dot{\theta}_6 \\ & + M_{356}\dot{\theta}_5\dot{\theta}_6 + G_3,\end{aligned} \tag{5.6.119}$$

$$\begin{aligned}\tau_4 = {} & M_{14}\ddot{\theta}_1 + M_{24}\ddot{\theta}_2 + M_{34}\ddot{\theta}_3 + M_{44}\ddot{\theta}_4 + M_{54}\ddot{\theta}_5 + M_{64}\ddot{\theta}_6 + M_{411}\dot{\theta}_1^2 \\ & + M_{422}\dot{\theta}_2^2 + M_{433}\dot{\theta}_3^2 + M_{444}\dot{\theta}_4^2 + M_{455}\dot{\theta}_5^2 + M_{466}\dot{\theta}_6^2 + M_{412}\dot{\theta}_1\dot{\theta}_2 \\ & + M_{413}\dot{\theta}_1\dot{\theta}_3 + M_{414}\dot{\theta}_1\dot{\theta}_4 + M_{415}\dot{\theta}_1\dot{\theta}_5 + M_{416}\dot{\theta}_1\dot{\theta}_6 + M_{423}\dot{\theta}_2\dot{\theta}_3 \\ & + M_{424}\dot{\theta}_2\dot{\theta}_4 + M_{425}\dot{\theta}_2\dot{\theta}_5 + M_{426}\dot{\theta}_2\dot{\theta}_6 + M_{434}\dot{\theta}_3\dot{\theta}_4 + M_{435}\dot{\theta}_3\dot{\theta}_5 \\ & + M_{436}\dot{\theta}_3\dot{\theta}_6 + M_{445}\dot{\theta}_4\dot{\theta}_5 + M_{446}\dot{\theta}_4\dot{\theta}_6 + M_{456}\dot{\theta}_5\dot{\theta}_6 + G_4,\end{aligned} \tag{5.6.120}$$

$$\begin{aligned}\tau_5 = {} & M_{15}\ddot{\theta}_1 + M_{25}\ddot{\theta}_2 + M_{35}\ddot{\theta}_3 + M_{45}\ddot{\theta}_4 + M_{55}\ddot{\theta}_5 + M_{65}\ddot{\theta}_6 + M_{511}\dot{\theta}_1^2 \\ & + M_{522}\dot{\theta}_2^2 + M_{533}\dot{\theta}_3^2 + M_{544}\dot{\theta}_4^2 + M_{555}\dot{\theta}_5^2 + M_{566}\dot{\theta}_6^2 + M_{512}\dot{\theta}_1\dot{\theta}_2 \\ & + M_{513}\dot{\theta}_1\dot{\theta}_3 + M_{514}\dot{\theta}_1\dot{\theta}_4 + M_{515}\dot{\theta}_1\dot{\theta}_5 + M_{516}\dot{\theta}_1\dot{\theta}_6 + M_{523}\dot{\theta}_2\dot{\theta}_3 \\ & + M_{524}\dot{\theta}_2\dot{\theta}_4 + M_{525}\dot{\theta}_2\dot{\theta}_5 + M_{526}\dot{\theta}_2\dot{\theta}_6 + M_{534}\dot{\theta}_3\dot{\theta}_4 + M_{535}\dot{\theta}_3\dot{\theta}_5 \\ & + M_{536}\dot{\theta}_3\dot{\theta}_6 + M_{545}\dot{\theta}_4\dot{\theta}_5 + M_{546}\dot{\theta}_4\dot{\theta}_6 + M_{556}\dot{\theta}_5\dot{\theta}_6 + G_6,\end{aligned} \tag{5.6.121}$$

and

$$\begin{aligned}\tau_6 = {} & M_{16}\ddot{\theta}_1 + M_{26}\ddot{\theta}_2 + M_{36}\ddot{\theta}_3 + M_{46}\ddot{\theta}_4 + M_{56}\ddot{\theta}_5 + M_{66}\ddot{\theta}_6 + M_{611}\dot{\theta}_1^2 \\ & + M_{622}\dot{\theta}_2^2 + M_{633}\dot{\theta}_3^2 + M_{644}\dot{\theta}_4^2 + M_{655}\dot{\theta}_5^2 + M_{666}\dot{\theta}_6^2 + M_{612}\dot{\theta}_1\dot{\theta}_2 \\ & + M_{613}\dot{\theta}_1\dot{\theta}_3 + M_{614}\dot{\theta}_1\dot{\theta}_4 + M_{615}\dot{\theta}_1\dot{\theta}_5 + M_{616}\dot{\theta}_1\dot{\theta}_6 + M_{623}\dot{\theta}_2\dot{\theta}_3 \\ & + M_{624}\dot{\theta}_2\dot{\theta}_4 + M_{625}\dot{\theta}_2\dot{\theta}_5 + M_{626}\dot{\theta}_2\dot{\theta}_6 + M_{634}\dot{\theta}_3\dot{\theta}_4 + M_{635}\dot{\theta}_3\dot{\theta}_5 \\ & + M_{636}\dot{\theta}_3\dot{\theta}_6 + M_{645}\dot{\theta}_4\dot{\theta}_5 + M_{646}\dot{\theta}_4\dot{\theta}_6 + M_{656}\dot{\theta}_5\dot{\theta}_6 + G_6.\end{aligned} \tag{5.6.122}$$

Note that in the above equations

$$\sum_{k=\max(i,j)}^{n} \mathrm{Tr}(\mathbf{D}_{kj}\mathbf{J}_k\mathbf{D}_{ki}^T) = \mathbf{M}_{ij} \tag{5.6.123}$$

and

$$\sum_{m=\max(i,j,k)}^{n} \mathrm{Tr}(\mathbf{D}_{mjk}\mathbf{J}_m\mathbf{D}_{mi}^T) = \mathbf{M}_{ikm}. \tag{5.6.124}$$

5.7 NEWTON–EULER ROBOTIC DYNAMICS EQUATIONS

We shall now use the Newton–Euler robotic dynamics equations to compute the generalized forces (torques) necessary for a given motion trajectory of a robotic manipulator. Initially, it is assumed that we know the position, velocity, and acceleration of each joint, that is, $\boldsymbol{\theta}$, $\dot{\boldsymbol{\theta}}$, and $\ddot{\boldsymbol{\theta}}$, and that we want to calculate the joint torques required to cause such time-dependent motions. As mentioned before, recursive type dynamic modeling is the answer to the current task. Before proceeding, however, let us write out the equations pertaining to the forces and torques acting on a robotic link. These are

$$\mathbf{f}_i = m_i\dot{\mathbf{v}}_{c(i)} \tag{5.7.1}$$

and

$$\boldsymbol{\tau}_i = {}^{C}\mathbf{I}_i\dot{\boldsymbol{\omega}}_i + \boldsymbol{\omega}_i \times ({}^{C}\mathbf{I}_i \cdot \boldsymbol{\omega}_i), \tag{5.7.2}$$

where $\mathbf{v}_{c(i)}$ is the velocity of the center of mass of the ith link, and ${}^{C}\mathbf{I}_i$ is the moment of inertia of the ith link about its center of mass with respect to frame **C**, which has its origin at the center of mass of link i. Here $\mathbf{f}_i$ and $\boldsymbol{\tau}_i$ are the net d'Alembert force and couple on the ith link and can be obtained by means of the

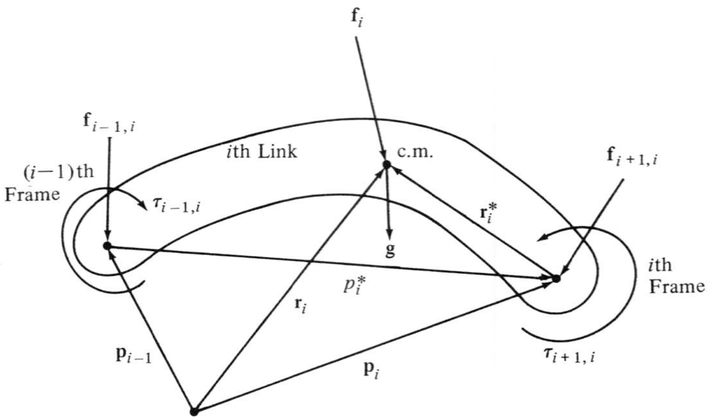

Figure 5.7.1 Forces and couples applied to the ith link.

balance equations for forces and couples applied to the ith link (Fig. 5.7.1), such that

$$\mathbf{f}_i = \mathbf{f}_{i-1,i} + \mathbf{f}_{i+1,i} + m_i\mathbf{f} \tag{5.7.3}$$

and

$$\tau_i = \boldsymbol{\tau}_{i-1,i} - \boldsymbol{\tau}_{i,i+1} + (\mathbf{p}_{i-1} - \mathbf{F}_i) \times \mathbf{f}_{i-1,i} - (\mathbf{p}_i - \mathbf{F}_i) \times \mathbf{f}_{i,i+1}, \tag{5.7.4}$$

where $\mathbf{f}_{i,j}$ and $\boldsymbol{\tau}_{i,j}$ are, respectively, the force and couple applied to the ith link by means of the adjoining jth link.

Now consider again two robotic coordinate systems $oxyz$ and $o^*x^*y^*z^*$ as shown in Figure 5.7.2. Let point p be described by a vector $\mathbf{r}$ with respect to the xyz system and $\mathbf{r}^*$ with respect to $x^*y^*z^*$ system. The two origins o and o^* are separated by a vector $\mathbf{a}$ with respect to the xyz system. It is then clear that

$$\mathbf{r} = \mathbf{r}^* + \mathbf{a} \quad \text{and} \quad \mathbf{r}^* = \mathbf{r} - \mathbf{a}, \tag{5.7.5}$$

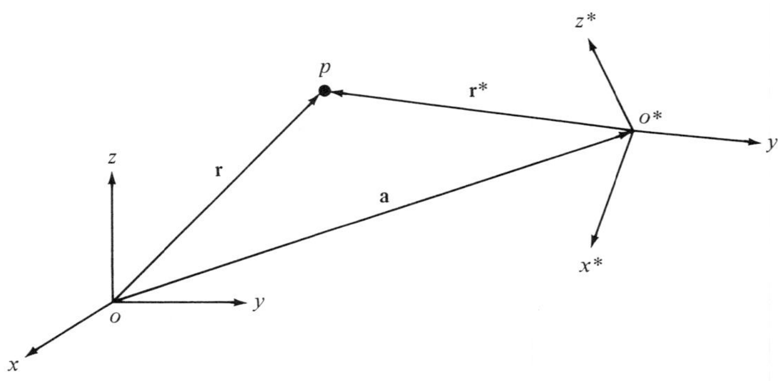

Figure 5.7.2 Moving coordinate systems for a robotic manipulator.

where $\mathbf{r} \equiv [x, y\, z]^T$, $\mathbf{r}^* \equiv [x^*\, y^*\, z^*]^T$, and $\mathbf{a} \equiv [a_x\, a_y\, a_z]^T$. If the origin o^* is moving with respect to the origin o, which is assumed to be fixed, then

$$\mathbf{v} = \frac{d\mathbf{r}}{dt} = \frac{d\mathbf{r}^*}{dt} + \frac{d\mathbf{a}}{dt} = \mathbf{v}^* + \mathbf{v}_a, \tag{5.7.6}$$

where $\mathbf{v}$ and $\mathbf{v}^*$ are the velocities of point p with respect to the xyz and $x^*y^*z^*$ frames, respectively. However, if the coordinate frame $x^*y^*z^*$ is moving and rotating with respect to the frame xyz, then

$$\frac{d\mathbf{r}}{dt} = \frac{d^*\mathbf{r}}{dt} + \boldsymbol{\omega} \times \mathbf{r}, \tag{5.7.7}$$

where $d^*\mathbf{r}/dt$ is the time derivative with respect to the $x^*y^*z^*$ coordinate system and $\boldsymbol{\omega}$ is the angular velocity vector of the $x^*y^*z^*$ coordinate system (Fig. 5.7.3).

Note that $d^*\mathbf{r}/dt$ is the time rate of change of $\mathbf{r}$ as seen by an observer sitting at the origin of the $x^*y^*z^*$ system, namely, o^*. From Equation (5.7.7), we then note that

$$\frac{d^2\mathbf{r}}{dt^2} = \frac{d}{dt}\left(\frac{d^*\mathbf{r}}{dt}\right) + \boldsymbol{\omega} \times \frac{d\mathbf{r}}{dt} + \frac{d\boldsymbol{\omega}}{dt} \times \mathbf{r} \tag{5.7.8}$$

$$= \frac{d^{*2}\mathbf{r}}{dt^2} + \boldsymbol{\omega} \times \frac{d^*\mathbf{r}}{dt} + \boldsymbol{\omega} \times \left(\frac{d^*\mathbf{r}}{dt} + \boldsymbol{\omega} \times \mathbf{r}\right) + \frac{d\boldsymbol{\omega}}{dt} \times \mathbf{r}. \tag{5.7.9}$$

Finally,

$$\frac{d^2\mathbf{r}}{dt^2} = \frac{d^{*2}\mathbf{r}}{dt^2} + 2\boldsymbol{\omega} \times \frac{d^*\mathbf{r}}{dt} + \boldsymbol{\omega} \times (\boldsymbol{\omega} \times \mathbf{r}) + \frac{d\boldsymbol{\omega}}{dt} \times \mathbf{r}. \tag{5.7.10}$$

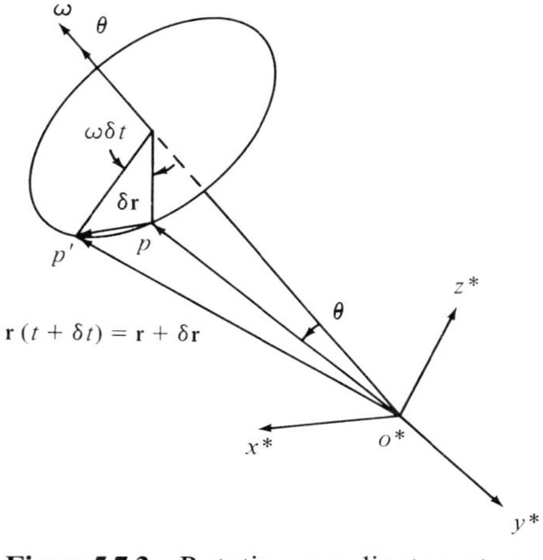

Figure 5.7.3 Rotating coordinate system.

Note that as discussed before

$$\frac{d^{*2}\mathbf{r}}{dt^2} \equiv \text{acceleration term as measured by an observer on the } x^*y^*z^* \text{ coordinate system,}$$

$$2\boldsymbol{\omega} \times \frac{d^*\mathbf{r}}{dt} \equiv \text{Coriolis acceleration,}$$

$$\boldsymbol{\omega} \times (\boldsymbol{\omega} \times \mathbf{r}) \equiv \text{centripetal (toward the center) acceleration,}$$

and

$$\frac{d\boldsymbol{\omega}}{dt} \times \mathbf{r} \equiv \text{angular acceleration.}$$

Now if the $x^*y^*z^*$ system is an inertial system—that is, its origin o^* has a uniform velocity with respect to the xyz system, which is assumed to be fixed—then one has

$$m\frac{d^{*2}\mathbf{r}}{dt^2} = \mathbf{f} - m\boldsymbol{\omega} \times (\boldsymbol{\omega} \times \mathbf{r}) - 2m\boldsymbol{\omega} \times \frac{d^*\mathbf{r}}{dt} - m\frac{d\boldsymbol{\omega}}{dt} \times \mathbf{r}, \tag{5.7.11}$$

where m is the mass associated with the point p and $\mathbf{f}$ is the resultant vector of all external forces applied to the mass m whose center is at p.

If the origin of the $x^*y^*z^*$ system is moving with a velocity $d\mathbf{a}/dt$ with respect to xyz, where $\mathbf{a}$ is the vector joining o and o^* (Fig. 5.7.2), then

$$\frac{d\mathbf{r}}{dt} = \frac{d^*\mathbf{r}^*}{dt} + \boldsymbol{\omega} \times \mathbf{r}^* + \frac{d\mathbf{a}}{dt} \tag{5.7.12}$$

and, furthermore,

$$\frac{d^2\mathbf{r}}{dt^2} = \frac{d^{*2}\mathbf{r}^*}{dt^2} + \boldsymbol{\omega} \times (\boldsymbol{\omega} \times \mathbf{r}^*) + 2\boldsymbol{\omega} \times \frac{d^*\mathbf{r}^*}{dt} + \frac{d\boldsymbol{\omega}}{dt} \times \mathbf{r}^* + \frac{d^2\mathbf{a}}{dt}. \tag{5.7.13}$$

Let us now consider the following generalized situation as depicted in Figure 5.7.4. Let us recall that an orthonormal coordinate system $(x_{i-1}, y_{i-1}, z_{i-1})$ is established at joint i and that (x_0, y_0, z_0) is the base coordinate system. Let us denote by $\mathbf{r}_i$ the position vector of o^1, the origin of the (x_i, y_i, z_i) system with respect to the base coordinate system. Let the position vector of o^*, the origin of the $(x_{i-1}, y_{i-1}, z_{i-1})$ system with respect to the base coordinate system, be denoted by $\mathbf{r}_{i-1}$. We assign $\mathbf{s}$ as the vector extending from o^* to o'. Note that

$$\mathbf{r}_{i-1} + \mathbf{s} = \mathbf{r}_i. \tag{5.7.14}$$

Let $\mathbf{v}_{i-1}$ and $\boldsymbol{\omega}_{i-1}$ be the linear and angular velocities, respectively, of the coordinate system $(x_{i-1}, y_{i-1}, z_{i-1})$ with respect to the base coordinate system (x_0, y_0, z_0). The linear velocity $\mathbf{v}_i$ of the coordinate system (x_i, y_i, z_i) with respect to the base coordinate system is

$$\mathbf{v}_i = \frac{d^*\mathbf{s}}{dt} + \boldsymbol{\omega}_{i-1} \times \mathbf{s} + \mathbf{v}_{i-1}, \tag{5.7.15}$$

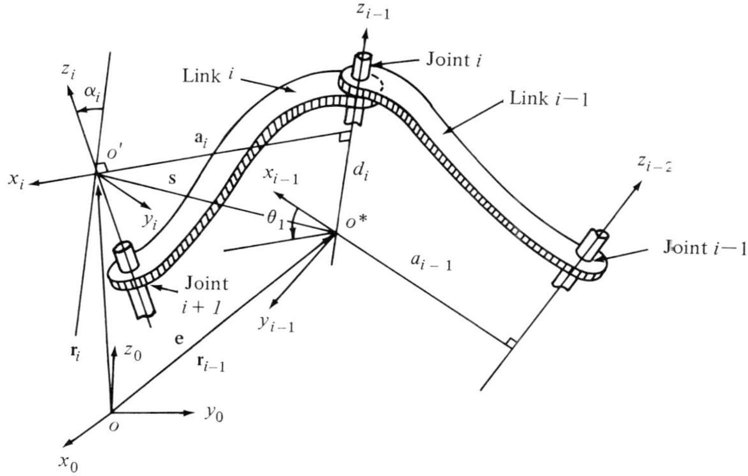

Figure 5.7.4 Generalized adjacent links situation for manipulators.

where $d^*(\ )/dt$ denotes the time derivative with respect to the moving/rotating coordinate system $(x_{i-1}, y_{i-1}, z_{i-1})$ and

$$\dot{\mathbf{v}}_i = \frac{d^{*2}\mathbf{s}}{dt^2} + \dot{\boldsymbol{\omega}}_{i-1} \times \mathbf{s} + 2\boldsymbol{\omega}_{i-1} \times \frac{d^*\mathbf{s}}{dt} + \boldsymbol{\omega}_{i-1} \times (\boldsymbol{\omega}_{i-1} \times \mathbf{s}) - \dot{\mathbf{v}}_{i-1}. \tag{5.7.16}$$

Let $\boldsymbol{\omega}_i$ and $\boldsymbol{\omega}_s$ be the angular velocities of (x_i, y_i, z_i) with respect to (x_0, y_0, z_0) and $(x_{i-1}, y_{i-1}, z_{i-1})$, respectively. Thus,

$$\boldsymbol{\omega}_i = \boldsymbol{\omega}_{i-1} + \boldsymbol{\omega}_s \tag{5.7.17}$$

and

$$\dot{\boldsymbol{\omega}}_i = \dot{\boldsymbol{\omega}}_{i-1} + \dot{\boldsymbol{\omega}}_s. \tag{5.7.18}$$

Note that

$$\dot{\boldsymbol{\omega}}_s = \frac{d^*\boldsymbol{\omega}_s}{dt} + \boldsymbol{\omega}_{i-1} \times \boldsymbol{\omega}_s, \tag{5.7.19}$$

and, therefore,

$$\dot{\boldsymbol{\omega}}_i = \dot{\boldsymbol{\omega}}_{i-1} + \frac{d^*\boldsymbol{\omega}_s}{dt} + \boldsymbol{\omega}_{i-1} \times \boldsymbol{\omega}_s. \tag{5.7.20}$$

We note that the angular velocity of the (x_i, y_i, z_i) system with respect to the $(x_{i-1}, y_{i-1}, z_{i-1})$ system, that is, $\boldsymbol{\omega}_s$, is the same as the angular velocity of link i about the z_{i-1} axis. Then

$$\boldsymbol{\omega}_s = \begin{cases} \mathbf{z}_{i-1}\dot{q}_i, & \text{if link } i \text{ is rotational,} \quad (5.7.21) \\ 0, & \text{if link } i \text{ is translational,} \quad (5.7.22) \end{cases}$$

where $\dot{q}_i$ is the generalized velocity of joint i or the magnitude of the angular velocity of link i with respect to the coordinate system $(x_{i-1}, y_{i-1}, z_{i-1})$, which, incidentally, is an orthonormal system. Thus,

$$\frac{d^*\boldsymbol{\omega}_s}{dt} = \begin{cases} \mathbf{z}_{i-1}\dot{q}_i, & \text{if link } i \text{ is rotational,} \quad (5.7.23) \\ 0, & \text{if link } i \text{ is translational.} \quad (5.7.24) \end{cases}$$

From Equations (5.7.17), (5.7.19), (5.7.21), (5.7.22), (5.7.23), and (5.7.24), we find that

$$\boldsymbol{\omega}_i = \begin{cases} \boldsymbol{\omega}_{i-1} + \mathbf{z}_{i-1}\dot{q}_i, & \text{if link } i \text{ is rotational,} \quad (5.7.25) \\ \boldsymbol{\omega}_{i-1}, & \text{if link } i \text{ is translational,} \quad (5.7.26) \end{cases}$$

and

$$\dot{\boldsymbol{\omega}}_i = \begin{cases} \dot{\boldsymbol{\omega}}_{i-1} + z_{i-1}\ddot{q}_i + \boldsymbol{\omega}_{i-1} \times (\mathbf{z}_{i-1}\dot{q}_i), & \text{if link } i \text{ is rotational,} \quad (5.7.27) \\ \dot{\boldsymbol{\omega}}_{i-1}, & \text{if link } i \text{ is translational.} \quad (5.7.28) \end{cases}$$

Let $\mathbf{s}$ be replaced by $\mathbf{p}_i^*$, where $\mathbf{p}_i^*$ is then the coordinate of the origin o' with respect to the $(x_{i-1}, y_{i-1}, z_{i-1})$ system. If link i is translational with respect to the $(x_{i-1}, y_{i-1}, z_{i-1})$ frame, it then travels in the direction of z_{i-1} with a joint velocity $\dot{q}_i$ relative to link $i-1$. If it is rotational in the $(x_{i-1}, y_{i-1}, z_{i-1})$ coordinate system, then it has an angular velocity $\boldsymbol{\omega}_s$ with respect to the $(x_{i-1}, y_{i-1}, z_{i-1})$ system. Thus,

$$\frac{d^*\mathbf{s}}{dt} = \begin{cases} \boldsymbol{\omega}_s \times \mathbf{p}_i^*, & \text{if link } i \text{ is rotational} \quad (5.7.29) \\ \mathbf{z}_{i-1}\dot{q}_i, & \text{if link } i \text{ is translational.} \quad (5.7.30) \end{cases}$$

Furthermore,

$$\frac{d^*\mathbf{s}^2}{dt^2} = \begin{cases} \dfrac{d^*\boldsymbol{\omega}_s}{dt} \times \mathbf{p}_i^* + \boldsymbol{\omega}_s \times (\boldsymbol{\omega}_s \times \mathbf{p}_i^*), & \text{if link } i \text{ is rotational} \quad (5.7.31) \\ \mathbf{z}_{i-1}\ddot{q}_i, & \text{if link } i \text{ is translational.} \quad (5.7.32) \end{cases}$$

Therefore, Equation (5.7.15) expands to

$$\mathbf{v}_i = \begin{cases} \boldsymbol{\omega}_i \times \mathbf{p}_i^* + \mathbf{v}_{i-1}, & \text{if link } i \text{ is rotational} \quad (5.7.33) \\ \mathbf{z}_{i-1}\dot{q}_i + \boldsymbol{\omega}_i \times \mathbf{p}_i^* + \mathbf{v}_{i-1}, & \text{if link } i \text{ is translational,} \quad (5.7.34) \end{cases}$$

and Equation (5.7.16) expands to

$$\dot{\mathbf{v}}_i = \begin{cases} \dot{\boldsymbol{\omega}}_i \times \mathbf{p}_i^* + \boldsymbol{\omega}_i \times (\boldsymbol{\omega}_i \times \mathbf{p}_i^*) + \dot{\mathbf{v}}_{i+1}, \\ \qquad \text{if link } i \text{ rotational} \quad (5.7.35) \\ \mathbf{z}_{i-1}\ddot{q}_i + \dot{\boldsymbol{\omega}}_i \times \mathbf{p}_i^* + 2\boldsymbol{\omega}_i \times (\mathbf{z}_{i-1}\dot{q}_i) + \boldsymbol{\omega}_i \times (\boldsymbol{\omega}_i \times \mathbf{p}_i^*) + \mathbf{v}_{i-1}, \\ \qquad \text{if link } i \text{ is translational.} \quad (5.7.36) \end{cases}$$

Note that, as explained before [Eq. (5.7.26)], in Equation (5.7.36) $\boldsymbol{\omega}_i = \boldsymbol{\omega}_{i-1}$ if link i is translational.

EXAMPLE 5.7.1 ■

Derive Equations (5.7.33) and (5.7.34) for $\mathbf{v}_i$.

Solution. Note from Equation (5.7.15) that we have to calculate $\mathbf{v}_i = d^*\mathbf{s}/dt + \boldsymbol{\omega}_{i-1} \times \mathbf{s} + \mathbf{v}_{i-1}$. From Equation (5.7.29) and (5.7.30), we see that

$$\frac{d^*\mathbf{s}^2}{dt} = \begin{cases} \boldsymbol{\omega}_s \times \mathbf{p}_i^* = (\boldsymbol{\omega}_i - \boldsymbol{\omega}_{i-1}) \times \mathbf{p}_i^*, & \text{if link } i \text{ is rotational,} \quad (5.7.37) \\ \mathbf{z}_{i-1}\dot{q}_i, & \text{if link } i \text{ is translational.} \quad (5.7.38) \end{cases}$$

Thus,

$$\mathbf{v}_i = \begin{cases} \boldsymbol{\omega}_i \times \mathbf{p}_i^* + \mathbf{v}_{i-1} = \boldsymbol{\omega} \times \mathbf{p}_i^* + \mathbf{v}_{i-1}, & \text{if link } i \text{ is rotational,} \quad (5.7.39) \\ \mathbf{z}_{i-1}\dot{q}_i + \boldsymbol{\omega}_i \times \mathbf{p}_i^* + \mathbf{v}_{i-1}, & \text{if link } i \text{ is translational.} \quad (5.7.40) \end{cases}$$

Note that in the translational case $\boldsymbol{\omega}_i = \boldsymbol{\omega}_{i-1}$ according to Equation (5.7.33). ■ ■

EXAMPLE 5.7.2 ■

Derive Equations (5.7.35) and (5.7.36) for $\dot{\mathbf{v}}_i$.

Solution. Note from Equation (5.7.16) that we must calculate $d^{2*}\mathbf{s}/dt^2$ and then simplify $\boldsymbol{\omega}_{i-1} \times (\boldsymbol{\omega}_{i-1} \times \mathbf{p}_i^*)$. From Equations (5.7.31) and (5.7.32) we have

$$\frac{d^{2*}\mathbf{s}}{dt^2} = \begin{cases} \mathbf{z}_{i-1}\ddot{q}_i \times \mathbf{p}_i^* + \boldsymbol{\omega}_s \times (\boldsymbol{\omega}_s \times \mathbf{p}_i^*), & \text{if link } i \text{ is rotational} \quad (5.7.41) \\ \mathbf{z}_{i-1}\ddot{q}_i, & \text{if link } i \text{ is translational.} \quad (5.7.42) \end{cases}$$

Note that we have the following vector identities:

$$\boldsymbol{\omega}_s \times (\boldsymbol{\omega}_s \times \mathbf{p}_i^*) = \boldsymbol{\omega}_s(\boldsymbol{\omega}_s \cdot \mathbf{p}_i^*) - \mathbf{p}_i^*(\boldsymbol{\omega}_s \cdot \boldsymbol{\omega}_s), \tag{5.7.43}$$

$$\boldsymbol{\omega}_{i-1} \times (\boldsymbol{\omega}_s \times \mathbf{p}_i^*) = \boldsymbol{\omega}_s(\boldsymbol{\omega}_{i-1} \cdot \mathbf{p}_i^*) - \mathbf{p}_i^*(\boldsymbol{\omega}_{i-1} \cdot \boldsymbol{\omega}_s), \tag{5.7.44}$$

and

$$\boldsymbol{\omega}_{i-1} \times (\boldsymbol{\omega}_{i-1} \times \mathbf{p}_i^*) = \boldsymbol{\omega}_{i-1}(\boldsymbol{\omega}_{i-1} \cdot \mathbf{p}_i^*) - \mathbf{p}_i^*(\boldsymbol{\omega}_{i-1} \cdot \boldsymbol{\omega}_{i-1}). \tag{5.7.45}$$

Thus, if link i is rotational,

$$\begin{aligned} \dot{\mathbf{v}}_i = {} & \mathbf{z}_{i-1}\ddot{q}_i \times \mathbf{p}_i^* + \dot{\boldsymbol{\omega}}_{i-1} \times \mathbf{p}_i^* + 2\boldsymbol{\omega}_{i-1} \times (\boldsymbol{\omega}_s \times \mathbf{p}_i^*) \\ & + \boldsymbol{\omega}_{i-1} \times (\boldsymbol{\omega}_{i-1} \times \mathbf{p}_i^*) + \dot{\mathbf{v}}_{i-1} + \boldsymbol{\omega}_s \times (\boldsymbol{\omega}_s \times \mathbf{p}_i^*). \end{aligned} \tag{5.7.46}$$

From Equation (5.7.27), for this case, we have

$$\mathbf{z}_{i-1}\ddot{q}_i = \dot{\boldsymbol{\omega}}_i - \dot{\boldsymbol{\omega}}_{i-1} - \dot{\boldsymbol{\omega}}_{i-1} \times (\mathbf{z}_{i-1}\dot{q}_i), \tag{5.7.47}$$

which in conjunction with Equation (5.7.25) yields

$$\begin{aligned} \mathbf{z}_{i-1}\ddot{q}_i &= \dot{\boldsymbol{\omega}}_i - \dot{\boldsymbol{\omega}}_{i-1} - \dot{\boldsymbol{\omega}}_{i-1} \times (\boldsymbol{\omega}_i - \boldsymbol{\omega}_{i-1}) \\ &= \dot{\boldsymbol{\omega}}_i - \dot{\boldsymbol{\omega}}_{i-1} - \boldsymbol{\omega}_{i-1} \times \boldsymbol{\omega}_i. \end{aligned} \tag{5.7.48}$$

Thus, Equation (5.7.46) reduces to

$$\begin{aligned}\dot{\mathbf{v}}_i = \dot{\boldsymbol{\omega}}_i \times \mathbf{p}_i^* - \boldsymbol{\omega}_{i-1} \times \mathbf{p}_i^* - \boldsymbol{\omega}_{i-1} \times \boldsymbol{\omega}_i \times \mathbf{p}_i^* + \dot{\boldsymbol{\omega}}_{i-1} \times \mathbf{p}_i^* \\ + 2\boldsymbol{\omega}_s(\boldsymbol{\omega}_{i-1} \cdot \mathbf{p}_i^*) - 2\mathbf{p}_i^*(\boldsymbol{\omega}_{i-1} \cdot \boldsymbol{\omega}_s) + \boldsymbol{\omega}_{i-1}(\boldsymbol{\omega}_{i-1} \cdot \mathbf{p}_i^*) \\ - \mathbf{p}_i^*(\boldsymbol{\omega}_{i-1} \cdot \boldsymbol{\omega}_{i-1}) + \dot{\mathbf{v}}_{i-1} + \boldsymbol{\omega}_s(\boldsymbol{\omega}_s \cdot \mathbf{p}_i^*) - \mathbf{p}_i^*(\boldsymbol{\omega}_{i-1} \cdot \boldsymbol{\omega}_s).\end{aligned} \tag{5.7.49}$$

Note that

$$\boldsymbol{\omega}_{i-1} \times \boldsymbol{\omega}_i \times \mathbf{p}_i^* = \boldsymbol{\omega}_i(\boldsymbol{\omega}_{i-1} \cdot \mathbf{p}_i^*) - \mathbf{p}_i^*(\boldsymbol{\omega}_i \cdot \boldsymbol{\omega}_{i-1}). \tag{5.7.50}$$

Thus,

$$\begin{aligned}\dot{\mathbf{v}}_i = \dot{\boldsymbol{\omega}}_i \times \mathbf{p}_i^* - \boldsymbol{\omega}_i(\boldsymbol{\omega}_{i-1} \cdot \mathbf{p}_i^*) + \mathbf{p}_i^*(\boldsymbol{\omega}_i \cdot \boldsymbol{\omega}_{i-1}) + 2\boldsymbol{\omega}_s(\boldsymbol{\omega}_{i-1} \cdot \mathbf{p}_i^*) \\ - 2\mathbf{p}_i^*(\boldsymbol{\omega}_{i-1} \cdot \boldsymbol{\omega}_s) + \boldsymbol{\omega}_{i-1}(\boldsymbol{\omega}_{i-1} \cdot \mathbf{p}_i^*) - \mathbf{p}_i^*(\boldsymbol{\omega}_{i-1} \cdot \boldsymbol{\omega}_{i-1}) + \dot{\mathbf{v}}_{i-1} \\ + \boldsymbol{\omega}_s(\boldsymbol{\omega}_s \cdot \mathbf{p}_i^*) - \mathbf{p}_i^*(\boldsymbol{\omega}_{i-1} \cdot \boldsymbol{\omega}_s).\end{aligned} \tag{5.7.51}$$

Finally, for this case (link i rotational),

$$\begin{aligned}\dot{\mathbf{v}}_i &= \dot{\boldsymbol{\omega}}_i \times \mathbf{p}_i^* + \boldsymbol{\omega}_i(\boldsymbol{\omega}_i \cdot \dot{\mathbf{p}}_i^*) - \mathbf{p}_i^*(\boldsymbol{\omega}_i \cdot \boldsymbol{\omega}_i) + \dot{\mathbf{v}}_{i-1} \\ &= \dot{\boldsymbol{\omega}}_i \times \mathbf{p}_i^* + \boldsymbol{\omega}_i(\boldsymbol{\omega}_i \cdot \mathbf{p}_i^*) + \dot{\mathbf{v}}_{i-1}.\end{aligned} \tag{5.7.52}$$

If link i is translational,

$$\begin{aligned}\dot{\mathbf{v}}_i &= \dot{\mathbf{z}}_{i-1}\ddot{q}_i + \dot{\boldsymbol{\omega}}_{i-1} \times \mathbf{p}_i^* + 2\boldsymbol{\omega}_{i-1} \times z_{i-1}\dot{q}_i + \boldsymbol{\omega}_{i-1} \times (\boldsymbol{\omega}_{i-1} \times \mathbf{p}_i^*) + \dot{\mathbf{v}}_{i-1} \\ &= \mathbf{z}_{i-1}\ddot{q}_i + \dot{\boldsymbol{\omega}}_i \times \mathbf{p}_i^* + 2\boldsymbol{\omega}_i \times (\mathbf{z}_{i-1}\dot{q}_i) + \boldsymbol{\omega}_i \times (\boldsymbol{\omega}_i \times \mathbf{p}_i^*) + \dot{\mathbf{v}}_{i-1}. \quad \blacksquare\blacksquare\end{aligned} \tag{5.7.53}$$

5.8 RECURSIVE DYNAMIC EQUATIONS FOR ROBOTIC MANIPULATORS

In this section we use the d'Alembert principle to derive a set of recursive equations of motion for robotic manipulators.

d'Alembert Principle The dynamic equilibrium of a multibody system is equivalent to a static equilibrium situation of the same system if only the inertial forces are treated as a set of independent forces in equilibrium with all other forces acting on the system at each instant of time. The inertial forces are defined as the negative of the product of the mass and the acceleration vector passing through the center of gravity of each link. Consider the ith link and let the origin o' be situated at its center of mass (see Fig. 5.8.1). Then $\mathbf{m}_i$ is the total mass of link i concentrated at o', $\bar{\mathbf{r}}_i$ is the position vector of the center of mass of link i with respect to the base coordinate system, $\bar{\mathbf{s}}_i$ is the position vector of the center of mass of link i with respect to the (x_i, y_i, z_i) system, $\bar{\mathbf{v}}_i = d\bar{\mathbf{r}}_i/dt$ is the linear velocity vector of the center of mass of link i with respect to the base coordinate system, $\mathbf{F}_i$ is the total external force (net d'Alembert force) vector exerted on link i with respect to the base coordinate system, $\boldsymbol{\tau}_i$ is the total external torque (net d'Alembert torque) exerted on link i with respect to the base coordinate system,

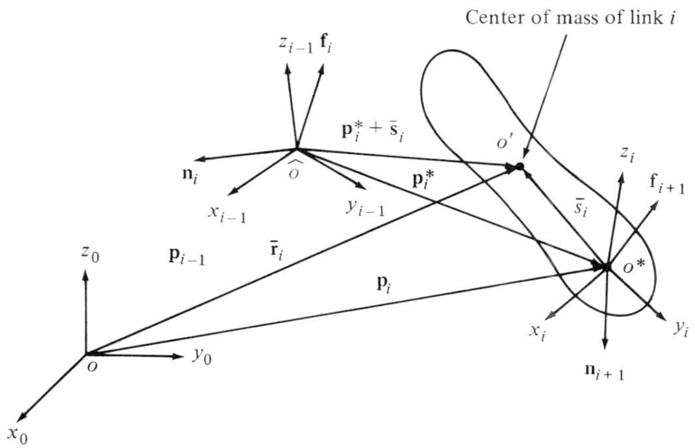

Figure 5.8.1 Forces and torques on the ith link.

$\mathbf{I}_i$ is the inertia matrix of link i about its center of mass with reference to the coordinate system (x_0, y_0, z_0), $\mathbf{f}_{i-1,i}$ is the force exerted on link i by link $i-1$, and $\boldsymbol{\tau}_{i-1,i}$ is the torque exerted on link i by link $i-1$.

Neglecting the effects of elasticities and viscous damping of the joints, we have

$$\mathbf{f}_i = m_i \bar{\mathbf{a}}_i, \tag{5.8.1}$$

$$\boldsymbol{\tau}_i = \frac{d}{dt}(\mathbf{I}_i \cdot \boldsymbol{\omega}_i) = \mathbf{I}_i \dot{\boldsymbol{\omega}}_i + \dot{\boldsymbol{\omega}}_i \times (\mathbf{I}_i \boldsymbol{\omega}_i), \tag{5.8.2}$$

$$\bar{\mathbf{v}}_i = \boldsymbol{\omega}_i \times \mathbf{s}_i + \mathbf{v}_i, \tag{5.8.3}$$

and

$$\bar{\mathbf{a}}_i = \dot{\boldsymbol{\omega}}_i \times \bar{\mathbf{s}}_i + \boldsymbol{\omega}_i \times (\boldsymbol{\omega}_i \times \bar{\mathbf{s}}_i) + \dot{\mathbf{v}}_i. \tag{5.8.4}$$

Clearly, the total external force $\mathbf{f}_i$ and moment $\boldsymbol{\tau}_i$ must be balanced, respectively, by the gravity and link forces from links $i-1$ and $i+1$. Thus,

$$\mathbf{f}_i = \mathbf{f}_{i-1,i} - \mathbf{f}_{i,i+1} + m_i \mathbf{g} \tag{5.8.5}$$

and

$$\boldsymbol{\tau}_i = \boldsymbol{\tau}_{i-1,i} - \boldsymbol{\tau}_{i,i+1} + (\mathbf{p}_{i-1} - \bar{\mathbf{r}}_i) \times \mathbf{f}_{i-1,i} - (\mathbf{p}_i - \bar{\mathbf{r}}_i) \times \mathbf{f}_{i,i+1}. \tag{5.8.6}$$

Note that, since $\bar{\mathbf{r}}_i - \mathbf{p}_{i-1} = (\mathbf{p}_i^* + \bar{\mathbf{s}}_i)$,

$$\mathbf{f}_{i-1,i} = \mathbf{f}_i + \mathbf{f}_{i,i+1} - m_i \mathbf{g} \tag{5.8.7}$$

and

$$\boldsymbol{\tau}_{i-1,i} = \boldsymbol{\tau}_{i,i+1} + \mathbf{p}_i^* \times \mathbf{f}_{i,i+1} + (\mathbf{p}_i^* + \bar{\mathbf{s}}_i) \times (\mathbf{f}_i - m_i \mathbf{g}). \tag{5.8.8}$$

Equations (5.8.7) and (5.8.8) are called the *recursive equations* for the dynamics of robotic manipulators. They can be used to compute $(\mathbf{f}_{i-1,i}, \boldsymbol{\tau}_{i-1,i})$, the force and couple applied to link i by link $i-1$. Note that $\mathbf{f}_{n,n+1}$ and $\boldsymbol{\tau}_{n,n+1}$ are, respectively, the force and torque exerted by the hand on the object being gripped. If joint i is rotational then it actually rotates q_i or θ_i radians about the $\mathbf{z}_{i-1}$ axis. Thus, the input torque at joint i is simply the sum of the projection $\mathbf{n}_i$ onto the $\mathbf{z}_{i-1}$ axis and the viscous damping and elasticity moments in that coordinate system. On the other hand, if joint i is translational, then it translates q_i units relative to (along) the $\mathbf{z}_{i-1}$ axis. In this case, the input torque $\boldsymbol{\tau}_i$ at the joint is the sum of the projection of $\mathbf{f}_i$ onto the $\mathbf{z}_{i-1}$ axis and the viscous damping and elasticity forces in that coordinate system. Thus, we finally have

$$\boldsymbol{\tau}_i^* = \begin{cases} (\boldsymbol{\tau}_{i-1,i} \cdot \mathbf{z}_{i-1})\mathbf{z}_{i-1} + [b_i\dot{q}_i + K_i(q_i)]\mathbf{z}_{i-1}, & (5.8.9) \\ (\mathbf{f}_{i-1,i} \cdot \mathbf{z}_{i-1,i})\mathbf{z}_{i-1,i} + [b_i\dot{q}_i + k_i(q_i)]\mathbf{z}_{i-1}, & (5.8.10) \end{cases}$$

where b_i is the viscosity damping coefficient, K_i is the rotational stiffness, k_i is the translational stiffness, and $\boldsymbol{\tau}_i^*$ is the joint torque. Clearly, if the base coordinate is bolted on a platform and link o is stationary, then $\boldsymbol{\omega}_0 = \dot{\boldsymbol{\omega}}_0 = \mathbf{v}_0 = 0$, $\dot{\mathbf{v}}_0 = \mathbf{a}_0 = [0, 0, g]^T$, where $g = 9.8062\ \text{m/s}^2$ is the ground-zero gravitational acceleration.

We therefore end up with a set of forward and backward recursion equations. These recursive equations are tabulated in Table 5.8.1. For the forward recursions, we start with linear velocity, linear acceleration, angular velocity, angular acceleration, and the total force and moment of each link and

TABLE 5.8.1 RECURSIVE EQUATIONS OF ROBOTICS DYNAMICS

Forward equations

$$\boldsymbol{\omega}_i = \begin{cases} \boldsymbol{\omega}_{i-1} + \mathbf{z}_{i-1}\dot{q}_i, & \text{if link } i \text{ is rotational} \quad (5.8.11) \\ \boldsymbol{\omega}_{i-1}, & \text{if link } i \text{ is translational} \quad (5.8.12) \end{cases}$$

$$\dot{\boldsymbol{\omega}}_i = \begin{cases} \dot{\boldsymbol{\omega}}_{i-1} + \mathbf{z}_{i-1}\ddot{q}_i + \dot{\boldsymbol{\omega}}_{i-1} \times (\mathbf{z}_{i-1}\dot{q}_i), & \text{if link } i \text{ is rotational} \quad (5.8.13) \\ \dot{\boldsymbol{\omega}}_{i-1}, & \text{if link } i \text{ is translational} \quad (5.8.14) \end{cases}$$

$$\mathbf{v}_i = \begin{cases} \boldsymbol{\omega}_i \times \mathbf{p}_i^* + \mathbf{v}_{i-1}, & \text{if link } i \text{ is rotational} \quad (5.8.15) \\ \mathbf{z}_{i-1}q_i + \boldsymbol{\omega}_i \times \mathbf{p}_i^* + \mathbf{v}_{i-1}, & \text{if link } i \text{ is translational} \quad (5.8.16) \end{cases}$$

$$\dot{\mathbf{v}}_i = \begin{cases} \dot{\boldsymbol{\omega}}_i \times \mathbf{p}_i^* + \boldsymbol{\omega}_i \times (\boldsymbol{\omega}_i \times \mathbf{p}_i^*) + \dot{\mathbf{v}}_{i-1}, & \text{if link } i \text{ is rotational} \quad (5.8.17) \\ \mathbf{z}_{i-1}\ddot{q}_i + \dot{\boldsymbol{\omega}}_i \times \mathbf{p}_i^* + 2\boldsymbol{\omega}_i \times (\mathbf{z}_{i-1}\dot{q}_i) + \boldsymbol{\omega}_i \times (\boldsymbol{\omega}_i \times \mathbf{p}_i^*) + \dot{\mathbf{v}}_{i-1}, & \text{if link } i \text{ is translational} \quad (5.8.18) \end{cases}$$

$$\bar{\mathbf{a}}_i = \dot{\boldsymbol{\omega}}_i \times \bar{\mathbf{s}}_i + \boldsymbol{\omega}_i \times (\boldsymbol{\omega}_i \times \bar{\mathbf{s}}_i) + \dot{\mathbf{v}}_i \quad (5.8.19)$$

$$\mathbf{f}_i = M_i\bar{\mathbf{a}}_i, \quad (5.8.20)$$

$$\boldsymbol{\tau}_i = \mathbf{I}_i\dot{\boldsymbol{\omega}}_i + \boldsymbol{\omega}_i \times (\mathbf{I}\boldsymbol{\omega}_i) \quad (5.8.21)$$

Backward equations

$$\mathbf{f}_{i-1,i} = \mathbf{f}_i + \mathbf{f}_{i,i+1} - m_i\mathbf{g}, \quad (5.8.22)$$

$$\boldsymbol{\tau}_{i-1,i} = \boldsymbol{\tau}_{i,i+1} + \mathbf{p}_i^* \times \mathbf{f}_{i,i+1} + (\mathbf{p}_i^* + \bar{\mathbf{s}}_i) \times \mathbf{f}_i + \boldsymbol{\tau}_i \quad (5.8.23)$$

$$\boldsymbol{\tau}_i^* = \begin{cases} (\boldsymbol{\tau}_{i-1,i} \cdot \mathbf{z}_{i-1})\mathbf{z}_{i-1} + [b_i\dot{q}_i + \mathbf{K}_i(q_i)]\mathbf{z}_{i-1} & (5.8.24) \\ (\mathbf{f}_{i-1,1} \cdot \mathbf{z}_{i-1})\mathbf{z}_{i-1} + [b_i\dot{q}_i + k(q_i)]\mathbf{z}_{i-1} & (5.8.25) \end{cases}$$

propagate from the base to the gripper or the hand. For the backward recursions, we start with the torques and forces exerted on each link propagated from the gripper to the base. Thus, the forward equations propagate the kinematics of each link from the base to the hand, while the backward equations compute the necessary torques and forces for each link from the gripper to the base.

5.9 RECURSIVE NEWTON–EULER EQUATIONS WITH RESPECT TO LOCAL LINK COORDINATE SYSTEM

Equations (5.8.11)–(5.8.25) are in fact with respect to the base coordinate system. It is more efficient to express the dynamic equations of each link with respect to its own coordinate system. Let $\mathbf{R}_{i-1}^{i}$ be the 3×3 rotation submatrix of $\mathbf{A}_{i-1}^{i}$. Recall that this matrix is an orthonormal one; that is, $(\mathbf{R}_{i-1}^{i})^{-1} = \mathbf{R}_{i}^{i-1} = (\mathbf{R}_{i-1}^{i})^{T}$, where

$$\mathbf{R}_{i-1}^{i} = \begin{bmatrix} \cos\theta_i & -\cos\alpha_i \sin\theta_i & \sin\alpha_i \sin\theta_i \\ \sin\theta_i & \cos\alpha_i(-\cos\theta_i) & -\sin\alpha_i \cos\theta_i \\ 0 & \sin\alpha_i & \cos\alpha_i \end{bmatrix} \tag{5.9.1}$$

and

$$(\mathbf{R}_{i-1}^{i})^{-1} = \mathbf{R}_{i}^{i-1} = (\mathbf{R}_{i-1}^{i})^{T} = \begin{bmatrix} \cos\theta_i & \sin\theta_i & 0 \\ -\cos\alpha_i \sin\theta_i & \cos\alpha_i \cos\theta_i & \sin\alpha_i \\ \sin\alpha_i \sin\theta_i & -\sin\alpha_i \cos\theta_i & \cos\alpha_i \end{bmatrix}. \tag{5.9.2}$$

Then Equations (5.8.11)–(5.8.25) will be reduced to the ones depicted in Table 5.9.1.

EXAMPLE 5.9.1 ■

Show that there are $135n - 48$ multiplications and $123n - 48$ additions involved in the Newton–Euler algorithm.

Solution. Breaking down Equations (5.9.3)–(5.9.17), we find for terms $\boldsymbol{\omega}_i$, $\mathbf{v}_i \dot{\boldsymbol{\omega}}_i$, $\dot{\mathbf{v}}_i$, $\bar{\mathbf{a}}_i$, $\mathbf{f}_i$, $\mathbf{f}_{i-1,i}$, $\boldsymbol{\tau}_i$, and $\boldsymbol{\tau}_{i-1,i}$ the number of multiplications and additions listed in Table 5.9.2. ■ ■

The following simple algorithm given for an n-jointed robotic manipulator generates the desired torque or force values for all joint actuators.

Algorithm 5.9.1. For the initial conditions we have

$$\boldsymbol{\omega}_0 = \dot{\boldsymbol{\omega}}_0 = \mathbf{v}_0 = 0 \quad \text{and} \quad \dot{\mathbf{v}}_0 = [0 \quad 0 \quad g]^T. \tag{5.9.18}$$

TABLE 5.9.1 EFFICIENT RECURSIVE EQUATIONS FOR ROBOTICS FORWARD DYNAMICS IN LOCAL COORDINATES

$$\mathbf{R}_i^0\boldsymbol{\omega}_i = \begin{cases} \mathbf{R}_i^{i-1}(\mathbf{R}_{i-1}^0\boldsymbol{\omega}_{i-1} + \mathbf{z}_0\dot{q}_i), & \text{if link } i \text{ is rotational} \quad (5.9.13) \\ \mathbf{R}_i^{i-1}(\mathbf{R}_{i-1}^0\boldsymbol{\omega}_{i-1}), & \text{if link } i \text{ is translational,} \quad (5.9.14) \end{cases}$$

$$\mathbf{R}_i^0\dot{\boldsymbol{\omega}}_i = \begin{cases} \mathbf{R}_i^{i-1}[\mathbf{R}_{i-1}^0\dot{\boldsymbol{\omega}}_{i-1} + \mathbf{z}_0\ddot{q}_i + (\mathbf{R}_{i-1}\boldsymbol{\omega}_{i-1}) \times \mathbf{z}_0\dot{q}_i], & \text{if link } i \text{ is rotational} \quad (5.9.15) \\ \mathbf{R}_i^{i-1}(\mathbf{R}_{i-1}^0\dot{\boldsymbol{\omega}}_{i-1}), & \text{if link } i \text{ is translational,} \quad (5.9.16) \end{cases}$$

$$\mathbf{R}_i^0\dot{\boldsymbol{\omega}}_i = \begin{cases} (\mathbf{R}_i^0\boldsymbol{\omega}_i) \times (\mathbf{R}_i^0\mathbf{p}_i^*) + \mathbf{R}_i^{i-1}(\mathbf{R}_{i-1}^0\mathbf{v}_{i-1}), & \text{if link } i \text{ is rotational} \quad (5.9.17) \\ \mathbf{R}_i^{i-1}(\mathbf{z}_0\dot{q}_i + \mathbf{R}_{i-1}^0\mathbf{v}_{i-1}) + (\mathbf{R}_i^0\boldsymbol{\omega}_i) \times (\mathbf{R}_i^0\mathbf{p}_i^*), & \text{if link } i \text{ is translational,} \quad (5.9.18) \end{cases}$$

$$\mathbf{R}_i^0\dot{\mathbf{v}}_i = \begin{cases} (\mathbf{R}_i^0\dot{\boldsymbol{\omega}}_i) \times (\mathbf{R}_i^0\mathbf{p}_i^*) + (\mathbf{R}_i\boldsymbol{\omega}_i) \times [(\mathbf{R}_i^0\boldsymbol{\omega}_i) \times (\mathbf{R}_i^0\mathbf{p}_i^*)] + \mathbf{R}_i^{i-1}(\mathbf{R}_{i-1}^0\dot{\mathbf{v}}_{i-1}), & \text{if link } i \text{ is rotational} \quad (5.9.19) \\ \mathbf{R}_i^{i-1}(\mathbf{z}_o\ddot{q}_i + \mathbf{R}_{i-1}^0\dot{\mathbf{v}}_{i-1}) + (\mathbf{R}_i^0\dot{\boldsymbol{\omega}}_i) \times (\mathbf{R}_i^0\mathbf{p}_i^*) \\ \quad + (\mathbf{R}_i^0\boldsymbol{\omega}_i) \times (\mathbf{R}_i^{i-1}\mathbf{z}_0\dot{q}_i) + \mathbf{R}_i^0\boldsymbol{\omega}_i \times [(\mathbf{R}_i^0\boldsymbol{\omega}_i) \times (\mathbf{R}_i^0\mathbf{p}_i^*)], & \text{if link } i \text{ is translational,} \quad (5.9.20) \end{cases}$$

$$\mathbf{R}_i^0\bar{\mathbf{a}}_i = (\mathbf{R}_i^0\dot{\boldsymbol{\omega}}_i) \times (\mathbf{R}_i^0\bar{\mathbf{s}}_i) + (\mathbf{R}_i^0\boldsymbol{\omega}_i) \times [(\mathbf{R}_i^0\boldsymbol{\omega}_i) \times (\mathbf{R}_i^0\bar{\mathbf{s}}_i)] + \mathbf{R}_i^0\dot{\mathbf{v}}_i \tag{5.9.21}$$

$$\mathbf{R}_i^0\mathbf{f}_i = m_i\mathbf{R}_i^0\bar{\mathbf{a}}_i \tag{5.9.22}$$

$$\mathbf{R}_i^0\boldsymbol{\tau}_i = (\mathbf{R}_i^0\mathbf{I}_i\mathbf{R}_0^i)(\mathbf{R}_i^0\dot{\boldsymbol{\omega}}_i) + (\mathbf{R}_i^0\boldsymbol{\omega}_i) \times [(\mathbf{R}_i^0\mathbf{I}_i\mathbf{R}_0^i)(\mathbf{R}_i^0\boldsymbol{\omega}_i)] \tag{5.9.23}$$

$$\mathbf{R}_i^0\mathbf{f}_{i-1,i} = \mathbf{R}_i^{i+1}(\mathbf{R}_{i+1}^0\mathbf{f}_{i,i+1}) + \mathbf{R}_i^0\mathbf{f}_i \tag{5.9.24}$$

$$\mathbf{R}_i^0\boldsymbol{\tau}_{i-1,i} = \mathbf{R}_i^{i+1}[\mathbf{R}_{i+1}^0\mathbf{n}_{i+1} + (\mathbf{R}_{i+1}^0\mathbf{p}_i^*) \times (\mathbf{R}_{i+1}^0\mathbf{f}_{i+1})] + (\mathbf{R}_i^0\mathbf{p}_i^* + \mathbf{R}_i^0\bar{\mathbf{s}}_i) \times (\mathbf{R}_i^0\mathbf{f}_i) + \mathbf{R}_i^0\boldsymbol{\tau}_i \tag{5.9.25}$$

$$\boldsymbol{\tau}_i^* = \begin{cases} (\mathbf{R}_i^0\boldsymbol{\tau}_{i-1,i})\cdot(\mathbf{R}_i^{i-1}\mathbf{z}_0)\mathbf{R}_i^{i-1}\mathbf{z}_0 + [b_i\dot{q}_i + k_i(q_i)]\mathbf{R}_i^{i-1}\mathbf{z}_0 & (5.9.26) \\ (\mathbf{R}_i^0\boldsymbol{\tau}_{i-1,i})\cdot(\mathbf{R}_i^{i-1}\mathbf{z}_0)\mathbf{R}_i^{i-1}\mathbf{z}_0 + [b_i\dot{q}_i + k_iq_i]\mathbf{R}_i^{i-1}\mathbf{z}_i, & (5.9.27) \end{cases}$$

where $\mathbf{z}_0 = [0\ 0\ 1]^T$ and $\dot{\mathbf{v}}_0 = [0\ 0\ \mathbf{g}]^T$, and $\boldsymbol{\tau}_i^*$ is the ith joint torque

TABLE 5.9.2 BREAKDOWN OF COMPUTATIONAL OPERATIONS IN THE NEWTON–EULER TECHNIQUE

Newton–Euler equations for robot	Multiplications	Additions
$\boldsymbol{\omega}_i$	$9(n-1)$	$9n-9$
$\mathbf{v}_i$	$9(n-1)$	$12n-9$
$\dot{\boldsymbol{\omega}}_i$	$12n-9$	$12n-9$
$\dot{\mathbf{v}}_i$	$27n$	$24n$
$\bar{\mathbf{a}}_i$	$18n$	$15n$
$\mathbf{f}_i$	$3n$	0
$\mathbf{f}_{i-1,i}$	$9(n-1)+3$	$9n-6$
$\boldsymbol{\tau}_i$	$27n$	$18n$
$\boldsymbol{\tau}_{i-1,i}$	$21n-15$	$24n-15$
Total	$135n-48$	$123n-48$

[a] n is the number of degrees of freedom for the robotic arm.

The joint variables are q_i, $\dot{q}_i$, and $\ddot{q}_i$, for $i = 1, 2, \ldots, n$; the link variables are i, $\mathbf{f}_i$, $\mathbf{f}_{i-1,i}$, and $\tau_{i-1,i}$.

Forward Iterations

1. Set $i = 1$.
2. Compute $\mathbf{R}_i\dot{\boldsymbol{\omega}}_i$, $\mathbf{R}_i^0\boldsymbol{\omega}_i^0$, $\mathbf{R}_i^0\mathbf{v}_i^0$, and $\mathbf{R}_i^0\mathbf{v}_i^0$ using Table 5.9.1.
3. Compute $\mathbf{R}_i^0\bar{\mathbf{a}}_i$, $\mathbf{R}_i^0\mathbf{f}_i$, and $\mathbf{R}_i^0\boldsymbol{\tau}_i$.
4. If $i = n$ go to the next step; otherwise, set $i = i + 1$ and return to step 2.

Backward Iterations

5. Set $\mathbf{f}_{n,n+1}$ and $\boldsymbol{\tau}_{n,n+1}$ to the required force and torque necessary for the gripper to carry the load.
6. Compute $\mathbf{R}_i^0\mathbf{f}_{i-1,i}$, $\mathbf{R}_i^0\boldsymbol{\tau}_{i-1,i}$, and $\boldsymbol{\tau}_i$ given $\mathbf{f}_{n,n+1}$ and $\boldsymbol{\tau}_{n,n+1}$.
7. If $i = 1$, stop; otherwise, set $i \rightarrow i - 1$ and go back to step 6.

EXAMPLE 5.9.2 ■

For the two-link nonplanar robotic manipulator shown in Figure 5.9.1, derive the dynamic equations of motion if all masses m_1 and m_2 are concentrated at the outermost end of each link.

Solution. Referring to Table 5.9.1, we can write the forward dynamics iterations in the following manner:

$$\boldsymbol{\omega}_0 = \dot{\boldsymbol{\omega}}_0 = \mathbf{v}_0 = 0, \qquad \dot{\mathbf{v}}_0 = [0 \quad 0 \quad -g]^T, \tag{5.9.19}$$

$$\mathbf{R}_1^0\boldsymbol{\omega}_1 = \mathbf{R}_1^0(\mathbf{R}_0\boldsymbol{\omega}_0 + \mathbf{z}_0\dot{\theta}_1) = \mathbf{R}_1^0\boldsymbol{\omega}_0 + \mathbf{z}_1\dot{\theta}_{11}, \tag{5.9.20}$$

or

$$\mathbf{R}_1^0\boldsymbol{\omega}_1 = [0 \quad 0 \quad \dot{\theta}_1]^T; \tag{5.9.21}$$

$$\mathbf{R}_1^0\dot{\boldsymbol{\omega}}_1 = \mathbf{R}_1^0[\mathbf{R}_0^0\dot{\boldsymbol{\omega}}_0 + \mathbf{z}_0\ddot{\theta}_1 + (\mathbf{R}_0^0\boldsymbol{\omega}_0) \times \mathbf{z}_0\dot{\theta}_1] \tag{5.9.22}$$

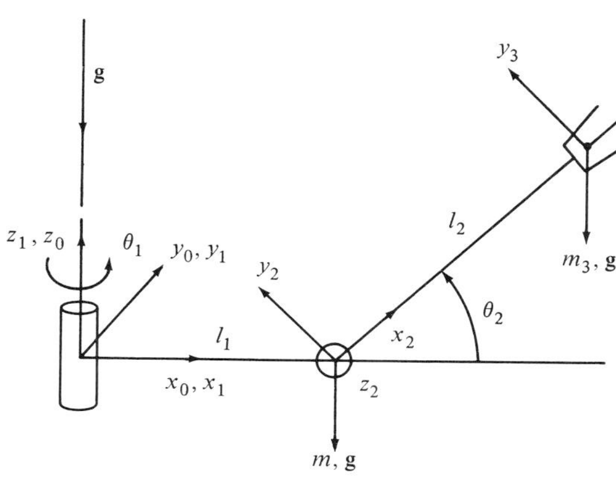

Figure 5.9.1 Two-link nonplanar manipulator.

or

$$\mathbf{R}_1^0\dot{\boldsymbol{\omega}}_1 = [0 \quad 0 \quad \ddot{\theta}_1]^T; \tag{5.9.23}$$

$$\mathbf{R}_1^0\mathbf{v}_1 = (\mathbf{R}_1^0\boldsymbol{\omega}_1) \times (\mathbf{R}_1^0\mathbf{p}_1^*)(\mathbf{R}_1^0(\mathbf{R}_0^0\mathbf{v}_0) \tag{5.9.24}$$

$$= \mathbf{R}_1^0[0 \quad 0 \quad 0] = [0 \quad 0 \quad 0]^T. \tag{5.9.25}$$

Furthermore, as can be seen from the Figure 5.9.1,

$$\mathbf{R}_0^1 = \begin{bmatrix} C_1 & -S_1 & 0 \\ S_1 & C_1 & 0 \\ 0 & 0 & 1 \end{bmatrix}, \tag{5.9.26}$$

$$\mathbf{R}_1^2 = \begin{bmatrix} C_2 & -S_2 & 0 \\ 0 & 0 & -1 \\ S_2 & C_2 & 0 \end{bmatrix}, \tag{5.9.27}$$

and

$$\mathbf{R}_2^3 = \begin{bmatrix} 1 & 0 & 0 \\ 0 & 1 & 0 \\ 0 & 0 & 1 \end{bmatrix}. \tag{5.9.28}$$

In addition,

$$\mathbf{p}_1^* = [0 \quad 0 \quad 0]^T, \qquad \mathbf{P}_2^* = [l_1 \quad 0 \quad 0], \qquad \mathbf{p}_3^* = [l_2 \quad 0 \quad 0], \tag{5.9.29}$$

$$\begin{aligned}\mathbf{R}_1^0\dot{\mathbf{v}}_1 &= \mathbf{R}_1^0\dot{\boldsymbol{\omega}}_1 \times \mathbf{R}_1^0\mathbf{p}_1^* + \mathbf{R}_1^0\boldsymbol{\omega}_1 \times (\mathbf{R}_1^0\boldsymbol{\omega}_1 \times \mathbf{R}_i^0\mathbf{p}_1^*) + \mathbf{R}_1^0(\mathbf{R}_0^0\mathbf{v}_0^0) \\ &= [0 \quad 0 \quad -q]^T,\end{aligned} \tag{5.9.30}$$

$$\begin{aligned}\mathbf{R}_1^0\bar{\mathbf{a}}_1 &= \mathbf{R}_1^0\boldsymbol{\omega}_1^0 \times \mathbf{R}_1^0\bar{\mathbf{s}}_1 + \mathbf{R}_1^0\boldsymbol{\omega}_1 \times (\mathbf{R}_1^0\boldsymbol{\omega}_1 \times \mathbf{R}_1^0\bar{\mathbf{s}}_1) + \mathbf{R}_1^0\mathbf{v}_1^0 \\ &= [-l_1\dot{\theta}_1^2 \quad l_1\ddot{\theta}_1 \quad -g]^T,\end{aligned} \tag{5.9.31}$$

$$\mathbf{R}_2^0\boldsymbol{\omega}_2 = \mathbf{R}_2^1(\mathbf{R}_1^0\boldsymbol{\omega}_1 + \mathbf{z}_0\dot{\theta}_2) = \mathbf{R}_2^1\mathbf{R}_1^0\boldsymbol{\omega}_1 + \mathbf{R}_2^1\mathbf{z}_0\dot{\theta}_2, \tag{5.9.32}$$

$$\mathbf{R}_2^0\boldsymbol{\omega}_2 + \begin{bmatrix} C_2 & 0 & S_2 & 0 \\ -S_2 & 0 & C_2 & 0 \\ 0 & -1 & 0 & \dot{\theta}_1 \end{bmatrix} + \begin{bmatrix} 0 \\ 0 \\ \dot{\theta}_2 \end{bmatrix} = \begin{bmatrix} S_2\dot{\theta}_1 \\ C_2\dot{\theta}_1 \\ \dot{\theta}_2 \end{bmatrix}, \tag{5.9.33}$$

$$\mathbf{R}_2^0\dot{\boldsymbol{\omega}}_2 = \mathbf{R}_2^1[\mathbf{R}_1^0\dot{\boldsymbol{\omega}}_1 + \mathbf{z}_0\ddot{\theta}_2 + (\mathbf{R}_2^0\boldsymbol{\omega}_1) \times \mathbf{z}_0\dot{\theta}_2], \tag{5.9.34}$$

$$\mathbf{R}_2^0\dot{\boldsymbol{\omega}}_2 = \mathbf{R}_2^1\mathbf{R}_1^0\dot{\boldsymbol{\omega}}_1 + \mathbf{z}_2\ddot{\theta}_2 + \mathbf{R}_2^1\mathbf{R}_2^0\boldsymbol{\omega}_1 \times \mathbf{z}_0\dot{\theta}_2, \tag{5.9.35}$$

$$\mathbf{R}_2^0\dot{\boldsymbol{\omega}}_2 = \begin{bmatrix} S_2\ddot{\theta}_2 \\ C_2\ddot{\theta}_1 \\ 0 \end{bmatrix} \times \begin{bmatrix} 0 \\ 0 \\ \dot{\theta}_2 \end{bmatrix} \to \begin{bmatrix} S_2\dot{\theta}_1 \\ C_2\dot{\theta}_1 \\ 0 \end{bmatrix} \times \begin{bmatrix} 0 \\ 0 \\ \dot{\theta}_2 \end{bmatrix} = \begin{bmatrix} S_2\ddot{\theta}_1 + C_2\dot{\theta}_1\dot{\theta}_2 \\ C_2\ddot{\theta}_1 - S_2\dot{\theta}_1\dot{\theta}_2 \\ \ddot{\theta}_2 \end{bmatrix}, \tag{5.9.36}$$

$$\mathbf{R}_2^0\dot{\mathbf{v}}_2 = \mathbf{R}_2^0\dot{\boldsymbol{\omega}}_2 \times \mathbf{R}_2^0\mathbf{p}_2^* + \mathbf{R}_2^0\boldsymbol{\omega}_2 \times [(\mathbf{R}_2^0\boldsymbol{\omega}_2) \times (\mathbf{R}_2^0\mathbf{p}_2^*)] + \mathbf{R}_2^1(\mathbf{R}_2^0\dot{\mathbf{v}}_2), \tag{5.9.37}$$

$$\mathbf{R}_2^0\dot{\mathbf{v}}_2 = \begin{bmatrix} C_2 & 0 & S_2 \\ -S_2 & 0 & C_2 \\ 0 & -1 & 0 \end{bmatrix} \begin{bmatrix} -l_1\dot{\theta}_1^2 \\ l_1\ddot{\theta}_1 \\ -g \end{bmatrix} = \begin{bmatrix} -l_1C_2\dot{\theta}_1^2 - S_2 g \\ l_1S_2\dot{\theta}_1^2 - C_2 g \\ -l_1\ddot{\theta}_1 \end{bmatrix}, \tag{5.9.38}$$

$$\mathbf{R}_2^0\bar{\mathbf{a}}_2 = \mathbf{R}_2^0\dot{\boldsymbol{\omega}}_2 \times \mathbf{R}_2^0\bar{S}_2 + (\mathbf{R}_2^0\boldsymbol{\omega}_2) \times [(\mathbf{R}_2^0\boldsymbol{\omega}_2 \times \mathbf{R}_2^0\bar{S}_2)] + \mathbf{R}_2^0\dot{\mathbf{v}}_2, \tag{5.9.39}$$

$$\mathbf{R}_2^0\bar{\mathbf{a}}_2 = \begin{bmatrix} S_2\ddot{\theta}_1 + C_1\dot{\theta}_1\dot{\theta}_2 \\ C_2\ddot{\theta}_1 - S_2\dot{\theta}_1\dot{\theta}_2 \\ \ddot{\theta}_2 \end{bmatrix} \times \begin{bmatrix} l_2 \\ 0 \\ 0 \end{bmatrix} + \begin{bmatrix} S_2\dot{\theta}_1 \\ C_2\dot{\theta}_1 \\ \dot{\theta}_2 \end{bmatrix} \times \begin{bmatrix} 0 \\ l_2\dot{\theta}_2 \\ -l_2S_2\dot{\theta}_1 \end{bmatrix} + \mathbf{R}_2^0\dot{\mathbf{v}}, \tag{5.9.40}$$

$$\mathbf{R}_2^0\bar{\mathbf{a}}_2 = \begin{bmatrix} 0 \\ l_2\ddot{\theta}_2 \\ -l_2C_2\ddot{\theta}_1 + l_2S_2\dot{\theta}_1\dot{\theta}_2 \end{bmatrix} + \begin{bmatrix} -l_2C_2^2\dot{\theta}_1^2 - l_2\dot{\theta}_2^2 \\ l_2S_2C_2\dot{\theta}_1^2 \\ l_2S_2\dot{\theta}_1\dot{\theta}_2 \end{bmatrix} + \begin{bmatrix} -l_1C_2\dot{\theta}_1^2 - S_2 g \\ l_1S_2\dot{\theta}_1^2 + C_2 g \\ -l_1\ddot{\theta}_1 \end{bmatrix}, \tag{5.9.41}$$

or, finally,

$$\mathbf{R}_2^0\bar{\mathbf{a}}_2 = \begin{bmatrix} -(l_1C_2 + l_2C_2^2)\dot{\theta}_1^2 - l_2\dot{\theta}_2^2 - S_2 g \\ (l_1S_2 + l_2S_2C_2)\dot{\theta}_1^2 - C_2 g + l_2\ddot{\theta}_2 \\ 2l_2S_2\dot{\theta}_1\dot{\theta}_2 - l_1\ddot{\theta}_1 - l_2C_2\ddot{\theta}_1 \end{bmatrix}. \tag{5.9.42}$$

The governing equations then become

$$\mathbf{R}_1^0\mathbf{f}_1 = m_1\mathbf{R}_1^0\bar{\mathbf{a}}_1 \tag{5.9.43}$$

or

$$\mathbf{f}_1 = \begin{bmatrix} -m_1l_1\dot{\theta}_1^2 \\ m_2l_1\ddot{\theta}_1 \\ m_1 g \end{bmatrix}, \tag{5.9.44}$$

and

$$\mathbf{R}_2^0\mathbf{f}_2 = m_2\mathbf{R}_2^0\bar{\mathbf{a}}_2 \tag{5.9.45}$$

or

$$\mathbf{R}_2^0\mathbf{f}_2 = \begin{bmatrix} -m_2(l_1 + l_2C_2)C_2\dot{\theta}_1^2 - m_2l_2\dot{\theta}_2^2 - m_2S_2 g \\ m_2(l_1 + l_2C_2)S_2\dot{\theta}_1^2 - m_2 gC_2 + m_2l_2\ddot{\theta}_2 \\ 2m_2l_2S_2\dot{\theta}_1\dot{\theta}_2 - m_2l_1\ddot{\theta}_1 - m_2l_2C_2\ddot{\theta}_1 \end{bmatrix}. \tag{5.9.46}$$

Now the torque equations become

$$\mathbf{R}_1^0\boldsymbol{\tau}_1 = (\mathbf{R}_1^0\mathbf{I}_1\mathbf{R}_0^1)(\mathbf{R}_1^0\dot{\boldsymbol{\omega}}_1) + \mathbf{R}_1^0\boldsymbol{\omega}_1 \times [(\mathbf{R}_1^0\mathbf{I}_1\mathbf{R}_0^1)(\mathbf{R}_1^0\boldsymbol{\omega}_1)] \tag{5.9.47}$$

and

$$\mathbf{R}_2^0\boldsymbol{\tau}_2 = (\mathbf{R}_2^0\mathbf{I}_2\mathbf{R}_0^2)(\mathbf{R}_2^0\dot{\boldsymbol{\omega}}_2) + (\mathbf{R}_2^0\boldsymbol{\omega}_2) \times [(\mathbf{R}_2^0\mathbf{I}_2\mathbf{R}_0^2)(\mathbf{R}_2^0\boldsymbol{\omega}_2)]. \tag{5.9.48}$$

Since both $\mathbf{I}_1$ and $\mathbf{I}_2$ are zero, it then follows that

$$\tau_1 = \tau_2 = 0. \tag{5.9.49}$$

Of course, $\mathbf{I}_1 = \mathbf{I}_2 = \mathbf{0}$ is an idealization because we are assuming that the masses are concentrated at a point at the center of gravity of the robotic structure.

For the backward force iterations from link 2 we have

$$\mathbf{R}_2^0\mathbf{f}_{1,2} = \mathbf{R}_2^3(\mathbf{R}_3^0\mathbf{f}_{2,3}) + \mathbf{R}_2^0\mathbf{f}_2 = \mathbf{R}_2^0\mathbf{f}_2 \quad \text{[see Eq. (5.9.46)]}, \tag{5.9.50}$$

$$\begin{aligned}\mathbf{R}_2^0\tau_{1,2} = {} & \mathbf{R}_2^3[\mathbf{R}_2^0\tau_{2,3} + (\mathbf{R}_3^0\mathbf{p}_2^*) \times (\mathbf{R}_3^0\mathbf{f}_{2,3})] \\ & + (\mathbf{R}_2^0\mathbf{p}_2^* + \mathbf{R}_2^0\bar{\mathbf{s}}_i) \times (\mathbf{R}_2^0\mathbf{f}_2) + \mathbf{R}_2^0\tau_2,\end{aligned} \tag{5.9.51}$$

$$\mathbf{R}_2^0\tau_{1,2} = \mathbf{R}_2^0\mathbf{p}_2^* \times \mathbf{R}_2^0\mathbf{f} = \begin{bmatrix} 0 \\ m_2 l_1 l_2 \ddot{\theta}_1 - 2m_2 l_2 S_2^2 \dot{\theta}_1 \dot{\theta}_2 + m_2 l_2^2 C_2 \ddot{\theta}_1 \\ m_2 l_2(l_1 + l_2 C_2) S_2 \dot{\theta}_1^2 - m_2 g l_2 C_2 + m_2 l_2^2 \ddot{\theta}_2 \end{bmatrix}, \tag{5.9.52}$$

$$\mathbf{R}_1^0\mathbf{f}_1 = \mathbf{R}_1^2(\mathbf{R}_2^0\mathbf{f}_{1,2}) + \mathbf{R}_1^0\mathbf{f}_1, \tag{5.9.53}$$

$$\begin{aligned}\mathbf{R}_1^0\tau_{0,1} = {} & \mathbf{R}_1^2[\mathbf{R}_2^0\tau_{1,2} + (\mathbf{R}_2^0\mathbf{p}_1^*) \times (\mathbf{R}_2^0\mathbf{f}_{1,2})] \\ & + (\mathbf{R}_1^0\mathbf{p}_1^* + \mathbf{R}_1^0\bar{\mathbf{s}}_1) \times (\mathbf{R}_1^0\mathbf{f}_1) + \mathbf{R}_1^0\tau_1.\end{aligned} \tag{5.9.54}$$

After simplifications using Equations (5.9.51), (5.9.52), and (5.9.54), one obtains

$$\mathbf{R}_1^0\tau_{0,1} = \begin{bmatrix} * \\ * \\ (m_1 l_1^2 + m_2 l_1^2 + 2m_2 l_1 l_2 C_2)\ddot{\theta}_1 - 2(l_1 + l_2 C_2) m_2 l_2 S_2 \dot{\theta}_1 \dot{\theta}_2 + m_2 l_2^2 C_2^2 \end{bmatrix}. \tag{5.9.55}$$

Finally, in the absence of joint friction and elasticity,

$$\tau_1^* = [m_1 l_1^2 + m_2(l_1 + l_2 C_2)^2]\ddot{\theta}_1 - 2(l_1 + l_2 C_2) m_2 l_2 S_2 \dot{\theta}_1 \dot{\theta}_2] \tag{5.9.56}$$

and

$$\tau_2^* = m_2 l_2^2 \ddot{\theta}_2 + (l_1 + l_2 C_2) m_2 l_2 S_2 \dot{\theta}_1 \dot{\theta}_2 - m_2 g l_2 C_2 \tag{5.9.57}$$

or, generally,

$$\tau = \mathbf{M}(\boldsymbol{\theta})\ddot{\boldsymbol{\theta}} + \mathbf{V}(\boldsymbol{\theta}, \dot{\boldsymbol{\theta}}) + \mathbf{G}(\boldsymbol{\theta}), \tag{5.9.58}$$

where

$$\mathbf{M}(\boldsymbol{\theta}) = \begin{bmatrix} m_1 l_1^2 + m_2(l_1 + l_2 C_2)^2 & 0 \\ 0 & m_2 l_2^2 \end{bmatrix}, \tag{5.9.59}$$

$$\mathbf{V}(\boldsymbol{\theta}, \dot{\boldsymbol{\theta}}) = \begin{bmatrix} -2(l_1 + l_2 C_2) m_2 l_2 S_2 \dot{\theta}_1 \dot{\theta}_2 \\ (l_1 + l_2 C_2) m_2 l_2 S_2 \dot{\theta}_1^2 \end{bmatrix}, \tag{5.9.60}$$

and

$$\mathbf{G}(\boldsymbol{\theta}) = \begin{bmatrix} 0 \\ -m_2 g l_2 C_2 \end{bmatrix}. \quad ■■ \tag{5.9.61}$$

5.10 CONCLUDING REMARKS

In this rather important chapter on robotic dyanamics, the different techniques for the derivations of the dynamic equations are described fully and a number of examples are solved to demonstrate the details involved.

It appears that the bond graph dynamic modeling of robotic manipulators offers advantages because it is applicable to highly complex robotic manipulators. However, even for a three-axis robot, both Lagrange–Euler dynamic modeling and Newton–Euler dynamic modeling become so involved and the equations become so complex that it is beyond the scope of this chapter to solve them as examples. In any event, they have been assigned as homework problems and their solutions can be found in the solutions manual.

BIBLIOGRAPHY

Bejczy, A. K. (1974). Robot Arm Dynamics and Control, NASA-JPL Tech. Memo Rep. 33-669 (Feb.).

Hollerbach, J. M. (1980). A Recursive Lagrangian Formulation of Maniplator Dynamics and a Comparative Study of Dynamics Formulation, *IEEE Trans. Syst. Man Cybern.*, **10**(11), 730–736.

Kahn, M. E., and Roth, B. (1971). The Near-Minimum-Time Control of Open Loop Kinematic Chains, *Trans. ASME J. Dyn. Syst. Meas. Control*. **93**, 164–172.

Lee, C. S. G. (1982). Robot Arm Kinematics, Dynamics and Control, *IEEE Trans. Computer*, 62–79 (Dec.)

Luh, J. Y. S., Walker, M. W., and Paul, R. P. (1980). On-Line Computational Scheme for Mechanical Manipulators, *Trans. ASME J. Dyn. Syst. Meas. Control*, **120**, 69–76.

Rosenberg, R. M. (1977). *Analytical Dynamics of Discrete Systems*, Mathematical Concepts and Methods in Science and Engineering, Angelo Miele (Ser. Ed.), Vol. 4, Plenum Press, New York.

Shahinpoor, M. (1983). *Bond Graph Dynamic Modeling of Robotic Manipulators*, Recent Developmental Applied Mathematics Series 1, I. G. Tadjbakhsh and F. F. Ling (Eds.) RPI Press, Troy, NY, pp. 176–186.

Shahinpoor, M. (1986). Robotics Dynamics, in *Encylopedia of Robotics*, Wiley, New York (to be published).

Silver, W. (1982). On the Equivalence of Lagrangian and Newton–Euler Dynamics for Manipulators, *Int. J. Robotics Res.*, **1**(2), 60–70.

Uicker, J. J., Jr. (1966). Dynamic Behaviour of Spatial Linkages, *Trans. ASME J. Appl. Mech.*, **5**(68), 1–15.

HOMEWORK PROBLEMS

1. Derive the governing dynamics equation for a one-legged jumping robotic manipulator (Fig. P5.1) by the bond graph modeling technique described in Section 5.2.

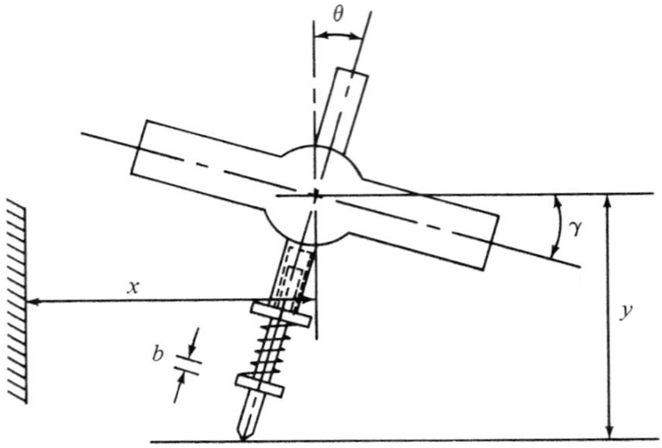

Figure P5.1 One-legged jumping robot (pogo stick).

2. Derive the same equations as for Problem 1, using Lagrange–Euler robotic dynamics.

3. Derive the same equations as for Problem 1, using Newton–Euler robotic dynamics.

4. Derive the inverse dyanamics equation for the torque $\boldsymbol{\tau}_{0,1}$ for the one-link robot shown in Figure P5.4 in terms of the tip forces and couples $\mathbf{r}_{1,2}$ and $\boldsymbol{\tau}_{1,2}$ as well as other pertinent parameters.

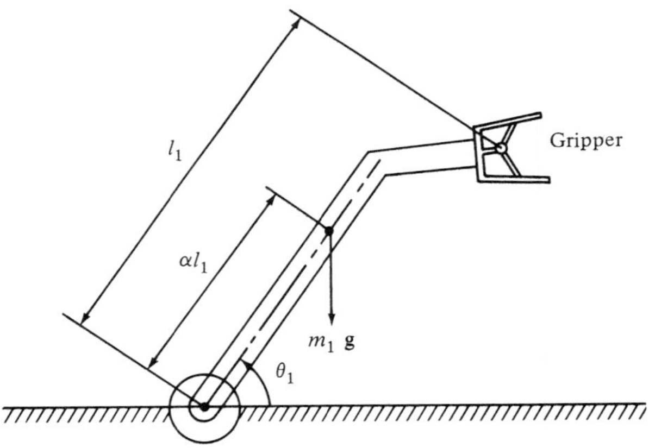

Figure P5.4 One-link robotic manipulator.

5. Show that for an n-axis robotic manipulator the Newton–Euler dynamics equations require $126n - 99$ multiplications and $106n - 92$ additions.

6. Show that for an n-axis robotic manipulator the Lagrange–Euler dynamics equations require $32n^4 + 86n^3 + 171n^2 + 53n - 128$ multiplications and $25n^4 + 66n^3 + 129n^2 + 42n - 96$ additions.

7. For the planar two-link robot shown in Figure P5.7, find the joint torques $\mathbf{n}_{1,2}$ and $\mathbf{n}_{0,1}$ using the Newton–Euler equations. Note that the links are of uniform diameter $2R$, lengths l_1 and l_2, and masses m_1 and m_2, respectively.

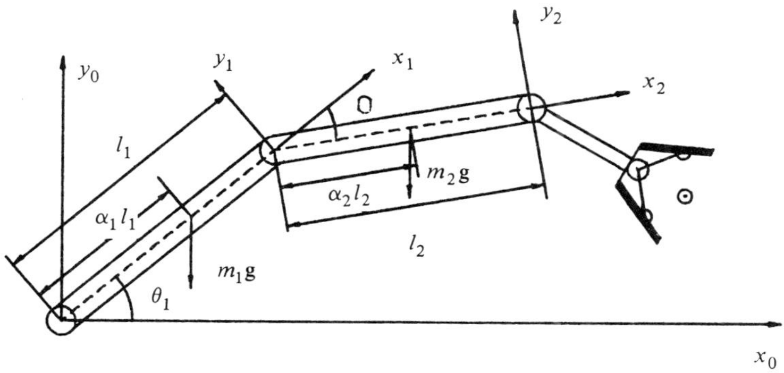

Figure P5.7

8. The following forces and moments (**f** and **τ**) are applied to the base of a robot in static equilibrium:

$$\mathbf{f} = 5\mathbf{i} + 6\mathbf{j} + 7\mathbf{k} \quad \text{and} \quad \boldsymbol{\tau} = 10\mathbf{i} + 20\mathbf{j} + 30\mathbf{k}.$$

What are the equivalent force and moment at the gripper if it is located in a frame **A** such that

$$\mathbf{A} = \begin{bmatrix} 0 & 0 & 1 & 20 \\ 1 & 0 & 0 & 30 \\ 0 & 1 & 0 & 0 \\ 0 & 0 & 0 & 1 \end{bmatrix}.$$

9. Derive the governing dynamic equations for a three-link planar robotic manipulator as shown in Figure P5.9.

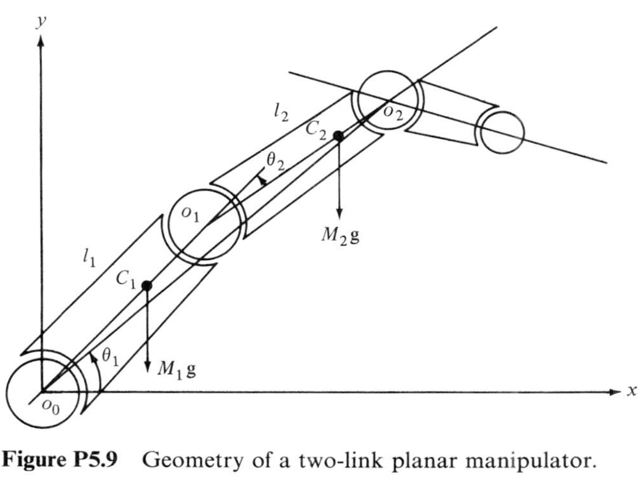

Figure P5.9 Geometry of a two-link planar manipulator.

10. Find the inertia tensor of the following objects with respect to a frame whose origin is at the center of gravity for each object:
    (a) A solid circular cylinder of uniform density.
    (b) A circular cylindrical tube of uniform density.
    (c) A tube with rectangular cross section.
11. Derive the explicit forms of the governing dynamics equations for a cylindrical polar robotic manipulator as described in Example 5.4.3 and Figure 5.4.5 by means of either (a) the Lagrange–Euler equations or (b) the Newton–Euler equations. (See also Fig. P5.11)

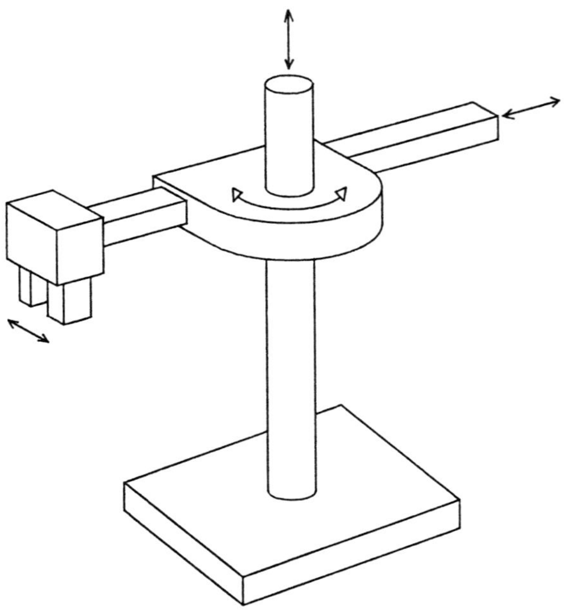

Figure P5.11

12. Derive the governing dynamics equations for a spherical robotic arm as described in Example 5.4.2 and Figure 5.4.3 using (a) the Lagrange–Euler equations and (b) the Newton–Euler equations. Compare your results with the ones obtained in Example 5.4.2. (See Fig. P5.12 for reference.)

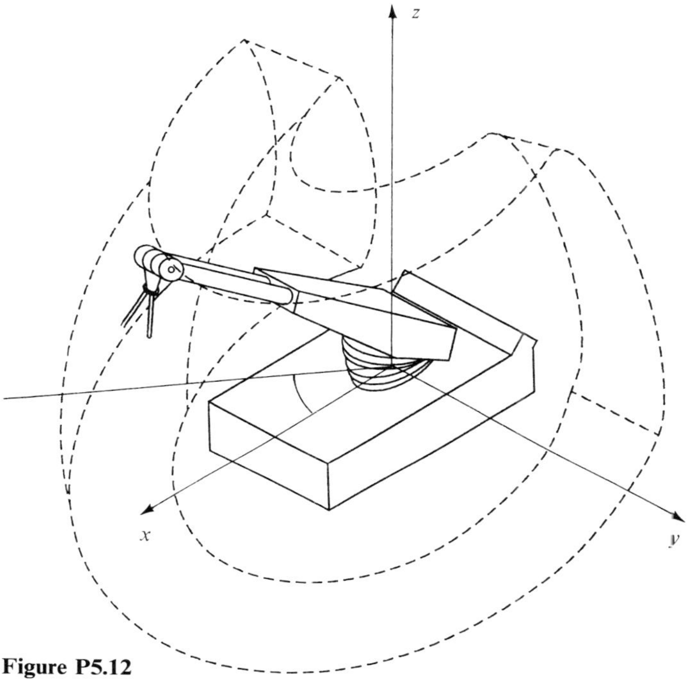

Figure P5.12

13. Derive the governing dynamics equations for a three-axis rectangular Cartesian robotic manipulator as described in the text (see Fig. P5.13).

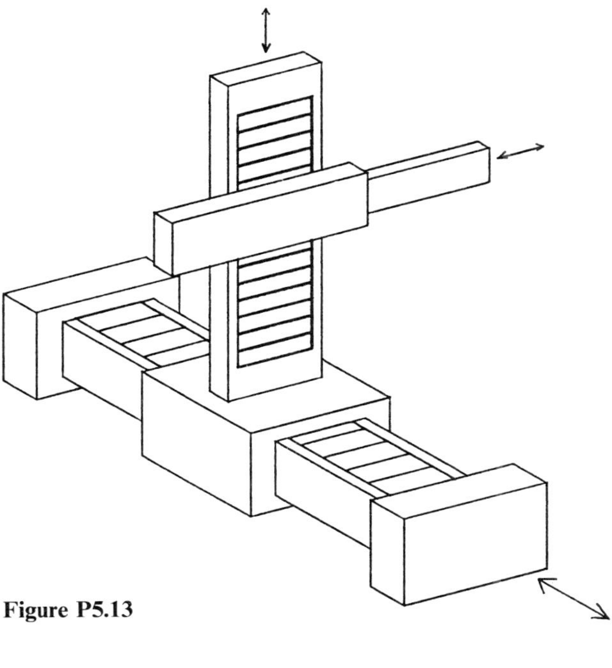

Figure P5.13

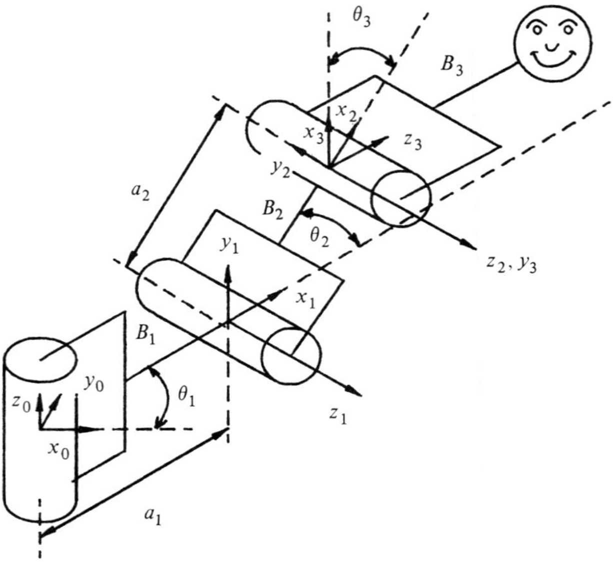

Figure P5.14

14. Derive the governing dynamics equations for a three-axis robotic manipulator as shown in Figure P5.14.

Chapter 6

Motion Control of Robotic Manipulators

6.1 INTRODUCTION

In this chapter we deal with the time-optimal control of an open-loop articulated kinematic chain in connection with robotic manipulators. Here we concentrate on the dynamic control of robotic motions and forces, that is, the forces or torques that must be exerted on the manipulator joints to move the manipulator from its present configuration to the desired configuration under the constraint of a particular trajectory. If the robot is to move very slowly, then no significant coupled dynamic forces act on the system. A simple solution to this problem is to control each of the joints independently by using a control system whose force output to each joint is a function of the position of the joint. This is called a *proportional control algorithm* and often suffices for most applications. If the robotic motions are rapid, the joints can no longer be controlled independently because the dynamic interactions between the links become extremely significant and a nonlinear or complex control algorithm is required. Generally, to construct such nonlinear control algorithms, the robot is split into two subsystems (minimal robot and a terminal robot). The dynamic variables (angles, velocities, accelerations, and torques) of the minimal configuration are calculated off-line and stored. During real-time control of the manipulator, the stored or prescribed dynamic data are imposed on the minimal configuration and then control torques for compensation are applied (calculated and commanded) on line for any task. Thus, the general objective for manipulator control can be stated as follows: *Find the joint motor torques* $\{\tau_i(t)\}$ *necessary to produce a given planned trajectory* $\{\theta_i(t)\}$ *or a planned set of forces* $\mathbf{f}(\mathbf{t})$ *and torques* $\boldsymbol{\tau}(t)$ *at the gripper.*

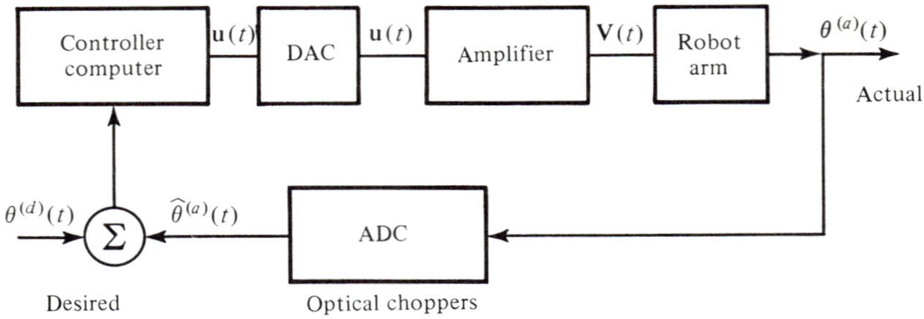

Figure 6.1.1 General block diagram for robotic arm control system.

In simple terms, the control problem is to find the necessary torques to actuate the joint motors at every instant of time in a desired manner. Obviously, because of the interactive effects of inertial, Coriolis, centrifugal, and gravitational loadings that vary continuously, the actual real-time control is extremely difficult if not impossible to achieve. We first treat the robotic arm as a continuous time-varying system and use the Laplace transform technique to simplify our analysis. The block diagram in Figure 6.1.1 depicts a general arrangement for a robotic control system.

Note that the controller is typically a small computer, the DAC is a digital analog converter, and the ADC is an analog to digital converter. The amplifier is the drive circuit to convert the low-level analog torque signal $\mathbf{u}(t)$ to $\mathbf{V}(t)$, which directly activates the joint motors. If the joint motors are dc motors (armature-controlled), the generated torque is proportional to the armature current. Therefore, the amplifier in Figure 6.1.1 may be an analog subsystem regulating the current through variations in the applied armature voltage. The ADC block detects the position (encoders) and the speed (tachometers) of the joint motors and converts them to a form readable by the digital controller. Thus, the sensors here may represent an encoded disk (optical encoder) type analog to digital converter. On the other hand, it might be a simple potentiometer and/or tachometer followed by an ADC circuit to detect the position and speed of the robot. Figure 6.1.2 displays the two main types of robotic motion control.

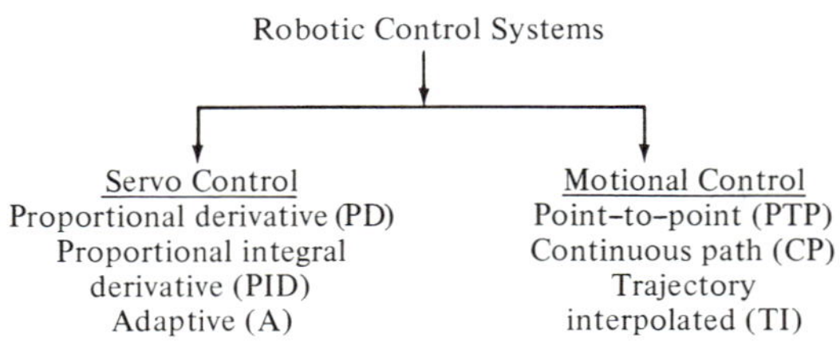

Figure 6.1.2 Robotic control system.

6.2 ROBOTIC OPEN- AND CLOSED-LOOP CONTROL SYSTEMS

Here we briefly recall the essential ideas behind the open- and closed-loop control systems. We express our analysis in terms of the Laplace transform variable (s). Note that the Laplace transform maps a time function into an s function such that

$$L[f(t)] = \int_0^{\infty} f(t)e^{-st}\, dt, \quad \text{with} \quad s = \sigma + jw. \tag{6.2.1}$$

The principal advantage of the Laplace transform is that ordinary differential equations in the time domain can be converted to algebraic equations in the Laplace transform domain. Referring to the robotic control systems in Figure 6.2.1, note that in both systems we want to find the system transfer functions, that is, $T(s)$. Here the variables are defined such that $I(s)$ is the input, $O(s)$ is the output, $G(s)$ is the transfer function, where for the open-loop system $G(s) = O(s)/I(s)$, $H(s)$ is the feedback transfer function, and $E(s)$ is the error signal. Note that the overall transfer functions $T_O(s)$ and $T_C(s)$ for open-loop and closed-loop systems, respectively, are

$$T_O(s) = \frac{O(s)}{I(s)} = G(s) \tag{6.2.2}$$

and

$$T_C(s) = \frac{O(s)}{I(s)} = \frac{G(s)}{1 + G(s)H(s)} \tag{6.2.3}$$

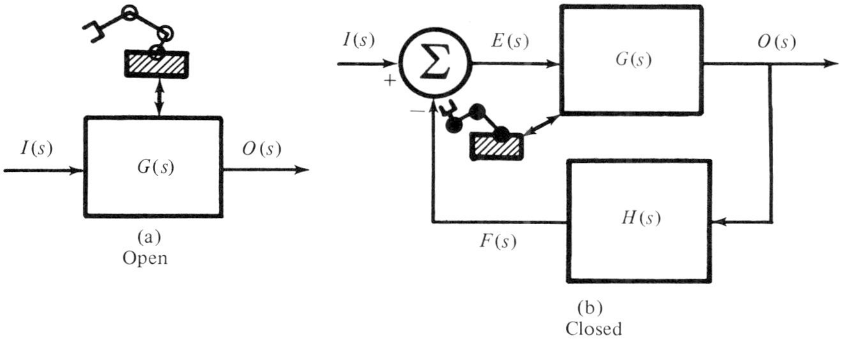

Figure 6.2.1 Open-loop and closed-loop robot control systems.

EXAMPLE 6.2.1 ■

Derive Equation (6.2.3) for closed-loop robotic control systems.

Solution. Note that

$$T_C(s) = \frac{O(s)}{I(s)} \quad \text{and} \quad G(s) = \frac{O(s)}{E(s)}; \tag{6.2.4}$$

but

$$E(s) = I(s) - F(s) \tag{6.2.5}$$

and

$$F(s) = H(s)O(s),$$

because $H(s) = F(s)/O(s)$. Therefore,

$$E(s) = I(s) - H(s)O(s) = I(s) - H(s)G(s)E(s), \tag{6.2.6}$$

or

$$I(s) = E(s)[1 + G(s)H(s)]. \tag{6.2.7}$$

Finally

$$T_C(s) = \frac{O(s)}{I(s)} = \frac{G(s)E(s)}{[1 + G(s)H(s)]E(s)} = \frac{G(s)}{1 + H(s)G(s)}. \quad ■■ \tag{6.2.8}$$

The dynamic stability of a robotic manipulator is an assessment of the response of a system to various inputs. In general, the robotic system is stable if for any given bounded input the output also is bounded. For example, for a particular disturbance the robot may oscillate violently as can be observed with a Rhino XR-2 robot. Oscillation is a mode of instability that may be either spontaneous or the result of some nonoscillating input. Another mode of instability is spiking an undamped or unbounded response to an input.

Note that the error signals of both systems are

$$E_O(s) = I(s) - O(s) = O(s)[G^{-1}(s) - 1] = I(s)[1 - G(s)] \tag{6.2.9}$$

and

$$E_C(s) = I(s)\left(\frac{1 - G(s) + G(s)H(s)}{1 + G(s)H(s)}\right). \tag{6.2.10}$$

In the following sections we describe a number of important areas of robotic control and, in particular, the second-order systems, error-controlled systems, and the general control problem.

6.3 SECOND-ORDER SYSTEMS

Consider a linear feedback second-order system characterized by $G(s) = K/s(s+p)$, such that

$$T(s) = \frac{G(s)}{1+G(s)} = \frac{K}{s^2+ps+K}. \tag{6.3.1}$$

Denoting $\omega_n^2 = K$ and $p = 2\xi\omega_n$, we can reduce Equation (6.3.1) to

$$T(s) = \frac{\omega_n^2}{s^2+2\xi\omega_n s+\omega_n^2}. \tag{6.3.2}$$

Traditionally, ξ is called the *damping factor* and ω_n is called the *natural frequency* of the system. The open-loop stability poles of the transfer function $T(s)$, which in this case is equal to $G(s)$, are at $s_1 = 0$ and $s_2 = -p$. However, for the closed-loop response the poles are at s_1 and s_2, such that

$$s_1, s_2 = -\xi\omega_n \pm j\omega_n\sqrt{1-\xi^2}. \tag{6.3.3}$$

There are four different cases to be considered, namely,

Case 1: $\xi = 0$, undamped, $s_1, s_2 = 0 \pm j\omega_n$. (6.3.4)

Case 2: $\xi > 1$, overdamped, $s_1, s_2 = -\xi\omega_n \pm j\omega_n\sqrt{1-\xi^2}$. (6.3.5)

Case 3: $\xi < 1$, $\xi > 0$, underdamped, $s_1, s_2 = -\xi\omega_n \pm j\omega_n\sqrt{1-\xi^2}$. (6.3.6)

Case 4: $\xi = 1$, critically damped, $s_1, s_2 = -\xi\omega_n$. (6.3.7)

Generally, for a unit-step input, the output for an underdamped system, $0 < \xi < 1$, is a damped sinusoidal function with overshoot, such that

$$C(t) = L^{-1}O(s) = 1 - \beta^{-1}e^{-\xi\omega_n t}\sin(\omega_n\beta t+\phi), \qquad t > 0, \tag{6.3.8}$$

where

$$\beta = \sqrt{1-\xi^2} \quad \text{and} \quad \phi = \tan^{-1}\xi.$$

To obtain a fast steady state or stable response in robotic manipulators, one should try to adjust the system parameters to obtain a critically damped response as depicted in Figure 6.3.1.

The following conventional techniques for the control of single-link robots are presented first. In these techniques, a combination of both feedback and feedforward controllers is used to ensure tracking and stability of desired input commands and the rejection of all undesirable disturbances. The control of multiaxis robots is discussed in Section 6.8 and various available techniques are outlined.

It must be noted that, although the actual control of any manipulator is by means of a digital computer in a discrete manner, the sampling rate is sufficiently large to justify the treatment of both the manipulator and its controller as continuous systems and to ignore the effect of sampling.

Figure 6.3.1 Response of a second-order critically damped system.

6.4 NONLINEAR CLOSED-LOOP EQUATIONS OF MOTION

Consider a simple one-dimensional motion of a single-axis robotic arm as shown in Figure 6.4.1. The equation governing the motion is

$$\tau_{0,1}(t) = mgl_c \cos\theta + ml_c^2\ddot{\theta} + b_\tau\dot{\theta}, \tag{6.4.1}$$

where b_τ is the rotational friction coefficient. Suppose we have sensors that can measure θ and $\dot{\theta}$ in terms of current, which can be amplified easily and turned to a torque, such that

$$\tau_{0,1}(t) = -k_1\theta - k_2\dot{\theta} + mgl_c \cos\theta. \tag{6.4.2}$$

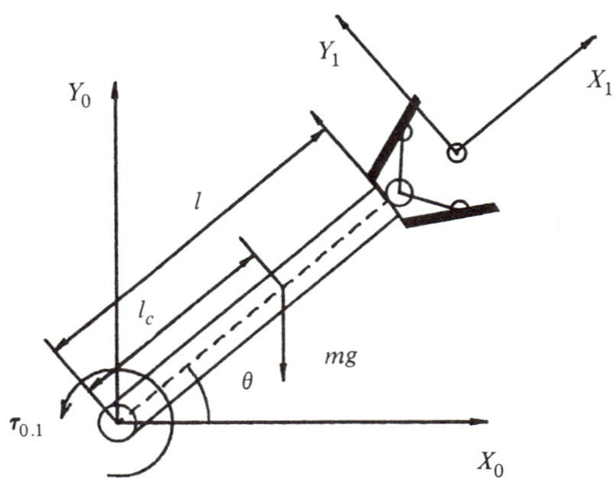

Figure 6.4.1 Single-axis robotic arm motion.

The calculated value of $\tau_{0,1}$ is the torque applied by the joint motor. Thus, the control law yields the following equation:

$$ml_c^2\ddot{\theta} + (b_\tau + k_2)\dot{\theta} + k_1\theta = 0. \tag{6.4.3}$$

Therefore, the system now acts like a linear second-order system with a stiffness of k_1 and a friction coefficient of $b_\tau + k_2$. Thus, we have a control algorithm that can regulate the angular stiffness and damping of the one-link robotic manipulator. This is the most basic extension of the nonlinear feedback control theory to a nonlinear mechanical system. If we want to control the position, we can choose a large value of k_1 so that the arm acts like an extremely stiff spring that holds a fixed rotational position (i.e., it does not rotate). This type of proportional control is called a *regulator*. As mentioned before, we may choose proper values of k_1 and k_2 to obtain a critically damped system for all disturbances about a nominal trajectory. We discuss these issues in the next few sections.

6.5 ERROR-CONTROLLED ROBOTIC DYNAMICS

Figure 6.5.1 displays a special computer-controlled manipulator system for a single-axis arm. The system outputs are $\theta(t)$ and $\dot{\theta}(t)$, recorded by a potentiometer and a tachometer, respectively, for each joint motor. The system inputs are the position set point and its derivatives, that is, $\theta^{(d)}(t)$, $\dot{\theta}^{(d)}(t)$, and $\ddot{\theta}^{(d)}(t)$, which are determined from planned joint motion trajectory equations as discussed in Chapter 4. Here, the control problem is analyzed as a continuous time system, such that errors due to A/D and D/A type quantizations are neglected. We set out to find the open-loop transfer function $G(s)$ of the robotic arm. Let us first consider the manipulator arm at each joint as an armature-controlled dc motor with an inertial load and negligible viscous friction. We are restricting ourselves with the armature-controlled dc motors because they are the type most often employed when an appreciable amount of shaft power is required. We further assume that the field flux is fixed. This means that the

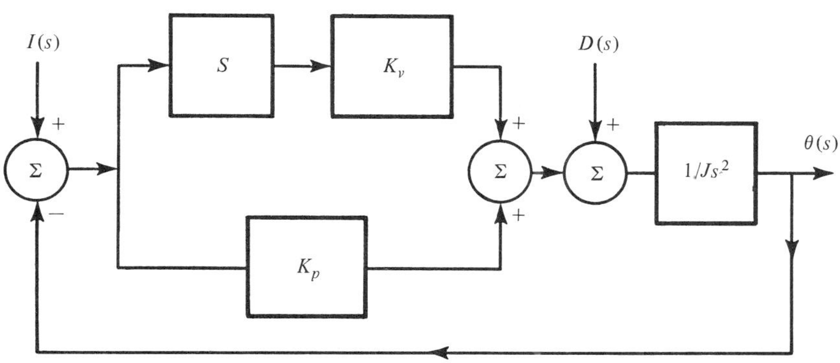

Figure 6.5.1 General block diagram for a computer control of a single-axis manipulator.

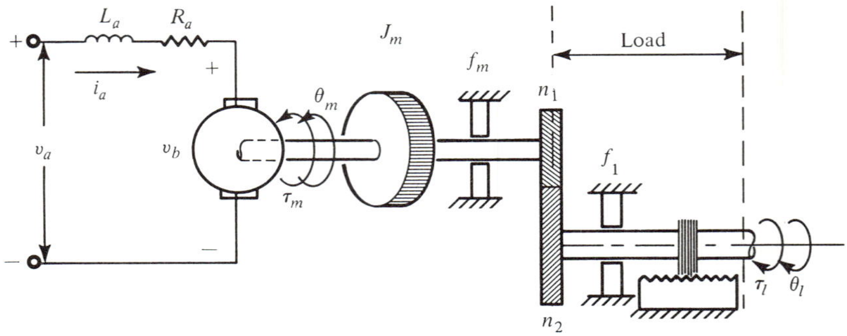

Figure 6.5.2 Structure of an armature-controlled dc motor in conjunction with load.

current in the field winding i_f is held constant. Alternatively, there may be no field winding if a permanent magnet motor is used. In any case, the model of such a dc motor is as shown in Figure 6.5.2.

Note that v_a is the applied armature voltage, i_a is the armature current, v_b is the back emf voltage, R_a is the armature winding resistance, L_a is the armature winding inductance, J_m and J are the polar inertias of the flywheels, f_m and f are bearing frictions, and n_1 and n_2 are the number of gear teeth. Using bond graph modeling, we can show that

$$L_a \frac{di_a}{dt} + R_a i_a + v_b = v_a. \tag{6.5.1}$$

The gear ratio between the pinion and the gear is denoted by $n = n_1/n_2$, such that the equivalent load polar moment of inertia and friction can be shown to be J and f, respectively:

$$J = J_m + n^2 J_l \tag{6.5.2}$$

and

$$f = f_m + n^2 f_l. \tag{6.5.3}$$

EXAMPLE 6.5.1 ■

Derive Equations (6.5.2) and (6.5.3) for the equivalent inertia and friction.

Solution. Employing the principle of conservation of energy, we have

$$\tfrac{1}{2} J_m \dot{\theta}_m^2 + \tfrac{1}{2} J_l \dot{\theta}_l^2 = \tfrac{1}{2} J \dot{\theta}_m^2. \tag{6.5.4}$$

But

$$n_2 \dot{\theta}_l = n_1 \dot{\theta}_m \quad \text{or} \quad \dot{\theta}_l = n\dot{\theta}_m; \qquad \dot{\theta}_l^2 = n^2 \dot{\theta}_m^2. \tag{6.5.5}$$

Thus,

$$\tfrac{1}{2} J_m \dot{\theta}_m^2 + \tfrac{1}{2} J_l n^2 \dot{\theta}_m^2 = \tfrac{1}{2} J \dot{\theta}_m^2 \tag{6.5.6}$$

or

$$J = J_m + n^2 J_l. \tag{6.5.7}$$

On the other hand, the total energy dissipated by frictional interactions is

$$\int f_m \dot{\theta}_m \dot{\theta}_m \, dt + \int f_l \dot{\theta}_l \dot{\theta}_l \, dt = \int f \dot{\theta}_m \dot{\theta}_m \, dt. \tag{6.5.8}$$

Again $\dot{\theta}_l^2 = n^2 \dot{\theta}_m^2$, such that

$$f = f_m + n^2 f_l. \quad \blacksquare\blacksquare \tag{6.5.9}$$

The armature current produces a torque τ_m that is delivered by the motor. Clearly, the equation of motion for this system without gravity is

$$\tau_m = J\ddot{\theta}_m + f\dot{\theta}_m. \tag{6.5.10}$$

Equations (6.5.1) and (6.5.10) are coupled with each other through the fact that the motor is a gyrator and the back emf v_b is related to the angular velocity $\dot{\theta}_m$ such that

$$v_b = k_b \dot{\theta}_m, \tag{6.5.11}$$

where k_b is a back emf constant or the gyration modulus in the context of the bond graph theory. Since the motor here is a gyrator, it is also true that

$$\tau_m = k i_a, \tag{6.5.12}$$

where k is a motor torque constant. Note that in the absence of any other kind of energy dissipation we must have

$$v_b i_a = \tau_m \dot{\theta}_m, \tag{6.5.13}$$

which means that $k \approx k_b$.

Combining Equations (6.5.1)–(6.5.13), we obtain

$$\frac{\theta_m(s)}{v_a(s)} = \frac{k}{s[L_a J s^2 + (L_a f + R_a J)s + (R_a f + k k_b)]}. \tag{6.5.14}$$

If we neglect L_a as being small,

$$\frac{\theta_m(s)}{v_a(s)} = \frac{k_m}{s(T_m s + 1)}, \tag{6.5.15}$$

where k_m and T_m are the motor gain constant and the motor torque constant, respectively. Setting $L_a \approx 0$ in (6.5.14) and equating terms with (6.5.15), we obtain

$$k_m = (R_a f + k k_b)^{-1} k \tag{6.5.16}$$

and

$$T_m = (R_a f + k k_b)^{-1} R_a J. \tag{6.5.17}$$

If we denote the ratio of $\theta_m(s)/v_a(s)$ by $T(s)$, the transfer function, it is then necessary to regulate the armature voltage $v_a(s)$ to control the torque. This is done indirectly by manipulating the applied armature current because the torque is dependent on the armature current i_a, that is, Equation (6.5.12). This process is performed by the amplifier, as explained in Section 6.6. However, at this stage it is convenient to express Equation (6.5.15) in terms of the input torque $T_m(s)$. Thus, from Equation (6.5.10) we conclude that

$$\frac{\theta_m(s)}{\tau_m(s)} = \frac{1}{s(Js + f)} \tag{6.5.18}$$

From Equations (6.5.15) and (6.5.18) it is clear that

$$\frac{\tau_m(s)}{v_a(s)} = \frac{k_m s(Js + f)}{s(T_m s + 1)}, \tag{6.5.19}$$

which is the torque–armature voltage transfer function, a desired function for robotics control.

Because of the complexity of the dynamic model, we consider a single-axis control problem first and then try to generalize to more complex situations. Consider Figure 6.1.1 again and assume that the sampling interval for the ADC measurements and the update interval for the DAC outputs are small compared with the fastest time constant of the amplifier. This assumption is necessary to treat the problem as a continuous time system. Let us further assume that the resolution of the ADC and DAC circuits is sufficiently high so that one may neglect the quantization error; that is, under such conditions the controller can sample $\boldsymbol{\theta}^{(d)}(t)$ and reconstruct a $\mathbf{u}(t)$ fast enough. This implies that $\hat{\mathbf{u}}(t) = \mathbf{u}(t)$ and $\hat{\theta}^{(a)}(t) = \theta^{(a)}(t)$ in Figure 6.1.1. In the next section we elaborate on the control system associated with the amplifier in Figure 6.1.1.

6.6 CONTROL STRUCTURE OF THE AMPLIFIER

Let us now consider the amplifier drive circuit in Figure 6.6.1. Note that $\hat{\mathbf{u}}(t)$ is an instantaneous torque command and thus the drive circuit must regulate the motor current using $\hat{\mathbf{u}}(t)$ as the set point. A typical diagram for such a control system using the pulse width modulation (PWM) technique and a transistor bridge is shown in Figure 6.6.1.

The power transistors Q_1, Q_2, Q_3, and Q_4 serve as elements of a bridge circuit such that Q_1 and Q_2 determine the direction of current flow through the armature winding of the dc motor. At each instant of time t, one of these transistors is on and the other is off. The lower-power transistors Q_3 and Q_4 determine the magnitude of the current flow through the armature winding. Thus, Q_3 and Q_4 control the magnitude of the torque by switching on and off with a duty cycle α, where ideally

$$\alpha = \frac{|\hat{\mathbf{u}}|}{\tau_{\max}}. \tag{6.6.1}$$

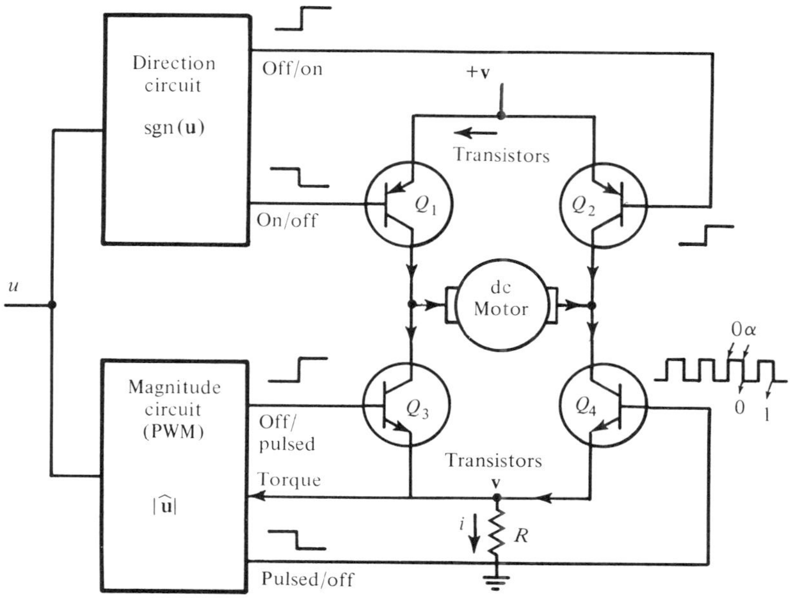

Figure 6.6.1 Bidirectional dc motor drive circuit.

Recall that $\tau_m = ki_a = k(v/R)$ according to Figure 6.6.1. Thus, $\tau_{\max}$ is the maximum possible torque that the motor can deliver. This corresponds to $Q_1 = Q_4$ full on or $Q_2 = Q_3$ full on. Therefore, by controlling the duty cycle α, various amounts of torque in the range $0 \le |\hat{\mathbf{u}}| \le \tau_{\max}$ can be produced. Note that varying α with $|\hat{\mathbf{u}}|$ is a form of pulse width modulation (PWM).

The period of the PWM signal $y(t)$ applied to Q_3 or Q_4 must be small compared to the time constants of the dc motor. Now consider the steady-state open-loop model of the drive circuit. The delivered torque τ in this case is proportional to the armature current i_a that in turn is proportional to the measured voltage drop $\mathbf{v}$ across the current-sensing resistor R. Thus, $\boldsymbol{\tau} = k_1(\mathbf{v}/R)$, where k_1 is a proportionality constant. The sensing resistor is typically very small ($R \ll 1$) and thus gives rise to high current and subsequently high power (Ri^2). Thus, it is typically a high-power resistor. From Equation (6.6.1) one notes that the desired torque $|\hat{\mathbf{u}}|$ is given by $|\hat{\mathbf{u}}| = \alpha\tau_{\max}$ and therefore $0 \le \alpha \le 1$. In the steady-state situation, the delivered torque matches the desired torque; that is, $\tau = |\hat{\mathbf{u}}|$. Thus, the open-loop transfer function of the drive circuit of the amplifier in the steady state ($s = 0$) is

$$\frac{\mathbf{v}}{\alpha} = \frac{R\tau_{\max}}{k_1}. \tag{6.6.2}$$

To develop a closed-loop drive circuit to respond to changes in the torque set point $|\hat{\mathbf{u}}|$, we can feedback the measured torque $k_1\mathbf{v}/R$ and use an integral error controller with gain k_2 as shown in Figure 6.6.2.

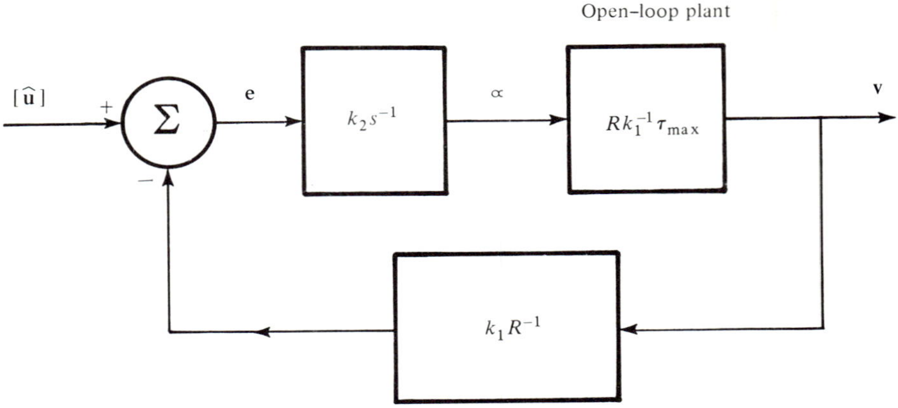

Figure 6.6.2 Closed-loop torque control.

Note that the integral control is needed to ensure zero steady-state tracking error to a step in $\hat{\mathbf{u}}$. The transfer function of this type 1 control system can be shown to be

$$\frac{V(s)}{W(s)} = Rk_1^{-1}\left(\frac{k_2\tau_{max}}{s + k_2\tau_{max}}\right), \tag{6.6.3}$$

where $V(s) = L\mathbf{v}$ and $W(s) = L|\hat{\mathbf{u}}|$. Now under steady-state conditions ($s = 0$) we have $v = R|u|k_1^{-1}$. Thus, the delivered torque $\boldsymbol{\tau}$ in this case equals the desired torque $|\hat{\mathbf{u}}|k_1^{-1}$. The delivered torque $\boldsymbol{\tau}$ in this case equals the desired torque $|\hat{\mathbf{u}}|$.

If we make the controller gain k_2 sufficiently large, the torque regulator time constant $1/k_2\tau_{max}$ can be made small compared to the time constants of the dc motor. Under these conditions, the torque drive circuit is sufficiently fast so that it can be modeled by its steady-state unity gain amplifier characteristic: $\tau = |\hat{\mathbf{u}}|$.

Table 6.6.1 displays two sets of values for two of the servomotors used for the Stanford arm.

TABLE 6.6.1 TYPICAL MOTOR PARAMETERS

Model	*A*	*B*
k_1 (oz·in./A)	6.1	14.4
J_0 (oz·in.·s^2/rad)	0.008	0.0033
B_m (oz·in.·s/rad)	0.01146	0.04297
k_b (V·s/rad)	0.04297	0.10123
L (μH)	100	100
$R(\Omega)$	1.025	0.91
k_t (V·s/rad)	0.02149	0.5062
f_m (oz·in.)	6.0	6.0
n	0.01	0.01

Figure 6.6.3 Robotic motion control in assembly processes.

6.7 CONTROL OF A SINGLE-AXIS ROBOTIC ARM

Consider a single-axis arm with a PD controller. This is shown schematically in Figure 6.7.1. Here $\boldsymbol{\theta}^{(d)}(t)$ is the desired family of trajectories and $\boldsymbol{\theta}^{(a)}(t)$ is the observed family of trajectories. Note that $\mathbf{u}(t)$ is a torque control vector signal and $\mathbf{d}(t)$ is the disturbance torque vector. As discussed before, the closed-loop transfer function under no disturbance ($\mathbf{d} = \mathbf{0}$) is

$$\frac{\theta^{(a)}(s)}{\theta^{(d)}(s)} = \frac{k_v s + k_p}{Js^2 + k_v s + k_p}. \tag{6.7.1}$$

EXAMPLE 6.7.1 ▪

Derive Equation (6.7.1).

Solution. Note from Equation (6.2.3) that

$$T(s) = \frac{\theta^{(a)}(s)}{\theta^{(d)}(s)} = \frac{G(s)}{1 + G(s)}$$

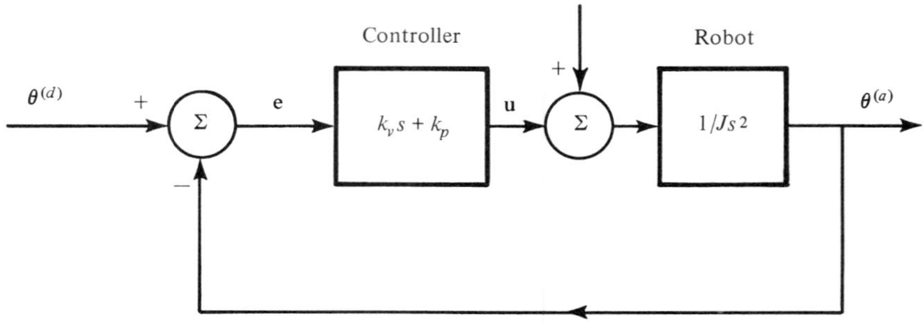

Figure 6.7.1 Proportional derivative (PD) controller.

because $H(s) = 1$. Now we note that $G(s) = (k_v s + k_p)/Js^2$. Thus,

$$\frac{\theta^{(a)}(s)}{\theta^{(d)}(s)} = \frac{(k_v s + k_p)/Js^2}{1 + (k_v s + k_p)/Js^2} = \frac{k_v s + k_p}{Js^2 + k_v s + k_p} \quad \text{Q.E.D.} \quad \blacksquare\blacksquare$$

Equation (6.7.1) can be rewritten in the following form:

$$\frac{\theta^{(a)}(s)}{\theta^{(d)}(s)} = \frac{(k_v/J)s + (k_p/J)}{s^2 + (k_v/J)s + k_p/J}. \tag{6.7.2}$$

It now can be seen by comparing Equation (6.3.2) and (6.7.2) that the natural frequency of this second-order system is $\omega_n = \sqrt{k_p/J}$, and the damping factor $\xi = k_v/2J\sqrt{k_p/J}$.

It is absolutely essential and necessary for a robot *not* to overshoot a command. For example, on a set down course of action, overshoot may cause surface penetration. Yet the arm should be moved as fast as possible for minimal task completion time. This implies that $\xi = 1$ is the most appropriate value. Thus, from Equation (6.7.2), the controller gains k_v and k_p should be chosen such that

$$k_p = J\omega_n^2 \tag{6.7.3}$$

and

$$k_v = 2\sqrt{Jk_p}. \tag{6.7.4}$$

This means that the system has a critically damped response. However, the difficulty with this type of response is that both controller gains k_p and k_v depend on the moment of inertia of the robotic arm. The difficulty arises because J is robot configuration dependent; for example, as the robot moves, J may change by as much as a factor of 10. Thus, we have to come up with a scheme of making the controller configuration-invariant. This can be achieved by scaling the controller gains k_v and k_p by J as shown in Figure 6.7.2.

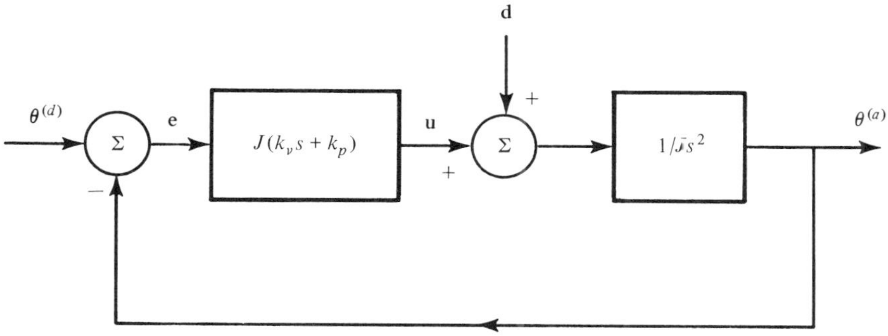

Figure 6.7.2 PD controller with configuration invariance.

EXAMPLE 6.7.2 ■

Show that the control system in Figure 6.7.2 has a transfer function independent of J.

Solution. Note that in this case and in the absence of a disturbance torque $d(t)$

$$G(s) = J(k_v s + k_p)\left(\frac{1}{Js^2}\right) = s^{-2}(k_v s + k_p).$$

Thus,

$$T(s) = \frac{G(s)}{1 + G(s)} = \frac{k_v s + k_p}{s^2 + k_v s + k_p}. \qquad ■■$$

Now let us apply a disturbance $\mathbf{d}(t)$ vector and set the desired $\theta^{(d)}(t) = 0$ and solve for $\theta^{(a)}(s)$ in terms of $d(s)$. We obtain the following disturbance transfer function $T_d(s)$:

$$T_d(s) = \frac{\theta^{(a)}(s)}{d(s)} = \frac{1}{Js^2 + Jk_v s + Jk_p}. \tag{6.7.5}$$

EXAMPLE 6.7.3 ■

Derive Equation (6.7.5) for the transfer function $T_d(s)$.

Solution. The block diagram for this case, where $\theta^{(d)} = 0$, is depicted in Figure 6.7.3. Our task is to find the transfer function $T_d(s)$, such that $[\theta^{(a)}(s)/d(s)] = T_d(s)$.

Note that this case is similar to the general feedback system shown in Figure 6.3.1, where

$$G(s) = \frac{1}{Js^2} \quad \text{and} \quad H(s) = J(k_v s + k_p). \tag{6.7.6}$$

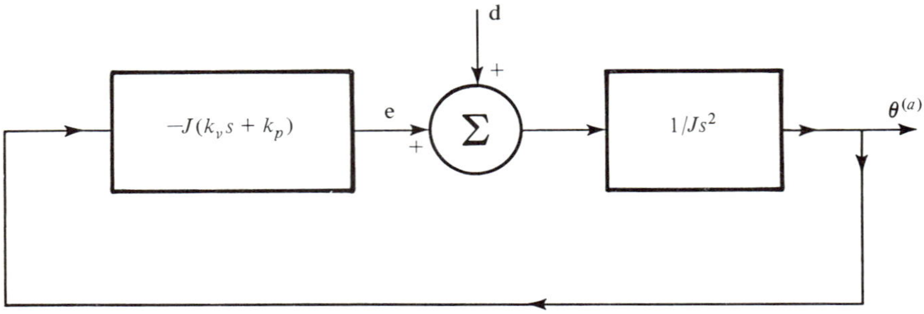

Figure 6.7.3 Zero-input controller with disturbance.

Since the signals **e** and **d** are added,

$$T_d(s) = \frac{G(s)}{1 + G(s)H(s)} = \frac{1/Js^2}{1 + J(k_v s + k_p)/Js^2}. \tag{6.7.7}$$

Finally,

$$T_d(s) = \frac{1}{Js^2 + Jk_v s + Jk_p}. \quad ■■ \tag{6.7.8}$$

Again in this case $\theta^{(d)} = 0$, and the response of the robotic arm to a torque disturbance may be critically damped because

$$Jk_p = J\omega_n^2 \to k_p = \omega_n^2 \tag{6.7.9}$$

and

$$Jk_v = 2J\delta\omega_n \to k_v = 2\delta\omega_n. \tag{6.7.10}$$

If $\delta = 1$, $k_v = 2\sqrt{k_p}$ can be achieved easily in the controller. Furthermore, for a unit step disturbance the response function is simply $(Jk_p)^{-1}$.

In the absence of mechanical design imperfections in the robotic arm, such as backlash, looseness, gear friction, eccentricities, imbalances, and inherent vibrations, the principal disturbance will be the result of gravitational shifts in the centers of gravities. Thus, gravity loading of the arm can create torque disturbances that have to be compensated. This is especially true for the shoulder and elbow joints when the arm gets extended. In this case, one has to supply a torque at the joint that is equal and opposite to the composite gravity loading d_g on the joint. This leads to a control system as shown in Figure 6.7.4.

EXAMPLE 6.7.4 ■

Show that the gravity disturbance torque may be represented by $s^{-1}d_g$.

Solution. Note that the gravity disturbance torque is given by d_g so that if we assume it to be independent of time, its Laplace transform will be

$$\int_0^\infty d_g e^{-st}\,dt = -\frac{d_g}{s}\int_0^\infty d^{-st} \int_0^\infty = \frac{d_g}{s} = s^{-1}\,d_g. \tag{6.7.11}$$

If the gravity disturbance torque is a function of time such as $d_g(t)$, then its actual transform $d_g(s)$ must be substituted for $s^{-1}\,d_g$. ■■

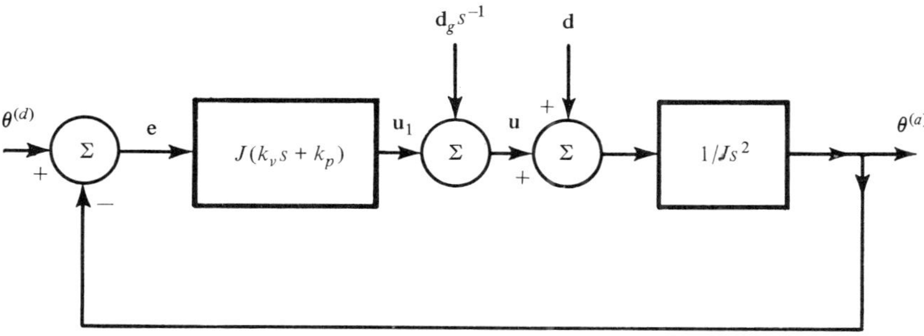

Figure 6.7.4 Controller with compensation for disturbance torques resulting from gravity (gravity-compensated robotic control).

The gravity-free disturbances (torque) are then represented by a residual torque $d_r(s)$ such that

$$\mathbf{d}_r(s) = \mathbf{d}(s) - \mathbf{d}_g(s). \tag{6.7.12}$$

Clearly, the error transform $e(s)$ is given by $e(s) = \theta^{(d)}(s) - \theta^{(a)}(s)$, such that

$$\mathbf{e}(s) = \boldsymbol{\theta}^{(d)}(s)\left(\frac{s^2}{s^2 + k_v s + k_p}\right) + \mathbf{d}_r(s)\left(\frac{1}{Js^2 + Jk_v s + Jk_p}\right). \tag{6.7.13}$$

The idea is to reduce the error $e(s)$ to zero in minimum time without causing any slowness (slag) or overshoot in the response. Thus, we have to cancel the effect of the first term on the right-hand side of Equation (6.7.13). To do this, we must adjoin a $s^2\theta^{(d)}(s)$ controller in parallel to the $k_v s + k_p$ controller in Figure 6.7.4 so that Figure 6.7.5 would result. Thus, Equation (6.7.14) simply reduces to

$$\mathbf{e}(s) = \mathbf{d}_r(s)]\left(\frac{1}{Js^2 + Jk_v s + Jk_p}\right). \tag{6.7.14}$$

Finally,

$$\left.\frac{\theta^{(a)}(s)}{\theta^{(d)}(s)}\right|_{\mathbf{d}_r = 0} = \frac{s^2 + k_v s + k_p}{s^2 + k_v s + k_p} = 1. \tag{6.7.15}$$

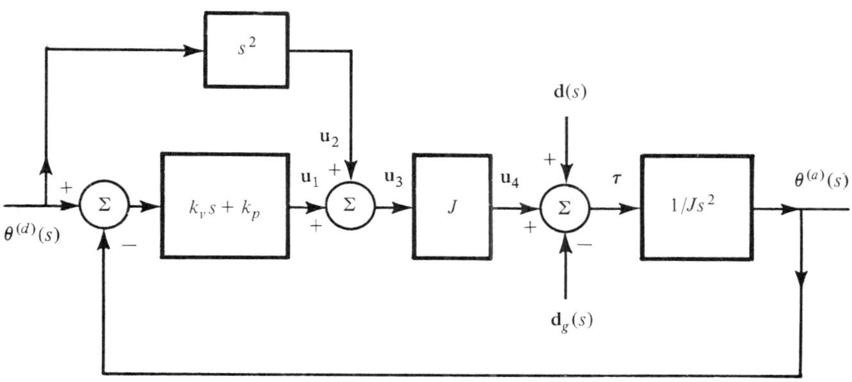

Figure 6.7.5 Controller with acceleration compensation.

Figure 6.7.6 Control of plasma cutting (courtesy: Cincinnati Milacron).

To summarize, a PD controller (Fig. 6.7.1) can be configuration-invariant (Fig. 6.7.2), can compensate for gravity loading (Fig. 6.7.3), and can include a feedforward acceleration compensation (Fig. 6.7.5). The control law that is associated with Figure 6.7.5 can be written

$$\tau(t) = J[\ddot{\theta}^{(d)}(t) + k_v(\dot{\theta}^{(d)} - \dot{\theta}^{(a)}) + k_p(\theta^{(d)} - \theta^{(a)})] + d_r(t). \tag{6.7.16}$$

The first term represents the acceleration compensation, the middle two terms represent the proportional derivative (PD), and the last term represents the gravitational torque. Parameters k_v and k_p are assumed to be constant, while J and d_g are essentially configuration dependent; that is, they depend on the θ_i's. Equation (6.7.16) actually represents a nonlinear control law because $J = J(\theta_i)$ and $d_r = d_r(\theta_i)$. For each joint i, Equation (6.7.16) may be rewritten in the form

$$\tau_i(t) = J_i[\ddot{\theta}_i^{(d)}(t) + k_v(\dot{\theta}_i^{(d)} - \dot{\theta}_i^{(a)}) + k_p(\theta_i^{(d)} - \theta_i^{(a)})] + d_{ri}(t). \tag{6.7.17}$$

6.8 GENERALIZED MOTION CONTROL LAWS FOR ROBOTIC MANIPULATORS

Recalling Equation (6.7.17) we note that the gains J and d_r are configuration-dependent variables. Thus, in the general case, one must consider the whole dynamics of the robotic arm. We can now rewrite the dynamic equations for multiaxis robotic manipulators in a compact form for the following general treatment of control:

$$\boldsymbol{\tau} = \mathbf{M}(\boldsymbol{\theta})\ddot{\boldsymbol{\theta}} + \mathbf{V}(\boldsymbol{\theta}, \dot{\boldsymbol{\theta}}) + \mathbf{G}(\boldsymbol{\theta}), \tag{6.8.1}$$

where $\mathbf{M}$ is a $n \times n$ mass matrix, $\mathbf{V}$ is an $n \times 1$ vector of centrifugal and Coriolis acceleration terms, and $\mathbf{G}$ is an $n \times 1$ vector of gravitational effects.

As discussed before, we choose to feedback the nonlinear terms, assuming for the time being negligible quantization, sampling, and computation time, in the following nonlinear feedback form:

$$\boldsymbol{\tau}^* = \mathbf{M}(\boldsymbol{\theta})[\ddot{\boldsymbol{\theta}}_{(d)} + \mathbf{k}_v(\dot{\boldsymbol{\theta}}_{(d)} - \dot{\boldsymbol{\theta}}) + \mathbf{k}_p(\boldsymbol{\theta}_{(d)} - \boldsymbol{\theta})] + \mathbf{V}(\boldsymbol{\theta}, \dot{\boldsymbol{\theta}}) + \mathbf{G}(\boldsymbol{\theta}), \tag{6.8.2}$$

where $\boldsymbol{\tau}^*$ is the feedback torque to the system and $\mathbf{k}_v$ and $\mathbf{k}_p$ are $n \times n$ diagonal matrices. The feedback term $\mathbf{V}(\boldsymbol{\theta}, \boldsymbol{\theta}) + \mathbf{G}(\boldsymbol{\theta})$ is the nonlinear feedback torque term that needs to be computed fast enough by a computer to make it an effective feedback control. This method is called the *computed torque technique* and the computation time for the nonlinear feedback term is one of the stumbling blocks in effective implementation of this control technique (Fig. 6.8.1).

The control (6.8.2) in conjunction with the dynamic equation (6.8.1) yields the following linearized dynamics for the control of robotic motion in terms of the error vector $\mathbf{e} = (\boldsymbol{\theta}_{(d)} - \boldsymbol{\theta})$:

$$\mathbf{M}(\ddot{\mathbf{e}} + \mathbf{k}_v\dot{\mathbf{e}} + \mathbf{k}_p\mathbf{e}) = 0. \tag{6.8.3}$$

Thus, any error can be critically damped, including that introduced by incorporating an acceleration feedforward turn as well as a typical PD controller as shown in Figure 6.8.1. Clearly, the desired trajectory $\theta_{(d)}(t)$ must be of class C_2 for $\ddot{\theta}_{(d)}(t)$ to exist. For example, if the desired trajectory must have a sudden change of velocity, then the above method fails because $\ddot{\theta}_{(d)}$ can increase dramatically and create uncontrollable noise in the feedforward term. The technique also requires knowledge on a complete set of structural and dynamic parameters such as link lengths and other dimensions, friction coefficients, and inertias. Since the underlying dynamic model is never perfect, the intended control algorithm will never be completely effective and midcourse corrections

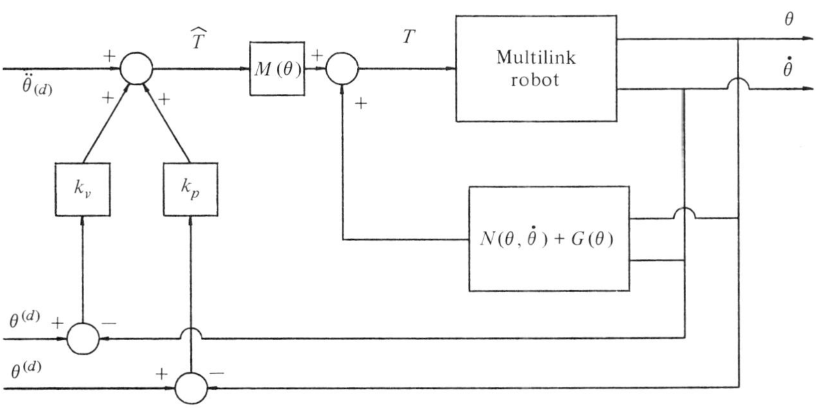

Figure 6.8.1 Computed torque technique.

must be made. To take care of such discrepancies, we may define a general vector of disturbances in the form of a torque $\boldsymbol{\tau}_D$, such that the control equation simply becomes

$$\ddot{\mathbf{e}} + \mathbf{k}_v\mathbf{e} + \mathbf{k}_p\mathbf{e} = \mathbf{M}^{-1}(\boldsymbol{\theta})\boldsymbol{\tau}_D. \tag{6.8.4}$$

This is correct only if the mass matrix $\mathbf{M}(\theta)$ has an inverse $\mathbf{M}^{-1}(\boldsymbol{\theta})$. Since $\mathbf{M}(\boldsymbol{\theta})$ is generally not symmetric, the disturbance in any particular joint can produce disturbances at other joints. Furthermore, some disturbances, such as picking up various parts and tools, actually affect the mass matrix $\mathbf{M}(\boldsymbol{\theta})$ and are hard to handle unless we adopt other control strategies.

Most manipulator control systems are performed in digital form and thus operate at a certain *sampling rate*. The motion sensors can be read only at discrete times and the actuator command would then be computed and sent to the joint motors. One therefore needs to analyze the effect of delay owing to computation and finite sampling rate by resorting to *discrete time control*. Only then can one analyze the problems of stability and pole placement. The memory size is generally very large in any control computational algorithm. For example, if the ranges of $\boldsymbol{\theta}$, $\dot{\boldsymbol{\theta}}$, and $\ddot{\boldsymbol{\theta}}$ are digitized by 100 points, which is very modest, then for a six-axis robot the memory space has 10^{36} cells. Each cell must be provided with six torque values, which amounts to 6×10^{36} memory cells. Thus, some simplification and reduction of computation time is highly desirable for robotic control systems.

6.9 COMMON CONTROL SYSTEMS FOR INDUSTRIAL ROBOTS

Since industrial robots must be economically viable, no complete control package is worth including because of the cost of the required computer time. This course of action is justified because of the lack of complete knowledge on various structural and dynamic parameters. Thus, many industrial robots have simple control algorithms that are generally error driven. In the following subsection we briefly describe some current industrial robot control algorithms.

6.9.1 Simple Robotic Manipulators

If the robotic structure is simple enough, such as SCARA robots, then the governing dynamic equations greatly simplify and decouple. For some other uncomplicated structures, the kinematics can be simplified and the inertia matrix may become diagonal, thus greatly simplifying the computational algorithms. Furthermore, removing nonlinear effects such as the backlash of speed reduction gears greatly simplifies the control problem. Therefore, designing simpler yet more effective robots is one way the industry is looking at simplifying the control tasks.

6.9.2 Independent Joint PID Control

The proportional integral derivative (PID) applied to an individual joint is a popular method of control for most industrial robots. Here, the servo torque is given by

$$\tau^* = \ddot{\boldsymbol{\theta}}_{(d)} + \mathbf{k}_v \dot{\mathbf{e}} + \mathbf{k}_p \mathbf{e} + \mathbf{k}_i \int \mathbf{e}\, dt. \tag{6.9.1}$$

Sometimes the values of $\ddot{\boldsymbol{\theta}}_{(d)}$ and $\dot{\theta}_{(d)}$ may not be available and the control law (6.9.1) reduces to

$$\tau^* = \mathbf{k}_v \dot{\boldsymbol{\theta}} + \mathbf{k}_p \mathbf{e} + \mathbf{k}_i \int \mathbf{e}\, dt. \tag{6.9.2}$$

The following example illustrates this.

EXAMPLE 6.9.1 ■

For the PID controller shown in Figure 6.9.1 present an analysis of the stability of the control algorithm.

Solution. The block diagram of the controller is shown in Figure 6.9.1 We consider a two-axis robot with two joint servos. The primary purpose of the proportional position feedback is to improve the speed-of-response. The integral feedback also improves speed-of-response and assures steady-state tracking of the reference input $\boldsymbol{\theta}_d$ in the face of disturbances. The derivative feedback enhances stability and reduces the tendency toward oscillation. These are precisely the features required for good response.

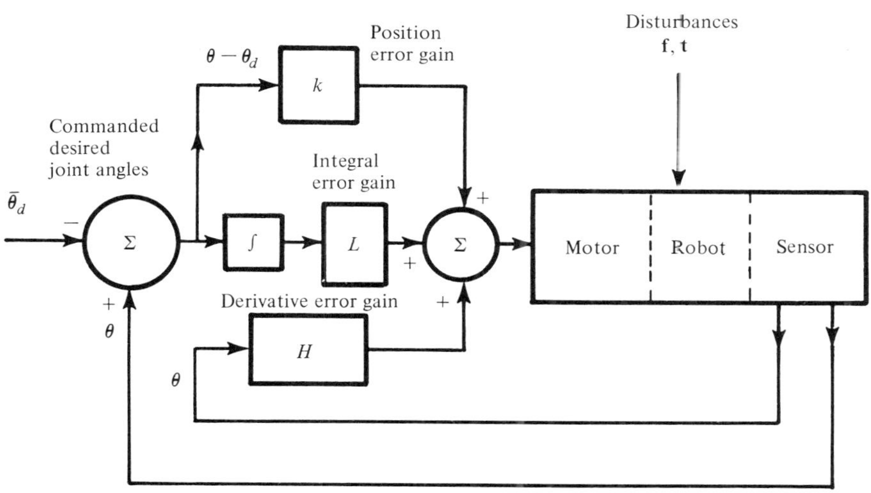

Figure 6.9.1 PID controller for a joint servo, including the block diagram of a typical joint servo.

The governing equations for the dynamics are

$$I_{11}\ddot{\theta}_1 + I_{12}\ddot{\theta}_2 = \tfrac{1}{2}m_2 l_1 l_2(\dot{\theta}_1 + \dot{\theta}_2)^2 \sin\theta_2 - (\tfrac{1}{2}m_1 + m_2)l_1 g\cos\theta_1 + \tau_{0,1} - \tau_{1,2} - (l_1\cos\theta_1)f_{2,3y} + (l_1\sin\theta_1)f_{2,3x} \tag{6.9.3}$$

and

$$I_{21}\ddot{\theta}_1 + I_{22}\ddot{\theta}_2 = -\tfrac{1}{2}m_2 l_1 l_2\dot{\theta}_1^2\sin\theta_2 - \tfrac{1}{2}m_2 l_2 g\cos(\theta_1+\theta_2) + \tau_{1,2} - \tau_{2,3} - [l_2\cos(\theta_1+\theta_2)]f_{2,3y} + [l_2\sin(\theta_1+\theta_2)]f_{2,3x}, \tag{6.9.4}$$

where $f_{2,3}$ and $\tau_{2,3}$ are the end effector vectors of forces and couples, $\boldsymbol{\tau} \equiv (\tau_{0,1}, \tau_{1,2})$ is the vector of the joint torques, and $\mathbf{I}$ is the inertia matrix. The control laws are then borrowed from Equation (6.9.2), such that

$$\tau_{0,1} = k_{p01}(\theta_1 - \bar{\theta}_{1d}) + k_{i01}e_1 + k_{v01}\dot{\theta}, \tag{6.9.5}$$

$$\dot{e}_1 = \theta_1 - \bar{\theta}_{1d} \quad\text{and}\quad \dot{e}_2 = \theta_2 - \bar{\theta}_{2d}, \tag{6.9.6}$$

and

$$\tau_{1,2} = k_{p12}(\theta_2 - \theta_{2d}) + k_{i12}e_2 + k_{v12}\dot{\theta}_2. \tag{6.9.7}$$

Now to analyze the stability of the proposed PID system, we linearize about a steady state $\bar{\boldsymbol{\theta}}_d$ and obtain from Equations (6.9.3) and (6.9.4) the following linearized equations:

$$I_{11}\delta\ddot{\theta}_1 + I_{12}\delta\ddot{\theta}_2 = k_{v01}\delta\dot{\theta}_1 - k_{v12}\delta\dot{\theta}_2 + (I_1^* + k_{p01})\delta\theta - k_{p12}\delta\theta_2 + k_{i01}\delta l_1 - k_{i12}\delta l_2 \tag{6.9.8}$$

and

$$I_{21}\delta\ddot{\theta}_1 + I_{22}\delta\ddot{\theta}_2 = k_{v12}\delta\dot{\theta}_2 + k_{i12}\delta l_2 + (I_2^* + k_{p12})\delta\theta_2 + I_2^*\delta\theta, \tag{6.9.9}$$

where

$$I_1^* = (\tfrac{1}{2}m_1 + m_2)l_1 g\sin\bar{\theta}_{1d} \tag{6.9.10}$$

and

$$I_2^* = \tfrac{1}{2}m_2 l_2 g\sin(\bar{\theta}_{1d} + \bar{\theta}_{2d}). \tag{6.9.11}$$

Taking the Laplace transform of Equations (6.9.8) and (6.9.9) and setting the determinant of coefficients of $\delta\theta_1$ and $\delta\theta_2$ equal to zero, we obtain the following characteristic polynomial:

$$\alpha(s) = \sum_{i=0}^{6} \alpha_i s^i = 0, \tag{6.9.12}$$

where

$$\alpha_0 = k_{i01}k_{i12}, \tag{6.9.13}$$

$$\alpha_1 = k_{i12}(k_{p01} + I_1^* + I_2^*) + k_{i01}(k_{p12} + I_2^*), \tag{6.9.14}$$

$$\alpha_2 = k_{i12}k_{v01} + k_{i01}k_{v12} + k_{p12}(k_{p01} + I_1^* + I_2^*) + (k_{p01} + I_1^*)I_2^*, \tag{6.9.15}$$

$$\alpha_3 = k_{p12}(k_{p01} + I_1^* + I_2^*) + k_{v01}(k_{p12} + I_2^*) - k_{i01}I_{22} - k_{i12}k_{i12}I_m, \tag{6.9.16}$$

$$\alpha_4 = k_{v01}k_{p12} - k_{p01}I_{22} - k_{p12}I_m - [I_1 + (\tfrac{1}{4}m_1 + m_2)l_1^2]I_1^2]I_2^* - I_1^*I_2^*, \tag{6.9.17}$$

$$\alpha_5 = -k_{v12}I_m - k_{v01}I_{22}, \tag{6.9.18}$$

$$\alpha_6 = I_1I_2 + (\tfrac{1}{4}m_1 + m_2)l_1^2l_2^2 + \tfrac{1}{16}m_1m_2l_1^2l_2^2 + \tfrac{1}{4}m_2^2l_1^2l_2^2 \sin \bar{\theta}_{2d} \tag{6.9.19}$$

and

$$I_m = I_1 + I_2 + (\tfrac{1}{4}m_1 + m_2)l_1^2 + \tfrac{1}{4}m_2l_2^2 + m_2l_1l_2 \cos \bar{\theta}_{2d}. \tag{6.9.20}$$

The Routh criterion provides the necessary and sufficient conditions for the roots of this polynomial to lie in the left half-plane and thus ensure stability. These are in the form of a set of inequalities involving functions α_0 to α_6. If those inequalities are satisfied simultaneously for some values of gain and some values of $(\bar{\theta}_{1d}, \bar{\theta}_{2d})$, the controller is locally and asymptotically stable in the sense that initial states ($\delta\theta_1$, $\delta\theta_2$, $\delta\dot{\theta}_2$, $\delta\dot{\theta}_2$, δ_1, and δ_2) sufficiently close to zero will decay to zero with time. Asymptotic stability may be generated for all set values of $\bar{\boldsymbol{\theta}}_d$ based on the above solutions. ■ ■

6.9.3 Independent Joint PID Control with Effective Joint Inertia

The servo law in this case is given by

$$\boldsymbol{\tau}^* = \mathbf{M}^*(\theta)\left(\ddot{\boldsymbol{\theta}}_d + \mathbf{k}_v\dot{\mathbf{e}} + \mathbf{k}_p\mathbf{e} + \mathbf{k}_i \int \mathbf{e}\, dt\right), \tag{6.9.21}$$

where $\mathbf{M}^*(\boldsymbol{\theta})$ is only the diagonal part of the proper mass matrix $\mathbf{M}(\boldsymbol{\theta})$. Thus, the joints in this model are independent and their associated inertias change with joint configurations.

This model may be modified by replacing $\mathbf{M}^*(\boldsymbol{\theta})$ with some other approximation or the actual inertia matrix $\mathbf{M}(\boldsymbol{\theta})$. Note that the closed-loop equation governing the error dynamics in this case is given by

$$\ddot{\mathbf{e}} + \mathbf{k}_v\dot{\mathbf{e}} + \mathbf{k}_p\mathbf{e} = \mathbf{M}^{-1}(\boldsymbol{\theta})(\mathbf{V} + \mathbf{G}). \tag{6.9.22}$$

Thus, one may introduce a gravity term such as $\mathbf{G}(\theta)$ in the servo law (6.9.21).

6.10 FORCE CONTROL OF ROBOTIC MANIPULATORS

In their path of motion, robotic manipulators commonly experience forces from the environment different from disturbances on the joint torques. Therefore, one has to consider ways of controlling such forces, without affecting the motion control already in progress, especially where motion and position control is not enough to perform a robotic task. For example, consider a robot in a surface-finishing operation. If the combined compliance of the robot, the robotic hand, and the work pieces is very low (or the combined stiffness is very high), it is hard to use the position feedback to control the forces necessary for a surface finish. Instead, it is much easier to sense the forces directly and use a force feedback to control the forces at the robotic hand that need to be maintained. Ultimately, it is best to control both the position and the force and thus to develop hybrid position/force controllers. To sense the forces, one needs to use six-axis force sensors or tactile sensors that output a voltage proportional to the force experienced. Then the voltage can be digitized and used in a computer to control the motion of the robot.

6.10.1 Types of Force Sensors

Force sensors make use of strain gauges, pressure transducers, and piezoelectric substrates, such as PZT piezocerams, that generate either a voltage or a current proportional to either deflection or force. There are three places where force or torque sensors can be placed on a manipulator. The first place is at the joint actuators, making them suitable for control rather than contact sensing. The second place is at the wrist because wrist force sensors can be used as three- to six-axis force or torque sensors. The third place is at the fingertips of a dexterous robotic hand to achieve tactile sensing.

6.10.2 Types of Force Control

There are essentially two types of force control: logic branching based on events and, continuous force control based on time differentials.

In logic branching the control algorithm senses an event, such as a collision, to assess that the magnitude of a force or moment has exceeded a certain limit. In this manner a change of motion (branching) is implemented.

The continuous force control is divided into two categories—*pure* and *hybrid*. In both categories the force is sensed first and the tool is moved next. Mathematically,

$$\begin{matrix} \dot{x} \\ \dot{y} \\ \dot{z} \\ \omega_x \\ \omega_y \\ \omega_z \end{matrix} \underset{6 \times 1}{\begin{bmatrix} \text{new} \\ \text{velocity} \\ \text{command} \\ \text{vector} \end{bmatrix}} = \underset{6 \times 6}{\begin{bmatrix} \text{force} \\ \text{feedback} \\ \text{matrix} \\ \text{(accommodation} \\ \text{matrix)} \end{bmatrix}} \underset{6 \times 1}{\begin{bmatrix} \text{sensed} \\ \text{force} \\ \text{vector} \end{bmatrix}} \begin{matrix} fx \\ fy \\ fz \\ \tau_x \\ \tau_y \\ \tau_z \end{matrix}$$

6.10.3 Torque Control of a Compliant Arm

Consider controlling the forces experienced by a single arm in the presence of some compliance either at the joint or in the hand (Fig. 6.10.1). Note that there exists an unknown disturbance force **f** that may originate because of collisions with other objects, friction, jamming in the mechanical structure, or any other effect. We wish to control the force **f*** acting on a compliant environment, that is,

$$\mathbf{f}^* = \mathbf{k}_e \mathbf{x}, \tag{6.10.1}$$

where $|x| = l^*\theta$.

The governing equation in the absence of friction is

$$\tau_{0,1} = ml^2\ddot{\theta} + mlg\cos\theta - k_e l^{2*}\theta - fl^*. \tag{6.10.2}$$

In terms of f^* we have

$$\tau_{0,1} = ml^2 k_e^{-1}(l^*)^{-1}\ddot{f}^* + mlg\cos[k_e^{-1}(l^*)^{-1}f^*] - f^*l^* - fl^*. \tag{6.10.3}$$

Suppose we apply a desired force f_d^* such that $e = f_d^* - f^*$ is the force error between the desired force and the sensed force. We may propose a servo law of the form

$$\begin{aligned}\tau_{0,1} &= ml^2 k_e^{-1}(l^*)^{-1}(\ddot{f}_d^* + k_v\dot{e} + k_p e)\\ &\quad + mlg\cos[ke^{-1}(l^*)^{-1}f^*] - f^*l^* - fl^*.\end{aligned} \tag{6.10.4}$$

Thus, the force error dynamics equation reduces to

$$\ddot{e} + k_v\dot{e} + k_p e = 0 \tag{6.10.5}$$

This system can be critically damped for any force error if we know the value of f at all times during a given interval of time.

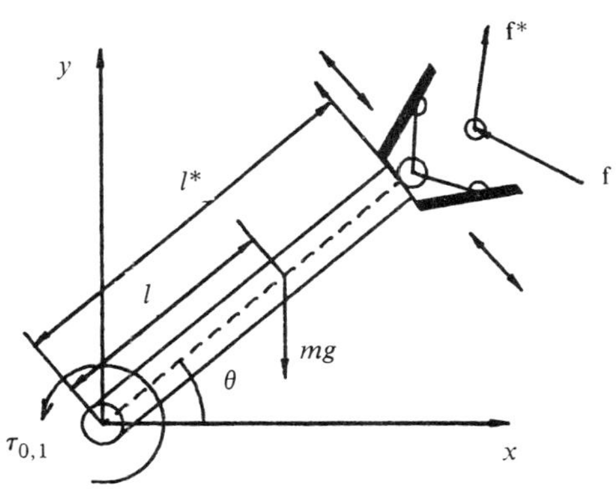

Figure 6.10.1 Compliant robotic arm.

6.11 TRACKING ERROR ANALYSIS

The tracking error is the difference between the planned trajectory $\boldsymbol{\theta}^{(d)}$ and the actual trajectory $\boldsymbol{\theta}^{(a)}$. Of course this also includes $\dot{\boldsymbol{\theta}}^{(d)}$, $\dot{\boldsymbol{\theta}}^{(a)}$, $\ddot{\boldsymbol{\theta}}^{(d)}$, and $\ddot{\boldsymbol{\theta}}^{(a)}$, When the manipulator arm is maintaining a fixed position, we may define a steady-state error as the difference between the planned end points and the actual end points. It must be emphasized that the major source of tracking results from inaccuracies encountered in the dynamic control equations. In this case,

$$\tau(t) = \mathbf{J} \cdot (\ddot{\boldsymbol{\theta}}^{(d)} - \ddot{\boldsymbol{\theta}}^{(a)}) + \mathbf{J}k_v \cdot (\dot{\boldsymbol{\theta}}^{(d)} - \dot{\boldsymbol{\theta}}^{(a)}) + \mathbf{J}k_p(\boldsymbol{\theta}^{(d)} - \boldsymbol{\theta}^{(a)}) + G(\boldsymbol{\theta}). \qquad (6.11.1)$$

In Equation (6.11.1) the effects of Coriolis and centrifugal accelerations are neglected. More accurate tracking can be achieved by retaining such acceleration terms, that is, to have

$$\tau(t) = \mathbf{J} \cdot (\ddot{\boldsymbol{\theta}}^{(d)} - \ddot{\boldsymbol{\theta}}^{(a)}) + \mathbf{J}k_v \cdot (\dot{\boldsymbol{\theta}}^{(d)} - \dot{\boldsymbol{\theta}}^{(a)}) + Jk_p \cdot (\boldsymbol{\theta}^{(d)} - \boldsymbol{\theta}^{(a)})$$
$$+ \mathbf{G}(\boldsymbol{\theta}) + \mathbf{H}(\boldsymbol{\theta}, \dot{\boldsymbol{\theta}}). \qquad (6.11.2)$$

The above equation is undesirable because with $\mathbf{H}(\boldsymbol{\theta}, \dot{\boldsymbol{\theta}})$ present the amount of computation is horrendous.

Using Equation (6.11.2), we state the following algorithm for robotic arm dynamic control.

Algorithm 6.11.1 To achieve robotic arm dynamic control we use Equation (6.11.2) as follows:

1. Design your trajectory as either a 4–3–4 or a 3–5–3.
2. Initiate the dynamic control by checking the value of $|\theta^{(d)} - \theta^{(d)}_{\text{final}}| < \varepsilon$.
3. Sample $\theta_i^{(a)}(t)$ and $\theta_i^{(d)}(t)$ for all i's.
4. Convert $\theta_i^{(a)}(t)$ and $\theta_i^{(a)}$ to joint angles in radians.
5. Calculate $\ddot{\boldsymbol{\theta}}_i^{(a)} = [\dot{\theta}^{(a)}(t) - \dot{\theta}^{(a)}(t-1)]/\Delta t$.
6. Calculate the error $\mathbf{e} = \boldsymbol{\theta}^{(d)} - \boldsymbol{\theta}^{(a)}$, $\dot{\mathbf{e}} = \dot{\boldsymbol{\theta}}^{(d)} - \dot{\boldsymbol{\theta}}^{(a)}$, and $\mathbf{e} = \ddot{\boldsymbol{\theta}}^{(d)} - \ddot{\boldsymbol{\theta}}^{(a)}$.
7. Compute the nonlinear terms such as $\mathbf{D}(\boldsymbol{\theta})$, $\mathbf{H}(\boldsymbol{\theta}, \dot{\boldsymbol{\theta}})$, $\mathbf{G}(\boldsymbol{\theta})$, k_p, and k_v for all i's.
8. Compute the required torque for each joint.
9. Convert the required torque to the required motor input voltage.
10. Output motor voltages to the D/A converter and current amplifier.
11. Increment time counter to next sampling time.
12. Compute $\boldsymbol{\theta}^{(d)}$, $\dot{\boldsymbol{\theta}}^{(d)}$, and $\ddot{\theta}^{(d)}$ for the new time.
13. Loop until you are done or the error tends to zero.
14. Go to step 2 (initiate).

6.12 COORDINATED MOVEMENT

The planned trajectory $\boldsymbol{\theta}^{(d)}(t)$, $t_0 \le t \le t_3$, creates a *continuous path motion* for the end effector to follow through the workspace. In applications such as arc welding, spray painting, and automatic cutting, the continuity of motion is essential. On the other hand, the velocity and acceleration of the end effector tip or tool tip also must be planned carefully. This requires accurate independent speed and acceleration control of all joint motors that are driving the axes. In applications such as spot welding, stacking, or material handling, a *point-to-point motion* is sufficient. However, in these applications one wants to minimize the transition time as the end effector moves between the points. To develop a coordinated movement of an end effector from $\boldsymbol{\theta}(t_0)$ to $\boldsymbol{\theta}(t_1)$ in a fast manner, we consider the following two basic cases.

1. *Variable Speed Controller.* Consider the joint motor at the ith joint. Suppose that the speed $\dot{\theta}_i(t)$ can be controlled independently by a controller. Let $w_{\max}$ be the *maximum* speed possible and let $(\Delta\theta)_{\max}$ be the maximum angle to be moved; that is,

$$(\Delta\theta)_{\max} = \max_{i=1}^{5} \{|\theta_i(t) - \theta_i(t_0)|\}. \tag{6.12.1}$$

The idea is to move the axis that has to travel the largest $\Delta\theta$ at the maximum possible speed $w_{\max}$ in order to minimize the travel time between the two points. To coordinate the motion, one must drive the other axes at slower speeds (in a proportional manner with respect to the $\Delta\theta$ values) so as to have all axes start and finish the motions simultaneously and respectively at t_0 and t_1. One may achieve this by requiring the following speed-control algorithm:

$$\dot{\theta}_i(t) = \frac{w_{\max}}{(\Delta\theta)_{\max}} [\theta_i(t_1) - \theta_i(t_0)]. \tag{6.12.2}$$

The speed controllers must provide different speeds for different joint motors, which is equivalent to having a first-degree polynomial for each joint motion. The controllers must be optimized to reach $w_{\max}$ initially in minimum time and to come to rest finally in minimum time.

2. *Constant Speed Controller.* If the controller is what is known as a *bang-bang controller*, then each joint motor can always be in any of the following states:

(a) Full speed forward.
(b) Full speed reverse.
(c) Full stop.

This is the case with the Rhino XR-2 controller board. Here the average speed is controlled by the fraction of the total task time that a given joint motor is supposed to be turned on. The axis that has the farthest angle to travel will be running full speed during the entire time interval $[t_0, t_1]$ to minimize the travel time. To obtain a coordinated motion, one wants to drive the other axes full speed for proportionally shorter subintervals of $[t_0, t_1]$ so that all axes complete the entire move in the required time. One may start all motors simultaneously and then turn them off individually as they complete their moves. On the other hand, one may want to delay the activation of each joint motor proportionally

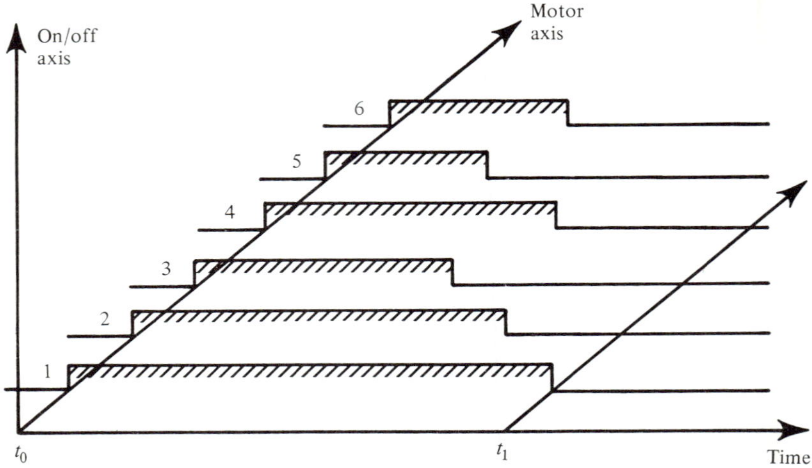

Figure 6.12.1 Bang-bang controller with start justification.

so as to have all axes finish the move simultaneously. Unfortunately, these techniques produce large current transients through the power supply and the controller. This is because all joint motors either are turned full on simultaneously or are turned from full on to off simultaneously. Another possibility is to drive each joint motor over a subinterval of $[t_0, t_1]$ such that the midpoint of each subinterval matches the midpoint of the whole interval $[t_0, t_1]$. In this manner the difficulties associated with current transients owing to large overall overshoots may be eliminated. The different possibilities for a six-axis manipulator are shown in Figure 6.12.1. We arbitrarily set $\theta_1(t_1) - \theta_1(t_0)$ to be the maximum $\Delta\theta$ traveled and thus assign the period $t_1 - t_0$ to joint motor number 1.

Finally, Figures 6.12.2 and 6.12.3 display typical situations corresponding to stop-justified coordinated movement strategies and symmetric movement strategies, respectively. As can be seen from Figure 6.12.3, various joint motors are turned on or off at different times and thus no high current transients are created. Here we can develop simple expressions for the turn on time $t_0^{(k)}$ and turn off time $t_1^{(k)}$ of the kth joint motor. These expressions are

$$t_0^{(k)} = t_0 + \left(\frac{1 - |\theta_k(t_1) - \theta_k(t_0)|/(\Delta\theta)_{\max}}{2}\right)(t_1 - t_0) \tag{6.12.3}$$

and

$$t_1^{(k)} = t_0 + \left(\frac{1 + |\theta_k(t_1) - \theta_k(t_0)|/(\Delta\theta)_{\max}}{2}\right)(t_1 - t_0). \tag{6.12.4}$$

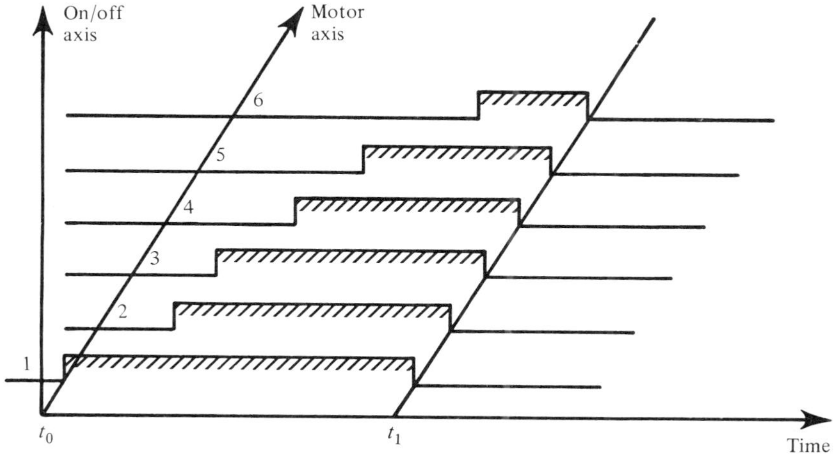

Figure 6.12.2 Bang-bang controller with stop justification.

Note that for the motor with maximum travel angle, $\theta_k(t_1) - \theta_k(t_0)$, $t_0^{(k)} = t_0$, and $t_1^{(k)} = t_1$. On the other hand, if for any joint motor $\theta_k(t_1) - \theta_k(t_0) = 0$, then the start and finish times from Equations (6.10.3) and (6.10.4) become

$$t_0^{(k)} = t_0 + \tfrac{1}{2}(t_1 - t_0) = \frac{t_1 + t_0}{2} \tag{6.12.5}$$

and

$$t_1^{(k)} = t_0 + \tfrac{1}{2}(t_1 - t_0) = \frac{t_1 + t_0}{2}, \tag{6.12.6}$$

and thus joint motor k will not move.

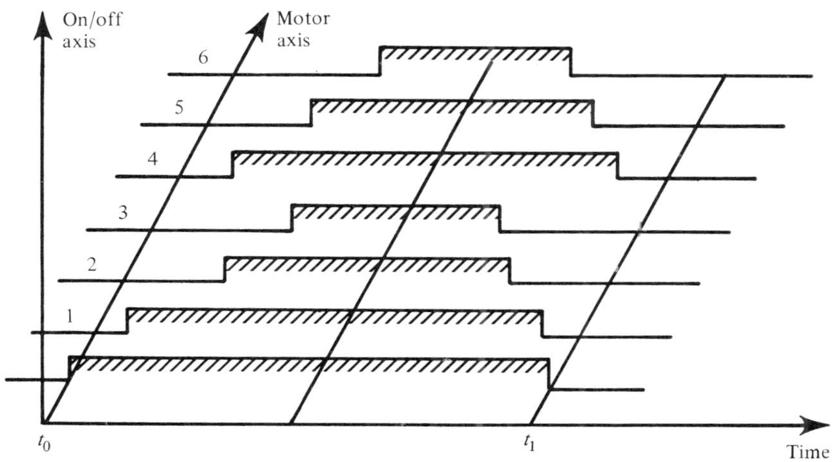

Figure 6.12.3 Bang-bang controller with symmetric movement justification.

BIBLIOGRAPHY

Bejczy, A. K. (1974). "Robot Arm Dynamics and Control," Technical Memorandum 33-669, NASA-Jet Propulsion Laboratory.

Koditschek, D. (1984). Natural Motion for Robot Arms, in *Proceedings of the 23rd Conference on Decision and Control*, Las Vegas, pp. 733–735.

Lee, C. S. G. and Chung, M. J. (1982). An Adaptive Control Strategy for Computer-based Manipulators, in *Proceedings of the 21st Conference on Decision and Control*, Tucson, pp. 95–100.

Luh, J. Y. S. (1983). Conventional Controller Design for Industrial Robots—A Tutorial, *IEEE Trans. Syst. Man Cybernet.*, **SMC 13**(3), 298–316.

Luh, J. Y. S., Walker, M. W., and Paul, R. P. (1980). Resolved Acceleration Control of Mechanical Manipulators, *IEEE Trans. Autom. Control*, **AC-25**(3), 468–474.

Markiewicz, B. R. (1973). "Analysis of the Computed Torque Drive Method and Comparison with Conventional Position Servo for a Computer-Controller Manipulator," Technical Memorandum 33-601, NASA-Jet Propulsion Laboratory.

Ogata, K. (1970). *Modern Control Engineering*, Prentice-Hall, Englewood Cliffs, NJ.

Paul, R. (1972). "Modeling, Trajectory Calculation, and Servoing of a Computer-Controller Arm," Stanford Artificial Intelligence Laboratory, A.I. Memo 177.

Paul, R. (1981). *Robot Manipulator: Mathematics, Programming and Control*, M.I.T. Press, Cambridge, MA.

Raibert, M. H., and Craig, J. J. (1981). Hybrid Position/Force Control of Manipulators, *ASME J. Dyn. Syst. Meas. Control*, **102**, 126–133.

Seraji, H., Shahinpoor, M., and Jamshidi, M. (1985). Perfect Tracking with Application to Robotics, in *Proceedings of the IASTED Symposium on Robotics*, Zurich.

Singh, S. N., and Schy, A. A. (1984). Invertibility and Robust Nonlinear Control of Robotic Systems, in *Proceedings of the 23rd Conference on Decision and Control*, Las Vegas, pp. 1058–1063.

Studenny, J., and Belanger, P. R. (1984). Robot Manipulator Control by Acceleration Feedback, in *Proceedings of the 23rd Conference on Decision and Control*, Las Vegas, pp. 1070–1072.

HOMEWORK PROBLEMS

1. What should be the transfer function $f(s)$ for the control system presented in Figure P6.1 with feedforward acceleration that would allow the error function to be zero; that is,

$$e(t) = \theta^{(d)}(t) - \theta^{(a)}(t) = 0.$$

2. Referring to the control system shown in Figure P6.2, show that the error signal is given by

$$e(s) = \theta^{(d)}(s)\left(\frac{s^2}{s^2 + k_v s + k_p}\right) - \theta^*(s)\left(\frac{1}{Js^2 + Jk_v s + Jk_p}\right).$$

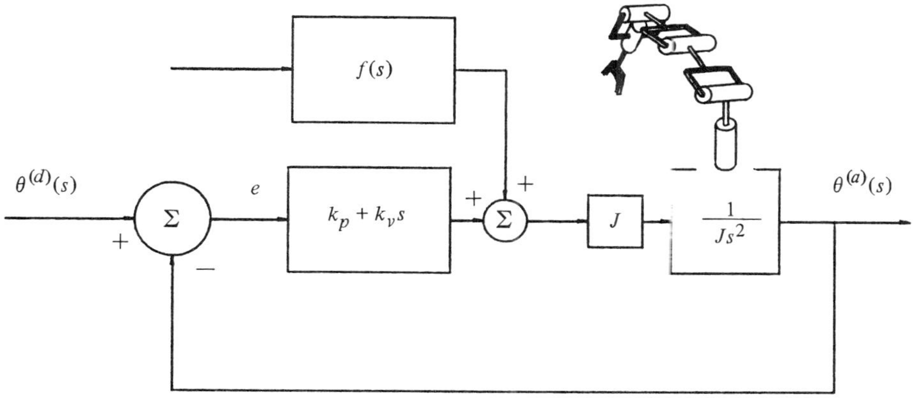

Figure P6.1 Acceleration compensation.

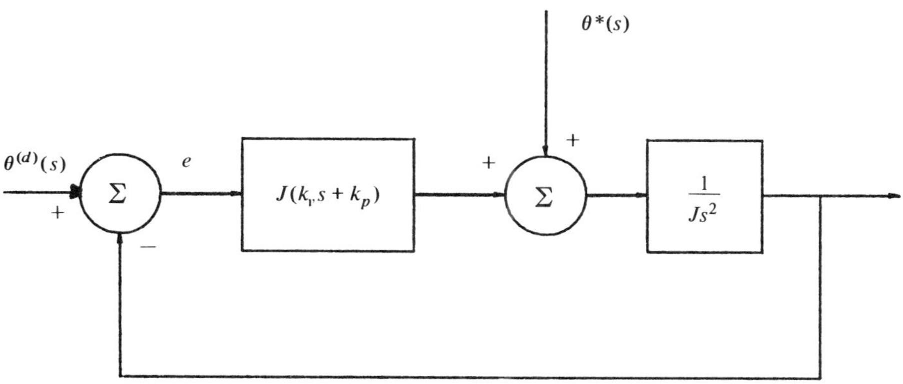

Figure P6.2 Error signal controller.

3. Consider a simple mass-spring-friction system whose governing dynamic equation is

$$m\ddot{x}(t) + b\dot{x}(t) + kx(t) = f(t).$$

To control the motion of the mass m in a straight line, we apply a position feedback control law (i.e., a servo force) of the form

$$f(t) = k_p[x_d(t) - x(t)],$$

where $x_d(t)$ is the desired position. Draw the control block diagram of the system and obtain the damping ratio and the undamped natural frequency ω_r in terms of k_p. Find the response of the system to a step as well as a ramp input $x_d(t)$. What is the response of the system under steady-state conditions?

4. For the system of Problem 3, apply the following servoforce and answer the same questions:

$$f(t) = k_p[x_d(t) - x(t)] - k_v\dot{x}(t).$$

5. For the system of Problem 3, apply the following servoforce and answer the same questions:

$$f(t) = k_p[x_d(t) - x(t)] - k_v\dot{x}(t) + k_i \int_0^t [x_d(t) - x(t)]dt.$$

6. Consider the single-link manipulator shown in Figure P6.6 having one rotational joint. Assume the mass to be concentrated at the tip and the friction to be present at the joint.

(a) Derive the governing dynamic equations.
(b) Linearize the equations about an operating point.
(c) Draw the control block diagram and find the transfer function.

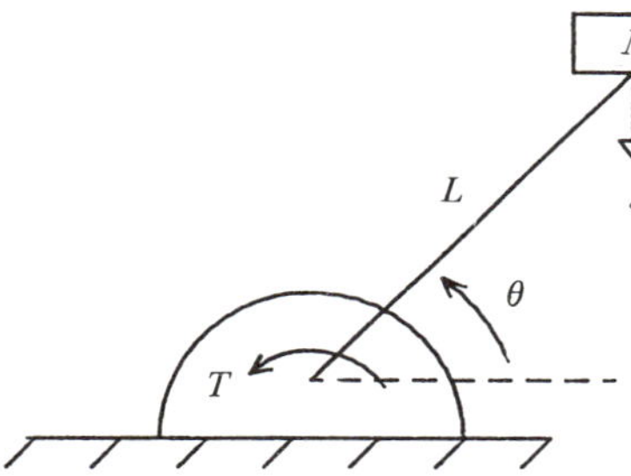

Figure P6.6 Single-link rotary manipulator.

7. In most industrial control systems, the PID-type control algorithms are used so that the system is insensitive to plant parameter variations. For the PID system depicted in Figure P6.7, show that steady-state tracking can be maintained despite plant disturbances.

8. For the control system shown in Figure P6.8, show that the transfer function $Q(s)$ of the feedforward controller may be chosen such that the tracking error is identically zero; namely,

$$E(s) = \theta_{(d)}(s) - \theta(s) = 0.$$

9. Consider the robot control system shown in Figure P6.9. Derive an expression for the tracking error $E(s)$ in terms of the desired position $\theta_{(d)}(s)$ and the disturbance torque $T^*(s)$. Find the steady-state error e_{ss} when $\theta_{(d)}$ and T^* are step functions.

10. An integral term is added to the feedback controller in Problem 7 to yield $k_p + k_v s + k_i/s$. Determine the steady-state error e_{ss} for a ramp command $\theta_{(d)}(s)$ and a step disturbance T^*. Obtain the closed-loop characteristic polynomial in terms of the gains k_p and k_v and the integral gain k_i.

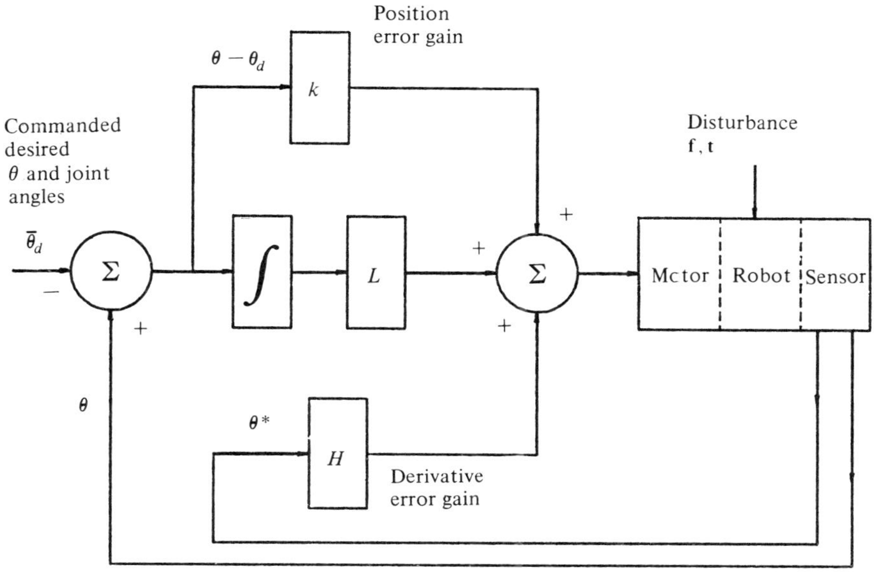

Figure P6.7 PID controller.

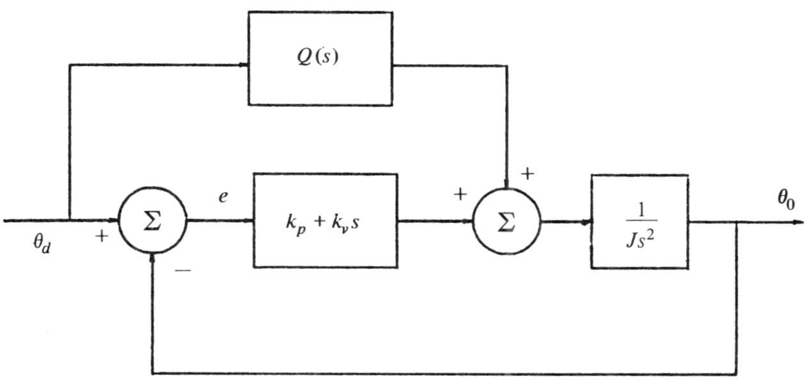

Figure P6.8 System of Problem 8.

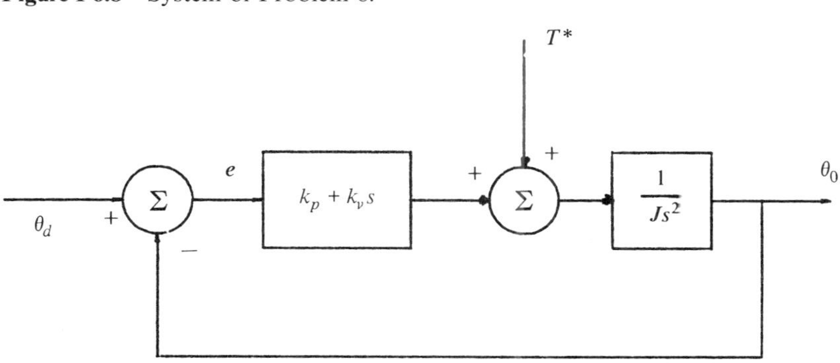

Figure P6.9 System of Problem 9.

11. Consider a nonlinear servocontrol law in the form

$$\tau(t) = (2\theta + 1)\ddot{\theta} + 4\dot{\theta}^2 - \cos\theta.$$

Discuss the conditions that will result in a critically damped system if $k_p = 8$.

12. For a generalized two-link robot, draw an appropriate block diagram of the robot control system based on the computed torque technique. Comment on the computational difficulties involved with such a technique.

13. Discuss the natural constraints on the insertion of a hexagonal peg in a hexagonal hole by sliding if the motion is to be controlled by a servoforce that is based on the motion feedback.

14. Figure P6.14 shows the block diagram of a typical positional servomechanism used in robot control applications. Obtain the closed-loop transfer function of the system, assuming the input and output are the reference shaft position $\theta_{(d)}(t)$ and the output shaft position $\theta_0(t)$, respectively. Assume the following numerical values for system parameters [1]:

k_1 = potentiometer gain = $24/\pi$ V/rad
k_p = amplifier gain = 10 V/V
R_a = armature resistance = 0.2 Ω
L_a = armature inductance ≈ 0
K_b = back-emf constant = 5.5×10^{-2} V·s/rad
K_a = motor torque constant = 6×10^{-5} lb·ft/A
J_m = motor moment of inertia = 1×10^{-5} lb·ft·s^2
f_m = motor viscous friction ≈ 0
J_l = load moment of inertia = 4.4×10^{-3} lb·ft·s^2
f_l = load viscous friction = 4×10^{-2} lb·ft/rad·s
n = gear ratio $N_1/N_2 = 1/10$

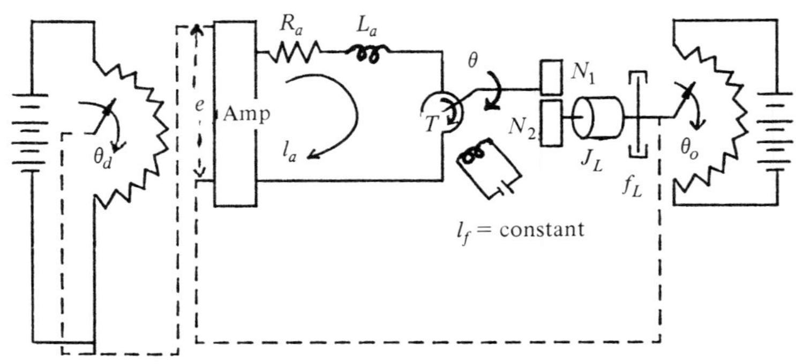

Figure P6.14 Positional servomechanism.

Chapter 7

Robotic Forces, Moments, and Compliance Analyses

7.1 INTRODUCTION

It is always necessary to know the magnitudes and directions of the various forces and moments that are at work in robotic manipulators. For one thing these forces and moments must satisfy either static or dynamic (D'Alembert) equilibria in various stages of robotic manipulations. Furthermore, the collective effect of all forces and moments must not cause any structural failure in robotic manipulators. Last, but not least, the analysis of forces and moments in robotic manipulators is absolutely essential in the understanding and control of compliance, that is, the ability of a robot to sense and react to external contact forces by modifying its trajectory. In this chapter we study the static and dynamic forces and moments, the equilibria of such forces and moments including body forces such as gravity, the force and moment transformations and the inverse static problem, the analysis of joint torques, design considerations on gripper loads, and finally the compliant motion of robotic manipulators. In the analysis of compliant motion of robotic manipulators, the subjects discussed are the force constrained motions, kinematically constrained motions, gripper stiffness control, stopping forces, exerting forces, wobbling motions, joint compliance analysis, and force sensing.

7.2 ROBOTIC FORCES AND MOMENTS: GENERALIZED PRESENTATIONS

For robotic manipulators in static or dynamic equilibrium, the forces and moments at any point can be represented by a six vector. In this sense, the first three components are the components of the force and the last three components are the components of the moments with respect to a local reference coordinate system. Let us denote the force vector and the moment vector at any point P_i by $\mathbf{f}_i$ and $\boldsymbol{\tau}_i$, respectively. The point P_i is assumed to be located in a coordinate frame i. Let us further denote the generalized six-dimensional force by $\mathbf{F}$, such that the following representation results:

$$\mathbf{F} = \begin{bmatrix} f_1 \\ f_2 \\ f_3 \\ \tau_1 \\ \tau_2 \\ \tau_3 \end{bmatrix} = [f_1 \quad f_2 \quad f_3 \quad \tau_1 \quad \tau_2 \quad \tau_3]^T. \tag{7.2.1}$$

The representation (7.2.1) of the generalized robotic forces is just a convenient and compact form. The forces and moments $\mathbf{f}_i$ and $\boldsymbol{\tau}_i$ can be considered as D'Alembert forces and moments, such that for an isolated robotic link (Fig. 7.2.1) we have the following relations:

$$\mathbf{f}_i = \mathbf{f}_{i-1,i} + \mathbf{f}_{i+1,i} + m_i\mathbf{g} \tag{7.2.2}$$

and

$$\boldsymbol{\tau}_i = \boldsymbol{\tau}_{i-1,i} + \boldsymbol{\tau}_{i+1,i} - (p_i^* + \mathbf{r}_i^*) \times \mathbf{f}_{i-1,i} - \mathbf{r}_i^* \times \mathbf{f}_{i+1,i}, \tag{7.2.3}$$

where the vectors $\mathbf{r}_i^*$ and $\mathbf{p}_i^*$ are defined, respectively, as the position vector of the center of mass of the ith link with respect to the origin of the ith coordinate system and the proximal-distal joint vector of the ith link, $\mathbf{f}_{i,j}$ is the force exerted from the ith link to the jth link, $\mathbf{g}$ is the gravitational vector, and $\boldsymbol{\tau}_{i,j}$ is the torque (moment) applied to the jth link by the ith link.

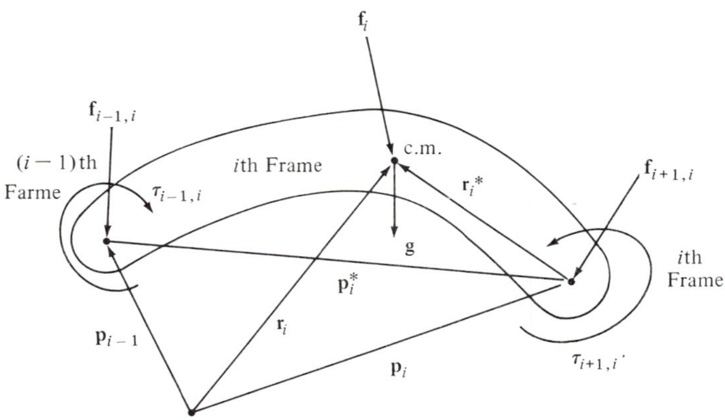

Figure 7.2.1 Forces and moments on an isolated robotic link.

Of course, the D'Alembert forces and moments $\mathbf{f}_i$ and $\boldsymbol{\tau}_i$ are given, respectively, as

$$\mathbf{f}_i = m_i \ddot{\mathbf{r}}_i \tag{7.2.4}$$

and

$$\boldsymbol{\tau}_i = \mathbf{I}_i \dot{\boldsymbol{\omega}}_i + \boldsymbol{\omega}_i \times (\mathbf{I}_i \boldsymbol{\omega}_i), \tag{7.2.5}$$

where $\ddot{\mathbf{r}}_i$ is the acceleration associated with the center of mass of the ith link $\mathbf{I}_i$ is the polar moment of inertia tensor of the ith link about its center of mass, $\boldsymbol{\omega}_i$ is the angular rotation of the ith link, and $\dot{\boldsymbol{\omega}}_i$ is the angular acceleration associated with the ith link. For static equilibrium, the net D'Alembert forces and moments are zero and thus

$$\mathbf{f}_{i-1,i} = \mathbf{f}_{i,i+1} - m_i \mathbf{g} \tag{7.2.6}$$

and

$$\boldsymbol{\tau}_{i-1,i} = \boldsymbol{\tau}_{i,i+1} + (\mathbf{p}_i^* + \mathbf{r}_i^*) \times \mathbf{f}_{i-1,i} + \mathbf{r}_i^* \times \mathbf{f}_{i+1,i}. \tag{7.2.7}$$

EXAMPLE 7.2.1 ■

For a two-axis planar manipulator, find the joint torques in terms of the tip force and moment $\mathbf{f}_{2,3}$ and $\boldsymbol{\tau}_{2,3}$ for static equilibrium.

Solution. From Equations (7.2.6) and (7.2.7) we have, respectively, for $i = 1, 2$

$$\mathbf{f}_{0,1} = \mathbf{f}_{1,2} - m_1 \mathbf{g}, \tag{7.2.8}$$

$$\mathbf{f}_{1,2} = \mathbf{f}_{2,3} - m_2 \mathbf{g}, \tag{7.2.9}$$

$$\boldsymbol{\tau}_{0,1} = \boldsymbol{\tau}_{1,2} + (\mathbf{p}_1^* + \mathbf{r}_1^*) \times \mathbf{f}_{0,1} - \mathbf{r}_1^* \times \mathbf{f}_{1,2}, \tag{7.2.10}$$

and

$$\boldsymbol{\tau}_{1,2} = \boldsymbol{\tau}_{2,3} + (\mathbf{p}_2^* + \mathbf{r}_2^*) \times \mathbf{f}_{1,2} - \mathbf{r}_2^* \times \mathbf{f}_{2,3}. \tag{7.2.11}$$

We have to find $\boldsymbol{\tau}_{0,1}$ and $\boldsymbol{\tau}_{1,2}$ in terms of $\mathbf{f}_{2,3}$. Thus, from Equations (7.2.8) and (7.2.9) we have

$$\mathbf{f}_{0,1} = \mathbf{f}_{2,3} - (m_1 + m_2)\mathbf{g}. \tag{7.2.12}$$

Now from Equation (7.2.11)

$$\boldsymbol{\tau}_{1,2} = \boldsymbol{\tau}_{2,3} + \mathbf{p}_2^* \times \mathbf{f}_{2,3} - m_2(\mathbf{p}_2^* + \mathbf{r}_2^*) \times \mathbf{g}. \tag{7.2.13}$$

From Equation (7.2.10)

$$\boldsymbol{\tau}_{0,1} = \boldsymbol{\tau}_{2,3} + (\mathbf{p}_1^* + \mathbf{p}_2^*) \times \mathbf{f}_{2,3} - m_2(\mathbf{p}_2^* + \mathbf{r}_2^*) \times \mathbf{g} - m_1(\mathbf{p}_1^* + \mathbf{r}_1^*) \times \mathbf{g} - m_2 \mathbf{p}_1^* \times \mathbf{g}. \quad ■■ \tag{7.2.14}$$

EXAMPLE 7.2.2 ■

Derive the inverse dynamic equation for the torque $\boldsymbol{\tau}_{0,1}$ for the one-link robot (Figs. 7.2.2 and 7.2.3) in terms of the tip forces and couples $\mathbf{f}_{1,2}$ and $\boldsymbol{\tau}_{1,2}$ as well as other pertinent parameters.

Solution. We note that in this case

$$\mathbf{p}_1^* = l_1\begin{bmatrix}\cos\theta_1\\ \sin\theta_1\end{bmatrix}, \tag{7.2.15}$$

$$\mathbf{r}_1 = \alpha_1\mathbf{p}_1^* = \alpha_1 l_1\begin{bmatrix}\cos\theta_1\\ \sin\theta_1\end{bmatrix}, \tag{7.2.16}$$

$$\mathbf{r}_1^* = (\alpha_1 - 1)\mathbf{p}_1^* \tag{7.2.17}$$

and

$$\mathbf{p}_1^* + \mathbf{r}_1^* = \alpha_1\mathbf{p}_1^*. \tag{7.2.18}$$

Thus, the forces and couples can be calculated from Equations (7.2.6)–(7.2.15) as

$$\mathbf{f}_1 = m_1\ddot{\mathbf{r}}_1 = \mathbf{f}_{0,1} - \mathbf{f}_{1,2} + m\mathbf{g} \Rightarrow \mathbf{f}_{0,1} = m_1\ddot{\mathbf{r}}_1 + \mathbf{f}_{1,2} + m\mathbf{g}, \tag{7.2.19}$$

$$\boldsymbol{\tau}_1 = I_1\ddot{\boldsymbol{\theta}}_1 = \boldsymbol{\tau}_{0,1} - \tau_{1,2} - \alpha_1\mathbf{p}_1^* \times \mathbf{f}_{0,1} + (\alpha_1 - 1)\mathbf{p}_1^* \times \mathbf{f}_{1,2}, \tag{7.2.20}$$

$$\begin{aligned}\tau_{0,1} = \tau_{1,2} &+ \alpha_1 l_1(f_{0,1y}\cos\theta_1 - f_{0,1x}\sin\theta_1)\\ &- (\alpha_1 - 1)l_1(f_{1,2y}\cos\theta_1 - f_{1,2y}\sin\theta_1) + I_1\ddot{\theta}_1.\end{aligned} \tag{7.2.21}$$

Now

$$f_{0,1x} = m_1\alpha_1 l_1(-\ddot{\theta}_1\sin\theta_1 - \dot{\theta}_1^2\cos\theta_1) + f_{1,2x} \tag{7.2.22}$$

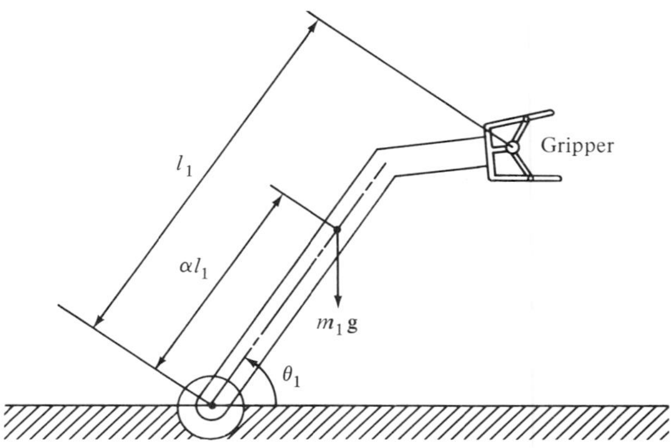

Figure 7.2.2 One-link robotic manipulator.

Figure 7.2.3 Robotic assembly requiring forces and torques. (Courtesy of Cincinnati Milacron.)

and

$$f_{0,1y} = m_1\alpha_1 l_1(\ddot{\theta}_1 \cos\theta_1 - \dot{\theta}_1^2 \sin\theta_1) + f_{1,2y} - mg_2. \qquad (7.2.23)$$

Therefore,

$$\begin{aligned}\tau_{0,1} &= \tau_{1,2} + \alpha_1 l_1(f_{1,2y}\cos\theta_1 - f_{1,2x}\sin\theta_1) + m_1\alpha_1^2 l_1^2\ddot{\theta}_1 \\ &\quad - \alpha_1 l_1 m g_2 C_1 - (\alpha_1 - 1)l_1(f_{1,2y}\cos\theta_1 - f_{1,2x}\sin\theta_1) + I_1\ddot{\theta}_1\end{aligned} \qquad (7.2.24)$$

and

$$\begin{aligned}\tau_{0,1} &= \tau_{1,2} + l_1 f_{1,2y}\cos\theta_1 - l_1 f_{1,2x}\sin\theta_1 + m_1\alpha_1^2 l_1^2\ddot{\theta}_1 \\ &\quad - \alpha_1 l_1 m g_2 \cos\theta_1 + I_1\ddot{\theta}_1.\end{aligned} \qquad (7.2.25)$$

7.3 FORCE AND MOMENT TRANSFORMATIONS

One of the fundamental problems in robotic force analysis is the determination of forces and couples experienced at any point by the robotic manipulator given the system of forces at a particular point. For example, if the gripper of a robot is experiencing a set of forces and couples, what are the forces and couples experienced at the base of the robot? To answer such questions, we resort to the principle of frame indifference of expressions for work performed by various sets of generalized forces in a robotic manipulator. Later we resort to the principle of conservation of energy in conjunction with the D'Alembert principle to relate the forces and the joint torques. Here we note that the work performed by a generalized force ${}^A\mathbf{F}$ expressed in a coordinate frame A is given by

$$ {}^AW = {}^A\mathbf{F}^T\,{}^A\mathbf{V}, \tag{7.3.1}$$

where ${}^A\mathbf{V}$ is a generalized velocity vector such that

$$ {}^A\mathbf{V} = [{}^Av_x \quad {}^Av_y \quad {}^Av_z \quad {}^Aw_x \quad {}^Aw_y \quad {}^Aw_z]^T \tag{7.3.2}$$

as discussed in Chapter 4. Note that this work must be invariant with respect to coordinate transformation; that is, the same work expressed with respect to another coordinate frame B must remain the same. Thus,

$$ {}^BW = {}^B\mathbf{F}^T\,{}^B\mathbf{V} = {}^AW = {}^A\mathbf{F}^T\,{}^A\mathbf{V}, \tag{7.3.3}$$

where ${}^B\mathbf{F}$ is the equivalent of ${}^A\mathbf{F}$ expressed in the B frame and ${}^B\mathbf{V}$ is the generalized velocity expressed in the B frame such that

$$ {}^B\mathbf{V} = [{}^Bv_x \quad {}^Bv_y \quad {}^Bv_z \quad {}^Bw_x \quad {}^Bw_y \quad {}^Bw_z]^T. \tag{7.3.4}$$

However, as discussed in Chapter 3, the two generalized velocities ${}^A\mathbf{V}$ and ${}^B\mathbf{V}$ are related by the Jacobian of transformation between the frames A and B, such that

$$ {}^B\mathbf{V} = {}^B\mathbf{J}_A\,{}^A\mathbf{V}, \tag{7.3.5}$$

where ${}^B\mathbf{J}_A$ is the Jacobian of transformation from the A frame to the B frame and is given explicitly by the following 6×6 matrix:

$$ {}^B\mathbf{J}_A = \begin{bmatrix} n_x & n_y & n_z & (\mathbf{p}\times\mathbf{n})_x & (\mathbf{p}\times\mathbf{n})_y & (\mathbf{p}\times\mathbf{n})_z \\ o_x & o_y & o_z & (\mathbf{p}\times\mathbf{o})_x & (\mathbf{p}\times\mathbf{o})_y & (\mathbf{p}\times\mathbf{o})_z \\ a_x & a_y & a_z & (\mathbf{p}\times\mathbf{a})_x & (\mathbf{p}\times\mathbf{a})_y & (\mathbf{p}\times\mathbf{a})_z \\ 0 & 0 & 0 & n_x & n_y & n_z \\ 0 & 0 & 0 & o_x & o_y & o_z \\ 0 & 0 & 0 & a_x & a_y & a_z \end{bmatrix}, \tag{7.3.6}$$

where $\mathbf{n}$, $\mathbf{o}$, $\mathbf{a}$, and $\mathbf{p}$ are, respectively, the normal, sliding, approach, and position vectors of the B with respect to the A frame, such that frame B with respect to frame A is described by

$$ {}^A\mathbf{B} = \begin{bmatrix} n_x & o_x & a_x & p_x \\ n_y & o_y & a_y & p_y \\ n_z & o_z & a_z & p_z \\ 0 & 0 & 0 & 1 \end{bmatrix}. \tag{7.3.7}$$

From Equations (7.3.3), (7.3.5), and (7.3.6) we obtain

$$^{B}\mathbf{F}^{T\,B}\mathbf{J}_A{}^{A}\mathbf{V} = {}^{A}\mathbf{F}^{T\,A}\mathbf{V} = {}^{A}\mathbf{F}^{T}\mathbf{I}\,{}^{A}\mathbf{V} \tag{7.3.8}$$

or

$$({}^{B}\mathbf{F}^{T\,B}\mathbf{J}_A - {}^{A}\mathbf{F}^{T}\mathbf{I}){}^{A}\mathbf{V} = 0. \tag{7.3.9}$$

Equation (7.3.9) must hold for all arbitrary generalized velocities ${}^{A}\mathbf{V}$ and thus

$$^{B}\mathbf{F}^{T\,B}\mathbf{J}_A = {}^{A}\mathbf{F}^{T} \tag{7.3.10}$$

or

$$^{A}\mathbf{F} = {}^{B}\mathbf{J}_A^{T\,B}\mathbf{F}. \tag{7.3.11}$$

In expanded form this equation reads

$$\begin{bmatrix} {}^{A}f_x \\ {}^{A}f_y \\ {}^{A}f_z \\ {}^{A}\tau_x \\ {}^{A}\tau_y \\ {}^{A}\tau_z \end{bmatrix} = \begin{bmatrix} n_x & o_x & a_x & 0 & 0 & 0 \\ n_y & o_y & a_y & 0 & 0 & 0 \\ n_z & o_z & a_z & 0 & 0 & 0 \\ (\mathbf{p}\times\mathbf{n})_x & (\mathbf{p}\times\mathbf{o})_x & (\mathbf{p}\times\mathbf{a})_x & n_x & o_x & a_x \\ (\mathbf{p}\times\mathbf{n})_y & (\mathbf{p}\times\mathbf{o})_y & (\mathbf{p}\times\mathbf{a})_y & n_y & o_y & a_y \\ (\mathbf{p}\times\mathbf{n})_z & (\mathbf{p}\times\mathbf{o})_z & (\mathbf{p}\times\mathbf{a})_z & n_a & o_z & a_z \end{bmatrix} \begin{bmatrix} {}^{B}f_x \\ {}^{B}f_y \\ {}^{B}f_z \\ {}^{B}\tau_x \\ {}^{B}\tau_y \\ {}^{B}\tau_z \end{bmatrix}. \tag{7.3.12}$$

Equation (7.3.12) may be inverted to yield

$$\begin{bmatrix} {}^{B}f_x \\ {}^{B}f_y \\ {}^{B}f_z \\ {}^{B}\tau_x \\ {}^{B}\tau_y \\ {}^{B}\tau_z \end{bmatrix} = \begin{bmatrix} n_x & n_x & n_x & 0 & 0 & 0 \\ o_x & o_y & o_z & 0 & 0 & 0 \\ a_x & a_y & a_z & 0 & 0 & 0 \\ (\mathbf{p}\times\mathbf{n})_x & (\mathbf{p}\times\mathbf{n})_y & (\mathbf{p}\times\mathbf{o})_z & n_x & n_y & n_z \\ (\mathbf{p}\times\mathbf{o})_a & (\mathbf{p}\times\mathbf{o})_y & (\mathbf{p}\times\mathbf{o})_z & o_x & o_y & o_z \\ (\mathbf{p}\times\mathbf{a})_x & (\mathbf{p}\times\mathbf{a})_y & (\mathbf{p}\times\mathbf{a})_z & a_x & a_y & a_z \end{bmatrix} \begin{bmatrix} {}^{A}f_x \\ {}^{A}f_y \\ {}^{A}f_z \\ {}^{A}\tau_y \\ {}^{A}\tau_z \\ {}^{A}\tau_z \end{bmatrix}. \tag{7.3.13}$$

The matrix equation (7.3.13) can be separated into the following individual relations for the computation of equivalent forces in various coordinate systems:

$$^{B}f_x = \mathbf{n}\cdot{}^{A}\mathbf{f}, \tag{7.3.14}$$

$$^{B}f_y = \mathbf{o}\cdot{}^{A}\mathbf{f}, \tag{7.3.15}$$

$$^{B}f_z = \mathbf{a}\cdot{}^{A}\mathbf{f}, \tag{7.3.16}$$

$$^{B}\tau_x = {}^{A}\mathbf{f}\cdot(\mathbf{p}\times\mathbf{n}) + \mathbf{n}\cdot{}^{A}\boldsymbol{\tau}, \tag{7.3.17}$$

$$^{B}\tau_y = {}^{A}\mathbf{f}\cdot(\mathbf{p}\times\mathbf{o}) + \mathbf{o}\cdot{}^{A}\boldsymbol{\tau}, \tag{7.3.18}$$

and

$$^{B}\tau_z = {}^{A}\mathbf{f}\cdot(\mathbf{p}\times\mathbf{a})\mathbf{a}\cdot{}^{A}\boldsymbol{\tau}. \tag{7.3.19}$$

We may simplify Equations (7.3.17)–(7.3.19) further rewrite them in the following forms:

$$^{B}\tau_x = \mathbf{n}\cdot[(^{A}\mathbf{f}\times\mathbf{p}) + {}^{A}\boldsymbol{\tau}], \tag{7.3.20}$$

$$^{B}\tau_y = \mathbf{o}\cdot[(^{A}\mathbf{f}\times\mathbf{p}) + {}^{A}\boldsymbol{\tau}], \tag{7.3.21}$$

and

$$^{B}\tau_z = \mathbf{a}\cdot[(^{A}\mathbf{f}\times\mathbf{p}) + {}^{A}\boldsymbol{\tau}], \tag{7.3.22}$$

which are computationally easier to handle.

EXAMPLE 7.3.1 ■

The following forces and moments **f** and τ are experienced at the base of a robot in static equilibrium:

$$\mathbf{f} = 5\mathbf{i} + 6\mathbf{j} + 7\mathbf{k} \tag{7.3.23}$$

and

$$\tau = 10\mathbf{i} + 20\mathbf{j} + 30\mathbf{k}. \tag{7.3.24}$$

What are the equivalent force and moment $^{A}\mathbf{f}$ and $^{A}\tau$ at the gripper if it is located in a frame **A** such that

$$\mathbf{A} = \begin{bmatrix} 0 & 0 & 1 & 20 \\ 1 & 0 & 0 & 30 \\ 0 & 1 & 0 & 0 \\ 0 & 0 & 0 & 1 \end{bmatrix}. \tag{7.3.25}$$

Solution. From the **A** matrix we find that

$$\mathbf{n} = \begin{bmatrix} 0 \\ 1 \\ 0 \end{bmatrix}, \tag{7.3.26}$$

$$\mathbf{o} = \begin{bmatrix} 0 \\ 0 \\ 1 \end{bmatrix}, \tag{7.3.27}$$

$$\mathbf{a} = \begin{bmatrix} 1 \\ 0 \\ 0 \end{bmatrix}, \tag{7.3.28}$$

and

$$\mathbf{p} = \begin{bmatrix} 20 \\ 30 \\ 0 \\ 1 \end{bmatrix}. \tag{7.3.29}$$

From expressions (7.3.14)–(7.3.16) and (7.3.20)–(7.3.22) we have

$$^{A}f_x = \mathbf{n} \cdot \mathbf{f}, \tag{7.3.30}$$

$$^{A}f_y = \mathbf{o} \cdot \mathbf{f}, \tag{7.3.31}$$

$$^{A}f_z = \mathbf{a} \cdot \mathbf{f}, \tag{7.3.32}$$

$$^{A}\tau_x = \mathbf{n} \cdot [(\mathbf{f} \times \mathbf{p}) + \boldsymbol{\tau}], \tag{7.3.33}$$

$$^{A}\tau_y = \mathbf{o} \cdot [(\mathbf{f} \times \mathbf{p}) + \boldsymbol{\tau}], \tag{7.3.34}$$

and

$$^{A}\tau_z = \mathbf{a} \cdot [(\mathbf{f} \times \mathbf{p}) + \boldsymbol{\tau}]. \tag{7.3.35}$$

Now

$$\mathbf{f} \times \mathbf{p} = \begin{bmatrix} \mathbf{i} & \mathbf{j} & \mathbf{k} \\ 5 & 6 & 7 \\ 20 & 30 & 0 \end{bmatrix} = \mathbf{i}(-210) - \mathbf{j}(-140) + \mathbf{k}(150 - 120) \tag{7.3.36}$$

or

$$\mathbf{f} \times \mathbf{p} = -210\mathbf{i} + 140\mathbf{j} + 30\mathbf{k} \tag{7.3.37}$$

and

$$(\mathbf{f} \times \mathbf{p}) + \boldsymbol{\tau} = -200\mathbf{i} + 160\mathbf{j} + 60\mathbf{k}. \tag{7.3.38}$$

Thus,

$$^{A}f_x = 6, \quad ^{A}f_y = 7, \quad \text{and} \quad ^{A}f_z = 5 \tag{7.3.39}$$

or

$$^{A}\mathbf{f} = 6\mathbf{i} + 7\mathbf{j} + 5\mathbf{k}. \tag{7.3.40}$$

$$^{A}\tau_x = 160, \quad ^{A}\tau_y = 60, \quad \text{and} \quad ^{A}\tau_z = -200 \tag{7.3.41}$$

or

$$^{A}\boldsymbol{\tau} = 160\mathbf{i} + 60\mathbf{j} - 200\mathbf{k}. \quad ■■ \tag{7.3.42}$$

EXAMPLE 7.3.2 ■

The gripper force and moment $^{G}\mathbf{f}$ and $^{G}\boldsymbol{\tau}$ of a robotic manipulator in static equilibrium are given by

$$^{G}\mathbf{f} = 10\mathbf{i} + 7\mathbf{j} + 8\mathbf{k} \tag{7.3.43}$$

and

$$^{G}\boldsymbol{\tau} = 200\mathbf{i} + 100\mathbf{j} - 100\mathbf{k}. \tag{7.3.44}$$

The gripper frame **G** with respect to the $\mathbf{T}_0^6$ frame is given by

$$^{T_0^6}\mathbf{G} = \begin{bmatrix} 1 & 0 & 0 & 1 \\ 0 & 1 & 0 & 2 \\ 0 & 0 & 1 & 3 \\ 0 & 0 & 0 & 1 \end{bmatrix}. \tag{7.3.45}$$

The $\mathbf{T}_0^6$ frame with respect to the base frame is given by

$$\mathbf{T}_0^6 = \begin{bmatrix} 0 & 0 & 1 & 10 \\ 1 & 0 & 0 & 20 \\ 0 & 1 & 0 & 10 \\ 0 & 0 & 0 & 1 \end{bmatrix}. \tag{7.3.46}$$

Find the equivalent force **f** and the equivalent moment $\boldsymbol{\tau}$ at the base of the robot due to gripper forces and moment.

Solution. Note that in this case we have

$$^{\text{Base}}\mathbf{G} = \mathbf{T}_0^6 \, {}^{T_0^6}\mathbf{G} = \begin{bmatrix} 0 & 0 & 1 & 10 \\ 1 & 0 & 0 & 20 \\ 0 & 1 & 0 & 10 \\ 0 & 0 & 0 & 1 \end{bmatrix} \begin{bmatrix} 1 & 0 & 0 & 1 \\ 0 & 1 & 0 & 2 \\ 0 & 0 & 1 & 3 \\ 0 & 0 & 0 & 1 \end{bmatrix} \tag{7.3.47}$$

or

$$^{\text{Base}}\mathbf{G} = \begin{bmatrix} 0 & 0 & 1 & 13 \\ 1 & 0 & 0 & 21 \\ 0 & 1 & 0 & 12 \\ 0 & 0 & 0 & 1 \end{bmatrix}. \tag{7.3.48}$$

Since we are to find the forces at the base, we need to find an expression for the base frame with respect to the gripper frame. This is obviously the inverse of $^{\text{Base}}\mathbf{G}$ or

$$^{G}\mathbf{Base} = (^{\text{Base}}\mathbf{G})^{-1} = \begin{bmatrix} 0 & 1 & 0 & -21 \\ 0 & 0 & 1 & -12 \\ 1 & 0 & 0 & -13 \\ 0 & 0 & 0 & 1 \end{bmatrix}. \tag{7.3.49}$$

Thus,

$$\mathbf{n} = 0\mathbf{i} + 0\mathbf{j} + 1\mathbf{k}, \tag{7.3.50}$$

$$\mathbf{o} = 1\mathbf{i} + 0\mathbf{j} + 0\mathbf{k}, \tag{7.3.51}$$

$$\mathbf{a} = 0\mathbf{i} + 1\mathbf{j} + 0\mathbf{k}, \tag{7.3.52}$$

and

$$\mathbf{p} = -21\mathbf{i} - 12\mathbf{j} - 13\mathbf{k}. \tag{7.3.53}$$

Now from expressions (7.3.14)-(7.3.16) and (7.3.20)-(7.3.22),

$$^{B}f_x = \mathbf{n}\cdot\mathbf{f} = 8, \tag{7.3.54}$$

$$^{B}f_y = \mathbf{o}\cdot\mathbf{f} = 10, \tag{7.3.55}$$

$$^{B}f_z = \mathbf{a}\cdot\mathbf{f} = 7, \tag{7.3.56}$$

$$\mathbf{f}\times\mathbf{p} = \begin{bmatrix} \mathbf{i} & \mathbf{j} & \mathbf{k} \\ 10 & 7 & 8 \\ -21 & -12 & -13 \end{bmatrix} = 5\mathbf{i} - 38\mathbf{j} + 27\mathbf{k}, \tag{7.3.57}$$

$$\mathbf{f}\times\mathbf{p} + \boldsymbol{\tau} = 205\mathbf{i} + 62\mathbf{j} - 73\mathbf{k}, \tag{7.3.58}$$

and

$$^{B}\tau_x = -73, \qquad ^{B}\tau_y = 205, \qquad ^{B}\tau_z = 62 \tag{7.3.59}$$

Thus,

$$^{B}\mathbf{f} = 8\mathbf{i} + 10\mathbf{j} + 7\mathbf{k} \tag{7.3.60}$$

and

$$^{B}\boldsymbol{\tau} = -73\mathbf{i} + 205\mathbf{j} + 62\mathbf{k}. \quad \blacksquare\blacksquare \tag{7.3.61}$$

7.4 GENERALIZED TIP FORCES AND EQUIVALENT JOINT TORQUES

In this section we employ the principle of conservation of energy to relate the generalized gripper (tip) forces to the joint torques that correspond to such forces. Here, by equivalence, it is implied that the D'Alembert joint torques necessary to compensate for gravitational shift owing to the weight of the robot itself is not included in the expressions for joint torques. The generalized expression for joint torques is given by Equation (7.2.3) whereby gravitational effects are included. Note that the rate of work performed by generalized gripper force $^{G}\mathbf{F}$ is $^{G}\mathbf{F}^{T}\,{}^{G}\mathbf{V}$. This is set equal to the total rate of work performed by all joint motors in the form $\boldsymbol{\tau}^{T}\dot{\boldsymbol{\theta}}$, where $\boldsymbol{\tau}$ is the vector of the joint torques and $\dot{\boldsymbol{\theta}}$ is the vector of the joint rotations. Thus,

$$^{G}\mathbf{F}^{T}\,{}^{G}\mathbf{V} = \boldsymbol{\tau}^{T}\dot{\boldsymbol{\theta}} = \mathbf{F}^{T}\mathbf{V}, \tag{7.4.1}$$

where $\mathbf{F}$ and $\mathbf{V}$ are the transformed $^{G}\mathbf{F}$ and $^{G}\mathbf{V}$ with respect to the base coordinate. Again, as discussed in Chapter 4, the two generalized vectors $^{G}\mathbf{V}$ and $\boldsymbol{\theta}$ are related by the Jacobian of transformation between the tip velocities and the joint rotations, such that

$$\mathbf{V} = \mathbf{J}\dot{\boldsymbol{\theta}}. \tag{7.4.2}$$

J generally is given by the following 6 × 6 matrix:

$$\mathbf{J} = \begin{bmatrix} d_{1x} & d_{2x} & d_{3x} & d_{4x} & d_{5x} & d_{6x} \\ d_{1y} & d_{2y} & d_{3y} & d_{4y} & d_{5y} & d_{6y} \\ d_{1z} & d_{2z} & d_{3z} & d_{4z} & d_{5z} & d_{6z} \\ \delta_{1x} & \delta_{2x} & \delta_{3x} & \delta_{4x} & \delta_{5x} & \delta_{6x} \\ \delta_{1y} & \delta_{2y} & \delta_{3y} & \delta_{4y} & \delta_{5y} & \delta_{6y} \\ \delta_{1z} & \delta_{2z} & \delta_{3z} & \delta_{4z} & \delta_{5z} & \delta_{6z} \end{bmatrix}, \tag{7.4.3}$$

where

$$d_{ix} = \mathbf{n}_i \cdot (\boldsymbol{\delta} \times \mathbf{p}_i), \tag{7.4.4}$$

$$d_{iy} = \mathbf{o}_i \cdot (\boldsymbol{\delta} \times \mathbf{p}_i), \tag{7.4.5}$$

$$d_{iz} = \mathbf{a}_i \cdot (\boldsymbol{\delta} \times \mathbf{p}_i), \tag{7.4.6}$$

$$\delta_{ix} = \mathbf{n}_i \cdot \boldsymbol{\delta}, \tag{7.4.7}$$

$$\delta_{iy} = \mathbf{o}_i \cdot \boldsymbol{\delta}, \tag{7.4.8}$$

and

$$\delta_{iz} = \mathbf{a}_i \cdot \boldsymbol{\delta}. \tag{7.4.9}$$

In the above equations i denotes the ith joint described by a coordinate frame A_i, such that

$$\mathbf{A}_i = \begin{bmatrix} \mathbf{n}_i & \mathbf{o}_i & \mathbf{a}_i & \mathbf{p}_i \\ 0 & 0 & 0 & 0 \end{bmatrix} \tag{7.4.10}$$

and $\boldsymbol{\delta} = \mathbf{k}$. From Equations (7.4.1) and (7.4.2), we find that

$$\boldsymbol{\tau}^T \dot{\boldsymbol{\theta}} = \mathbf{F}^T \mathbf{J} \dot{\boldsymbol{\theta}}. \tag{7.4.11}$$

This equation should hold for all arbitrary joint rotations $\dot{\boldsymbol{\theta}}$ and thus

$$\boldsymbol{\tau}^T = \mathbf{F}^T \mathbf{J}. \tag{7.4.12}$$

Finally, from Equation (7.4.12), we obtain

$$\boldsymbol{\tau} = \mathbf{J}^T \mathbf{F}, \tag{7.4.13}$$

relating the joint torques to the generalized gripper force.

EXAMPLE 7.4.1 ■

For a two-link planar manipulator (Figs. 7.4.1 and 7.4.2), find the equivalent joint torques in terms of the gripper forces and couples (inverse static problem).

Solution. The tip location is characterized by x and y, such that

$$\begin{bmatrix} x \\ y \end{bmatrix} = \begin{bmatrix} l_1 \cos\theta_1 + l_2 \cos(\theta_1 + \theta_2) \\ l_1 \sin\theta_1 + l_2 \sin(\theta_1 + \theta_2) \end{bmatrix}. \tag{7.4.14}$$

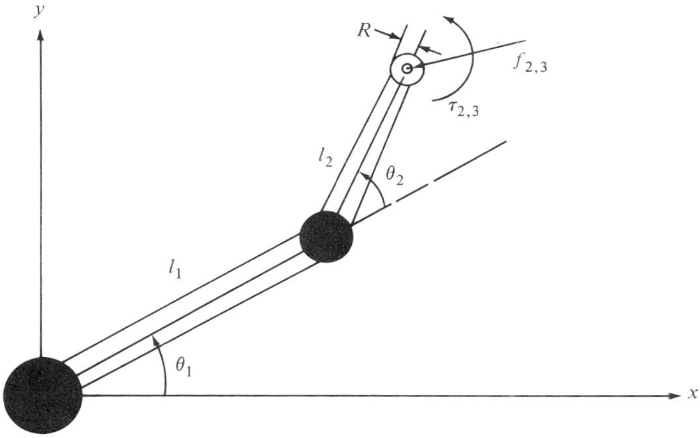

Figure 7.4.1 Two-link planar robotic manipulator with rotary joints.

Figure 7.4.2 Overhead gantry robots for manipulation of heavy loads. (Courtesy of Cincinnati Milacron.)

Differentiating the above expression with respect to time yields the following relation:

$$\begin{bmatrix} \dot{x} \\ \dot{y} \end{bmatrix} = \begin{bmatrix} -l_1 \sin \theta_1 - l_2 \sin(\theta_1 + \theta_2) - l_2 \sin(\theta_1 + \theta_2) \\ l_1 \cos \theta_1 + l_2 \cos(\theta_1 + \theta_2) l_2 \cos(\theta_1 + \theta_2) \end{bmatrix} \begin{bmatrix} \dot{\theta}_1 \\ \dot{\theta}_2 \end{bmatrix} \tag{7.4.15}$$

Note that the generalized tip velocity $\mathbf{V} = [\dot{x}\ \dot{y}]^I = \mathbf{J}\dot{\boldsymbol{\theta}}$, $w = \dot{\theta}_1 + \dot{\theta}_2$, and $\dot{\boldsymbol{\theta}} = [\dot{\theta}_1\ \dot{\theta}_2]^T$, so that the Jacobian of velocity transformation is determined completely. Thus, according to Equation (7.4.13), the vector of the joint torques $\tau_{0,1}$ and $\tau_{1,2}$ is given by

$$\tau \begin{bmatrix} \tau_{0,1} \\ \tau_{1,2} \end{bmatrix} = \mathbf{J}^T\mathbf{F} = \mathbf{J}^T \begin{bmatrix} f_{2,3x} \\ f_{2,3y} \\ \tau_{2,3} \end{bmatrix} \tag{7.4.16}$$

or

$$\begin{aligned} \tau_{0,1} = & -[(l_1 \sin \theta_1 + l_2 \sin(\theta_1 + \theta_2)]f_{2,3x} \\ & + [(l_1 \cos \theta_1 + l_2 \cos(\theta_1 + \theta_2)]f_{2,3y} + \tau_{2,3} \end{aligned} \tag{7.4.17}$$

and

$$\tau_{1,2} = -l_2 \sin(\theta_1 + \theta_2)f_{2,3x} + l_2 \cos(\theta_1 + \theta_2)f_{2,3y} + \tau_{2,3}, \tag{7.4.18}$$

where

$$\mathbf{J}^T = \begin{bmatrix} -l_1 \sin \theta_1 - l_2 \sin(\theta_1 + \theta_2) & l_1 \cos \theta_1 + l_2 \cos(\theta_1 + \theta_2) & 1 \\ -l_2 \sin(\theta_1 + \theta_2) & l_2 \cos(\theta_1 + \theta_2) & 1 \end{bmatrix}. \quad \blacksquare\blacksquare \tag{7.4.19}$$

7.5 VIRTUAL WORK AND THE D'ALEMBERT PRINCIPLE

The principle of virtual work along with the D'Alembert principle on dynamic equilibrium provide a powerful tool for the analysis of generalized robotic forces, motions, and deformations. A complete discussion on robotic deformation is beyond the scope of this chapter. These deformations may be elastic, plastic, hyperelastic, viscoelastic, or pure rigid body motions in nature. Here we present examples pertaining only to robotic rigid body motions.

Let there be N locations (joints) in a robot where forces are being applied and let us denote these forces by $\mathbf{F}_j$, $j = 1, 2, \ldots, N$. Let the associated motions be denoted by $\mathbf{q}_j$, $j = 1, 2, \ldots, N$. Then the principle of virtual work states that

$$\sum_{j=1}^{N} \mathbf{F}_j \cdot \delta\mathbf{q}_j = U, \tag{7.5.1}$$

where δ denotes the incremental change and U is the energy stored in the robot. For rigid body motions only the energy stored is equal to zero and the principle would read

$$\sum_{j=1}^{N} \mathbf{F}_j \cdot \delta \mathbf{q}_j = 0. \tag{7.5.2}$$

If one is dealing with a dynamic situation, the D'Alembert principle can be incorporated in the virtual work principle and Equation (7.5.1) becomes

$$\sum_{j=1}^{N} (\mathbf{F}_j - \mathbf{I}_j \ddot{\mathbf{q}}_j) \cdot \delta \mathbf{q}_j = U, \tag{7.5.3}$$

where $\mathbf{I}_j$ is the inertia tensor associated with the jth location on the robot.

EXAMPLE 7.5.1 ■

For the pendulum shown in Figure 7.5.1, determine θ_1 and θ_2 if P is known.

Solution.

$$\delta W = \sum_{j=1}^{3} \mathbf{F}_j \cdot \delta \mathbf{r}_j, \tag{7.5.4}$$

$$\mathbf{r}_1 = l(\sin\theta_1\, \mathbf{i} + \cos\theta_1\, \mathbf{j}), \tag{7.5.5}$$

$$\mathbf{r}_2 = l(\sin\theta_1 + \sin\theta_2)\mathbf{i} + l(\cos\theta_1 + \cos\theta_2)\mathbf{j}, \tag{7.5.6}$$

$$\mathbf{r}_3 = \mathbf{r}_2, \tag{7.5.7}$$

$$\mathbf{F}_1 = m_1 g \mathbf{j}, \tag{7.5.8}$$

$$\mathbf{F}_2 = m_2 g \mathbf{j}, \tag{7.5.9}$$

$$\mathbf{F}_3 = P\mathbf{i}, \tag{7.5.10}$$

$$\delta \mathbf{r}_1 = \frac{\partial \mathbf{r}_1}{\delta\theta_1}\delta\theta_1 + \frac{\partial \mathbf{r}_1}{\delta\theta_2}\delta\theta_2 = l(\cos\theta_1 \mathbf{i} - \sin\theta_1 \mathbf{j})\delta\theta, \tag{7.5.11}$$

$$\delta \mathbf{r}_2 = l(\cos\theta_1\, \delta\theta_1 + \cos\theta_2\, \delta\theta_2)\mathbf{i} - l(\sin\theta_1\, \delta\theta_1 - \sin\theta_2\, \delta\theta_2)\mathbf{j}, \tag{7.5.12}$$

and

$$\delta \mathbf{r}_3 = \delta \mathbf{r}_2. \tag{7.5.13}$$

Therefore,

$$\delta W = [Pl\cos\theta_1 - (m_1 + m_2)gl\sin\theta_1]\delta\theta_1 + [Pl\cos\theta_2 - m_2 gl\sin\theta_2]\delta\theta_2 = 0. \tag{7.5.14}$$

Since $\delta\theta_1$ and $\delta\theta_2$ are arbitrary,

$$Pl\cos\theta_1 - (m_1 + m_2)gl\sin\theta_1 = 0 \Rightarrow \tan\theta_1 = \frac{P}{(m_1 + m_2)g} \tag{7.5.15}$$

and

$$Pl\cos\theta_2 - m_2 gl\sin\theta_2 = 0 \Rightarrow \tan\theta_2 = \frac{P}{m_2 g}. \quad ■■ \tag{7.5.16}$$

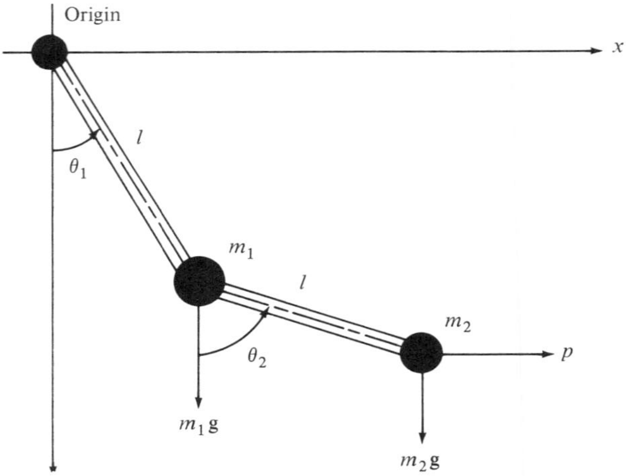

Figure 7.5.1 Double pendulum in static equilibrium.

EXAMPLE 7.5.2 ■

For the single pendulum shown in Figure 7.5.2, derive the governing equations for dynamic equilibrium using the principle of virtual work along with the D'Alembert principle.

Solution.

$$\delta W = \sum_{j=1}^{1} (\mathbf{F}_j - m_j \ddot{\mathbf{r}}_j) \cdot \delta \mathbf{r}_j = 0, \tag{7.5.17}$$

$$\mathbf{r} = \mathbf{r}_j = l \sin \theta \, \mathbf{i} + l \cos \theta \, \mathbf{j}, \tag{7.5.18}$$

$$\mathbf{F}_j = \mathbf{F} = mg\mathbf{j}, \tag{7.5.19}$$

$$\dot{\mathbf{r}} = (l \cos \theta)\dot{\theta}\mathbf{i} - (l \sin \theta)\dot{\theta}\mathbf{j}, \tag{7.5.20}$$

$$\ddot{\mathbf{r}} = [(-l \sin \theta)\dot{\theta}^2 + (l \cos \theta)\ddot{\theta}]\mathbf{i} - [(l \cos \theta)\dot{\theta}^2 + (l \sin \theta)\theta]\mathbf{j}, \tag{7.5.21}$$

$$\delta \mathbf{r} = [(l \cos \theta)\mathbf{i} - (l \sin \theta)\mathbf{j}]\delta\theta, \tag{7.5.22}$$

and

$$F - m\ddot{\mathbf{r}} = + m[(l \sin \theta)\dot{\theta}^2 - (l \cos \theta)\ddot{\theta}]\mathbf{i} + m[g + (l \cos \theta)\dot{\theta}^2 + (l \sin \theta)\ddot{\theta}]\mathbf{j}. \tag{7.5.23}$$

Therefore,

$$\delta W = [(ml^2 \sin \theta \cos \theta)\dot{\theta}^2 - (ml^2 \cos^2 \theta)\ddot{\theta} - mgl \sin \theta - (ml^2 \sin \theta \cos \theta)\dot{\theta}^2 - (ml^2 \sin^2 \theta)\ddot{\theta}]\delta\theta = 0. \tag{7.5.24}$$

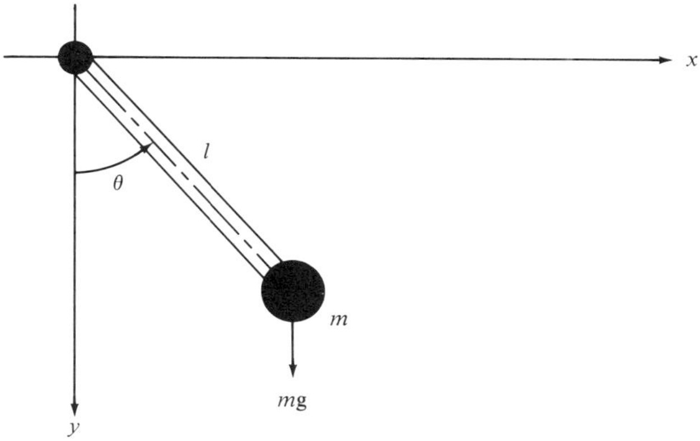

Figure 7.5.2 Simple pendulum.

This should hold for all arbitrary $\delta\theta$ and thus

$$\ddot{\theta} + \left(\frac{g}{l}\right)\sin\theta = 0. \tag{7.5.25}$$

EXAMPLE 7.5.3 ■

For the two-link robot shown in Figures 7.5.3 and 7.5.4, determine the necessary joint torques to maintain equilibrium.

Solution.

$$\delta W = \sum_{j=1}^{5} \mathbf{F}_j \cdot \delta \mathbf{q}_j, \tag{7.5.26}$$

$$\mathbf{q}_1 = \mathbf{r}_1 = l_1(\cos\theta_1\,\mathbf{i} + \sin\theta_1\,\mathbf{j}), \tag{7.5.27}$$

$$\mathbf{q}_2 = \mathbf{r}_2 = \mathbf{r}_1 + l_2[\cos(\theta_1 + \theta_2)\mathbf{i} + \sin(\theta_1 + \theta_2)\mathbf{j}], \tag{7.5.28}$$

$$\mathbf{q}_3 = \mathbf{r}_2, \tag{7.5.29}$$

$$\mathbf{q}_4 = -\theta_1\mathbf{k}, \tag{7.5.30}$$

$$\mathbf{q}_5 = -\theta_2\mathbf{k}, \tag{7.5.31}$$

$$\mathbf{F}_1 = -m_1 g\mathbf{j}, \tag{7.5.32}$$

$$\mathbf{F}_2 = -m_2 g\mathbf{j}, \tag{7.5.33}$$

$$\mathbf{F}_3 = f_{2,3x}\mathbf{i} + f_{2,3y}\mathbf{j}, \tag{7.5.34}$$

$$\mathbf{F}_4 = \tau_{0,1}\mathbf{k}, \tag{7.5.35}$$

$$\mathbf{F}_5 = \tau_{1,2}\mathbf{k}, \tag{7.5.36}$$

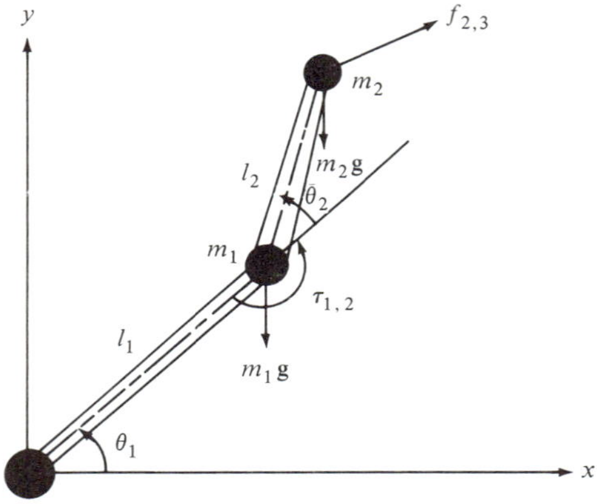

Figure 7.5.3 Two-link robotic manipulator.

Figure 7.5.4 Robotic application requiring handling of heavy loads. (Courtesy of Cincinnati Milacron.)

$$\delta \mathbf{q}_1 = \delta \mathbf{r}_1 = \frac{\partial \mathbf{r}_1}{\partial \theta_1}\delta\theta_1 + \frac{\partial \mathbf{r}_2}{\partial \theta_1}\delta\theta_2 = l_1(-\sin\theta_1\,\mathbf{i} + \cos\theta_2\,\mathbf{j})\delta\theta_1, \tag{7.5.37}$$

$$\delta \mathbf{q}_2 = \delta \mathbf{r}_2 = \frac{\partial \mathbf{r}_2}{\partial \theta_1}\delta\theta_1 + \frac{\partial \mathbf{r}_2}{\partial \theta_2}\delta\theta_2, \tag{7.5.38}$$

$$\begin{aligned}\delta \mathbf{q}_2 = \delta \mathbf{r}_2 = {} & \{[-l_1\sin\theta_1 - l_2\sin(\theta_1+\theta_2)]\delta\theta_1 - l_2\sin(\theta_1+\theta_2)\delta\theta_2\}\mathbf{i} \\ & + \{[l_1\cos\theta_1 + l_2\cos(\theta_1+\theta_2)]\delta\theta_1 \\ & + l_2\cos(\theta_1+\theta_2)\delta\theta_2\}\mathbf{j},\end{aligned} \tag{7.5.39}$$

$$\delta \mathbf{q}_3 = \delta \mathbf{r}_2, \tag{7.5.40}$$

$$\delta \mathbf{q}_4 = -\delta\theta_1\mathbf{k}, \tag{7.5.41}$$

and

$$\delta \mathbf{q}_5 = -\delta\theta_2\mathbf{k}. \tag{7.5.42}$$

Therefore, the principle of virtual work is reduced to

$$\delta W = \mathbf{F}_1\cdot\delta\mathbf{r}_1 + \mathbf{F}_2\cdot\delta\mathbf{r}_2 + \mathbf{F}_3\cdot\delta\mathbf{r}_3 - \tau_{0,1}\delta\theta_1 - \tau_{1,2}\delta\theta_2 \tag{7.5.43}$$

or

$$\begin{aligned}\delta W = {} & (-m_1 g l_1\cos\theta_1)\delta\theta_1 - [m_2 g l_1\cos\theta_1 + m_2 l_2 g\cos(\theta_1+\theta_2)]\delta\theta_1 \\ & - m_2 l_2 g\cos(\theta_1+\theta_2)\delta\theta_2 + f_{2,3x}\{[-l_1\sin\theta_1 - l_2\sin(\theta_1+\theta_2)]\delta\theta_1 \\ & - l_2\sin(\theta_1+\theta_2)\delta\theta_2\} + f_{2,3y}\{[l_1\cos\theta_1 + l_2\cos(\theta_1+\theta_2)]\delta\theta_1 \\ & + l_2\cos(\theta_1+\theta_2)\delta\theta_2\} - \tau_{0,1}\delta\theta_1 - \tau_{1,2}\delta\theta_2 = 0.\end{aligned} \tag{7.5.44}$$

Thus,

$$\begin{aligned}& \delta\theta_1\{-(m_1+m_2)gl_1\cos\theta_1 - m_2 g l_2\cos(\theta_1+\theta_2) \\ & \quad + f_{2,3x}[-l_1\sin\theta_1 - l_2\sin(\theta_1+\theta_2)] \\ & \quad + f_{2,3y}[l_1\cos\theta_1 + l_2\cos(\theta_1+\theta_2)] - \tau_{0,1}\} \\ & \quad + \delta\theta_2\{-m_2 l_2 g\cos(\theta_1+\theta_2) - f_{2,3x}l_2\sin(\theta_1+\theta_2) \\ & \quad + f_{2,3y}l_2\cos(\theta_1+\theta_2) - \tau_{1,2}\} = 0.\end{aligned} \tag{7.5.45}$$

Since $\delta\theta_1$ and $\delta\theta_2$ are arbitrary, it is concluded that

$$\tau_{1,2} = -m_2 l_2 g\cos(\theta_1+\theta_2) - f_{2,3x}l_2\sin(\theta_1+\theta_2) + f_{2,3y}l_2\cos(\theta_1+\theta_2) \tag{7.5.46}$$

and

$$\begin{aligned}\tau_{0,1} = {} & -(m_1+m_2)gl_1\cos\theta_1 - m_2 g l_2\cos(\theta_1+\theta_2) \\ & + f_{2,3x}[-l_1\sin\theta_1 - l_2\sin(\theta_1+\theta_2)] \\ & + f_{2,3y}[l_1\cos\theta_1 + l_2\cos(\theta_1+\theta_2).\end{aligned} \tag{7.5.47}$$

7.6 COMPLIANCE ANALYSIS AND COMPLIANT MOTION

Compliance is the ability of a robotic manipulator to react to contact forces and tactile stimuli throughout the motion. The most common applications of robotic manipulators are welding, pick-and-place, and spray painting. Generally, these tasks do not require highly sophisticated tactile and force sensing. However, as the robotic tasks become more complex, it will be necessary to monitor the force stimuli continuously throughout the motion of the robotic manipulator. These applications, that is, compliant motion, are particularly significant in automatic assembly and complex manufacturing processes. One of the main features of robotic motion is that it is stiff and has a blindlike response. Equipping a robot with a controllable compliance gives it the ability to sense and to react to external contact forces and subsequently to modify its trajectory according to a prescribed strategy. Thus, compliance makes the robotic manipulator to perform complex manufacturing tasks. Contact forces generally are created by the surfaces that come into contact with the gripped object rather than the gripper itself as the robot is performing its task. For most cases, the exact position or force relations between the gripped objects and the external surfaces are not known a priori. In other words, the force–displacement relations or the constitutive equations may change even if one assumes common linear relations. For example, the linear stiffness coefficient may vary from place to place as the operation proceeds. Note that the gross motion of a manipulator may have excellent repeatability (high resolution and accuracy) while the fine motion for each cycle may be entirely different in nature to suit the randomly varying fine geometry of external surfaces of the mating parts (i.e., tolerances).

EXAMPLE 7.6.1 ■

Present some examples of robotic tasks requiring compliant motion.

Solution. There are numerous examples that can be cited here. Any automatic assembly type operation requires compliant robotic motion. For example, one may cite the following specific cases:

1. Inserting a peg in a hole.
2. Wiping a pane of glass or a window pane.
3. Cutting a roast beef.
4. Tuning a door knob.
5. Placing an object on a surface.
6. Tuning a guitar or a piano.
7. Playing a musical instrument.
8. Handling delicate objects.
9. Inspecting the roughness of a surface by touching it. ■ ■

Generally, compliance is provided by either passive compliance inherent in the robot (e.g., elasticities) or proper programming of the control system to react to random tactile stimuli. The primary issue of determining the compliance is the functional specification of mechanisms responsible for compliant motion. This essentially means the mathematical analysis of force interactions between the robot and its environment and the complete specification of constraint equations. It is important to know the nature of compliant motion to identify the contact forces that are to be computer controlled during the robotic task execution. We now present the two basic ways of analyzing the compliant motion.

1. We can use explicit feedback by means of linear constitutive equations and generalized stiffness matrices, such that

$$\mathbf{f} = \mathbf{k}\boldsymbol{\delta}, \tag{7.6.1}$$

where **f** is the generalized force vector including the torques, **k** is a generalized stiffness matrix, and $\boldsymbol{\delta}$ is the generalized displacement vector that includes rotations and indicates the deviation in position and orientation of the manipulator from the prescribed path. Incidentally, the inverse of the stiffness matrix is called the *mechanical compliance* **C**, such that

$$\boldsymbol{\delta} = \mathbf{C}\mathbf{f}. \tag{7.6.2}$$

Thus, knowing the mechanical compliance **C** and having measured the contact force **f** by means of force sensors, we can determine the actual displacement of the contact surfaces from Equation (7.6.2) and proper feedback control action may be executed to correct for such displacements.

2. We also can use a hybrid controller that allows force to be commanded along certain degrees of freedom while position is commanded along the remaining degrees of freedom. Thus, mathematically, equations (7.5.1) and (7.5.2) are used simultaneously.

An important concept inherent in both of the above methods is the *center of compliance*, that is, a point in the object where the generalized force and generalized displacement are colinear and the generalized stiffness matrix is completely diagonal. This can be represented as

$$\begin{bmatrix} F_1 \\ F_2 \\ F_3 \\ \tau_1 \\ \tau_2 \\ \tau_3 \end{bmatrix} \begin{bmatrix} k_1 & 0 & 0 & 0 & 0 & 0 \\ 0 & k_2 & 0 & 0 & 0 & 0 \\ 0 & 0 & k_3 & 0 & 0 & 0 \\ 0 & 0 & 0 & k_4 & 0 & 0 \\ 0 & 0 & 0 & 0 & k_5 & 0 \\ 0 & 0 & 0 & 0 & 0 & k_6 \end{bmatrix} \begin{bmatrix} \delta_1 \\ \delta_2 \\ \delta_3 \\ \delta_4 \\ \delta_5 \\ \delta_6 \end{bmatrix}. \tag{7.6.3}$$

Generally, the explicit feedback is a simpler approach and is particularly suitable for special compliant motions, while the hybrid controller is more suitable for robotic task planning.

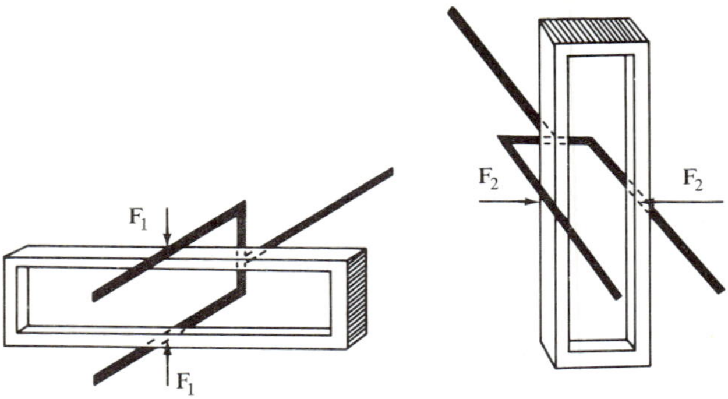

Figure 7.6.1 Gripper holding and rotating a slender object.

EXAMPLE 7.6.2 ■

At first a gripper is holding a slender object horizontally as shown in Figure 7.6.1a. Then the gripper rotates the body, which weighs W pounds, clockwise until the body is vertical (Fig. 7.6.1b). Determine the additional gripper force necessary for the object not to slip if the coefficient of friction between the object and gripper is f.

Solution. The additional gripper force $F_2 - F_1 = \delta F$ necessary to keep the part from sliding down is simply equal to the force necessary to create enough mobilized friction force to compensate for the weight W of the object. Thus,

$$W = 2f(F_2 - F_1) \tag{7.6.4}$$

or

$$F_2 - F_1 = \delta F = \frac{W}{2f}. \quad ■■ \tag{7.6.5}$$

7.7 CONCEPT OF REMOTE CENTER OF COMPLIANCE (RCC)

As discussed before, a center of compliance (COC) is a point in the workspace where the constitutive equations for force/moment–displacement/rotation relations become decoupled. Thus, the stiffness or compliance matrix of the robotic structure is diagonal at the COC. As is well known from the theory of structures, this concept is related to the center of mass/inertia as well as the fundamental frequencies of vibrations of the robotic structures. During parts assembly it is necessary to have the COC as close as possible to the initial points of contact. Under these circumstances the concept of a remote center of compliance (RCC) emerges because the robotic joints providing the necessary articulation at the tip are generally at an appreciable distance from the tip while providing the compliant motion. The compliant motion is provided by the controller sensing

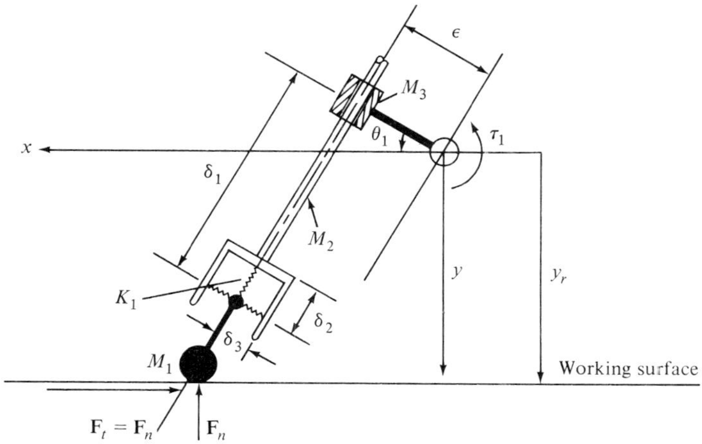

Figure 7.7.1 Compliant motion of a cylindrical polar robot on a working surface.

the magnitude and direction of the contact forces at the contact points (where presumably the COC is located) and immediately translating that into moving (away or toward) the contact surfaces in the same direction (contact force).

EXAMPLE 7.7.1 ■

For the surface-finishing robotic manipulator shown in Figure 7.7.1, the normal force F_n is not to exceed a certain amount while the tangential force F_t is to be equal to the product of the friction factor μ and the normal force F_n. Completely model the robot–surface instruction for compliant motion.

Solution. Bond graph modeling yields the following governing equations:

$$\ddot{\theta}_1 = \frac{1}{I_1(\delta_1)}[\tau_1 + F_1(\dot{\theta}_1) + \delta_2 k_1 \varepsilon + k_1 \delta_3 \delta_1], \tag{7.7.1}$$

$$\ddot{\delta}_1 = \frac{1}{(M_1 + M_2)}[F_s + F_2(\dot{\delta}_1) + k_1 \delta_2], \tag{7.7.2}$$

$$\ddot{\delta}_2 = \frac{1}{M_1}[-k_1 \delta_2 - F_n \cos\theta_1 - \mu F_n \sin\theta_1, \tag{7.7.3}$$

and

$$\ddot{\delta}_3 = \frac{1}{M_1}(-k_1 \delta_3 + F_n \sin\theta_1 + \mu F_n \cos\theta_1], \tag{7.7.4}$$

where k_1 is the stiffness of the wrist sensor, M_1, M_2, and M_3 are the masses associated with the gripper and the two other links, F_1 and F_2 are joint friction forces, τ_1 is the joint torque, F_s is the sliding joint force, δ_2 and δ_3 are the spring deflections of the wrist, δ_1 is the sliding joint displacement,

θ_1 is the joint rotation, and I_1 is the moment of inertia of the robot about the rotary joint and is the link offset. The reaction force F_n can be modeled linearly in terms of the surface stiffness k_s; namely,

$$F_n = k_s (y - y_r), \tag{7.7.5}$$

where y_r is the location of the reaction surface. ■ ■

EXAMPLE 7.7.2 ■

Analyze the planar compliant motion of a peg being inserted into a hole (Fig. 7.7.2). Determine the various distinct stages of compliant motion and the relevant governing equations.

Solution. As depicted in Figure 7.7.2, there are four distinct stages of compliant motion:

(a) peg approaching the hole;
(b) peg crossing the rounded corner of the hole;
(c) peg in a one-point contact with the wall of the hole; and
(d) peg in a two-point contact with the wall of the hole.

In each of the above distinct stages there are a number of geometrical and equilibrium equations that have to be satisfied. In part (a) as the peg is being moved toward the hole (Fig. 7.7.3) the governing equilibrium (D'Alembert) equations would be those pertaining to small vibrations of the peg about the compliance center at the wrist such that

$$M\ddot{\delta}_x - k_x\delta_x - b_x\dot{\delta}_x = 0, \tag{7.7.6}$$

$$M\ddot{\delta}_y + k_y\delta_y - b_y\dot{\delta}_y = 0, \tag{7.7.7}$$

and

$$I\ddot{\theta} + k_\theta\theta - b_y\dot{\theta} = 0, \tag{7.7.8}$$

where k_x, k_y, and k_θ are the stiffness coefficients of the wrist, while b_x, b_y, and b_θ are the linear damping coefficients of the compliant wrist. The governing geometric equation is

$$d_0 = \epsilon_0 + l\theta_0. \tag{7.7.9}$$

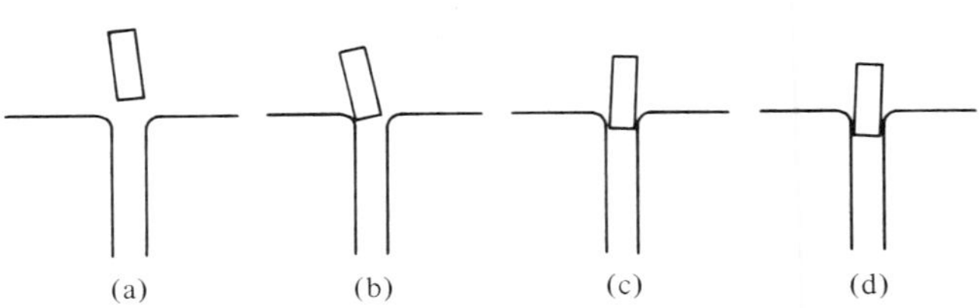

Figure 7.7.2 Planar compliant motion of a peg being inserted into a hole.

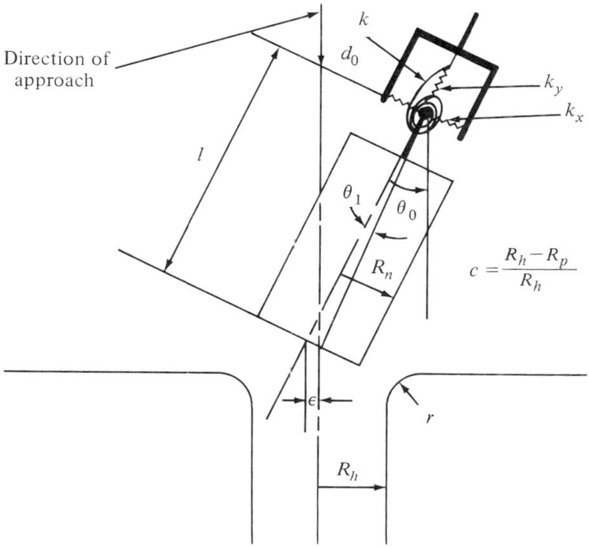

Figure 7.7.3 Peg compliantly attached to the wrist by spring actions k_x, k_y, and k_θ.

When the peg makes a one-point contact with the rounded corner (Fig. 7.7.4), we may assume a quasistatic situation and write down the governing equations for the balance of forces and couples as well as the geometric conditions.

The governing geometric relation is

$$d = l\theta + \epsilon. \tag{7.7.10}$$

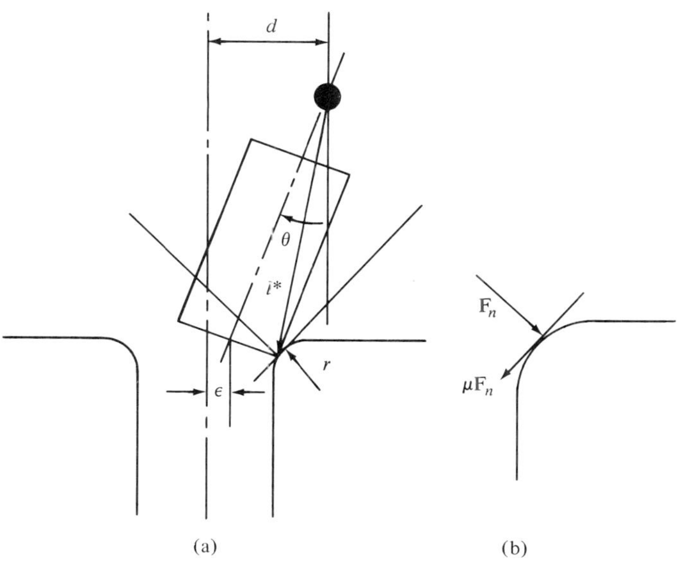

Figure 7.7.4 Peg compliantly making contact with the rounded corner.

The rounded corner experiences a normal force $\mathbf{F}_n$ and a friction force $\mathbf{F}_\mu = \mu\mathbf{F}_n$ as shown in Figure 7.7.4b. These forces must balance the forces and couples at the COC of the wrist. Thus,

$$|\mathbf{l}^* \times (\mathbf{F}_n - \mathbf{F}_\mu)| + k_\theta(\theta - \theta_0) = 0 \qquad (7.7.11)$$

and

$$\mathbf{F}_n + \mathbf{F}_\mu + k_x\boldsymbol{\delta}_x + k_y\boldsymbol{\delta}_y = 0, \qquad (7.7.12)$$

where $\mathbf{l}^*$ is the position vector extended from the COC to the point of contact. Similar relations apply to the case where there is a one-point contact with the wall of the hole. The only change is the contact position vector $\mathbf{l}^*$.

For a two-point contact the situation becomes slightly more involved, as shown in Figure 7.7.5. The balance of the forces and moments in this case becomes

$$|\mathbf{l}_1^* \times (\mathbf{F}_{n_1} - \mathbf{F}_{\mu_1})| + |\mathbf{l}_2^*(\mathbf{F}_{\mu_2} - \mathbf{F}_{n_2})| + k_\theta(\theta - \theta_0) = 0 \qquad (7.7.13)$$

and

$$\mathbf{F}_{n_1} + \mathbf{F}_{\mu_1} + \mathbf{F}_{n_2} + \mathbf{F}_{\mu_2} + k_x\boldsymbol{\delta}_x + k_y\boldsymbol{\delta}_y = 0. \qquad (7.7.14).$$

These equations can be expanded and elaborated on quite readily. ■ ■

In a similar manner one may discuss problems associated with wedging and jamming in a compliant parts assembly.

The geometric and force/couple–deformation/rotation analyses pertaining to parts assembly by robotic manipulators play an important role in the design and control of robotic manipulators. There are standard devices that provide

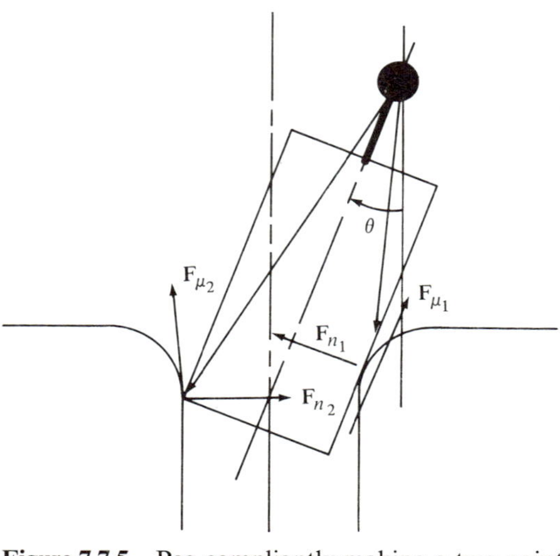

Figure 7.7.5 Peg compliantly making a two-point contact with the wall of the hole.

efficient compliant motions and avoid wedging and jamming of parts to be assembled. One well-known device is the remote center of compliance (RCC), which has the ability to reduce mating forces and prevent jamming.

7.8 SIMUNOVIC'S JAMMING CRITERION

The insertion of a peg in a hole has been studied quite extensively in the robotics literature. This is mainly because almost 35 percent of all assembly processes involve some kind of insertion operation involving pegs, round shafts, and holes. Here we present a simple analysis of a two-dimensional peg-in-hole insertion process following Simunovic's (1979) approach. The two-dimensional situation is chosen because of its simple geometry and ease of mathematical and geometrical manipulations. Certainly, a general geometry of a peg in a hole can be analyzed in a similar but more complicated manner. Here, the goal is to study the effect of friction and geometrical parameters of both the peg and the hole in the insertion process and the associated jamming phenomena. Furthermore, the goal is to obtain a set of design criteria for the creation of a passing impedance in the system to enable the forces generated to correct the error on the relative position between the parts. In the analysis that follows, we assume all processes to be quasistatic; that is, inertial forces are negligible. We further assume that the mating parts are rigid and static and the kinematic coefficients of friction are equal.

Consider Figure 7.8.1. Let $\mathbf{F} = [F_x \ F_y \ 0]$ and $\mathbf{M} = [0 \ 0 \ M_0]$ be the components of the force and the moment vectors, all with respect to a reference rectangular Cartesian frame XYZ. Note that the moment $\mathbf{M}_0$ is created by the force system $\mathbf{F}_x$, $\mathbf{F}_y$ about the Z axis because the center of action of the F_x, F_y system is not necessarily coincident with the origin of the XYZ system. The moment of $\mathbf{F}$ about any point P is $\mathbf{M}_P$, such that

$$\mathbf{M}_P = \mathbf{M} - \mathbf{OP} \times \mathbf{F}. \tag{7.8.1}$$

If $\mathbf{M}_P$ is set equal to zero, Equation (7.8.1) yields the focus of all points along which the applied force system creates no moments; that is,

$$X_P F_y - Y_P F_x = M_0, \tag{7.8.2}$$

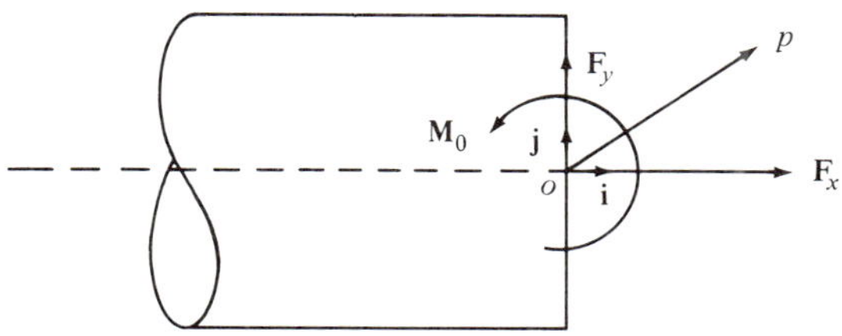

Figure 7.8.1 Structure of forces applied to a peg.

where X_p and Y_p are the X and Y coordinates of all such points P, respectively. We may nondimensionalize all variables by defining the following quantities:

$$x = \frac{X_P}{r}, \quad y = \frac{Y_P}{r}, \quad f_x = \frac{RF_x}{M_0}, \quad \text{and} \quad f_y = \frac{rF_y}{M_0}, \tag{7.8.3}$$

where r is the peg radius. Equation (7.8.2) is now reduced to

$$f_y x - f_x y = 1. \tag{7.8.4}$$

With no loss of generality we may replace $\mathbf{F}$ by $\mathbf{f} = [f_x \; f_y \; 0]$ in the subsequent analysis.

Now consider Figure 7.8.2. Let A and B be the points of contact between the peg and the wall of the hole, $\mathbf{N}_A$ and $\mathbf{N}_B$ be the normal reaction forces, and $\mathbf{T}_A$ and $\mathbf{T}_B$ be the frictional forces on these points. Let the angle between the peg axis and the hole axis be θ.

From Figure 7.8.2, it is clear that

$$N_A \geq 0, \qquad N_B \geq 0 \tag{7.8.5}$$

$$\frac{T_A}{N_A} \leq \mu_A = \tan \phi_A, \tag{7.8.6}$$

and

$$\frac{T_B}{N_B} \leq \mu_A = \tan \phi_B, \tag{7.8.7}$$

where μ_A and μ_B are the coefficients of friction at points A and B, respectively. The line of action of the force at A has a slope of s_A, such that its equation is given by

$$y_A = s_A x_A + r \tag{7.8.8}$$

or in a nondimensional form

$$y = s_A x + 1. \tag{7.8.9}$$

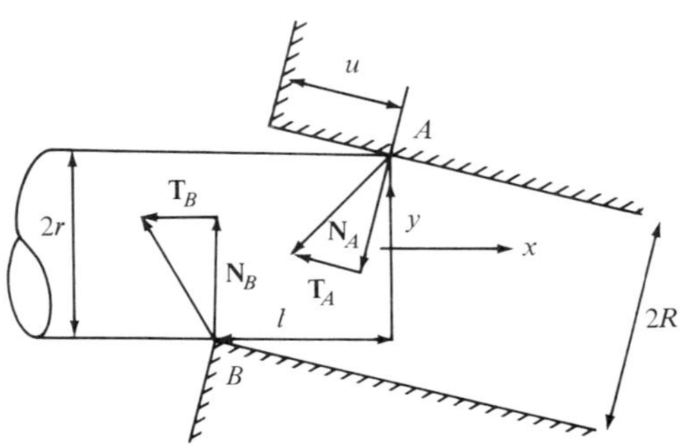

Figure 7.8.2 Structure of reaction forces.

It is clear that the slope s_A is bounded by $\cotan(\phi_A - \theta)$ and $\cotan(\phi_A + \theta)$ such that

$$\cotan(\phi_A + \theta) < s_A < \cotan(\phi_A - \theta). \tag{7.8.10}$$

Similarly, for point B

$$y + 1 = s_B\left(X + \frac{1}{r}\right), \tag{7.8.11}$$

where S_B is the slope of the line of action of $\mathbf{F}_B$, such that

$$|s_B| \geq \mu_B^{-1}. \tag{7.8.12}$$

Note that from the geometry of Figure 7.8.2

$$l^2 + (2r)^2 = U^2 + (2R)^2. \tag{7.8.13}$$

Jamming could occur if any of the contact points is outside the friction cone of the other contact point. Clearly, this can happen if

$$\frac{1}{2r} \geq \mu + C\left(\frac{1 + \mu^2}{\mu}\right), \tag{7.8.14}$$

where $C = (R - r)/R$ is the clearance to diameter ratio. Condition (7.8.14) is known as *Simunovic's jamming criterion.* (Refer to Homework Problem 11.)

BIBLIOGRAPHY

Arai, T. and Kinoshita, N. (1981). The Part Mating Forces that Arise When Using a Worktable with Compliance, *Assembly Autom.* **VI**(4), 204–210.

Drake, S. H. (1977). The Use of Compliance in a Robot Assembly System, in *IFAC Symposium*, Tokyo, Oct. 1977; "Using Compliance in Lieu of Sensory Feedback for Automatic Assembly," Ph.D. thesis, M.I.T., Mechanical Engineering Department.

Nevins, J. L. and Associates (1973–1978). "Exploratory Research in Industrial Modular Assembly," C. S. Draper Lab. Rept. No. R-764 (1973), R-800 (1974), R-850 (1975), R-996 (1976), R-1111 (1977), R-1218 (1978).

Nevins, J. L. and Whitney, D. E. (1978). Computer-Controlled Assembly, *Sci. Am.*, **238**(2), 62–74.

Nevins, J. L. and Whitney, D. E. (1980). Assembly Research, *Indus. Robot*, **7**(1), 27–43.

Ohwovoriole, M. S. (1980). "An Extension of Screw Theory and Its Application to the Automation of Industrial Assemblies," Ph.D. thesis, Stanford University, Department of Computer Science.

Simunovic, S. (1975). Force Information in Assembly Processor, in *Proceedings of the 5th International Symposium on Industrial Robots*, Vol 7, No. 1, pp. 415–431.

Simunovic, S. (1979). "An Information Approach to Parts Mating," Sc.D. thesis, M.I.T., Mechanical Engineering Department and C. S. Draper Lab. Rept. T-690.

Whitney, D. E. (1982). Quasi-static Assembly of Compliantly Supported Rigid Parts, *Trans. ASME, J. Dyn. Syst. Meas. Control*, **104**(3), 65–77.

HOMEWORK PROBLEMS

1. The following forces and moments **f** and τ are experienced at the base of a robotic manipulator in static equilibrium:

$$\mathbf{f} = 8\mathbf{i} + 16\mathbf{j} + 9\mathbf{k} \quad \text{and} \quad \boldsymbol{\tau} = 8\mathbf{i} + 3\mathbf{j} + 41\mathbf{k}.$$

What is the equivalent system of forces and couples, respectively, at the gripper if it is located in a frame **G**, such that **G** with respect to the base frame is given by

$$\mathbf{G} = \begin{bmatrix} 0 & 0 & 1 & 14 \\ 1 & 0 & 0 & 29 \\ 0 & 1 & 0 & 10 \\ 0 & 0 & 0 & 1 \end{bmatrix}.$$

2. The gripper force and moment ${}^{G}\mathbf{f}$ and ${}^{G}\boldsymbol{\tau}$ of a robotic manipulator in static equilibrium are given by

$${}^{G}\mathbf{f} = 100\mathbf{i} + 17\mathbf{j} + 18\mathbf{k} \quad \text{and} \quad {}^{G}\boldsymbol{\tau} = 20\mathbf{i} + 150\mathbf{j} - 80\mathbf{k}.$$

The gripper frame **G** with respect to the $\mathbf{T}_0^6$ frame and the $\mathbf{T}_0^6$ frame are given by

$${}^{T_0^6}\mathbf{G} = \begin{bmatrix} 1 & 0 & 0 & 1 \\ 0 & 1 & 0 & 2 \\ 0 & 0 & 1 & 1 \\ 0 & 0 & 0 & 1 \end{bmatrix} \quad \text{and} \quad \mathbf{T}_0^6 = \begin{bmatrix} 0 & 0 & 1 & 15 \\ 1 & 0 & 0 & 14 \\ 0 & 1 & 0 & 16 \\ 0 & 0 & 0 & 1 \end{bmatrix}.$$

Find the force and moment **f** and τ at the base of the robot.

3. For a PUMA 560 robotic manipulator, determine the general expressions for the force and moment vectors felt at the base if the gripper is subjected to a general force of $\mathbf{F} = [f_x\ f_y\ f_z\ \tau_x\ \tau_y\ \tau_z]^T$. (Hint: If the gripper coordinate is located with a transform ${}^{6}\mathbf{G}$ from the $\mathbf{T}_0^6$ coordinate, find the equivalent force and torque between the coordinate $\mathbf{A} = \mathbf{T}_0^6\ {}^{6}\mathbf{G}$ and the base coordinate. Express your results in terms of the joint angles θ_i as well as other structural parameters.)

4. The gripper force and moment ${}^{G}\mathbf{f}$ and ${}^{G}\boldsymbol{\tau}$ of a robot in static equilibrium are given by

$${}^{G}\mathbf{f} = 10\mathbf{i} + 7\mathbf{j} + 8\mathbf{k} \quad \text{and} \quad {}^{G}\boldsymbol{\tau} = 200\mathbf{i} + 100\mathbf{j} - 100\mathbf{k}.$$

The gripper frame **G** with respect to the $\mathbf{T}_0^6$ frame and the $\mathbf{T}_0^6$ frame are given by

$${}^{T_0^6}\mathbf{G} = \begin{bmatrix} 1 & 0 & 0 & 1 \\ 0 & 1 & 0 & 2 \\ 0 & 0 & 1 & 3 \\ 0 & 0 & 0 & 1 \end{bmatrix} \quad \text{and} \quad \mathbf{T}_0^6 = \begin{bmatrix} 0 & 0 & 1 & 10 \\ 1 & 0 & 0 & 20 \\ 0 & 1 & 0 & 10 \\ 0 & 0 & 0 & 1 \end{bmatrix}.$$

Find the force and moment **f** and τ at the base of the robot.

5. Derive the inverse dynamics equation for the torque $\tau_{0,1}$ for the one-link robot (Fig. P7.5) in terms of the tip forces and couples $\mathbf{f}_{1,2}$ and $\boldsymbol{\tau}_{1,2}$ as well as other pertinent parameters.

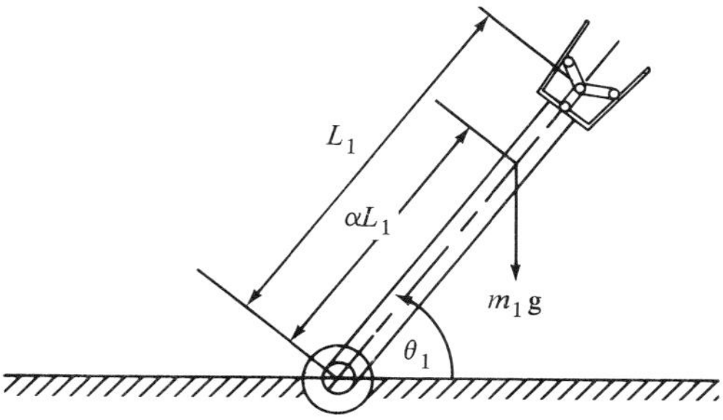

Figure P7.5

6. The following forces and moments $\mathbf{F}$ and τ are applied to the base of a robot in static equilibrium:

$$\mathbf{F} = 5\mathbf{i} + 6\mathbf{j} + 7\mathbf{k} \quad \text{and} \quad \tau = 10\mathbf{i} + 20\mathbf{j} + 30\mathbf{k}.$$

What are the equivalent force and moment at the gripper if it is located in a frame **A** such that

$$\mathbf{A} = \begin{bmatrix} 0 & 0 & 1 & 20 \\ 1 & 0 & 0 & 30 \\ 0 & 1 & 0 & 0 \\ 0 & 0 & 0 & 1 \end{bmatrix}.$$

7. Find the Jacobian of transformations J_1 and J_2 for the Rhino XR-2 robot as described in the text. J_1 and J_2 have the same definitions as given in the text by Equations (7.3.6) and (7.4.3), respectively (Fig. P7.7).

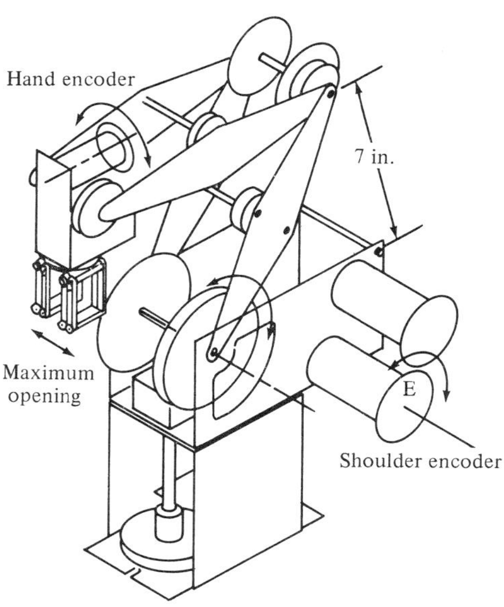

Figure P7.7 Rhino XR-2 robot.

8. For the robotic peg-in-hole insertion process shown in Figure P7.8, show that jamming will occur if

$$2\left(\frac{1}{d}\right) + \frac{C(1 - \mu^2)}{\mu} = 0,$$

where $C = (D - d)/d$ and μ is the coefficient of friction.

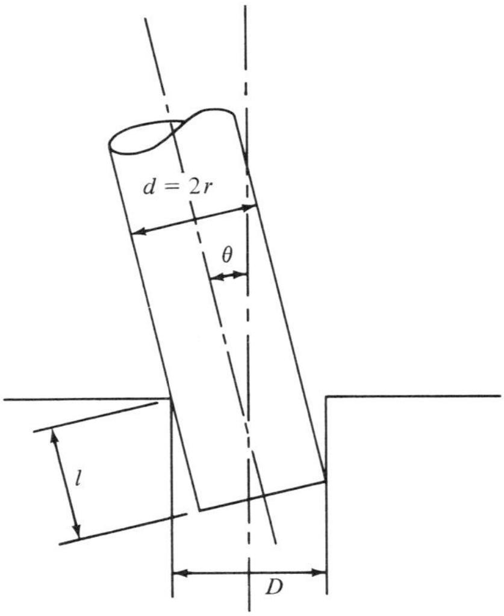

Figure P7.8 Rigid peg-in-hole insertion process.

9. Repeat Problem 8 for a peg that is linearly elastic with a Young modulus of elasticity E, a Poisson ratio V, and a shear modulus G.

10. Find the Jacobian of transformations J_1 and J_2 for the polar cylindrical robot (Stanford arm) as shown in Figure P7.10. Note that the definitions of J_1 and J_2 are the same as the ones described in the text.

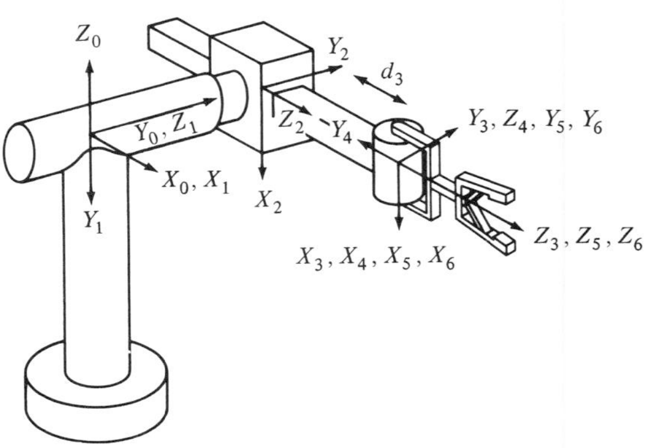

Figure P7.10 Cylindrical polar robotic manipulator.

11. Derive the Simunovich jamming criterion as given by Equation (7.8.14).

Chapter 8

Robotic Sensing and Vision

8.1 INTRODUCTION

The main objective of equipping robotic manipulators with intelligent sensing is to enable them to work in nonstructured and random environments. If robots are not equipped with intelligent sensors, they will be unable to sense and to respond to any change in their work environment. A robotic gripper will always tend to grip an object that it was programmed to grip at a specified point, even if the object is not there. Furthermore, it will move in its path toward the object even if there is momentarily an obstacle in its path. Thus, it will collide with the obstacle rather than intelligently sense the obstacle and move around it to reach the object to be grabbed. Therefore, for robotic manipulators to operate effectively and intelligently they must be equipped with an array of sensors, and the associated robotic software must be augmented accordingly so that the computer running the robot can receive data from the sensors and process the necessary real-time commands and information for decision making. Robots capable of performing such operations are called *intelligent robots.*

A general control diagram for a computer-based intelligent robot is shown in Figure 8.1.1.

As seen from Figure 8.1.1, the system consists of multilevel hierarchical closed-loop control. At the lower levels, the position, velocity, and possibly acceleration are fed back by means of optical encoders, tachometers, or acceleration gauges, respectively. At the higher levels of hierarchical control, the system has a series of more sophisticated sensors to sense the robot environment and feed back the information.

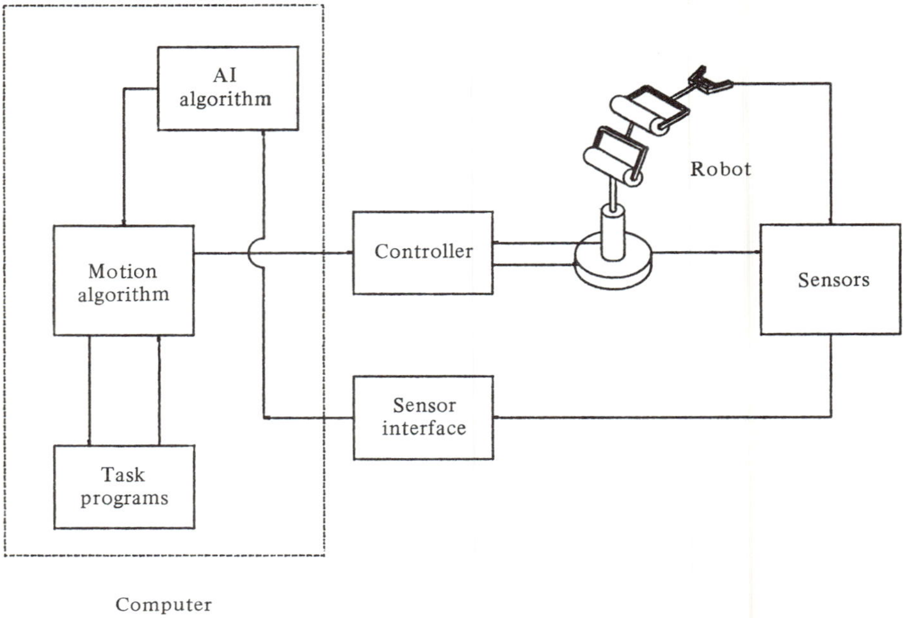

Figure 8.1.1 General schematic diagram of a computer-based intelligent robotic manipulator.

Generally, sensors are destined to have the following functions performed regularly:

1. To provide a priori information about the position and/or orientation of objects to be articulated and of the work pieces, as well as of the robot itself.
2. To correct position and/or orientation errors of both the robotic gripper and the work pieces.
3. To provide a posteriori information for detecting faults and failures in the finished products by visual inspection, that is, computer vision, or by ultrasonic or sonic means.
4. To provide continuous information about the forces between the gripper and the work pieces as well as force sensing of the surfaces of the work pieces by means of tactile sensing.
5. To provide continuous information about the other aspects of the robot environment such as the presence of thermal gradients, chemicals, and waves.

Next we present a broad classification of robotic sensors.

8.2 CLASSIFICATION OF ROBOTIC SENSORS

It is convenient to classify robotic sensors into two distinct and yet broad groups, that is, internal and external, as depicted in Figure 8.2.1. Note that the external group is further subdivided into contact and noncontact sensors. The division and subcategorization is to some extent arbitrary because, for example, tactile sensing appears to be *internal*; that is, sensors are mounted permanently in the robotic hand. However, their function would be to provide a rough outline of the external part being grasped or its texture (surface). A simple tactile sensor is a microswitch that senses the presence of barriers, obstacles, or surfaces beyond which the robot is not permitted to move. Thus, it can be used to avoid collision or to signal the robot that the target has been reached or to measure the dimensions of parts or objects to be manipulated or assembled. On the other hand, force and torque sensors, which are normally mounted somewhere between the end effector and the wrist joints, measure the reaction forces and torques. They may consist of simple strain gauges that are bonded onto the end effector or the robotic hand or they may be piezoelectric transducers mounted onto the robotic gripper.

The other important group of robotic sensors is the noncontact sensors, which includes visual sensors, proximity sensors, acoustic sensors, range detectors, temperature sensors, and chemical detectors.

EXAMPLE 8.2.1 ■

Give brief descriptions of some noncontact sensors, their advantages, and disadvantages.

Solution. Let us start with visual sensors and note that these are essentially based on digital TV cameras or laser beam scanners. The

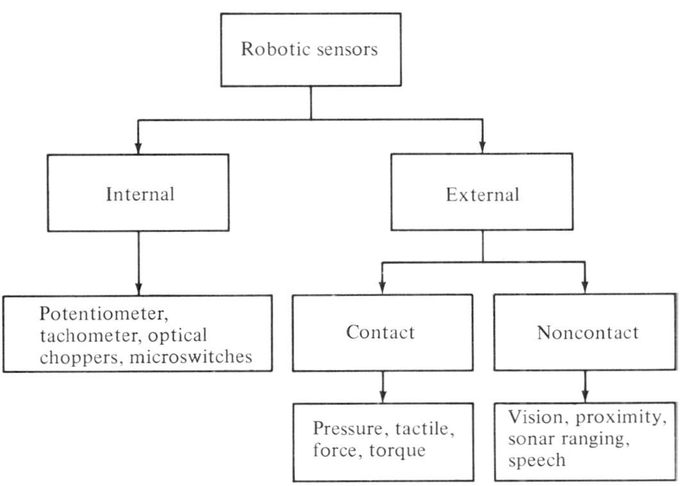

Figure 8.2.1 Classification of robotic sensors.

digitized signals of the cameras are fed at a rate of about 50 frames per second (FPS) into a central computer, which in turn analyzes the data and extracts the necessary and useful information. This procedure is discussed in more detail later in this chapter. At this point we discuss another class of noncontact sensors, namely, the class of proximity sensors. These types of sensor sense the presence of an object within a fixed workspace near the sensors themselves. One good example of a proximity sensor is the eddy-current sensor that can maintain a constant distance between the robot and a steel plate and is thus useful for welding or surface-finishing purposes. Another example of a simple proximity sensor is a light-emitting diode (LED) and an associated light-absorbing diode (photodiode) that can receive the reflection of a light beam from a surface and thus signal its proximity. The intensity of the received signal depends strongly on the reflectance of the surfaces to be examined for proximity, which presents a drawback for such sensors. Acoustic detectors include devices such as special microphones capable of recognizing isolated words or the presence of flaws and cracks in materials.

Ranging is another technique of noncontact sensing whereby information on objects' shapes or distribution in a space may be obtained by techniques such as triangulation. This is discussed in more detail later in this chapter. ■ ■

There are two basic techniques of using robotic sensors: closed-loop sensing and static sensing. In closed-loop sensing the robotic manipulator is controlled by the sensor during the actual motion. For example, most vision systems are in a closed-loop mode, such that the camera monitors the error between the end effector's actual kinematic attitude and the desired kinematic attitude. The information is used by the computer to move the robot effectively to achieve the right kinematic attitude for the gripper. A major difficulty of this technique is the long computation time for the computer to analyze the received images and accordingly command the robot to correct its kinematic attitude.

In static sensing the robot remains stationary while the sensors do their job of sensing the environment and giving feedback to the computer. After this process the robot is commanded to move in a blind fashion (Fig. 8.2.2).

EXAMPLE 8.2.2 ■

Consider a closed-loop robotic vision system as shown in Figure 8.2.3. The camera is monitoring the motion of an object as it moves while attached to link 5 of the manipulator. The transforms $^5\mathbf{C}$ describing the camera frame with respect to the $\mathbf{T}_5$ frame and the gripper frame with respect to the $\mathbf{T}_5$ frame, that is, $\mathbf{A}_5$, are given by

$$^5\mathbf{C} = \begin{bmatrix} 0 & 0 & -1 & 8 \\ 0 & -1 & 0 & 0 \\ -1 & 0 & 0 & 5 \\ 0 & 0 & 0 & 1 \end{bmatrix} \tag{8.2.1}$$

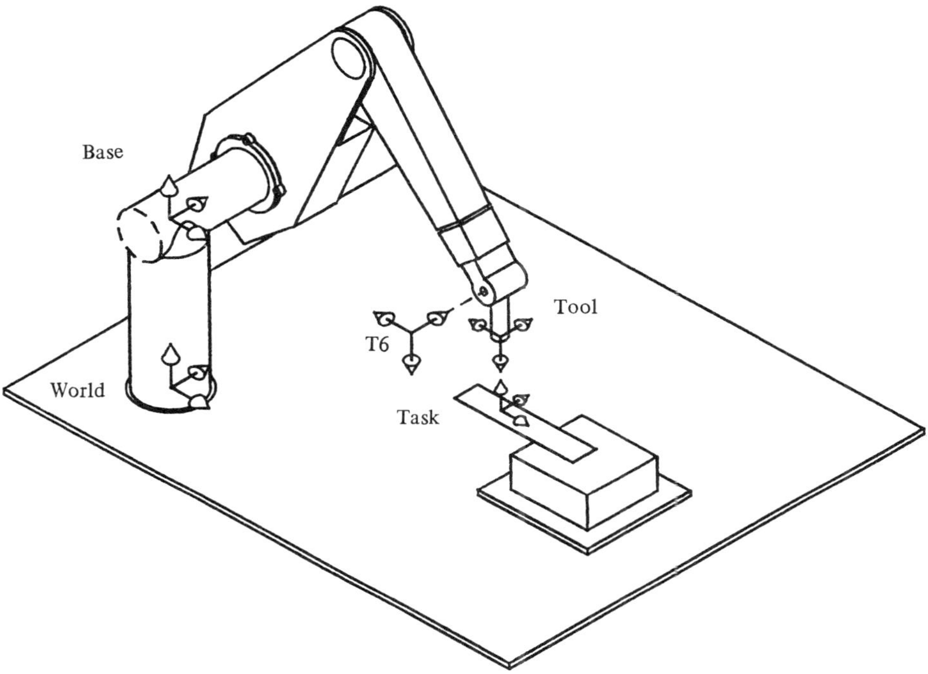

Figure 8.2.2 Static sensing in robotic manipulation.

and

$$\mathbf{A}_5^6 = \begin{bmatrix} 0 & -1 & 0 & 0 \\ 1 & 0 & 0 & 0 \\ 0 & 0 & 0 & 10 \\ 0 & 0 & 0 & 1 \end{bmatrix}. \tag{8.2.2}$$

The camera frame observes the object frame at ${}^c\mathbf{O}$. The object frame is at ${}^6\mathbf{O}$ from the end-effector frame 6. Determine the generalized differential vector of the end-effector frame with respect to the object frame if the generalized differential vector of the object frame with respect to the camera frame is given by

$${}^c\mathbf{D} = [-0.2 \quad 0.2 \quad 0 \quad 0 \quad 0 \quad 0.2]^T. \tag{8.2.3}$$

Solution. Let us refer to Figure 8.2.3 and note that the gripper frame with respect to the camera frame ${}^c\mathbf{G}$ is given by

$${}^c\mathbf{G} = {}^5\mathbf{C}^{-1}\mathbf{A}_5^6. \tag{8.2.4}$$

Note that

$${}^5\mathbf{C}^{-1} = \begin{bmatrix} 0 & 0 & -1 & 5 \\ 0 & -1 & 0 & 0 \\ -1 & 0 & 0 & 8 \\ 0 & 0 & 0 & 1 \end{bmatrix}. \tag{8.2.5}$$

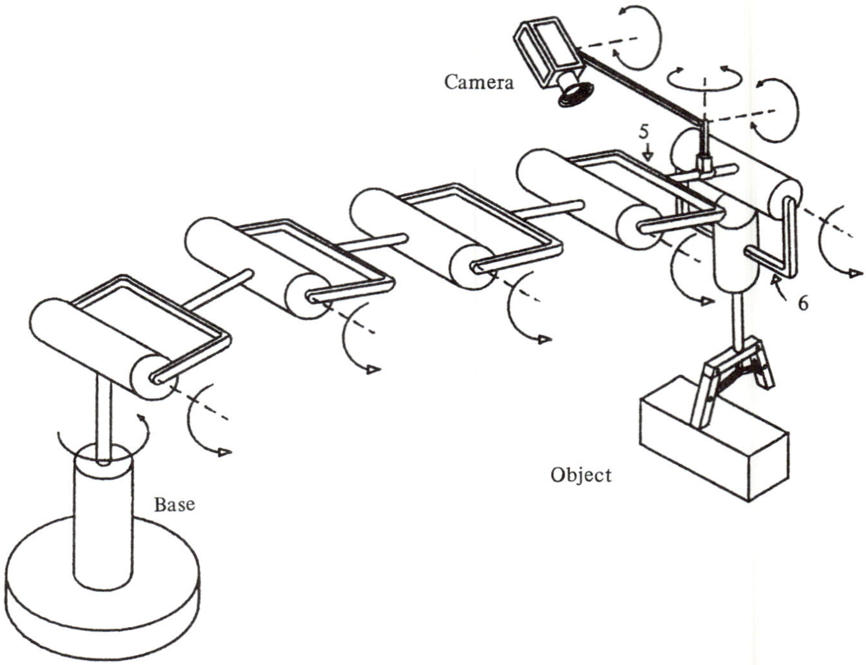

Figure 8.2.3 Closed-loop vision system mounted on a robotic manipulator.

Thus, ${}^c\mathbf{G}$ is obtained as

$$ {}^c\mathbf{G} = \mathrm{CAM}^{-1}\mathbf{A}_6 = \begin{bmatrix} 0 & 0 & -1 & 5 \\ 0 & -1 & 0 & 0 \\ -1 & 0 & 0 & 8 \\ 0 & 0 & 0 & 1 \end{bmatrix} \begin{bmatrix} 0 & -1 & 0 & 0 \\ 1 & 0 & 0 & 0 \\ 0 & 0 & 1 & 10 \\ 0 & 0 & 0 & 1 \end{bmatrix} \tag{8.2.6} $$

or

$$ {}^c\mathbf{G} = \begin{bmatrix} n_x & o_x & a_x & p_x \\ n_y & o_y & a_y & p_y \\ n_z & o_z & a_z & p_z \\ 0 & 0 & 0 & 1 \end{bmatrix} = \begin{bmatrix} 0 & 0 & -1 & -5 \\ -1 & 0 & 0 & 0 \\ 0 & 1 & 0 & 8 \\ 0 & 0 & 0 & 1 \end{bmatrix}. \tag{8.2.7} $$

Thus, $\mathbf{n} = 0\mathbf{i} - 0\mathbf{j} + 0\mathbf{k}$, $\mathbf{o} = 0\mathbf{i} + 0\mathbf{j} + \mathbf{k}$, $\mathbf{a} = \mathbf{i} + 0\mathbf{j} + 0\mathbf{k}$, and $\mathbf{p} = 5\mathbf{i} + 0\mathbf{j} + 8\mathbf{k}$. Also note that

$$ {}^c\mathbf{d} = -0.2\mathbf{i} + 0.2\mathbf{j} + 0\mathbf{k} \tag{8.2.8} $$

and

$$ {}^c\boldsymbol{\delta} = 0\mathbf{i} + 0\mathbf{j} + 0.2\mathbf{k}. \tag{8.2.9} $$

Therefore,

$$^{T}d_x = \boldsymbol{\delta}\cdot(\mathbf{p}\times\mathbf{n}) + \mathbf{d}\cdot\mathbf{n} = \mathbf{n}\cdot(\boldsymbol{\delta}\times\mathbf{p}+\mathbf{d}), \tag{8.2.10}$$

$$^{T}d_y = \boldsymbol{\delta}\cdot(\mathbf{p}\times\mathbf{o}) + \mathbf{d}\cdot\mathbf{o} = \mathbf{o}\cdot(\boldsymbol{\delta}\times\mathbf{p}+\mathbf{d}), \tag{8.2.11}$$

$$^{T}d_z = \boldsymbol{\delta}\cdot(\mathbf{p}\times\mathbf{a}) + \mathbf{d}\cdot\mathbf{a} = \mathbf{a}\cdot(\boldsymbol{\delta}\times\mathbf{p}+\mathbf{d}), \tag{8.2.12}$$

$$^{T}\boldsymbol{\delta}_x = \mathbf{n}\cdot\boldsymbol{\delta}, \tag{8.2.13}$$

$$^{T}\boldsymbol{\delta}_y = \mathbf{o}\cdot\boldsymbol{\delta}, \tag{8.2.14}$$

and

$$^{T}\boldsymbol{\delta}_z = \mathbf{a}\cdot\boldsymbol{\delta}, \tag{8.2.15}$$

so that $\boldsymbol{\delta}\times\mathbf{p} = 0\mathbf{i} + 1\mathbf{j} + 0\mathbf{k}$. Thus,

$$^{6}\mathbf{d} = -1.2\mathbf{i} + 0\mathbf{j} + 0\mathbf{k} \tag{8.2.16}$$

and

$$\boldsymbol{\delta} = 0\mathbf{i} + 0.2\mathbf{j} + 0\mathbf{k}. \tag{8.2.17}$$

Finally,

$$^{6}\mathbf{D} = [-1.2 \quad 0 \quad 0 \quad 0 \quad 0.2 \quad 0]^{T}. \quad ■■ \tag{8.2.18}$$

8.3 ROBOTIC VISION SYSTEMS

Robotic vision systems generally are equipped with a number of TV cameras capable of digitizing and processing the acquired digital information. A central processor thus analyzes the information and defines the object, the visual situation, or the scene for any task-oriented robotic manipulation and assembly. The main applications of robotic vision systems are in assembly processes, quality control, and parts handling and classification.

A typical robotic vision system consists of a TV camera interfaced with a computer vision system through a frame grabber and a visual information preprocessor. The frame grabber acts as a video buffer. The video signal can be sent to a TV monitor to display continuously the scene to be analyzed. The scene visual information processing is generally very slow and thus a preprocessor normally is employed to retain only the useful information and reduce the processing time. Images also must be put into a contrast with their background and thus a threshold circuit is used to set the intensities of the various parts of the scene to either white or black. Figure 8.3.1 depicts a general diagram of a typical robotic vision system.

The TV cameras can be mounted either on the robot (Fig. 8.2.3) or separately. As was discussed in Example 8.2.2, under these circumstances the robot's movement is based on the differential error signal between the object's location and the gripper or camera location. A disadvantage of this system is that it requires a rugged camera because it is subjected to rough handling.

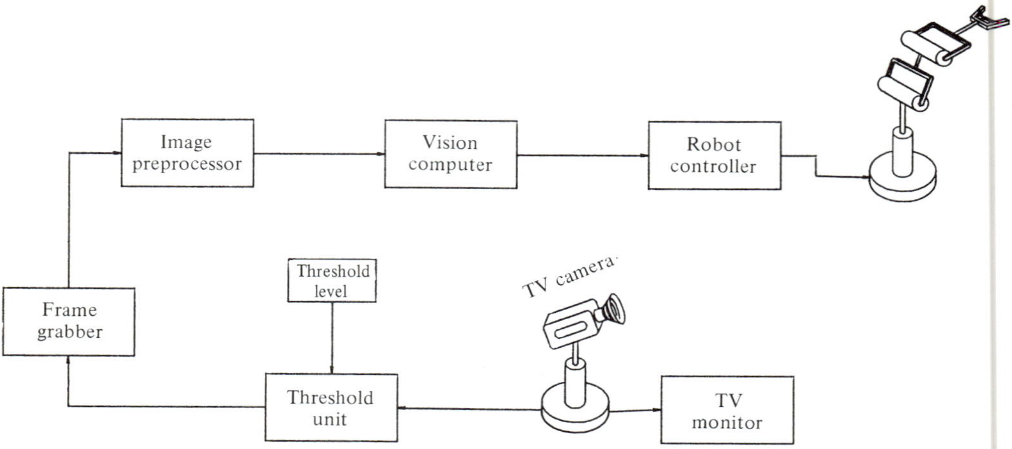

Figure 8.3.1 Typical robotic vision system.

Furthermore, the added weight of the camera on the robotic arm is undesirable and requires additional control algorithms and control computation time. On the other hand, one may mount the vision system outside the robot system (Fig. 8.3.1); however, this requires a complete three-dimensional knowledge of the robotic scene, which may cause control and computational problems. The advantage of the first scheme (Fig. 8.2.3) is that the robot is able to move the camera around the object and get a much better set of visual information while the setup in Figure 8.3.1, for example, gives us only a fixed location view of the three-dimensional scene of activities.

8.3.1 Robotic Eyes—TV Cameras

Robotic vision system cameras essentially fall in the domain of TV cameras. In these cameras that play the role of robotic eyes, the images are real time and already in digital form. The camera's electronics is capable of scanning the entire image of over 500 scan lines (525 lines in the United States), which is called a *frame*. Each frame consists of two *fields* each made up of alternate scan lines from the frame. The camera's electronics generates and transmits the fields sequentially. For instance, the transmitted image is interlaced with odd-numbered scan lines being drawn on the screen alternating with even-numbered scan lines. There are presently seven different robotic camera systems used to generate TV images of live scenes.

Image Orthicon Tube This is a storage type tube, since it is based on the neutralization of positive charges by a scanning electron beam. The image orthicon is divided into an imaging and readout section. In the imaging section, light from the scene is focused onto a semitransparent photocathode. The photocathode emits electrons that are focused magnetically by a coil and then

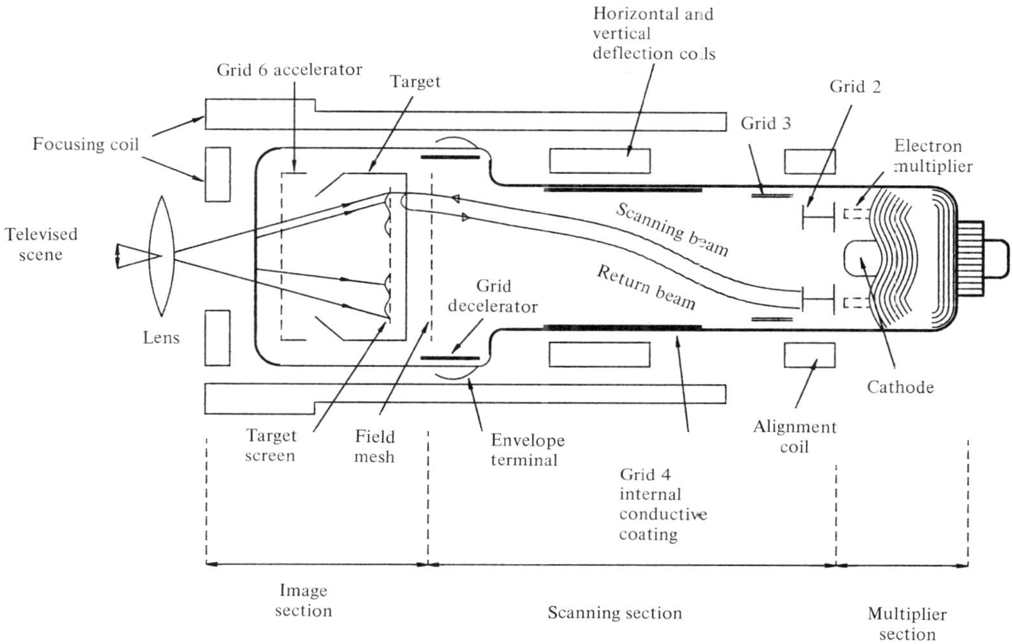

Figure 8.3.2 Orthicon tube.

accelerated toward a positively charged target. This target is made up of a thin disk of glass with a fine-wire-mesh screen on it that faces the cathode. When electrons strike the screen, secondary electrons are emitted from the glass disk and the scanning side of the glass disk gets positively charged. The pattern of these positively charged regions corresponds to the pattern of light intensity in the scene being examined. The back of the glass disk is then scanned by an electron beam from an electron gun (Fig. 8.3.2) at the rear of the orthicon. The final image is represented by the amplitude-modulated intensity of the beam.

Vidicon Tube The vidicon tube is generally more rugged, smaller, lighter, and portable compared to the orthicon tube. The target in the vidicon tube is coated with a transparent conducting film that acts as an electrode for video signals. The film is deposited on a thin photosensitive layer that is composed of a large number of tiny resistive elements. These elements are such that their resistance is decreased by illumination. It is this layer that is scanned in raster fashion by an electron beam from the electron gun at the rear of the tube. Electrons are deposited on the layer and reduce its potential. The two target surfaces essentially form a capacitor, while the scanning action of the beam produces a capacitive current at the video signal electrode, which is in fact the video signal (Fig. 8.3.3).

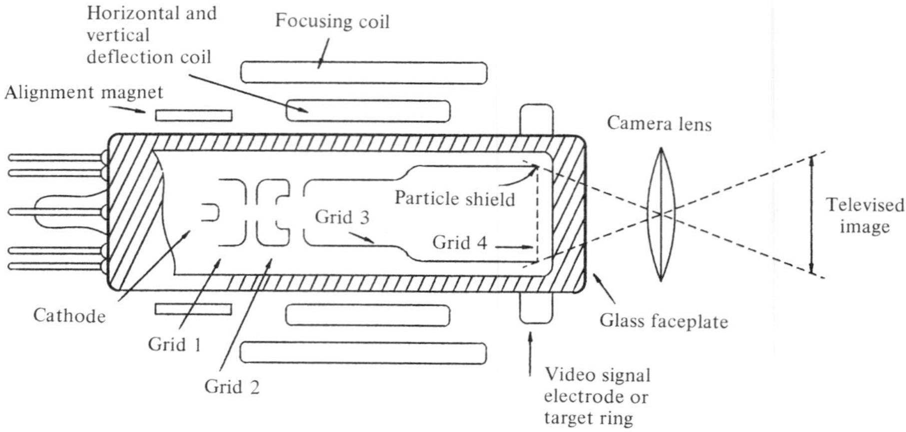

Figure 8.3.3 Vidicon tube.

Plumbicon Tube The plumbicon tube is basically the same as the vidicon tube except that the photosensitive layer is made up of lead oxide. Thus, it offers some advantages over the vidicon tube. Among the advantages are lower dark current, higher sensitivity, and negligible lag.

Iconoscope Tube The iconoscope tube is composed of an electron beam that scans a target. The target consists of a thin mica sheet of mosaic that is coated with a photosensitive layer. In the iconoscope tube both the electron beam and light beam strike the same side of the target surface, in sharp contrast to the orthicon, vidicon, and plumbicon tubes. The back of the thin mica sheet is covered with a conductive film that is connected to an output load. The electronic assembly is similar to a matrix of small capacitors that discharge through a common connection or lead. This tube is no longer in use in modern image processing.

Image Dissector Tube In an image dissector tube light from the scene to be examined is focused on a cathode coated with a photosentitive layer. In this sense the cathode emits electrons whose number is proportional to the amount of light striking the cathode. These electrons are accelerated toward a target by the anode. The target is covered by a small aperture that allows only a small part of the *electron image* emitted by the cathode to reach the target. Thus, in this sense, it is an electron multiplier, and the whole apparatus operates on instantaneous scanning instead of secondary neutralization of positive charges (Fig. 8.3.4).

A focusing coil is used to focus the electron image and produce an axial magnetic field. The electron image is then scanned past the target aperture by the deflection coils. This operation causes the electron multiplier to produce a variable voltage that represents the video signals. Thus, the image is "dissected" as it is scanned past the target.

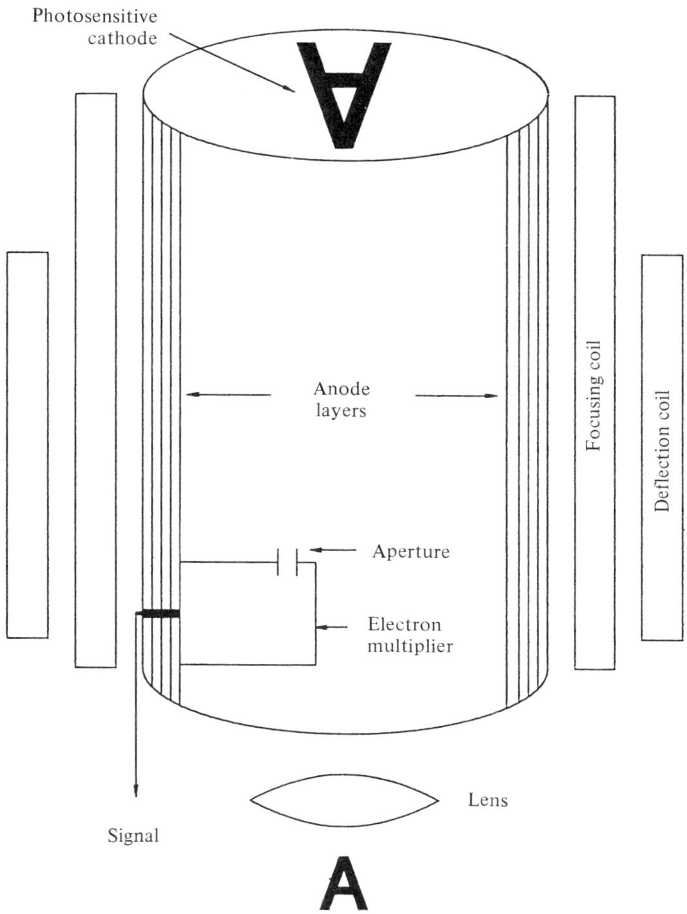

Figure 8.3.4 Image dissector tube.

Charge-Coupled Device (CCD) The CCD type robotic cameras fall in the general category of charge-transfer devices or CTDs. The CCD robotic camera operates in a manner similar to a metal-oxide semiconductor field-effect transistor because the CCD camera has a *source* region and a *drain* region coupled by a *depletion*-region channel (Fig. 8.3.5). In the imaging modus operandi the CCD camera is considered a monolithic array of closely spaced metal-oxide semiconductor (MOS) capacitors that form a shift register. A series of clocking pulses are applied to a row of electrodes between the source and the drain. This causes the charges in the depletion region to be transferred to the output. A series of charges are generated on the CCD array by photons incident on the MOS (Fig. 8.3.6). The generated charges are transferred directly to an output register either one line at a time or an area at a time. These are called, respectively, *line transfer* and *frame transfer*.

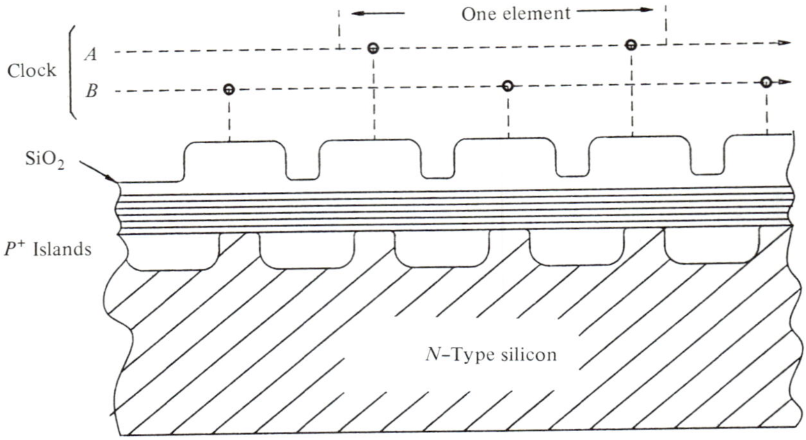

Figure 8.3.5 Charge-coupled device (CCD).

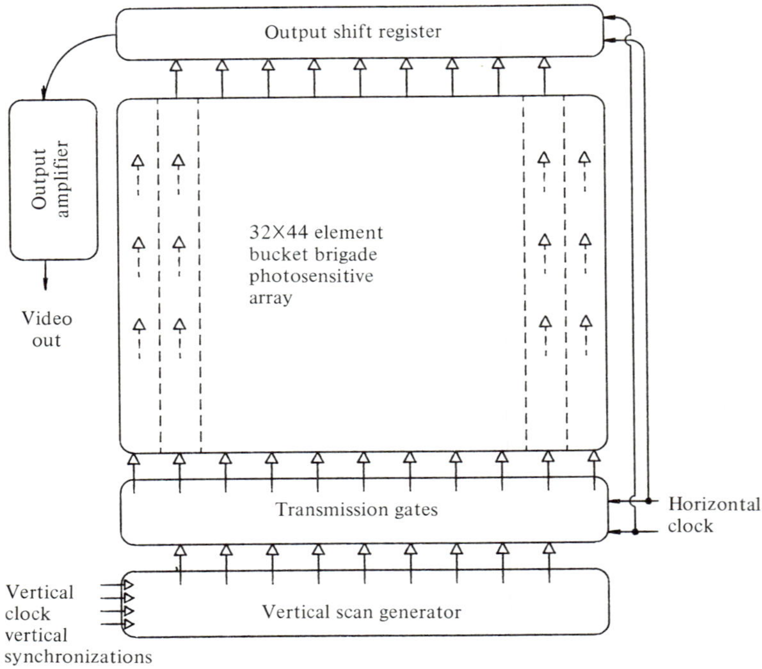

Figure 8.3.6 CCD line transfer array.

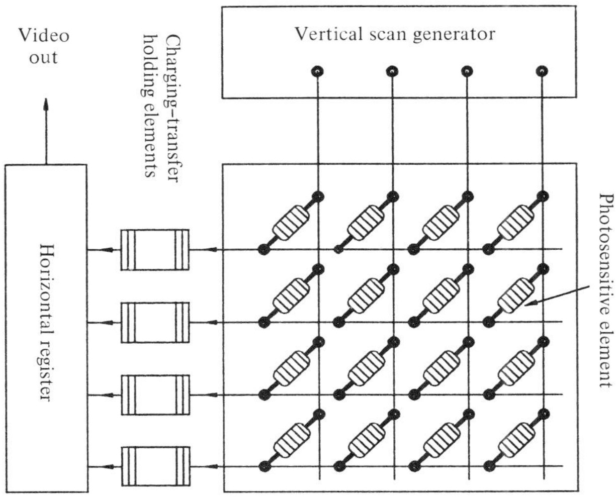

Figure 8.3.7 Charge-injected device array.

Charge-Injected Devices (CIDs) The CID type robotic cameras again fall into the general category of charge-transfer devices or CTDs. The CIDs generally resemble the CCDs except that during image sensing the charge is confined to the image site where it was generated. The charges are read by a technique similar to the X-Y addressing technique used in computer memories. Essentially, the stored charge is injected into the substrate, resulting in a displacement current that subsequently is interpreted as the video signal (Fig. 8.3.7).

The CIDs have a number of advantages over the CCDs in that they have less defects, simple mechanization, less charge-transfer losses, and minimal blooming.

8.3.2 Fundamentals of Image Processing

The representation of an image as an analog electrical signal is usually an unavoidable precursor to further image processing. Thus, the first step is to digitize the video information that is normally in the form of an analog voltage as discussed before (Fig. 8.3.8). Now the function of an analog to digital (A/D) converter is to take the input voltage, namely, the video signal, and to produce an output that is binary and thus suitable for reading by a digital computer via a suitable interface (Fig. 8.3.9). The quality of an A/D converter is tested against its temporal resolution or the speed at which it can perform the necessary conversions as well as the accuracy of its digital output. Presently, these A/D converters are produced as IC chips with a resolution of 8–16 bits.

It is quite possible to digitize an entire TV camera raster image (frame), which is almost 525 scan lines, by 300 samples along each scan line in a single frame time, that is 1/30 second. Such fast digitization of images is possible by

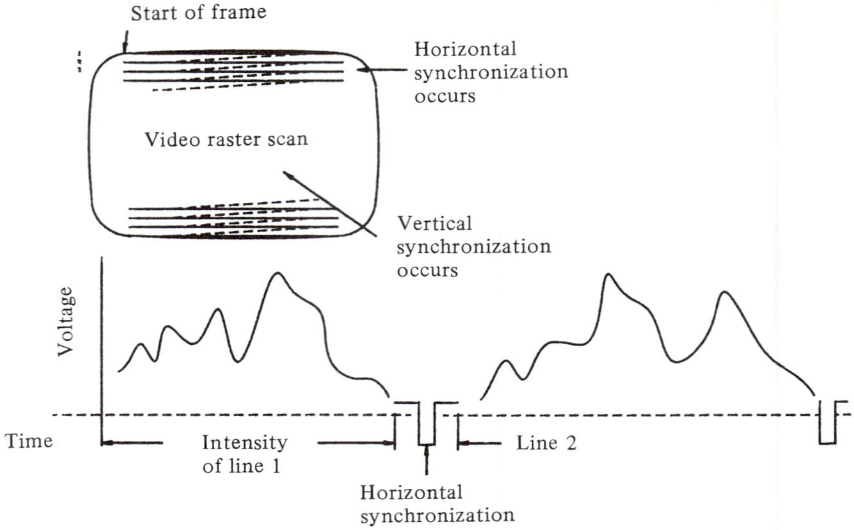

Figure 8.3.8 Standard analog video signal. Each line of the raster scan is proportional to the brightness along the line.

existing commercial systems. They do it by digitizing into a *frame buffer* memory, along with raster graphics display capabilities from the same frame buffer, and by *video rate processing* of the digital data.

The digitized image of the frame is stored in a memory of the vision computer and subsequently processed. The computer can then determine the types of object and their orientations and transfer the information to the robot computer. There are two forms in which such information may be processed: the binary image and the gray-scale image.

The digitized image consists of pixels or picture elements. The brightness of each is represented by a number. In a binary vision system, the brightness is represented by either a 0 for background or a 1 for the object silhouette or 0 for the black and 1 for the white. In a gray-scale system, the shades of an image or the gray levels can take a range of values. For instance, if 8-bit controllers are used, then each pixels can have a gray-level value from 0 (black) to 255 (white). It is clear that binary image processing systems are more cost effective but they impose restrictions on the scene to be examined. For example, generally the

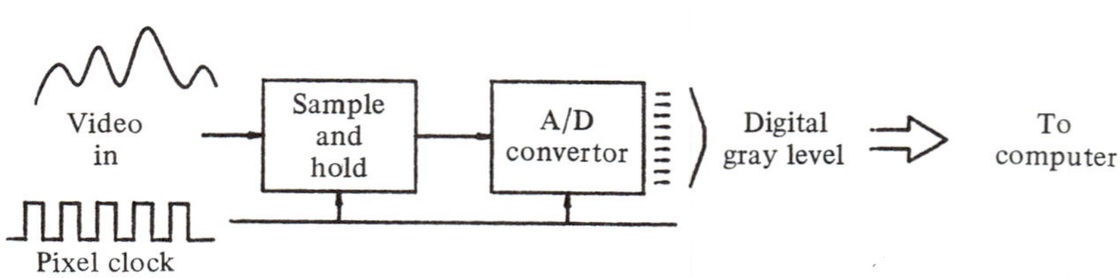

Figure 8.3.9 Conversion from analog video to digital gray levels.

depth of the objects cannot be detected using binary image processing, while gray-scale image processing allows the detection of curved surfaces, depths, and the presence of overlapping parts. However, the gray-scale system also requires more hardware, software, and computing power.

Some important areas of application are acquisition of randomly located parts on moving conveyer belts, the random bin picking problem, visual inspection of assembled parts, chasing of moving objects to be grabbed by the robot, and collision avoidance problems.

The essential steps in visual processing of images with a computer are outlined in Figure 8.3.10.

Next we describe in detail the aforementioned essential steps in visual data processing.

8.3.3 Visual Data Acquisition

Visual data acquisition consists of reading either a series of horizontal raster scans of the image from top to bottom and side to side or the output of a solid-state imaging array based on CCD technology. In either case, the raw image

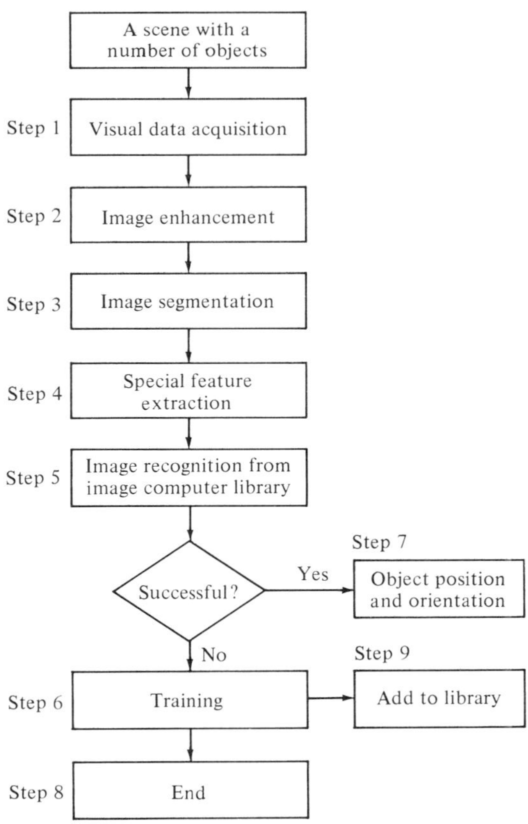

Figure 8.3.10 Essential steps in visual inspection.

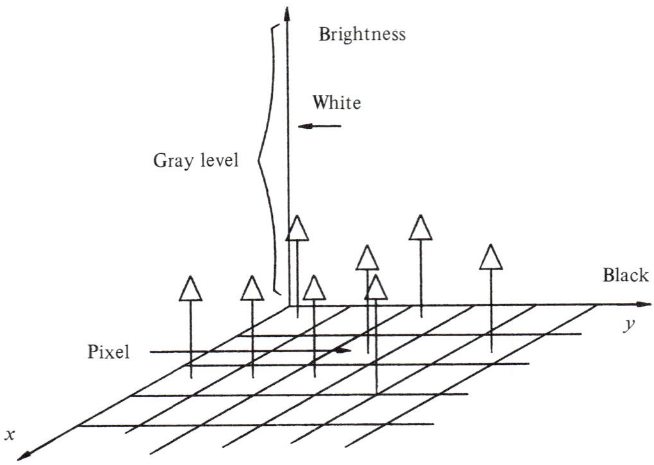

Figure 8.3.11 Image digitization by means of pixels.

takes the form of a two-dimensional array of gray levels, where each array element corresponding to a pixel (picture element) is as shown in Figure 8.3.11.

Here the average value of the light intensity over each pixel is the gray level associated with that particular pixel. Obviously, the size of a pixel is a measure of the resolution of the TV camera under consideration.

There are two purposes in obtaining visual data in the form of gray levels: (1) to measure the spatial parameters on the objects to be examined visually and (2) to recognize the objects under consideration. The signal processing operations are based on an averaging (integral) technique applied to the visual signal. However, the differential operations are also necessary to create contrast and recognition for the image contours.

By applying the integral operation to visual data a whole series of useful information is obtained. Some of the important features are:

1. Largeness.
2. Center of area (gravity).
3. Inclination (orientation).
4. Features (roundness, slenderness, etc.).
5. Similarity (other figures).

To describe the above features let us consider Figure 8.3.12. In the context of binary image processing,

$$I(x, y) = \begin{cases} 1 & \text{for all } x, y \in C \\ 0 & \text{for all } x, y \notin C. \end{cases} \tag{8.3.1}$$

The above functional definition shows that the figure has uniform brightness everywhere and can be distinguished clearly from the background. If one

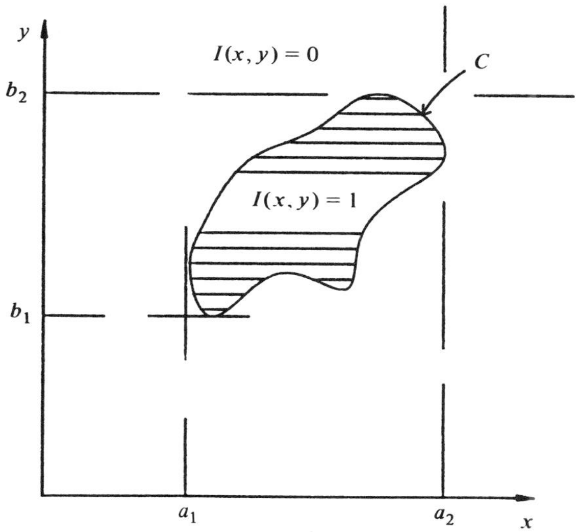

Figure 8.3.12 Binary intensity function $I(x, y)$.

integrates the function $I(x, y)$ over the entire visual field, one obtains the largeness A, which represents the brightness area of the planar image:

$$A = \int_{a_1}^{a_2} \int_{b_1}^{b_2} I(x, y)\, dx\, dy. \tag{8.3.2}$$

To obtain the center area or gravity of the image, which in a way identifies its position, two integral representations have been proposed as follows.

The Center of Area (x_A, y_A) This is represented by

$$\int_{a_1}^{a_2} \int_{b_1}^{b_2} sg_n(x - x_A) I(x, y)\, dx\, dy = 0 \tag{8.3.3}$$

and

$$\int_{a_1}^{a_2} \int_{b_1}^{b_2} sg_n(y - y_A) I(x, y)\, dx\, dy = 0. \tag{8.3.4}$$

The Center of Gravity (x_g, y_g) This is represented by

$$x_g = A^{-1} \int_{a_1}^{a_2} \int_{b_1}^{b_2} xI(x, y)\, dx\, dy \tag{8.3.5}$$

and

$$y_g = A^{-1} \int_{a_1}^{a_2} \int_{b_1}^{b_2} yI(x, y)\, dx\, dy. \tag{8.3.6}$$

To obtain the orientation or the inclination of the object consider an arbitrary direction in the body represented by a line L in Figure 8.3.13. Integrating the square of distance l from a point $P(x, y)$ in the image to a straight line that passes through the center of gravity of the binary image making an angle θ with the x axis, one obtains the M_2 about the line L such that

$$M_2 = A^{-1} \int_{a_1}^{a_2} \int_{b_1}^{b_2} \frac{[(y - y_g) - (x - x_g) \tan \theta]^2 I(x, y)}{1 + \tan^2 \theta} \, dx \, dy. \tag{8.3.7}$$

The value of θ that makes M_2 a minimum is defined as the inclination of the binary image with respect to the x axis.

To obtain information on other features of the binary image, such as roundness and slenderness, one can use higher-order moments such as

$$\bar{x}^2 = A^{-1} \int_{a_1}^{a_2} \int_{b_1}^{b_2} (x - x_g)^2 I(x, y) \, dx \, dy. \tag{8.3.8}$$

$$\bar{y}^2 = A^{-1} \int_{a_1}^{a_2} \int_{b_1}^{b_2} (y - y_g)^2 I(x, y) \, dx \, dy. \tag{8.3.9}$$

and

$$\overline{xy} = A^{-1} \int_{a_1}^{a_2} \int_{b_1}^{b_2} (x - x_g)(y - y_g) I(x, y) \, dx \, dy. \tag{8.3.10}$$

These are additional information that can be communicated to the vision computer memory.

The similarity between two binary images is obtained by a procedure called *template matching*, which is discussed later. Essentially, the various information obtained for a binary image is compared to the same information

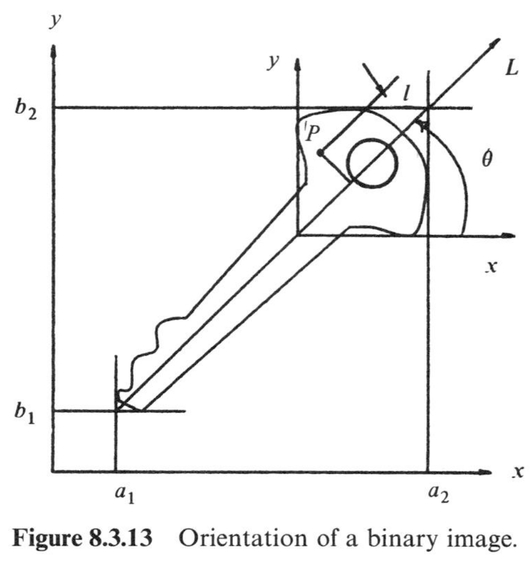

Figure 8.3.13 Orientation of a binary image.

for another binary image and the differences are spelled out by the vision computer.

EXAMPLE 8.3.1 ■

Derive Equation (8.3.7) and graphically find the value of θ that makes M_2 a minimum.

Solution. Note that

$$M_2 = A^{-1} \int_{a_1}^{a_2} \int_{b_1}^{b_2} l^2 I(x, y)\, dx\, dy. \tag{8.3.11}$$

Figure 8.3.14 is a representation of the situation. We observe that

$$a^* = (y - y_g) - b^*, \tag{8.3.12}$$

$$b^* = (x - x_g) \tan \theta, \tag{8.3.13}$$

and

$$l = a^* \cos \theta = a^* \left(\frac{1}{1 + \tan^2 \theta} \right)^{1/2}. \tag{8.3.14}$$

Thus,

$$l^2 = \frac{[(y - y_g) - (x - x_y) \tan \theta]^2}{1 + \tan^2 \theta} \tag{8.3.15}$$

and

$$M_2 = A^{-1} \int_{a_1}^{a_2} \int_{b_1}^{b_2} \frac{[(y - y_g) - (x - x_g) \tan \theta]^2}{1 + \tan^2 \theta} I(x, y)\, dx\, dy. \tag{8.3.16}$$

Figure 8.3.14 depicts the minimum solution.

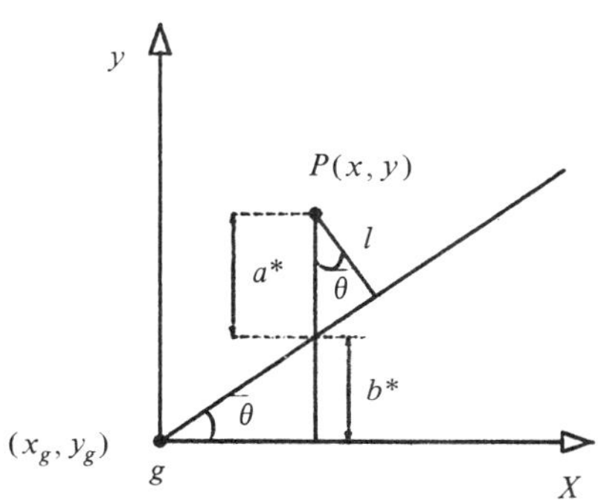

Figure 8.3.14

Based on the definitions of x, y, and xy, one may define the following measures of *roundness* and *slenderness*.

$$\text{Roundness } R : R = A^{-1}(\bar{x}^2 + \bar{y}^2). \tag{8.3.17}$$

$$\text{Slenderness } S : S = \frac{[(\bar{y}^2 - \bar{x}^2)^2 + 4(\bar{x}\bar{y})^2]^{1/2}}{\bar{x}^2 + \bar{y}^2}. \quad ■■ \tag{8.3.18}$$

8.3.4 Image Enhancement

In this phase of visual data processing the original binary or gray-scale image is used and processed to produce a second image of much improved quality and resolution by eliminating the obviously observed noise or other artifacts. Basically, a video preprocessor is employed to improve the signal-to-noise ratio of the image, which in turn reduces the amount of information sent to the computer and improves the quality of the picture for analysis. One of the most important features of the preprocessor is that it clearly brings out the edges of the image by amplification of the difference in the light intensity observed at an edge of the image compared to the light intensity of the background. This process is sometimes referred to as *edge detection*.

8.3.5 Image Segmentation

The next essential step in visual processing is to locate regions in the image that correspond to certain geometrical objects as distinct from the background image on which the objects are placed. In Figure 8.3.15 the total number of dots per unit area could be regarded as the gray level of the image scene. This process of separating objects from the background is called *image segmentation*. It is clear that careful structured lighting in an industrial environment can simplify the process of image segmentation considerably and thus increase its reliability.

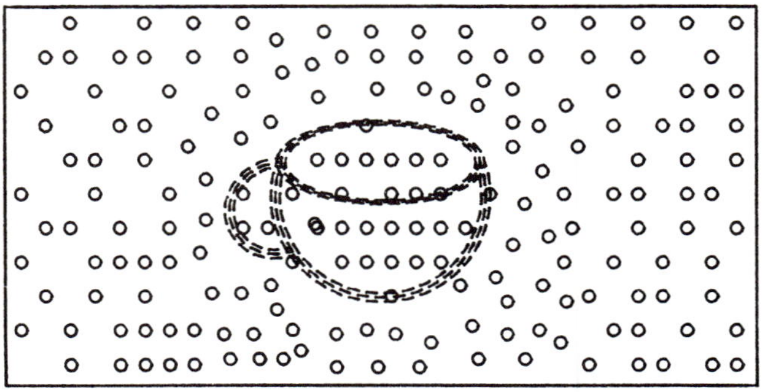

Figure 8.3.15 Image segmentation.

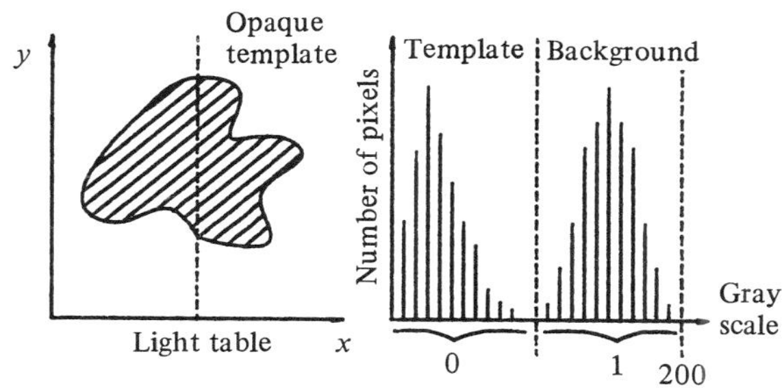

Figure 8.3.16 Threshold image segmentation.

For example, the parts may be placed on translucent tables with diffused backlighting to obtain sharp contrasts. In the case of foreground lighting the object should be placed in such a way that a minimum amount of image shadows are produced. The use of a background color that contrasts sharply with the color of the foreground objects is also helpful for improved image contrast. The segmentation process operates on the gray-scale image to produce a binary (two-level gray-scale) image. For example, a logical 0 might represent an object while a logical 1 might represent the background. The most commonly used technique for the segmentation of an image by gray scaling is the *threshold image segmentation* technique.

Generally, an intensity histogram is used to select an optimal gray-scale value for the threshold that separates the object from the background. An opaque disk, for example, placed on a light table might produce an intensity histogram as shown in Figure 8.3.16, where the threshold might be picked in a manner similar to drawing the line between letter grades in a histogram of student scores or grades.

8.3.6 Image Extraction and Recognition

After completion of the process of segmenting a binary image, the pattern it represents must be classified. The most common and straightforward technique of pattern classification is to compare a pattern with stored models of known patterns to find the best match. These stored models are known as *templates*. Here we define a template as a geometrical object or image produced or generated by "training" the vision system using a common representative member of the class of objects. Consider an object in the form of a wrench and consider another pair of objects such as a quarter coin and a safety pin. Let us place these three objects in the center of the image viewing area. Now the segmented images of these three objects might be stored in an *object library* as representative templates of such classes of objects. When an unknown object is placed in the viewing area, its segmented image $I(i, j)$ is obtained. This is the intensity of the pixel at the location (i, j). The kth stored template $T_k(i, j)$ is then obtained from the library and moved over the image until the best match is obtained. If $T_k(i - x, j - y)$ represents a translation of the template $T_k(i, j)$ by (x, y), then the problem is to find (x, y) to minimize an error $E_k(x, y)$ where, by definition, we have

$$E_k(x, y) = \sum_j \sum_i [I(i, j) - T_k(i - x, j - y)]^2, \qquad k = 1, 2, \ldots, n - 1. \tag{8.3.19}$$

Here the summations with respect to i and j are taken over overlapping portions of the images $I(i, j)$ and $T_k(i, j)$. The minimum value of $E_k(x, y)$ is calculated for $k = 1$ to n, where n is the number of objects in the library. For example, suppose that $\hat{x}$, $\hat{y}$, and l are found such that

$$E_l(\hat{x}, \hat{y}) < E_k(x, y) \tag{8.3.20}$$

for all k's in the library and for all x and y in the field of view. If the optimal $E_l(\hat{x}, \hat{y})$ is sufficiently small, that is, $E_l(\hat{x}, \hat{y}) \approx 0$, then it is a perfect match and the image $I(i, j)$ represents the object. Furthermore, the object representation $I(i, j)$ would be translated by $(\hat{x}, \hat{y})$ in comparison with the position of the object in template k. If the optimal $E_l(\hat{x}, \hat{y})$ is too large, then a match has not been found. Thus, under these circumstances, the object that has generated the image $I(i, j)$ should be classified as unknown.

A better measure, known as the *normalized cross-correlation measure*, has been defined by

$$m_k(x, y) = \frac{\sum_j \sum_i I(i, j) T_k(i - x, j - y)}{((\sum_j \sum_i I^2(i, j))(\sum_j \sum_i T_k^2(i - x, j - y)))^{1/2}}. \tag{8.3.21}$$

This measure is insensitive to the absolute intensity and contrast, and if it achieves a maximum value of 1, it then indicates an exact match between the templates.

Sometimes it is beneficial to represent templates by a collection of subtemplates with some specific spatial relations. Then it is easier to match the subtemplates independent of each other. Under those circumstances, a further measure of the spatial relations in the input pattern should be used to evaluate the template matching process. Template matching based on the index defined in Equation (8.3.19) is invariant with respect to translation of the object as long as the entire object remains in the field of view. However, the measure is not invariant with respect to rotation, scaling (size), or perspective transformations. These additional transformations could be incorporated into the template matching procedure by simply applying them to either the templates or the image in the vision computer since we are concerned with two-dimensional images only.

A typical camera consists of a lens and a plane on which the image is to be formed. Typically, the situation with an image is as depicted in Figure 8.3.17, where there is a lens center C, an image plane I, and a distance f that is the focal length of the imaging system from C.

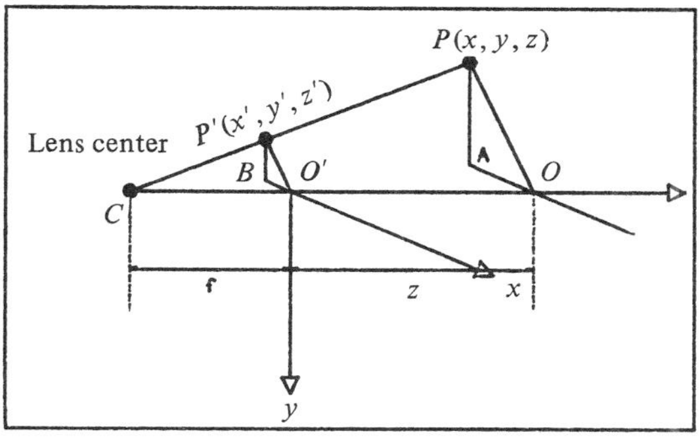

Figure 8.3.17 Planar TV camera image scene.

EXAMPLE 8.3.2 ■

Show that the points x', y', and z' are given by

$$x' = \frac{fx}{f+z}, \quad y' = \frac{fy}{f+z}, \quad \text{and} \quad z' = 0. \tag{8.3.22}$$

Solution. From similar triangles PAO and $P'BO'$, it is clear that

$$\frac{P'B}{PA} = \frac{x'}{x} = \frac{P'O'}{PO} = \frac{BO'}{AO} = \frac{y'}{y}. \tag{8.3.23}$$

From similar triangles $P'O'C$ and POC it is clear that

$$\frac{P'O'}{PO} = \frac{f}{f+z}, \tag{8.3.24}$$

and thus from Equations (8.3.23) and (8.3.24), we can derive Equations (8.3.22) easily. ■ ■

Transformations (8.3.22) are called *perspective transformations* and can be employed in the vision computer memory to make it more intelligent. However, they are nonlinear and need to be linearized for algebraic matrix type operations. As discussed in Chapter 2, these transformations can be linearized by use of homogeneous coordinates and transformations. For example, let us consider the following homogeneous transformation:

$$\begin{bmatrix} x'_n \\ y'_n \\ z'_n \\ w'_n \end{bmatrix} = \begin{bmatrix} 1 & 0 & 0 & 0 \\ 0 & 1 & 0 & 0 \\ 0 & 0 & 0 & 0 \\ 0 & 0 & f^{-1} & 1 \end{bmatrix} \begin{bmatrix} x_h \\ y_h \\ z_h \\ w_h \end{bmatrix}, \tag{8.3.25}$$

where the homogeneous coordinates are defined as

$$x_h w_h^{-1} = x, \quad x'_n w_h'^{-1} = x', \quad \text{and so on.} \qquad (8.3.26)_{1,2}$$

Obviously, the relation between the primed and unprimed coordinates become

$$x'_h = x_h, \quad y'_h = y_h, \quad z'_h = 0, \quad \text{and} \quad w'_h = f^{-1}z_h + w_h. \tag{8.3.27}$$

By dividing Equations (8.3.27) by w_h, we obtain

$$x' = w_h^{-1}x'_h = \frac{fx}{f+z}, \tag{8.3.28}$$

$$y' = y'_h w_h^{-1} = \frac{fy}{f+z}, \tag{8.3.29}$$

and

$$z' = 0. \tag{8.3.30}$$

Thus, the identical perspective transformations are obtained. The above transformations do not give any unique information about the object depth given the image. Thus, defining an augmented transformation T_A such that

$$T_A = \begin{bmatrix} 1 & 0 & 0 & 0 \\ 0 & 1 & 0 & 0 \\ 0 & 0 & 1 & 0 \\ 0 & 0 & f^{-1} & 1 \end{bmatrix} \tag{8.3.31}$$

will yield a z' such that

$$z' = \frac{fz}{f+z}. \tag{8.3.32}$$

This can be used to yield additional information on the object such as hidden lines or shades.

As discussed before, other types of transformation can be used to improve the process of image recognition. These transformations include rotation R, translation T, and scaling S and were discussed in Chapter 2. For example, a rotation transformation is represented by

$$\mathbf{R} = \begin{bmatrix} & \mathbf{R} & & 0 \\ & (3 \times 3) & & 0 \\ & & & 0 \\ 0 & 0 & 0 & 1 \end{bmatrix} \quad \text{and} \quad \mathbf{R}'^T = \mathbf{R}'^{-1}. \tag{8.3.33}$$

A translation transformation $\mathbf{T}$ is represented by

$$\mathbf{T} = \begin{bmatrix} 1 & 0 & 0 & x_0 \\ 0 & 1 & 0 & y_0 \\ 0 & 0 & 1 & z_0 \\ 0 & 0 & 0 & 1 \end{bmatrix} \tag{8.3.34}$$

and a scaling transformation S is represented by

$$\mathbf{S} = \begin{bmatrix} 1 & 0 & 0 & 0 \\ 0 & 1 & 0 & 0 \\ 0 & 0 & 1 & 0 \\ 0 & 0 & 0 & s \end{bmatrix}. \tag{8.3.35}$$

8.3.7 Object–Model Matching

Having obtained an image of an object in the vision computer library we seek to choose a model m such that the calculated image of it under some transformation T_m has minimal difference with the observed image of the object. Thus, one must choose the transformation T_m to minimize the error e_m between a set of object image points v_k and the model image points v'_k from a set of model points v_m. These are defined as vertices such that

$$V_p = \{v_{ip}, i = 1, 2, \ldots, n\}, \tag{8.3.36}$$

$$V'_p = \{v'_{ip}, i = 1, 2, \ldots, n\}, \tag{8.3.37}$$

and

$$V_m = \{v_{im}, i = 1, 2, \ldots, n\}, \tag{8.3.38}$$

where v'_k are the predicted vertices of the model with n vertices under transformation T_m. Generally, one wants to minimize the error e_m such that

$$e_m = \sum_{i=1}^{n} |v_{ip} - v'_{ip}|. \tag{8.3.39}$$

Now let a model point, v_{im}, be represented as a homogeneous point; that is,

$$v_{im} \equiv [x_{im} \quad y_{im} \quad z_{im} \quad w_{im}]^T. \tag{8.3.40}$$

Similarly, an object image point v_{ip} can be represented by

$$v_{ip} \equiv [X_{ip} \quad Y_{ip} \quad W_{ip}]^T. \tag{8.3.41}$$

The two coordinate systems are different in the sense that the model points are in a reference three-dimensional coordinate system while the image points are in a two-dimensional coordinate system in the image plane. Now let V_k, V'_k, and V_m represent matrices corresponding to the picture points, image points, and model points, respectively. Thus, V_k and V'_k are $3 \times n$ matrices while V_m is a $4 \times n$ matrix. Suppose there exists a 3×4 transformation H such that once applied to V_m it would transform the model $\mathbf{V}_m$ to its image $\mathbf{V}_k$ pending correction for scaling of each homogeneous point. With zero error, that is, $e_m = 0$, one has

$$\mathbf{HV}_m = \mathbf{V}_p \boldsymbol{\lambda}, \tag{8.3.42}$$

where $\boldsymbol{\lambda}$ is a $n \times n$ diagonal matrix with n unknowns, that is, $[\lambda_1 \; \lambda_2, \ldots, \lambda_n]$.

EXAMPLE 8.3.3 ■

Find a solution for $\mathbf{H}$ if $\lambda = \mathbf{I}$, where $\mathbf{I}$ is the identity matrix.

Solution. From Equation (8.3.42) it is clear that

$$\mathbf{H}\mathbf{V}_m\mathbf{V}_m^T = \mathbf{V}_p\lambda\mathbf{V}_m^T. \tag{8.3.43}$$

Now the matrix $\mathbf{V}_m\mathbf{V}_m^T$ is a 4×4 matrix and can be inverted easily such that

$$\mathbf{H}(\mathbf{V}_m\mathbf{V}_m^T)(\mathbf{V}_m\mathbf{V}_m^T)^{-1} = \mathbf{V}_p\lambda\mathbf{V}_m^T(\mathbf{V}_m\mathbf{V}_m^T)^{-1} \tag{8.3.44}$$

or

$$\mathbf{H}\mathbf{I} = \mathbf{H} = \mathbf{V}_p\mathbf{V}_m^T(\mathbf{V}_m\mathbf{V}_m^T)^{-1}. \quad ■■ \tag{8.3.45}$$

8.3.8 Image Rotation Estimation

Determining the angular rotation of the image of an object is one of the fundamental computational tasks for robotic vision. This is especially important when robots are charged with the task of acquiring randomly placed parts with a gripper, such as a parallel jaw gripper, that must have the right orientation relative to the part. Here we define a number of integral measures, namely, moments, to estimate the degree of rotation of the image.

Suppose that the scene intensity function $I(x, y)$ is given as either a binary image or gray-scale image such that

$$I_b(x, y) = \begin{cases} 1, & \text{if } I(x, y) \geq T_b \quad (8.3.46) \\ 0, & \text{otherwise,} \quad (8.3.47) \end{cases}$$

or

$$I_g(x, y) = \begin{cases} I(x, y), & \text{if } I(x, y) \geq T_g \quad (8.3.48) \\ 0, & \text{otherwise,} \quad (8.3.49) \end{cases}$$

where T_b and T_g are the thresholds (background suppressing or separating).

The mixed (rectangular) moments of an image M_{mn} are defined such that

$$M_{mn} = \iint x^m y^n I(x + x_g, y + y_g)\, dx\, dy \tag{8.3.50}$$

where (x_g, y_g) are the coordinates of the center of area of the image. The polar moments of the image are defined as

$$S_{mn} = \iint r^{n+1}(\cos m\theta)I(r \cos m\theta + y_c r \sin m\theta + y_0)\, dr\, d\theta \tag{8.3.51}$$

and

$$R(i, j) = \mathbf{V}I(i, j)_F. \tag{8.3.52}$$

The polar moment vector $\mathbf{P}_{mn}$ is defined as

$$\mathbf{P}_{mn} = \begin{bmatrix} C_{mn} \\ S_{mn} \end{bmatrix} \tag{8.3.53}$$

and remains length invariant if the image is rotated through an angle θ^*, while $\mathbf{P}_{mn}$ is rotated by $m\theta^*$.

8.3.9 Camera Calibration

The transformations that relate the coordinates of the objects in a world coordinate system to the image coordinates are known as the *camera transformations.* Commonly, the camera is fixed in a coordinate system and it is useful to measure or calibrate the parameters that affect the camera transformations. The procedure is to observe a known three-dimensional object in a known position and orientation and measure points in the image corresponding to known points on the object. One can then compute the necessary transformation based on the equations in Section 8.3.6.

Camera calibration supplies the values of the model parameters that are needed to compute the line of sight ray in space corresponding to a point in the image plane. Given the three-dimensional coordinates of a point in the field of view, the camera transforms allow the image coordinates to be computed. In the inverse camera transformation, the line of sight ray in space is determined from the image coordinates. In robotics it is the inverse transform that is used when an object location has to be computed from an image to allow for object acquisition manipulation.

For stereo vision analysis, two cameras are used and the calibration of the two cameras becomes important and poses additional problems. In the pertinent literature (see the bibilography at the end of this chapter), two methods for camera calibrations have been reported. These are essentially based on the *pinhole* model and the *multiplane* model.

8.3.10 Image Extraction

Image extraction is an important part of robot vision. For example, extraction of lines that correspond to object edges is of primary importance for obvious reasons. Normally, the extraction is based on the assumption that the intensity of light is either constant or smoothly varying over the image or a face of an object and only can suffer a discontinuity at the edge of the face where it intersects with another page.

In a continuous image plane, the gradient of the intensity function is generally finite and is given by Robert's gradient function $R(i, j)$ such that

$$R(i, j) = \nabla I(i, j)_F \tag{8.3.54}$$

or

$$R(i, j) = \{[I(i+1, j+1) - I(i, j)]^2 + [I(i, j+1) - I(i+1, j)]^2\}^{1/2}, \tag{8.3.55}$$

where $I(i, j)$ is the image intensity at pixel point (i, j). The direction of the gradient is defined by α such that

$$\alpha = -\frac{\pi}{4} + \tan^{-1}\left(\frac{I(i, j+1) - I(i+1, j)}{I(i+1, j+1) - I(i, j)}\right). \tag{8.3.55}$$

An edge is claimed to be present at pixel point (i, j) if Robert's cross-operator $R(i, j) > r^*$, where r^* is the chosen threshold for edge extraction. This method,

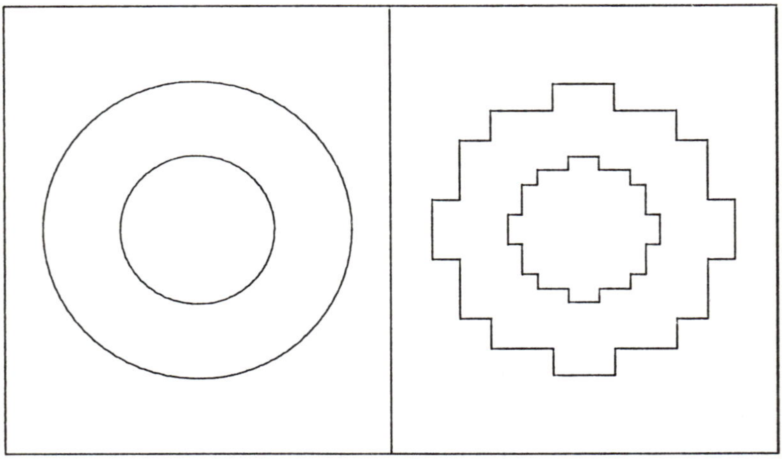

Figure 8.3.18 An object and its digitized binary image.

however, shows all markings on the surface or the texture of the faces of the image and thus careful consideration is necessary.

Recursive algorithms can be developed for contour extractions and the recording of all external and internal contour information. Sometimes image extraction must be preceded by removing ambiguities that arise because of the digitization of images. For example, consider a ring and its digitized binary image in Figure 8.3.18. One may select a number of parameters to compare the two and extract the correct form of the object based on its digitized image. These parameters may include the following:

1. total area,
2. various moments,
3. maximum and minimum radii,
4. location of the center of areas,
5. perimeter length,
6. number of holes,

or any other measurable feature.

8.3.11 Some Typical Vision Systems

In this section three randomly selected commercially available vision systems are described briefly to provide a look at the features that they offer. These are the SRI robotic vision system, the GE optimation II vision system, and the Unimation Univision II robotic vision system.

The SRI (Stanford Research Institute) Robotic Vision System The SRI robotic vision system has been developed as part of the SRI Industrial

Automation Program, sponsored jointly by the National Science Foundation and industrial companies affiliated with that program. The SRI vision module is designed as a research tool for robotic and machine vision to identify, position, and orient machine parts for assembly processes and noncontact inspection and quality control. It is used also for vision-controlled manipulation relative to stationary or moving targets. In fact, all three vision systems introduced here are capable of performing the same functions more or less.

The SRI Mark II Vision Module consists of the three components:

1. A modified GE TN2200, 128 × 128 element solid-state camera, with a 10-meter long cable.
2. A preprocessor interface between the camera and microcomputer, performing general-purpose functions in hardware.
3. A DEC LSI-11/2 microcomputer with 30K word random access memory; this stores the SRI library of basic machine-vision subroutines and some useful application programs, such as "training by showing" for determination of part identity, position, and orientation.

The above system can be interfaced with peripherals that normally are not supplied by SRI but are available from other commercial vendors. These peripherals include a Tektronix 475 oscilloscope capable of displaying the gray scale or the binary images of a part seen by the TV camera, Tektronix 4012 graphics terminal for communicating with the LSI-11/2 computer and for displaying the computer image of a part, and an external microcomputer.

The GE Optimation II Robotic Vision System The GE optimation II robotic vision system is equipped with a TN2500 CID camera with high resolution (244 × 248 pixels) as well as a PN23304 decision processor. This system can perform noncontact quality inspection of up to 900 parts per minute and its software has complex image analysis capability in terms of geometric measurement such as the centroids, ruler, and caliper. The multiprocessor is equipped with 8086, 2901 and 8748 microprocessors. It can operate with up to four cameras and it uses a VPL (Vision Programming Language) that is similar to BASIC.

The Unimation Univision II Robotic Vision System The Univision II robotic vision system is primarily a seam tracking system suitable for automated arc welding purposes. However, it is also capable of other noncontact optical sensing and real-time image processing. Univision II consists of a microprocessor-based image processor, a quick-change hand mechanism for the weld torch, a fiber optics sensing head, a solid-state TV camera, and a software and hardware interface for PUMA robots. The system can adjust to a path to accommodate seam deviations of up to 10 mm in the vertical and horizontal planes. The operation concept is depicted in Figure 8.3.19.

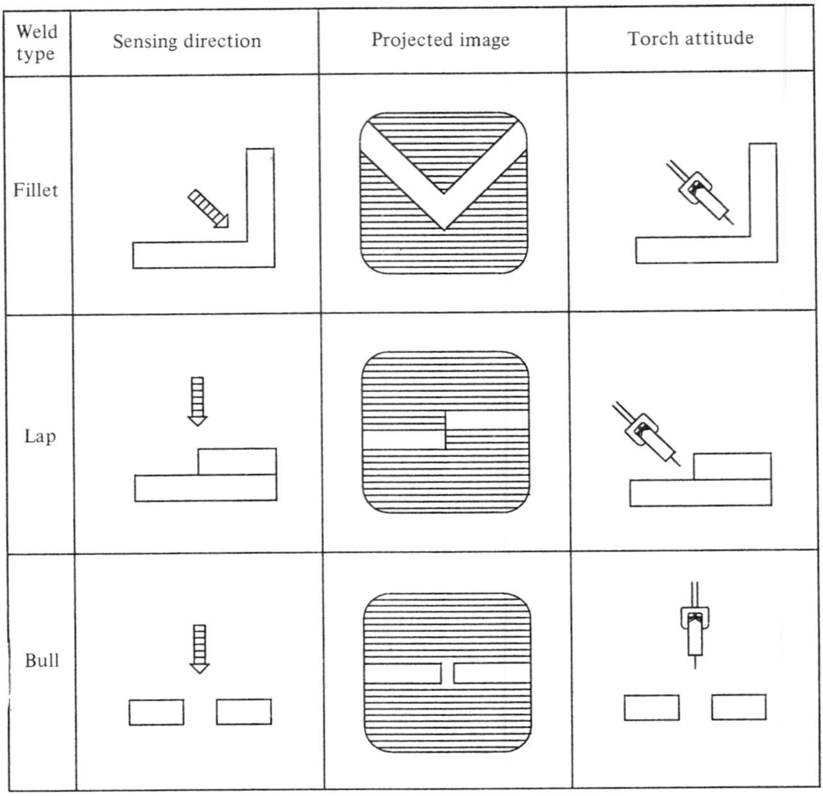

Figure 8.3.19 Seam tracking concept.

8.4 ROBOTIC ASSEMBLY SENSORS

Assembly is a highly intricate and sophisticated human task that is often taken for granted. When one starts to think of robotic manipulators to duplicate various assembly tasks, one begins to appreciate the complexity of the whole process. Furthermore, one is convinced immediately that only intelligent robots can perform assembly processes. Most assembly processes require high precision because they involve mating of parts with complex geometries. The tolerances of parts make the process much more complicated. High-precision robots with high resolutions are generally very expensive and fit only a limited range of applications. Thus, assembly robots must be equipped with suitable sensors for intelligent operations. Generally speaking, the assembly sensors fall into two categories: *passive assembly sensors* and *active assembly sensors*.

The passive assembly sensors are devices such as a remote center of compliance (RCC) and the active assembly sensors are devices such as a six-axis generalized force sensor in which the contact forces and torques are measured and fed back to the robot controller to trigger corrective commands. Both devices compensate for attitude (position and orientation) inaccuracies of the end effector (robotic hand). However, the passive devices do not induce any change in the attitude of the hand but only the tool, while the active devices employ a search strategy by means of which the attitude of the robotic hand is changed.

The assembly process can be considered a positioning problem and one may show schematically the flow of information for a general type configuration of an assembly system. Referring to Figure 8.4.1 we can see the nominal relative position between parts as inputted to a robotic manipulator.

Disturbances to the nominal position because of errors such as jigging, grasping, mechanical inaccuracies of the robot, and resolution problems, and other disturbances such as tolerances of the parts further alter the nominal position. Forces and torques generated by the interaction of parts also alter this position through physical compliance of the robot and parts system. In Chapters 6 and 7, we discussed the various ramifications of disturbances and the jamming conditions pertaining to peg-in-hole assembly processes. Peg-in-hole and screw insertion assembly processes constitute some 35 percent of all assembly tasks. Furthermore, in assembly processes almost 70 percent of parts arrive from one direction, which makes it possible to assemble them with devices with less than six degrees of freedom, such as three-component force sensors or the RCC devices.

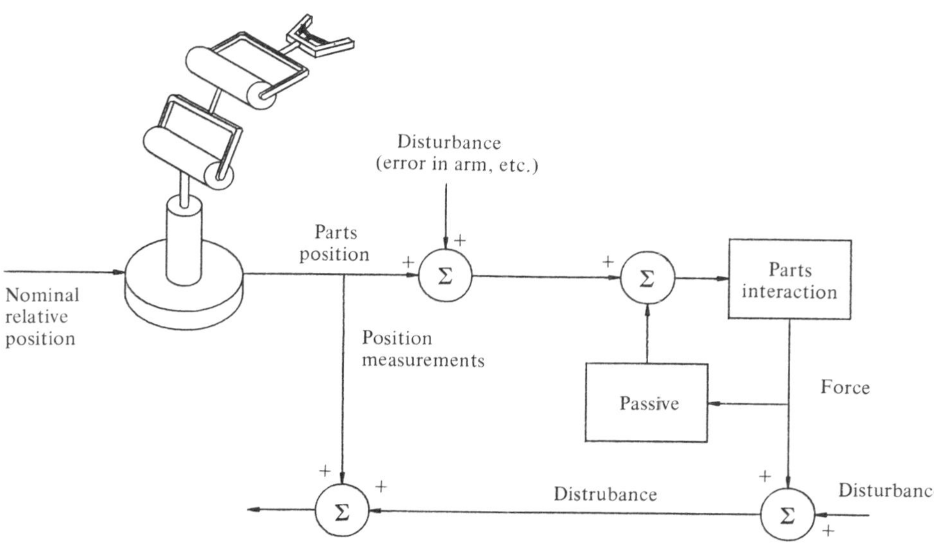

Figure 8.4.1 Assembly information flow.

8.4.1 Remote Center of Compliance (RCC) as an Assembly Aid

The remote center of compliance (RCC) is used in the insertion processes of a peg or a screw in a chamfered hole that can be threaded. An extreme example of such a process is the mounting of the front plate of a clock mechanism with all the drive spindles and shafts in the correct holes; this requires some mechanical manipulative skill for the correct orientation and insertion. Another extreme example is the insertion of a fine screw into a threaded hole, which requires a degree of sophistication in sight and fine force sensing. Insertion of a simple peg in a hole is perhaps the simplest assembly process because the insertion of a screw into a tapped hole requires much greater accuracy of alignment for threads to mesh correctly. Thus, the robot has to make sure that the peg is in a vertical position and that the hole is also vertical. The human hand is capable of inserting a screw into a tapped hole at almost any inclination to the vertical. This of course is done with a lot of help from our vision as well as touch sensors to ensure alignment and proper orientation. Thus, in a typical assembly process, the peg should be pivoted from the top and should be capable of touching the edges (chamfers) of the hole as shown in Figure 8.4.2. The chamfer at the top of the hole or on the end of the peg is assumed to be slightly greater than the maximum displacement error so that the end of the peg can reach the chamfer and slide right down into the hole.

To continue the insertion process, the peg must be rotated easily in all possible ways and translated laterally in all possible ways. This operation can be performed automatically and passively by a device called the RCC originally developed at the M.I.T. Charles Draper Laboratory by Nevin and Whitney in 1977. The device is now commercially available by a number of vendors such as Lord Kinematics. The basic structure of the RCC is shown in Figure 8.4.3.

Normally, the RCC is attached to the end of the last link of the manipulator and acts as a gripper holding the peg to be inserted. As seen in Figure 8.4.3, it consists of two sets of links—one for easy rotations and one for easy translation of the peg without unwanted rotation. The RCC therefore facilitates assembly of close-fitting parts and improves part and tool alignment in sequential machining operations. Basically, the function of the RCC is to provide the necessary *yield* to avoid wedging, jamming, galling, scratching, pitting, and general surface damage, while exerting the necessary force for assembly or machining.

We now describe the basic operation of the RCC with sketches describing how the assembly or insertion takes place using a RCC. Basically, the positional error between the hole and the shaft or the peg creates a force on the leading end of the peg because of the presence of the chamfer. This force has two components—a horizontal force and a vertical force (Fig. 8.4.4a). The horizontal force causes the peg to translate laterally into the hole for easy insertion if the axis of the hole is aligned with the axis of the shaft (Fig. 8.4.4b). If it is not (Fig. 8.4.4c), then a jamming couple is created that causes the RCC links to rotate easily to align the peg passively for easy insertion (Fig. 8.4.4d). By combining the

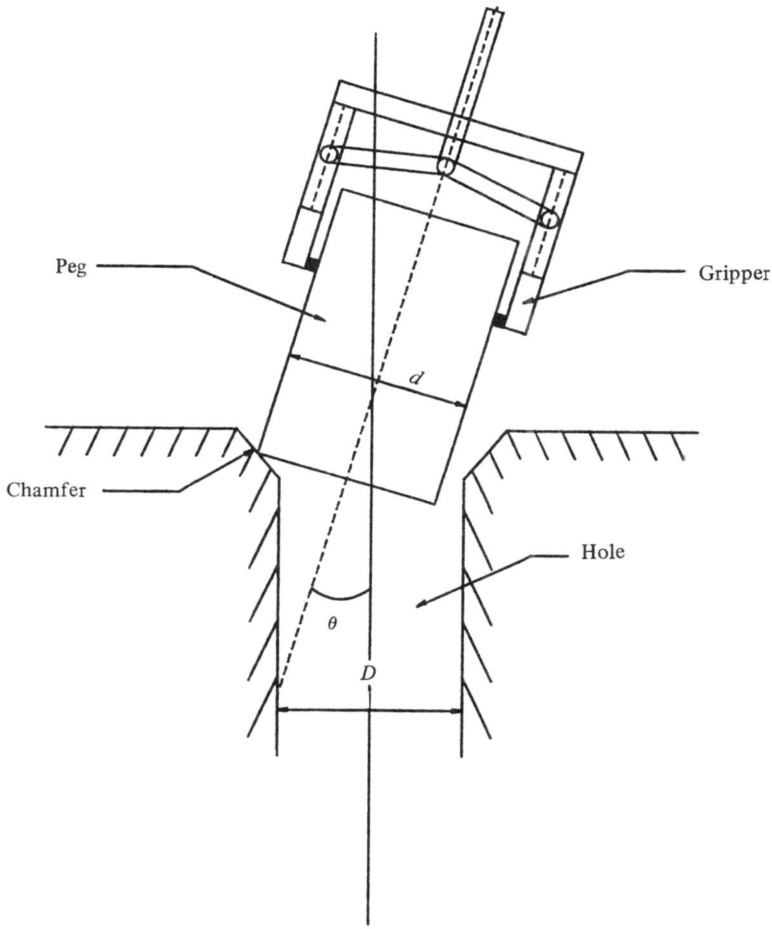

Figure 8.4.2 Peg-in-hole assembly.

two modes of operation, a versatile compliant device is developed (Fig. 8.4.3). Thus, the RCC possesses two outstanding characteristics: (1) controlled elastic center projection or remote center of compliance and (2) controlled flexibility or stiffness.

Controlled flexibility is obtained by means of laminated elastomeric pads and metal shim components. These components are much stiffer in compression than in lateral shear and facilitate controlled elastic center projection and flexibility. By changing the number of pads, elastomeric material, and geometry, a large range of flexibilities can be covered for assembly operations involving fastening or insertion processes. Figure 8.4.5 shows a diagram of a RCC called an accommodator that is commercially available from ASTEK Engineering, Inc. (recently acquired by Barry Wright Corporation).

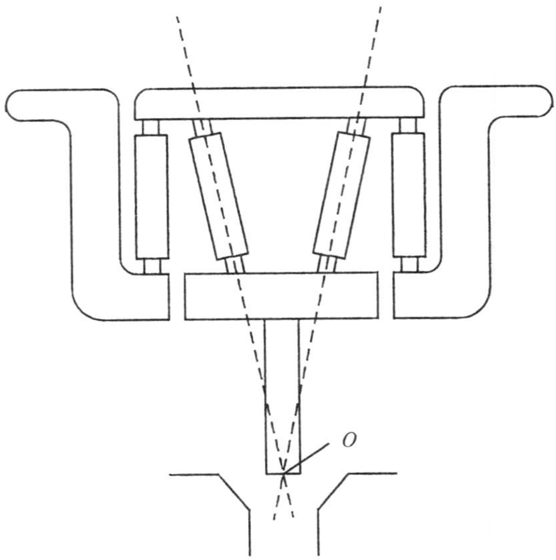

Figure 8.4.3 Basic structure of a RCC wrist.

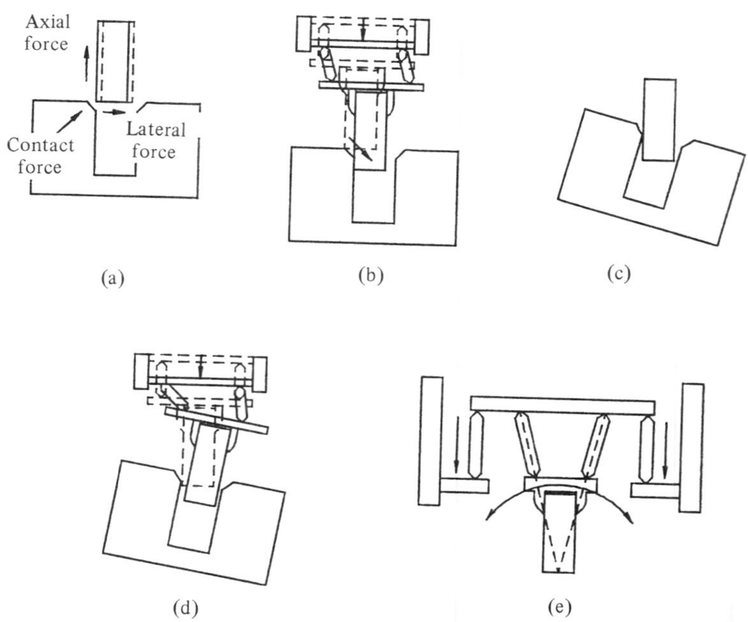

Figure 8.4.4 Insertion operational phases of a RCC.

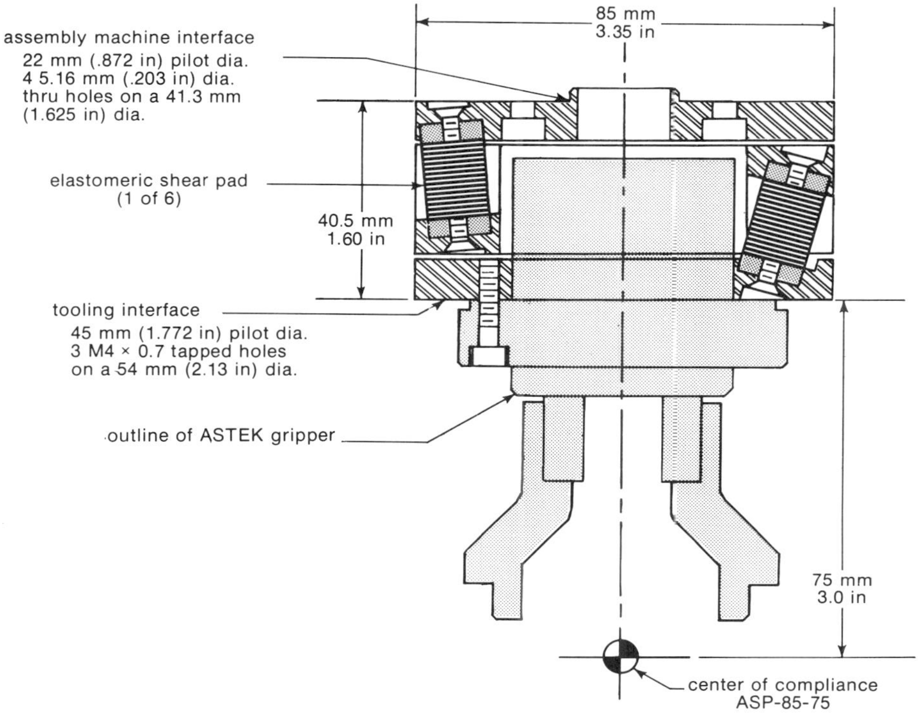

Figure 8.4.5 View of a RCC from ASTEK Engineering, Inc.

RCC devices have a number of disadvantages: the size range of operations is quite limited, the assembly of horizontal pegs is rather difficult, the device does not function effectively in the absence of chamfers, and the device does not allow for dynamic effects in high-speed operations, that is, it is inherently slow.

To overcome these difficulties, one may resort to force and couple sensors that can perform assembly processes even in the absence of RCC devices. This is discussed next.

8.4.2 Robotic Force and Torque Sensors

As discussed in Chapter 7, the generalized force experienced by a robotic gripper is a six-dimensional vector consisting of three components of the force vector and three components of the couple vector at the center of the coordinate frame of the gripper. To sense these six components of the forces and the couples, suitable transducers are mounted properly somewhere between the last robotic link joint and the robotic gripper or hand. These transducers are called *wrist force sensors* because they measure the components of the force and the torque by sensing the deflections of the compliant section associated with those sensors.

A force feedback algorithm is generally capable of accepting force and motion commands, as well as sensed forces and kinematic attitudes, and can

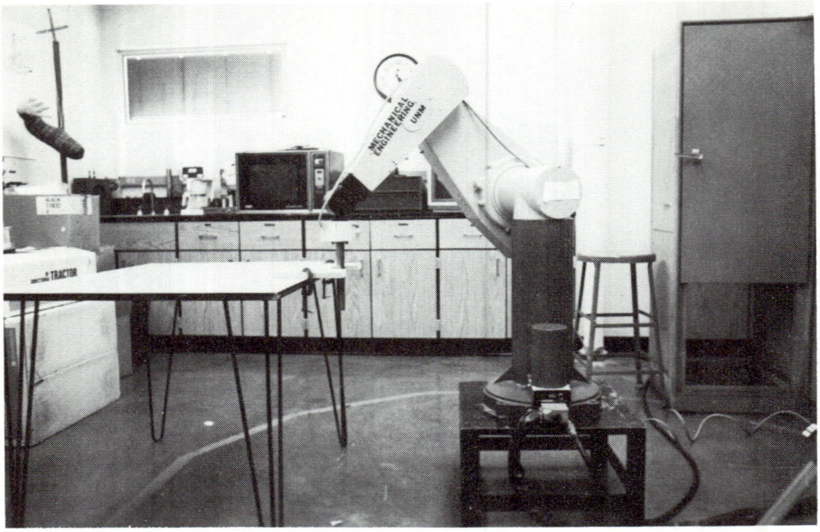

Figure 8.4.6 Six-axis force sensor (ASTEK) on the PUMA 560 hand at the Robotics Research Laboratory, University of New Mexico.

generate motion commands. The force sensors that are used to sense the forces and torques are made from either piezoelectric transducers or typical strain gauges that can be bonded onto sensitive sections of the compliant devices. There are three places where force and torque gauges may be mounted on a robotic arm:

1. At the joints where actuators are at work to measure the force and torque actually put out by the joint actuators.
2. Somewhere between the last joint and the robotic hand, that is, the wrist sensors.
3. At the fingertips of a multifingered robotic hand for fingertip dexterity and hand articulation.

Figure 8.4.6 shows a six-axis force sensor on a PUMA 560 robot at the Robotics Research Laboratory in the University of New Mexico's Department of Mechanical Engineering.

The first robotic wrist force sensor was developed at SRI and consisted of strain gauges cemented on elastic beams as shown in Figure 8.4.7.

The strain gauges should have a high signal-to-noise ratio and low hysteresis to be accurate enough for sensitive force sensing. The structure of the sensor is such that it can measure the forces and torques separately. Referring to the structure of the SRI's six-axis force sensor, we see that there are eight beams—four of them are parallel to the approach axis z and four are in a plane perpendicular to the z direction. Clearly, when a generalized six-axis force $\mathbf{F} = [f_x \; f_y \; f_z \; T_x \; T_y \; T_z]^T$ is applied to a center of compliance of the robotic wrist,

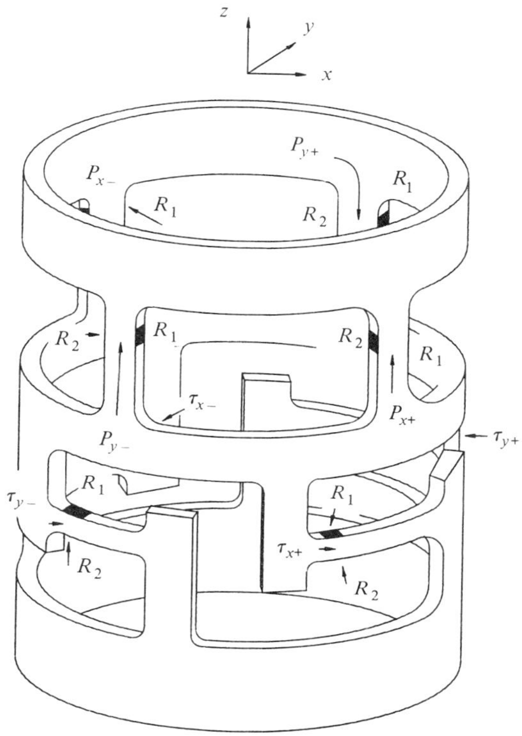

Figure 8.4.7 Six-axis force sensor based on a design at SRI.

the various voltages put out by the strain gauges, that is, v_{x-}, v_{x+}, v_{y-}, v_{y+}, q_{x-}, q_{x+}, q_{y-}, and q_{y+}, are related to the components of **F** by the following relations:

$$f_x = a_x(v_{y+} + v_{y-}), \tag{8.4.1}$$

$$f_y = a_y(v_{x+} + v_{x-}), \tag{8.4.2}$$

$$f_z = a_x(q_{x+} + q_{x-} + q_{y+} + q_{y-}), \tag{8.4.3}$$

$$\tau_x = b_x(q_{y+} - q_{y-}), \tag{8.4.4}$$

$$\tau_y = b_y(q_{x-} - q_{x+}), \tag{8.4.5}$$

and

$$\tau_z = b_z(v_{x+} - v_{x-} - v_{y+} + v_{y-}). \tag{8.4.6}$$

It must be noted that similar types of six-axis force sensor have been developed at the Charles Draper Laboratory and other robotics research laboratories. Three-axis force sensors have been around for quite a while now and sometimes offer advantages over the six-axis force sensors because of their simplicity.

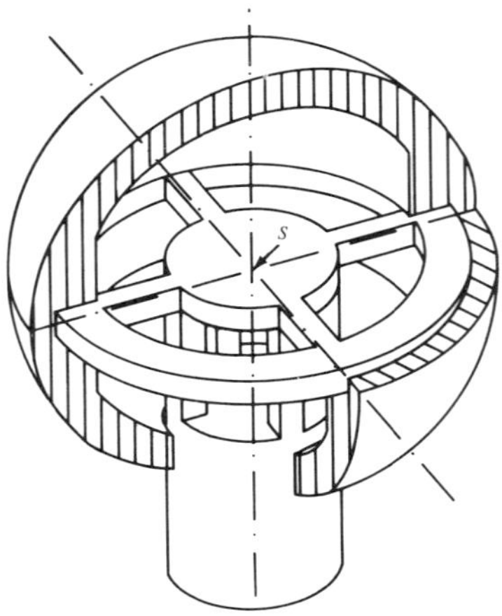

Figure 8.4.8 Sensor ball developed at DFVLR, West Germany.

Force and torque perception can be taught to a robot by what is called a sensor ball as developed in West Germany by DFVLR. This is shown in Figure 8.4.8.

8.5 OTHER TYPES OF SENSOR FOR INTELLIGENT ROBOTS

In this section we discuss other types of intelligent robotic sensor that are presently in use in robotics research and automated/flexible manufacturing operations. These types of sensor generally fall into three distinct categories:

1. Proximity sensors.
2. Range sensors.
3. Tactile sensors.

We discuss each of the above separately.

8.5.1 Proximity Sensors

Proximity sensors are used quite frequently in robotic applications as devices that sense when the end effector, the robotic hand, or any other part of the robot body is close to other objects in its workspace. The distances can vary anywhere from a few inches to a few millimeters. To sense larger distances, one may use range finders, which are discussed later. Most proximity sensors indicate only the presence or absence of an object within their workspace (sensing domain). Few proximity sensors can give information regarding the distance between the sensor and the object to be sensed. The following list provides a brief description of a variety of these sensors.

1. Optical proximity sensors measure the amount of light reflected from an object. The IR sensors with incandescent light source are quite

common. Light emitting diodes (LEDs) are quite reliable as solid-state devices because they are fairly insensitive to shock and vibration. Laser diodes also have been used as optical proximity sensors; however, they are expensive compared to LEDs.

2. Magnetic-field sensors basically measure the presence of an object that can alter magnetic flux and induce effects such as the Hall effect and magnetoresistive effects.
3. Eddy-current proximity sensors produce an alternating magnetic field in the tip of the robotic end effector or probe. This alternating field induces an eddy current in any conductive body that penetrates the field. These proximity sensors are especially useful for very close approaches and proximities in the range of 1 mm.
4. Electrostatic-effect proximity sensors can be built and operated in just the same manner as the electromagnetic-effect proximity sensors. The problem with such electrostatic-effect sensors is that they are overly sensitive to stray fields radiated by electronic equipment and to fields from static charges produced by friction or by spraying.
5. Touch-sensitive and limit-switch type proximity sensors are in fact very close approach and touch detectors. These devices normally react to contact with a large conductive body, such as a person. However, by attaching a conducting body to the contact point, these devices can respond to objects at a distance by means of their own capacitance.
6. Acoustic proximity sensors have been used successfully for robotic applications recently. Essentially, the presence of an object changes the sound pressure received by a microphone in such devices.

A typical proximity sensor operation is shown in Figure 8.5.1. In addition, there are proximity sensors based on fluid mechanical and chemical phenomena, but we do not discuss them here.

8.5.2 Robotic Range Detectors

A range sensor is a device that can provide precise measurement of the distance between the sensor itself and an object. Commonly, they require a single line of sight in order to look along the optical axis of a camera lens or to hear along the sonic axis of an ultrasonic device such as a sonar. One problem with acoustic ranging devices is that their spatial resolution as a whole is inferior to optical range finders. It is generally difficult to produce a narrow beam of sound.

Range detection devices basically measure the reflection time of a laser or sonar beam to the object and back to the source to measure the object's distance from the beam source. In doing so, they operate on the principle of beam triangulation (Fig. 8.5.2)

From Figure 8.5.2 we see that d_i is the distance of the ith light detector and x is the distance of the object from the detector. It can be shown that

$$x = \frac{dh}{d_i - d}. \tag{8.5.1}$$

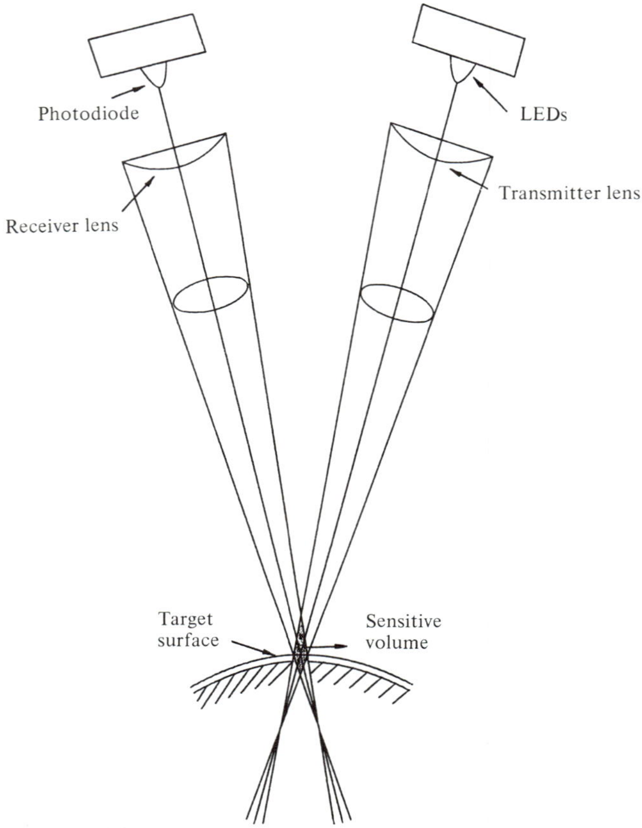

Figure 8.5.1 Operation of a typical proximity sensor.

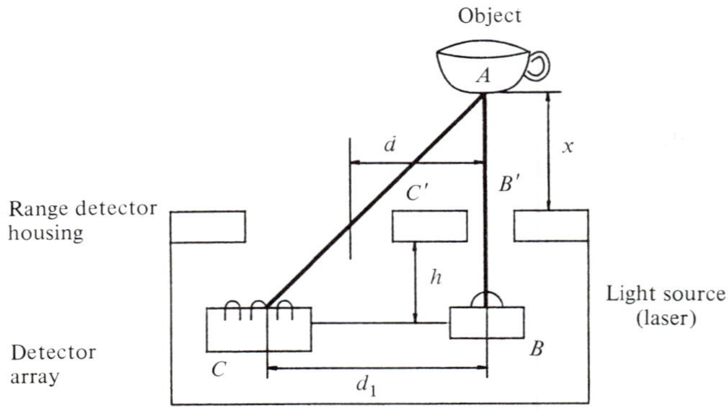

Figure 8.5.2 Triangulation principle used in range detectors.

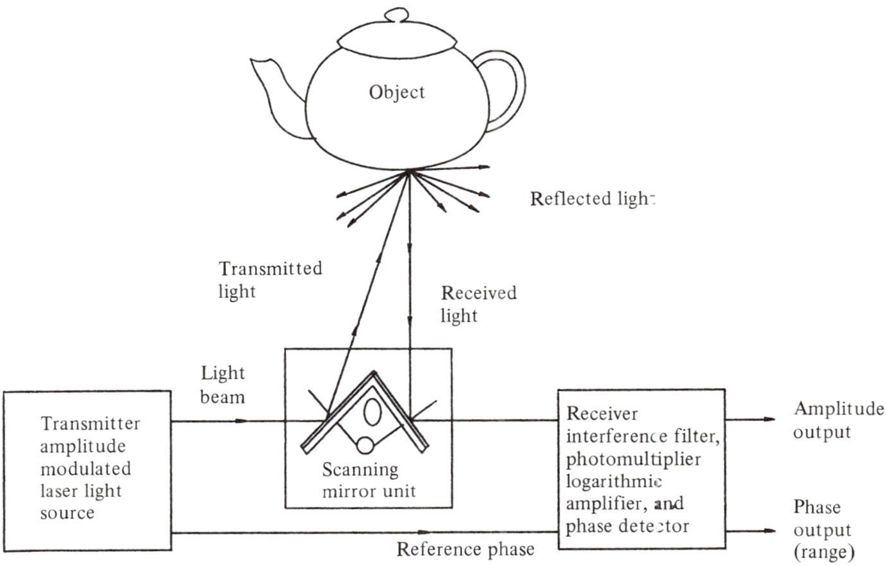

Figure 8.5.3 Simplified block diagram of a range imaging sensor.

EXAMPLE 8.5.1 ■

Show that the distance x between an object and the range detector in Figure 8.5.2 is given by Equation (8.5.1).

Solution. Referring to Figure 8.5.2, consider the two similar triangles ABC and $AB'C'$. Note that in this case

$$\frac{x}{x+h} = \frac{d}{d_i}. \tag{8.5.2}$$

Equation (8.5.1) easily follows. ■ ■

Scanning lasers on autonomous and mobile robotic manipulators also have been used to determine the location of the robot in a crowded workspace. Again the principle of triangulation is used for many objects to determine the location of the robot with respect to all other objects present.

A simplified block diagram of a range detector is shown in Figure 8.5.3.

8.5.3 Tactile Sensors for Robotic Applications

A lot of work has been done in robotics to provide the equivalent of the human touch and muscular sensors of force, roughness, dryness, hotness, surface texture, and so on. Thus, tactile sensors are made to respond to contact forces that arise between the sensors themselves and the surfaces of solid objects. Tactile forces are either of single touch type or of contact point matrix type, giving information about the topography and texture of contact surfaces.

Generally, they consist of either an array of individual points or *tactels* or some variety of pulsed devices using time domain processes to determine the state of any particular contact point and to construct a pressure image of the surface.

Unlike proximity sensors, tactile sensors do not indicate the presence of objects until an object actually comes into contact with a sensor. Thus, a useful combination of sensors in a robotic end effector is a proximity sensor working in conjunction with a touch sensor because the proximity sensor can detect the proximity of an object that is destined to be examined by the tactile sensor and can move the end effector safely toward that object. The signal from the proximity sensor can command the controller to move the end effector quickly toward the object and then to slow down when it gets close to the object in order to avoid collision. Tactile sensors also can be categorized as either touch sensors or stress sensors.

Touch sensors produce a binary output signal just like the binary image of a vision sensor depending on whether or not they are in contact with an object, while stress tactile sensors produce a signal that indicates the magnitude and distribution of contact forces similar to a gray-scale image of a vision sensor. By suitable geometrical arrangement of tactile sensing arrays, it is possible to measure the forces and couples in two or three directions. The following different types of tactile sensor for robotic applications are presently in use:

1. Microswitches, which are the simplest and least expensive tactile sensing devices of the binary output type.
2. Silver-doped rubber pads that become conductive if pressed because the conducting granules embedded in the pads contact each other as a result of the high elasticity of the rubber pad.
3. Strain gauges, which are used also as force and couple sensors and were extensively discussed in Section 8.4.2. In this mode of operation they respond to contact forces and couples experienced by the robotic gripper in the form of a wrist sensor as shown in Figure 8.5.4. The sensors are such that the individual strain gauges are attached to the roots of cantilever beams milled into the solid blocks of the metallic frame. The orientation of the beams and the connections between them are such that the six components of the generalized force, that is, F_x, F_y, F_z, M_x, M_y, and M_z, can be measured as shown in Figure 8.5.4b and can be related to the contact forces experienced by the gripper. For example, surface roughness or friction can be translated into various components of force and couple, and in this sense the wrist sensor may act as a tactile sensor.
4. Resistive and capacitive array tactile sensors that react to applied contact forces by changing their resistivity or capacitance as more points contact each other.
5. Magnetoelastic tactile sensors that sense the magnitude of contact forces by elastic deformation changes in magnetostriction permeability and the resulting secondary inductions.

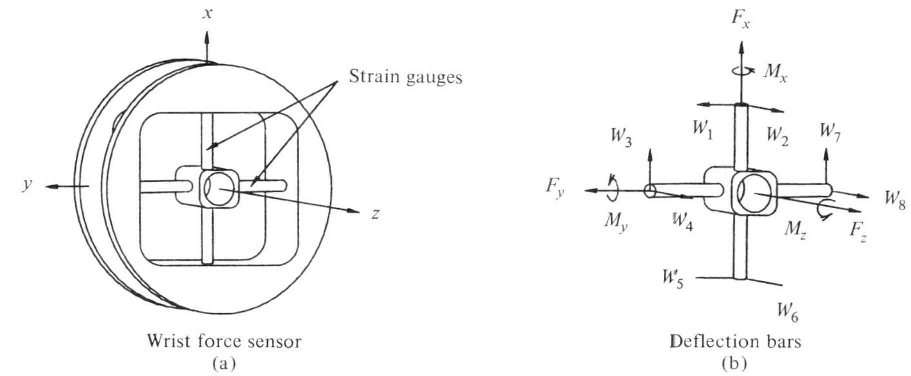

Figure 8.5.4 Wrist sensor as a tactile sensor.

6. Piezoelectric tactile sensors that respond to contact forces by creating charges that can be converted to either a current or a voltage. Two commonly used crystals are the PZTs (lead–zirconate–titanate) piezoceramics and lithium niobate ($LiNbO_3$) because of their linear responses.
7. Electrooptical tactile sensors in which the magnitude of contact forces is converted into an optical signal by either mechanical displacement or other electromechanical effects.

Figure 8.5.5 displays a robotic gripper configuration with a pair of tactile sensing arrays as well as a six-axis wrist sensor.

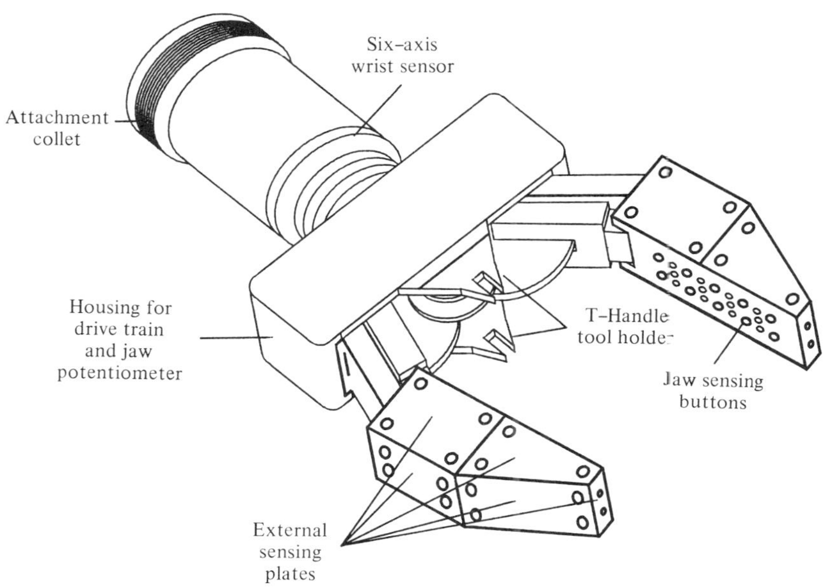

Figure 8.5.5 Hand with wrist force sensor and analog touch sensors.

BIBLIOGRAPHY

Birk, J., Kelley, R., Barron, T., Crouch, J., Duncan, D., Hall, J., Kolanko, F., Mak, A., Martins, H., Mehta, R. Spirito, A., Tella, R., and Vinci, A. (1980). "General Methods to Enable Robots with Vision to Acquire, Orient and Transport Work Pieces," University of Rhode Island, National Technical Information Services (Aug.).

Nevins J. L. and Whitney, D. E. (1978). Computer-Controlled Assembly, *Sci. Am.*, **238**(2), 62–73.

Nevins, J. L. and Whitney, D. E. (1980). Assembly Research, *Indus. Robot*, **7**(1), 27–43.

Roberts, L. G. (1968). Machine Perception of Three-Dimensional Solids, in *Optical and Electro-Optical Information Processing*, J. T. Tippet et al. (Ed.), M.I.T. Press, Cambridge, MA, pp. 159–197.

Simunovic, S. (1975). Force Information on Assembly Processes, in *5th International Symposium on Industrial Robots*, Chicago, IL, pp. 415–431 (Sept.).

Toepperwein, L. L., Blackman, M. T., Park, W. T., Tanner, W. R., and Adolfson, W. F. (1983). "Robotics Applications for Industry: A Practical Guide," Neyes Data Corp., Dark Ridge, NJ.

HOMEWORK PROBLEMS

1. Give brief descriptions of proximity sensors that are based on inductive or capacitive properties of materials.
2. Give brief descriptions of some acoustic sensors.
3. Referring to Example 8.2.2, find the generalized differential vector of the object frame with respect to the gripper frame if the generalized differential vector of the gripper frame with respect to the camera frame is given by

$$^{c}\mathbf{D}_g = [-0.1 \quad 0.1 \quad 0 \quad 0.1 \quad 0 \quad 1]^T$$

and the gripper and the camera frames with respect to the frame of link 5 are given by

$$^{5}\mathbf{G} = \begin{bmatrix} 1 & 1 & 0 & 12 \\ -1 & 0 & 0 & 10 \\ 0 & 0 & 1 & 8 \\ 0 & 0 & 0 & 1 \end{bmatrix} \quad \text{and} \quad ^{5}\mathbf{C} = \begin{bmatrix} 1 & 0 & 0 & 5 \\ 0 & 1 & 0 & 0 \\ 0 & 0 & 1 & -15 \\ 0 & 0 & 0 & 1 \end{bmatrix}.$$

4. Find the inclination of the binary image shown in Figure P8.4.

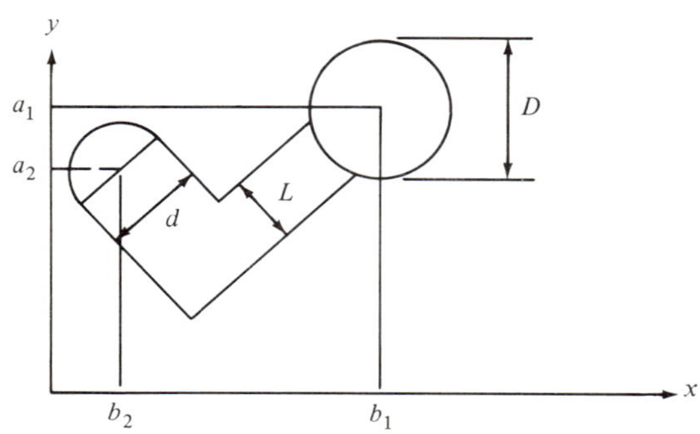

Figure P8.4 Inclined binary image.

5. Referring to Example 8.3.3 and Equation (8.3.42), find a set of solutions for **H** and λ when $\lambda \neq \mathbf{I}$.

6. Referring to Figure 8.4.4, explain the various phases (a), (b), (c), and (d) of an RCC device in terms of the forces and moments applied to the tip of the peg.

7. Describe a proximity sensor that could operate based on changes in a fluid velocity or pressure.

8. Consider the laser range-finding device shown in Figure P8.8. Show that the range Z of the target is given by

$$Z = \frac{(E + M)F}{M + F}.$$

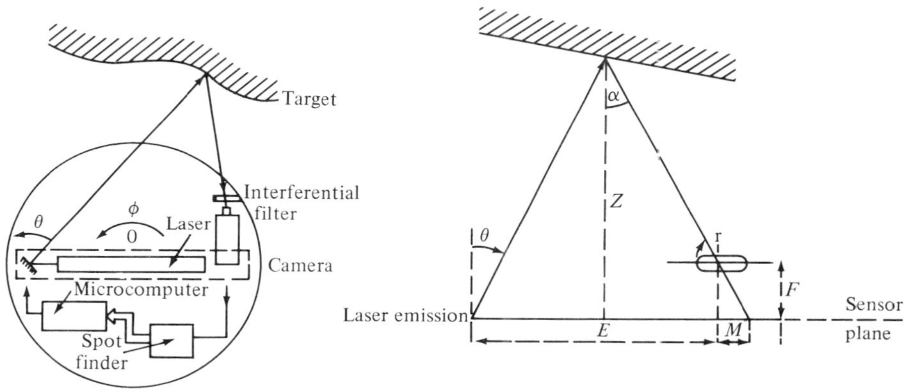

Figure P8.8

Chapter 9

Robotic Motion Programming

9.1 INTRODUCTION

In this chapter we address some important issues in robotic motion programming. The programming of multiaxis robotic manipulators is much more involved than the programming of NC machines. Many robotic systems employ what is called on-line programming in which the robot itself is used at various programming stages. This is also known as the *show-and-teach* technique of robot programming in which the operator moves the robotic arm through a set of required points or trajectory in the three-dimensional workspace by means of push buttons, joysticks, mice, control handles, or the computer keyboard. Thus, the computer remembers all the points and details of motion trajectory so that it can command the robot to reproduce the motion exactly at a later time. On the other hand, off-line programming techniques employ programs that command the robot without originally using the robot to write the program. In this case, the software must be able to solve the inverse kinematics problem in general and to manipulate the robotic joints to achieve any desired kinematic attitude for the gripper. Thus, there are three programming techniques employed in robotics:

1. *Manual teaching* or *show-and-teach.*
2. *Lead-and-teach* or *guiding.*
3. Programming languages.

The first two techniques fall in the domain of on-line programming because they require the robot itself to implement.

9.2 MANUAL TEACHING OR SHOW-AND-TEACH

This normally is done by means of a teach box on which a number of buttons and joysticks enable the programmer to move the robot from one kinematic attitude (i.e., position and orientation) to another kinematic attitude and subsequently save every move in the memory of a computer and finally create a file of motion that can be replayed by the robot.

In the programming section of this chapter, a teach program that converts the entire keyboard of an IBM PC-compatible computer to a sophisticated teach pendant is presented to illustrate the manual teaching technique (Fig. 9.2.1). In addition to manual teaching, operations such as closing the gripper or activating a welding tip are prescribed at intermediate positions and recorded.

9.3 LEAD-AND-TEACH OR GUIDING

This involves physically grasping the robot end effector and leading it through any desired path at the required speed and possibly acceleration while simultaneously storing and recording the continuous positions and orientations. This method can be applied only to a certain class of robots for which the motors can be moved manually while feedback encoders keep track of the motion trajectories, thus recording such motions. Another solution is to have a second robot or a robotic simulator (teaching arm), whereby motion can be taught to the robotic simulator and that in turn recorded for use on the robot itself. In this case, the simulator is grasped manually by the operator and is led through the desired motion trajectory and the motion is computer recorded by sampling at a certain frequency. Because of a number of disadvantages with this lead-and-teach technique it is not widely used. However, for highly articulated robotic tip manipulation such as special welding, spray painting of complex auto parts, and complex motion trajectories, it is still used.

9.4 PROGRAMMING LANGUAGES

Some robotic systems are equipped with computer programming languages capable of accessing information from the sensors mounted on the robotic hand as well as specifying the robotic motion. These languages are referred to as *explicit* or *robot-level* programming languages. The most important advantage of an explicit programming language for robotics is that it enables the use of external data from various robotic sensors, such as force, vision, and motion, in the modification of various robotic tasks and manipulations (Fig. 9.4.1). Obviously, under these circumstances, the robot can cope with a greater degree of randomness and uncertainty in its workspace and thus enlarge its sphere of application. The main disadvantage of robot-level programming language as compared to show-and-teach or lead-and-teach is that it requires the robot

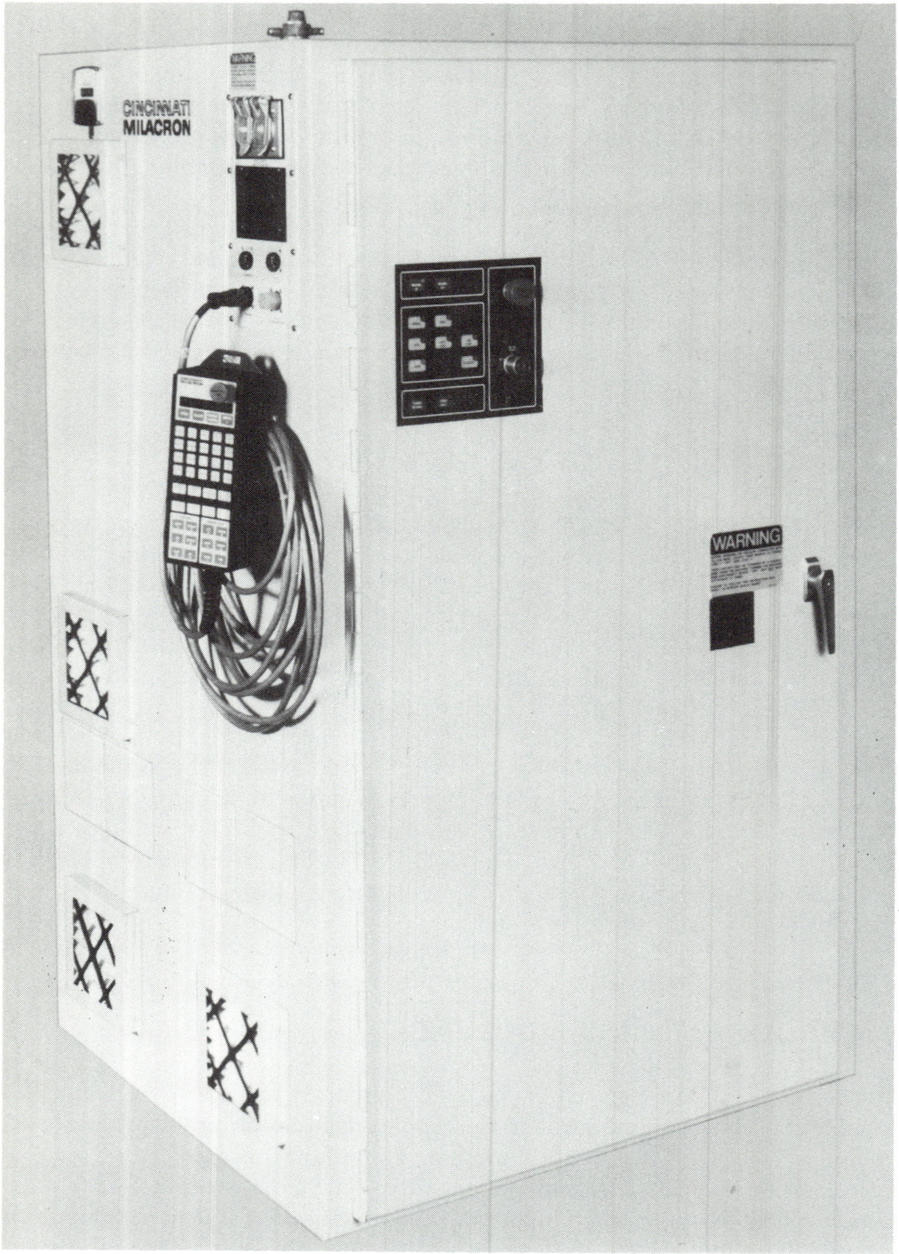

Figure 9.2.1 Typical robot control and teach pendant. (Courtesy of Cincinnati Milacron.)

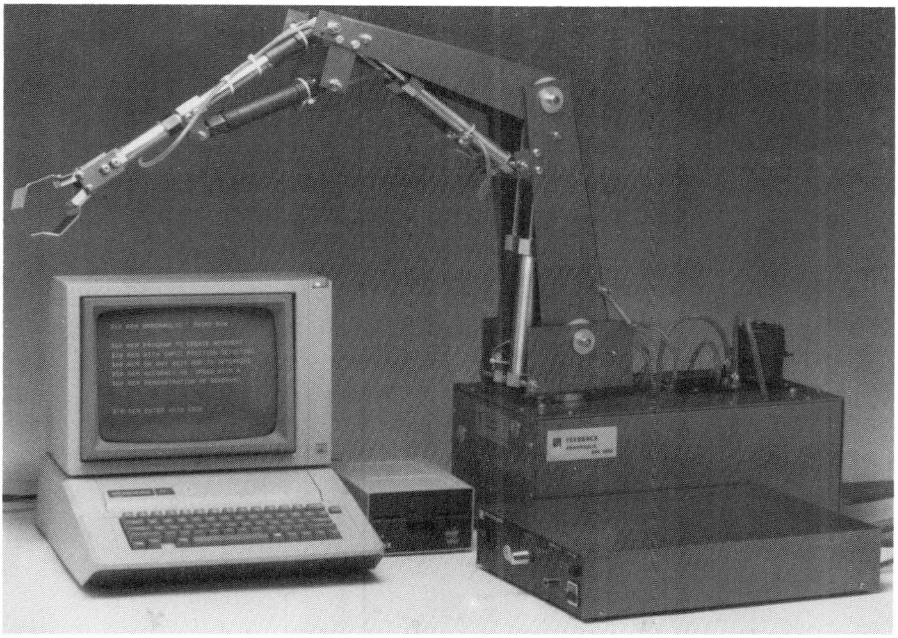

Figure 9.4.1 Typical robot + computer + controller. (Courtesy of Feedback Corp. USA.)

programmer to be an expert in programming and the design of robotic sensor-based motion strategies. A typical factory-floor worker is not equipped with such expertise. One way of circumventing this difficulty is to resort to *task-level* programming. In task-level programming the programmer does not have to be concerned about the details of every action that the robot is to take but rather is concerned with the commands that achieve the desired subgoals of the task directly. For example, if the command is to "place the coffee cup on the table," then the system automatically plans an optimum path for the manipulator that avoids collision with any obstacles in its path and places the cup on the table in an appropriate manner. Task-level programming is under development and should be available soon.

9.5 SURVEY OF ROBOT-LEVEL PROGRAMMING LANGUAGES

There are two excellent survey papers on robot programming languages that have appeared in the recent literature. One is "Robot Programming" by T. Lozano-Perez (1983) and the other is "Industrial Robot Programming Languages: A Comparative Evaluation" by W. A. Gruver et al. (1984). Readers are referred to these papers for further details. Here we briefly review the pertinent literature and present some typical peg-in-hole programs.

The first robot-level programming language was developed at M.I.T. by Ernst (1961). It was named a *mechanical hand interpreter* or MHI. The programming structure of MHI is modeled around *guarded moves*; that is, the hand moves until a sensor detects a particular situation. A typical program would involve commands such as "move" to indicate a direction and a speed, "until" to test a sensor, "ifgoto" to indicate a logic branching, and "ifcontinue" to continue the program if the desired conditions are met.

The next robotic language was developed at M.I.T. by Silver (1973). This was named MINI and was basically an extension of the LISP system. The serious limitation of MINI was that it could not control the robotic joints dependently. It was used in conjunction with a rectangular robot with a six-degree-of-freedom force-sensing wrist with a sensitivity of about 1 oz. The main part of an insertion for a peg-in-hole program reads as follows:

```
PEG-IN-HOLE INSERTION PROGRAM IN MINI
(DEFUN MOVE-ABOVE(P OFFSET))
;set x, y, z goals and wait until they are reached
(X = (X-LOCATION P))
(Y = (Y-LOCATION P))
(Z = (PLUS(Z-LOCATION P)OFFSET))
(WAIT'(AND(?X)(?Y)(?Z)))
(DEFUN INSERT(HOLE))
(MOVE-ABOVE HOLE 0.25)
;define a target 1 inch below current position
(SET Q ZTARGET (DIFFERENCE(GETM ZPOS) 1.0))
;move down until a contact force is met or until
;the position target is met
(FZ=LANDING-FORCE)
(WAIT'(OR(?FZ)(SEQ(GETM ZPOS) ZTARGET)))
(COND((SEQ(GETM ZPOS)ZTARGET)))
     ;if the position goal was met, i.e., no surface
      encountered
     ;comply with lateral forces
     (FX=0)(FY=0)
     ;and push down until enough resistance is met.
     (FZ=INSERTION-FORCE)
     (WAIT'(FZ)))
     (T;if a surface was encountered
     (ERROR INSERT)
```

The next generation of robotic software was developed in the Stanford Artificial Intelligence Laboratory (SAIL) by Paul (1977) and was an experimental research software designed as a general purpose robot programming language (WAVE). The syntax of WAVE was modeled after the assembly language of PDP-10 and thus was a low-level syntax. Despite that, the software provided many high-level functions enabling the programmer to describe the kinematic attitude (position + orientation) of the robotic gripper and to coordinate joint motions for continuity in velocities and accelerations. Most important of all was the ability of WAVE to specify the robotic joint and hand compliance. The WAVE software was also quite successful in robotic motion trajectory design because small deviations from the designed trajectory could be observed during execution. A typical program for the insertion of a peg in a hole given by Paul (1977) reads as follows:

```
PEG-IN-HOLE INSERTION PROGRAM IN WAVE
NOTE:  USE OF GUARDED MOVES
NOTE:  USE OF COMPLIANCE
    TRANSPIN...                  ;Location of pin
    TRANSHOLE...                 ;Location of hole
    ASSIGN TRIES 2               ;Number of pickup attempts
    MOVE PIN                     ;Move to PIN, MOVE first
                                  moves in + Z
                                 ;then to a point above PIN,
                                  then -Z
PICKUP:
    CLOSE 1                      ;Pickup pin
    SKIPE 2                      ;Skip next instruction if
                                  Error 2 occurs
                                 ;(Error 2: fingers closed
                                  beyond argument
                                 ;to CLOSE)
    JUMP OK                      ;Error did not occur, go to
                                  OK
    OPEN 5                       ;Error did occur, open the
                                  fingers
    CHANGE Z,-1.NIL,0,0          ;Move down 1 inch
    SOJG TRIES, PICKUP           ;Decrement TRIES, if not
                                  negative
                                 ;jump to PICKUP
    WAIT NO PIN                  ;Print "NO PIN" and wait
                                  for the operator
    JUMP PICKUP                  ;Try again when operator
                                  types
                                 PROCEED
OK:
    MOVE HOLE                    ;Move above hole
    STOP FV.NIL                  ;Stop on 50 oz.
    CHANGE Z,-1.NIL,0,0          ;Try to go down one inch
    SKIPE 23                     ;Error 23, failed to stop
    JUMP NOHOLE                  ;Error did not occur (pin
                                  hit surface)
```

```
    FREE 2,X,Y                  ;Proceed with insertion by
                                 complying
                                ;with forces along x and y
    SPIN 2,X,Y                  ;Also comply with torques
                                 about x and y
    STOP FV.NIL                 ;Stop on 50 oz.
    CHANGE Z,-2,NIL.0,0         ;Make the insertion
NOHOLE:
    WAIT NO HOLE                ;Failed
```

Later, a more advanced robot programming language based on ALGOL software was developed at SAIL and was named AL [see Finkel et al. (1974) and Mujtaba and Goldman (1979)]. This new language had the capability of programming assembly operations using force-sensing devices. AL is suitable for both robot-level and task-level programming. Language statements include motion instructions such as "Move Arm To" (position) and "Open Hand To" (value) and conditional commands such as "On Force (value) Do (operation)". A typical peg-in-hole insertion program in AL reads as follows:

```
PEG-IN-HOLE INSERTION PROGRAM IN AL
BEGIN "insert peg into hole"
    FRAME peg_bottom,peg_grasp,hole_bottom,hole_top;
    {The coordinate frames represent actual positions
    of object features, not hand positions}
    peg_bottom←FRAME(nilrot,VECTOR(20,30,0)*inches);
    hole_bottom←FRAME(nilrot,VECTOR(25,35,0)*inches);
    {Grasping position relative to peg_bottom}
    peg_grasp←FRAME(ROT(xhat,180*degrees).3*zhat*inches);
    tries←2;
    grasped←FALSE;
    [the top of the hole is defined to have a fixed
    relation to the bottom]
    AFFIX hole_top to hole_bottom RIGIDLY
              AT TRANS(nilrot,3*zhat*inches);
    OPEN bhand TO peg_diameter + 1*inches;
    {Initiate the motion to the peg, note the
    destination frame}
    MOVE barm TO peg_bottom*peg_grasp;
    WHILE NOT grasped AND i<tries DO
         BEGIN "Attempt grasp"
         CLOSE bhand<TO 0*inches;
         IF bhand<peg_diameter/2
           THEN BEGIN "No object in grasp"
             OPEN bhand TO peg_diameter+1*inches;
             MOVE barm TO ⊗ -1*inches; {⊗indicates
             current location}
             END
```

```
            ELSE grasped ← TRUE;
          i←i+1;
          END
     IF NOT grasped THEN ABORT ("Failed to grasp the
     peg");
     {Establish a fixed relation between arm and peg.}
     AFFIX peg_bottom TO barm RIGIDLY;
     {Note that we move the peg_bottom, not barm}
     MOVE peg_bottom TO hole_top;

     {Test if a hole is below us}
     MOVE barm TO ⊗ -1*inches
          ON FORCE(zhat)>10*inches DO ABORT("No Hole");

     {Exert downward force, while complying to side
     forces}
     MOVE peg_bottom to hole_bottom DIRECTLY
     WITH FORCE_FRAME = station IN WORLD
     WITH FORCE(zhat) =-10*ounces
     WITH FORCE(xhat) = 0*ounces
     WITH FORCE(yhat) = 0*ounces
     SLOWLY;
END "insert peg in hole"
```

One of the best-known commercial languages for robotic motion programming and control is the VAL software developed by Shimano (1979) and Unimation Inc. (1980, 1983). The robotic software that runs on a DEC LSI-11 computer to control the robot consists of both coordinate and motion specifications. It provides a subset of WAVE on a stand-alone minicomputer. It supports relative motion, which in conjunction with the capability of stopping a motion in response to a signal provides the ability for guarded moves. VAL is similar to BASIC in the sense of basic commands but does not support floating-point numbers or character strings and arithmetic on position data. Neither can VAL support compliant motion and thus a typical peg-in-hole insertion program in VAL assumes that the clearance between the peg and the hole is greater than the robot's accuracy. To circumvent this problem one may mount a multiaxis passive force sensor or a compliance device on the robot's end effector. A typical program in VAL for the insertion of a peg in a hole reads as follows:

```
PEG-IN-HOLE INSERTION PROGRAM IN VAL
     SET1      TRIES=2

     REMARK    If the hand closes to less than 100 mm,
               go to statement labeled 20.
10   GRASP     100,20
     REMARK    Otherwise continue at statement 30.
20   GOTO      30
```

```
        REMARK     Open the fingers, displace down along
                   world Z axis and try again.
20      OPEN1      500
        DRAW       0.0,-200
        SET1       TRIES=TRIES-1
        IF         TRIES GE 0 THEN 10
        TYPE       NO PIN
        STOP
        REMARK     Move 300 mm above HOLE following a
                   straight line.
30      APPROS     HOLE,300
        REMARK     Monitor signal line 3 and call procedure
                   ENDIT to STOP the program.
        REMARK     if the signal is activated during the
                   next motion.
        REACT1     3,ENDIT
        APPROS     HOLE,200
        REMARK     Did not feel force, so continue to HOLE.
        MOVES      HOLE
```

The following program in VAL-II moves a PUMA 560 robotic manipulator in a random manner depending on the range of numbers given. The program could be used for waving a flag.

```
PROGRAM random
        HERE #ref
        DECOMPOSE h[] = #ref
        FOR number = 1 TO count
             FOR i = 0 TO 5
                  j[i] = (RANDOM*2-1)* mag+h[i]
             END
             MOVE #PPOINT(j[0], j[1], j[2], j[3], j[4],
             j[5])
        END
        RETURN
```

IBM in the meantime was developing a much more advanced robotic software such as EMILY by Evans et al. (1976), ML by Will and Grossman (1975), and finally AML (a manufacturing language) by Taylor et al. (1982), IBM (1981, 1982), and Grossman (1982). AML is basically a general computer programming language that can be adopted for robotic applications. It supports arithmetic and Boolean operations, data aggregates used to implement operations on vectors and matrices and thus coordinate transformations and motion trajectory design for position, velocity, and accelerations, as well as sensor monitoring for guarded moves. Recently, a version of AML called AML/E has been developed for AML/Entry that runs on an IBM PC or compatibles such as the Zenith 151 microcomputer. However, the original AML runs on the famous series 1 (RS-1) computer. A typical AML program for peg-in-hole insertion reads as follows:

```
PEG-IN-HOLE INSERTION PROGRAM IN AML
PICKUP:SUBR(PART_DATA,TRIES);
    MOVE(GRIPPER,DIAMETER(PART_DATA)+0.2);
    MOVE(<1,2,3>,XYZ_POSITION(PART_DATA)+<0,0,1>);
    TRY_PICKUP(PART_DATA,TRIES);
    END;

TRY_PICKUP:SUBR(PART_DATA,TRIES);
    IF TRIES LT 1 THEN RETURN('NO PART');
    DMOVE(3,-1.0);
    IF GRASP(DIAMETER(PART_DATA)='NO PART'
       THEN TRY_PICKUP(PART_DATA,TRIES-1);
    END;

GRASP:SUBR(DIAMETER,F);
    FMONS:NEW APPLY($MONITOR,PINCH_FORCE(F));
    MOVE(GRIPPER,O,FMONS);
    RETURN(IF QPOSITION(GRIPPER)LE DIAMETER/2
         THEN 'NO PART'
         ELSE 'PART');
    END;
INSERT:SUBR(PART_DATA,HOLE);
  FMONS:NEW APPLY(SMONITOR,
              TIP_FORCE(LANDING_FORCE));
    MOVE (<1,2,3>,HOLE+<0,0,25>);
    DMOVE(3,-1.0.FMONS);
    IF QMONITOR(FMONS)=1
      THEN RETURN ('NO HOLE');
    MOVE(3,HOLE(3) + PART_LENGTH(PART_DATA));
    END;

PART_IN_HOLE:SUBR(PART_DATA,HOLE);
    PICKUP(PART_DATA,2.);
    INSERT(PART_DATA,HOLE);
    END;
```

The following program is written in AML for an IBM 7565 robotic manipulator just to warm up the six degrees of freedom motion as well as the grip motion and the speed sensors. It moves the robot to seven different kinematic attitudes and sets the distance between the jaws of the gripper and the speed.

```
JOINT MOTION WARMUP FOR AN IBM 7565 (WHITE CLOUD)
ROBOTIC MANIPULATOR

EXERCISE:SUBR[N,S];
  N:NEW IF ? N THEN N ELSE 10;
  S:NEW IF ? S THEN S ELSE 1.0;
10  OLDS:NEW SPEED;
    STARTUP;
    DELAY[5];
    SPEED[S];
    CLEANUP[$3XCLN];
    MOVE[ARM,0];
    WHILE <N:=N-1> GE 0 DO(
         MOVE[ARM,<5,18,8,134,89,135,3>],
         MOVE[ARM,<-2,-3,-4,-134,-89,-134,0>],
         MOVE[ARM,<8,15,8,124,79,105,2>],
         MOVE[ARM,<-6,-24,-7,-57,-78,-90,3>],
         MOVE[ARM,<7,24,7,120,70,120,2.8>],
         MOVE[ARM,<-7,-24,-7,-120,-50,-120,2>],
         MOVE[ARM,<-3,10,-7,-120,-64,-120,0>]);
    EXCLN:SUBR;
    MOVE[ARM,0];
    SPEED[OLDS];
    END;
END;

TYPE[EOL, 'EXERCISE[REPS,SPEED];',EOL,
    'CAUTION:MOVE ARM TO SAFE LOCATION FOR MOVE TO
    ZERO]',EOL];
```

Among the other existing recent software developed for robotic motion manipulations and control one should mention TEACH by Ruoff (1979, 1980) for the Bendix Corporation, PAL by Takase et al. (1979), which is capable of executing multiple coordinate transformations, JARS by the Robotics and Teleoperators Group of NASA's Jet Propulsion Laboratory [see Craig (1980)] as a high-level programming software for robotics control, MAPLE by Darringer and Blasgen (1975) for IBM, SIGLA by Salmon (1978) for Olivetti's SIGMA Robots, MAL by Gini et al. (1979) of the Milan Polytechnic in Italy, LAMA-S by Falek and Parent (1980) developed in France, LM by Latomobe and Mazer (1981) of IMAG in France, MCL by McDonnel-Douglas (1981) as an extension of the APT programming language for numerically controlled machines, RAIL by Franklin and Vanderburg (1982) of Automation Inc. to control both vision and motion of robotic manipulators, RPS by Park (1979, 1981) for SRI International for control of assembly operations using multiple robots, and HELP by the General Electric Company (1981, 1982) to control the multiaxis Cartesian robots.

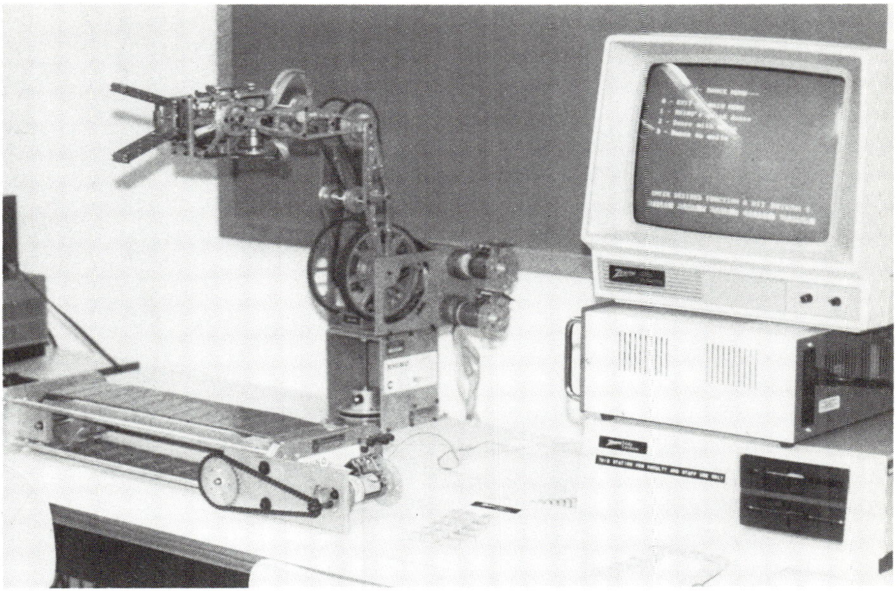

Figure 9.5.1 Typical educational robotic cell at the University of New Mexico's Robotic Instructional Laboratory.

Again, the reader is referred to the already mentioned extensive reviews by Lozano-Perez (1983) and Gruver et al. (1984) on robot programming. More recent developments indicate that BASIC and C are emerging as bases for robotic software in the future and thus we devote the rest of this chapter to programming robotic motion in BASIC.

In what follows we present a set of complete source code programs in BASIC for various task implementations of manipulations and assembly processes for a class of general robots and in particular for Rhino XR-2 educational robotic cells. Figure 9.5.1 displays a typical robotic educational cell comprising a Rhino XR-2 five-axis robot in conjunction with a conveyer belt, a carousel, and the associated controller computer.

9.6 COMPLETE SET OF ROBOTIC TEACH-ASSEMBLY PROGRAMS

These programs are specifically applicable to Rhino XR-2 robotic manipulators. However, they are general enough for any robot whose controller can communicate with a host computer by means of ASCII codes over an asynchronous port (serial RS-232). The ASCII codes do not allow for the advanced features in today's computers and robotic manipulators. Thus, robot manufacturers have to adjust the codes to suit their own needs, which means we may often be faced with compatibility problems between the robotic manipulators and the computers. The following specific commands are used on conjunction with the BASIC software:

LPRINT "An" Move motor A by n steps (resolutions)

LPRINT "AX" Stop motor A

LPRINT "Aϕ" Read the position of a limit switch for motor A

LPRINT "I"; X = INP (PORT)-32 Send control codes to the serial port

LPRINT "A?" Read the position of motor A

X = INP(PORT)-3 Receive error message from the port

If the following programs are to be used with robots other than the Rhino XR-2, then the above library of commands must be replaced in the programs with an appropriate library of commands for the robots under consideration. Similar robot programming languages have been developed by starting with a popular computer language such as BASIC or PASCAL and adding a library of robot-specific subroutines. The user then writes a BASIC program making use of frequent calls to the predefined subroutine package for robot-specific needs. Examples include the AR-BASIC by the American Robot Corporation [see Gilbert et al. (1984)] and ROBOT-BASIC by Intelledex [see Intelledex (1983)]. Of course, NASA's JPL JARS robotic software is another example of a robot programming language using PASCAL.

9.6.1 The RESET Program

This program resets a five-axis robot to a home position that is basically arbitrary and whose vector of the joint angles is given by the array in line 1080. It does so by moving all joints so as to sense a number of tactile sensors (limit switches in the case of Rhino XR-2). The motion stops and changes direction and the joint motion is resumed until the reference configuration is reached:

```
1000 PROGRAM RESET (HOME)
1010
1020 PURPOSE-RESET THE XR-2 TO THE "HOME" POSITION
1030
1040 DEFINT A-Z
1050 DIM OFFSET(8),MAP(8),RDIR(8),IPOS(8),NET(8)
1060
1070 DATA 8,7,4,3,2,1,6,5
1080 DATA 0,0,-45,+278,-322,-870,24,25
1090 DATA 1,1,-1,-1,1,1,1,-1
1100
1110 NCON = 8
1120 NAXIS = 6
```

```
1130 ISTEP = 25
1140 PORT = &H3F8 (changes for other computers)
1150
1160 FOR I = 1 TO NCON : READ MAP(I) : NEXT I
1170 FOR I = 1 TO NCON : READ OFFSET(I) : NEXT I
1180 FOR I = 1 TO NCON : READ RDIR(I) : NEXT I
1190 FOR I = 1 TO NCON : NET(I) = 0 : NEXT I
1200
1210 MR$="HGFEDCBA"
1220 I$ = "I"
1230
1240 DISPLAY MENU
1250
1260 K$ = INKEYS 'CLEAR KEYBOARD
1270 CLS          'CLEAR SCREEN
1280 GOSUB 2360   'MOTORS QUIET
1290
1300 PRINT "RESET PROGRAM:"
1310 PRINT
1320 PRINT "    [1] = RESET ALL JOINTS MICROSWITCH"
1330 PRINT
1340 PRINT "          [2] = RESET BASE"
1350 PRINT "          [3] = RESET SHOULDER"
1360 PRINT "          [4] = RESET ELBOW"
1370 PRINT "          [5] = RESET WRIST PITCH"
1380 PRINT "          [6] = RESET WRIST ROLL"
1390 PRINT "          [7] = RESET FINGERS"
1400
1410 PRINT "    [8] = RETURN TO TEST PROGRAM"
1440 LOCATE 15,1
1450
1460 PRINT "PLEASE CHOOSE ONE OF THE ABOVE"
1470
1480 K$ = INKEY$ : IF K$ = " " THEN 1480
1490 X = VAL(K$)
1500 JLOW = 2 : JHIGH = NAXIS + 1
1510 IF K$ = "1" THEN GOSUB 1600 : GOTO 1440
1520 FOR I = 2 TO 7
1530   IF X <> I THEN 1550
1540   JLOW = I : JHIGH = I
1550 NEXT I
1560 IF X > 1 AND X < 8 THEN GOSUB 1600 : GOTO 1440
1570 IF K$ = "8" THEN CHAIN "TEST"
1580 IF K$ = "9" THEN CHAIN "TEST"
1590 GOTO 1440
1600
```

```
1610 EXECUTE THE SELECTED RESET COMMAND
1620
1630 FOR J = JHIGH TO JLOW STEP -1
1640   I = J + 1
1650   MASK = 2^(MAP(I) - 1)
1660   M$ = MID$(MR$,I,1)
1670   SO$ = M$ + "O"
1680   SX$ = M$ + "X"
1690   SQ$ = M$ + "?"
1700   SIGN = RDIR(I)
1710   K$ = INKEY$
1720   LOCATE 2+I,9 : PRINT "->"
1730   LOCATE 2+I,49 : PRINT "<-"
1740   ZSTEP = SIGN*ISTEP : IF I = 8 THEN ZSTEP = SIGN
1750   GOSUB 1840 'ROUGH SEARCH
1760   ZSTEP = -RDIR(I)
1770   GOSUB 2190 'ROUGH OFFSET
1780   ZSTEP = RDIR(I)
1790   GOSUB 1840 'FINE SEARCH
1800   GOSUB 2020 'OFFSET
1810   IPOS(I) = O
1820 NEXT J
1830 RETURN
1840
1850  MOVE FORWARD TO CLOSE MICROSWITCH USING RESOLUTION
      ZSTEP
1860
1870 SZ$ = M$ + STR$(ZSTEP)
1880 LPRINT SO$
1890  LPRINT SO$
1900  X = (INP(PORT) - 32) AND MASK
1910  LOCATE 2+I,42
1920  IF X = 0 THEN PRINT "CLOSED" : BEEP : GOTO 1990
1930   IF X <> O THEN PRINT "OPEN"
1940 LPRINT SQ$
1950   IF INP(PORT) - 32 > ISTEP THEN 1890
1960   LPRINT SZ$
1970   NET(I) = NET (I) + ZSTEP
1980 GOTO 1880
1990 LPRINT M$ + STR$(-3*ZSTEP)
2000 NET(I) = NET(I) - 3*ZSTEP
2010 RETURN
2020
2030 OFFSET ROUTINE
2040
2050 LOCATE 2+I,42 : PRINT "OPEN" : BEEP
```

```
2060 JHOLE = ABS(OFFSET(I))
2070 JSIGN = SGN(OFFSET(I))
2080 SJ$ = M$ + STR$(JSIGN*ISTEP)
2090 LPRINT SQ$
2100   IF INP(PORT) - 32 > ISTEP THEN 2090
2110   IF JHOLE < ISTEP THEN 2150
2120   LPRINT SJ$
2130   JHOLE = JHOLE - ISTEP
2140 GOTO 2090
2150 LPRINT M$ + STR$(JSIGN*JHOLE)
2160 NET(I) = NET(I) + OFFSET(I)
2170 GOSUB 2430
2180 RETURN
2190
2200 ROUGH OFFSET TO OPEN THE MICROSWITCH
2210
2220 SZ$ = M$ + STR$(ZSTEP)
2230 LPRINT SO$
2240  LPRINT I$
2250  X = (INP(PORT) - 32) AND MASK
2260  LOCATE 2+I,42
2270  IF X = 0 THEN PRINT "CLOSED" : BEEP
2280  IF X <> 0 THEN PRINT "OPEN"
2290  IF X <> 0 THEN 2350
2300  LPRINT SQ$
2310 IF INP(PORT) - 32 > ISTEP THEN 2240
2320  LPRINT SZ$
2330 NET(I) = NET(I) + ZSTEP
2340 GOTO 2230
2350 RETURN
2360
2370 TURN OFF ALL MOTORS
2380
2390 FOR J = 1 TO NCON
2400 LPRINT MID$(MR$,J,1) + "X"
2410 NEXT J
2420 RETURN
2430
2440 UPDATE MOVE DISPLAY
2450
2460 LOCATE 2+I,9 : PRINT " "
2470 LOCATE 2+I, 42 ; PRINT " "
2480 RETURN
2490
2500 END
```

9.6.2 Program TEST

This program sets the stage for a teach program and chains a number of other programs to it. The other programs include RESET, CMOV, PMOVE, EMOVE, and INKIN.

```
100 PROGRAM TEST
110
115 CLOSE
120 PURPOSE--TEACH THE XR-2 TO FOLLOW A SERIES OF MOVES
130
140 DIM IPOS(8),PMIN(8),PMAX(8)
150
160 NCON = 8
170 NAXIS = 6
180
190 DATA -1000,-1000,-1300,-1000,-1000,-1000,-1000,-1000
200 DATA 1000,1000,1300,1000,1000,1000,1000,1000
210
220 FOR I = 1 TO NCON : READ PMIN(I) : NEXT I
230 FOR I = 1 TO NCON : READ PMAX(I) : NEXT I
240
250 K$ = INKEY$   'CLEAR KEYBOARD
260 CLS           'CLEAR SCREEN
270
280 PRINT "TEST PROGRAM:"
290 PRINT
300 PRINT "[1] = RESET ROBOT TO 'HOME' POSITION"
310 PRINT "[2] = CREATE A NEW MOVE FILE"
320 PRINT "[3] = PLAYBACK AN EXISTING MOVE FILE"
330 PRINT "[4] = EDIT AN EXISTING MOVE FILE"
340 PRINT "[5] = EXIT THE TEST PROGRAM"
341 PRINT "[6] = INVERSE KINEMATICS FILE"
350 PRINT
360 PRINT "PLEASE SELECT ONE OF THE ABOVE"
370 PRINT
380 K$ = INKEY$ : IF K$ = " " THEN 380
390 X = VAL(K$)
400
410 IF X = 1 THEN CHAIN "HOME"
420 IF X = 2 THEN CHAIN "CHU"
430 IF X = 3 THEN CHAIN "PHU"
440 IF X = 4 THEN CHAIN "EHU"
450 IF X = 5 THEN GOTO 500
451 IF X = 6 THEN CHAIN "INVERSE"
460
```

```
470 GOTO 260
480
500 INPUT "ARE YOU SURE? [Y OR N]";K$
510 IF K$='Y' THEN GOTO 540
520 GOTO290
540 END
```

9.6.3 Program CREATE A MOVE FILE

This program enables a five-axis robotic arm to be taught an elaborate set of moves for assembly processes, in conjunction with a conveyer belt. The program also allows the robotic hand to grip and release objects at desired times and also to reset itself to a reference hand configuration. The entire set of taught moves can be stored in the memory of the computer and can be played back in any fashion and for any number of cycles.

```
610 PORT = &H3F8
1000 PROGRAM CMOVE(Create a move file)
1010
1020 PURPOSE - CREATE AN XR-2 MOVE FILE
1030
1040 DEFINT A-Z
1050 DIM STEPS(8),MK$(4),IPOS(8),JPOS(8),DMAP(8)
1060
1070 DATA 100,10,-10,-100,10,1,-1,-10
1080 DATA 0,0,0,0,0,0,0,0
1090 DATA 0,0,0,0,0,0,0,0
1100 DATA 1,1,-1,-1,1,1,-1,-1
1110
1111 BLOCK - 1
1112 FLAG1 = O, SET GRIPPER IN OPEN STATUS
1120 NCON = 8
1130 NAXIS = 6
1140 ISTEP = 40
1150 PORT = &H3F8
1160
1170 FOR I = 1 TO 8 : READ STEPS(I) : NEXT I
1180 FOR I = 1 TO NCON : READ IPOS(I) : NEXT I
1190 FOR I = 1 TO NCON : READ JPOS(I) : NEXT I
1200 FOR I = 1 TO NCON : READ DMAP(I) : NEXT I
1210
1220 MR$="HAFEDCBG"
1230 MK$(1) = "YI123456"
1240 MK$(2) = "UIQWERTY"
```

```
1250 MK$(3) = "JKASDFGH"
1260 MK$(4) = "M,ZXCVBN"
1270
1280 K$ = INKEY$ 'CLEAR KEYBOARD
1290 CLS         'CLEAR SCREEN
1300
1310 PRINT "CMOVE PROGRAM:"
1320 PRINT
1330 INPUT "ENTER FILENAME";MFILE$
1340
1350 OPEN "O",1,MFILE$
1360 PRINT CHR$(27);"x"; "<" 'DISABLE KEYBOARD AUTO REPEAT
1370
1380 CLS
1390 PRINT "----------------------------------------------------------------------"
1400 PRINT ":   BASE    : SHOULDER :  ELBOW   :  PITCH   :   ROLL   : CONVEYER :"
1410 PRINT "----------------------------------------------------------------------"
1420 PRINT ":    [1]    :    [2]   :   [3]    :   [4]    :   [5]    :   [6]    :"
1430 PRINT ": CW   100  : UP   100 : UP   100 : UP   100 : CW   100 : RH   100 :"
1440 PRINT "----------------------------------------------------------------------"
1450 PRINT ":    [Q]    :    [W]   :   [E]    :   [R]    :   [T]    :   [Y]    :"
1460 PRINT ": CW   10   : UP   10  : UP   10  : UP   10  : CW   10  : RH   10  :"
1470 PRINT "----------------------------------------------------------------------"
1480 PRINT ":    [A]    :    [S]   :   [D]    :   [F]    :   [G]    :   [H]    :"
1490 PRINT ":CCW  -10   : DN  -10  : DN  -10  : DN  -10  : CCW -10  : LF  -10  :"
1500 PRINT "----------------------------------------------------------------------"
1510 PRINT ":    [Z]    :    [X]   :   [C]    :   [V]    :   [B]    :   [N]    :"
1520 PRINT ":CCW  -100  : DN  -100 : DN  -100 : DN  -100 : CCW -100 : LF  -100 :"
1530 PRINT "----------------------------------------------------------------------"
1540 PRINT ":           :          :          :          :          :          :"
1550 PRINT "----------------------------------------------------------------------"
1560 PRINT ":           :          :          :          :          :          :"
1570 PRINT "----------------------------------------------------------------------"
1571 PRINT " BLOCK #"
1580 PRINT
1590 PRINT "[L] TOGGLES THE SENSITIVITY"
1600 PRINT "[P] ENDS A MOVE BLOCK [O] ENDS THE FILE"
1610 PRINT "[M] RESETS GRIPPER  [K] GRIPS AN OBJECT [I]
     RELEASES AN OBJECT"
1620
1630 FOR I = 3 TO NCON
1640   LOCATE 17,11*I-25 : PRINT " "
1650   LOCATE 17,11*I-28 : PRINT IPOS(I)
1660 NEXT I
1670
1680 FOR I = 3 TO NCON
1690   LOCATE 15,11*I-25 : PRINT " "
1700   LOCATE 15,11*I-28 : PRINT JPOS(I)
1710 NEXT I
1720
1721 LOCATE 19,36 :PRINT BLOCK
```

```
1730 K$ = INKEY$ : IF K$ = " " THEN 1730
1731 LOCATE 19,36 : PRINT BLOCK
1740 IF K$ = "L" THEN GOSUB 2340 : GOTO 1730
1750 IF K$ = "P" THEN GOSUB 2030 : GOTO 1680
1760 IF K$ = "O" THEN GOSUB 2140 : GOTO 1730
1761 IF K$ = "K" THEN GOSUB 2600 : GOSUB 2030 :
     GOTO 1730
1762 IF K$ = "I" THEN GOSUB 2721 : GOTO 1730
1763 IF K$ = "M" THEN GOSUB 2480 : GOTO 1730
1770
1780 FOR I = 3 TO NCON
1790 FOR J = 1 TO 4
1800  IF K$ = MID$(MK$(J),I,1) THEN 1850
1810  NEXT J
1820 NEXT I
1830 GOTO 1730
1840
1850 GOSUB 1870 'MOVE MOTOR
1860 GOTO 1730
1870
1880 'MOVE MOTOR STEPS(J) HOLES
1881 COLOR 0,7,6
1890
1891 LOCATE 3*J+1,11*I-27 : PRINT STEPS(J)
1900 SIZE = ABS(STEPS(J))
1910 SIGN = DMAP(I)*SGN(STEPS(J))
1920 M$ = MID$(MR$,I,1)
1930 LPRINT M$ + "?"
1940  IF INP(PORT) - 32> ISTEP THEN 1930
1950  IF SIZE < ISTEP THEN 1990
1960  LPRINT M$ + STR$(SIGN*ISTEP)
1970  SIZE = SIZE - ISTEP
1971 IPOS(I)=IPOS(I)+ISTEP*SIGN*DMAP(I)
1972 JPOS(I)=JPOS(I)+ISTEP*SIGN*DMAP(I)
1973 COLOR 7,0,6
1975 LOCATE 17,11*I-28 : PRINT IPOS(I)
1976 LOCATE 15,11*I-28 : PRINT JPOS(I)
1980 GOTO 1930
1990 LPRINT M$ + STR$(SIGN*SIZE)
1991 COLOR 7,0,6
1992 LOCATE 3*J+1,11*I-27 :PRINT STEPS(J)
2001 IPOS(I)=IPOS(I)+SIZE*SIGN*DMAP(I)
2002 JPOS(I)=JPOS(I)+SIZE*SIGN*DMAP(I)
2203 LOCATE 17,11*I-28 : PRINT IPOS(I)
2204 LOCATE 15,11*8-28 : PRINT JPOS(I)
```

```
2020 RETURN
2030
2040 'SAVE MOVE ON DISK : [P]
2050
2060 FOR L = 1 TO NOCON : WRITE #1,JPOS(L) : NEXT L
2061 WRITE#1 ,BLOCK:BLOCK=BLOCK +1
2065 JPOS(2)=0
2070 FOR L = 3 TO NCON
2080  JPOS(L) = 0
2090  LOCATE 17,11*L-25 : PRINT " "
2100  LOCATE 17,11*L-28 : PRINT IPOS(L)
2110 NEXT L
2120 RETURN
2130
2140 'CLOSE FILE : [0]
2150
2160 CLOSE#1
2170 CLS
2180 K$ = INKEY$
2190 PRINT "THE FILE";MFILE$ : "HAS BEEN CREATED"
2200 PRINT
2210 PRINT "[1] = CREATE ANOTHER MOVE FILE"
2220 PRINT "[2] = RETURN TO THE TEST PROGRAM"
2240 PRINT
2250 PRINT "PLEASE SELECT ONE OF THE ABOVE"
2260 PRINT
2270 K$ = INKEY$ : IF K$ = " " THEN 2270
2280
2290 IF K$ = "1" THEN 1310
2300 PRINT CHR$(27) ; "y"; "<" 'ENABLE KEYBOARD AUTO
     REPEAT
2310 IF K$ = "2" THEN CHAIN "TEST"
2330 GOTO 2270
2340
2350 CHANGE THE SENSITIVITY OF THE KEYS : [I]
2360
2370 FOR L = 1 TO 4
2380  X = STEPS(L)
2390  STEP(L) = STEPS (L+4)
2400  STEPS (L+4) = X
2410  FOR K = 1 TO NAXIS
2420   LOCATE 1+3*L,11*K-4
2430   PRINT ABS(STEPS(L))
2440  NEXT K
2450 NEXT L
2460 RETURN
```

```
2470
2480 'GRIPPER RESET ROUTINE
2490 PORT = &H3F8
2500 LPRINT "A126"
2510 FOR I= 1 TO 10 :I=I+1 : NEXT
2520 LPRINT "A?"
2530 X1=INP(PORT)-32
2540 2550 FOR I= 1 TO 10 : I-I+1 : NEXT
2550 LPRINT "A?"
2560 X1=INP(PORT)-32
2570 IF X1<>X2 THEN GOTO 2520
2580 LPRINT "AX"
2590 LPRINT "A-60"
2591 FLAG1=0
2592 JPOS(2)=3
2595 RETURN
2600 'GRIP ROUTINE
2601 IF FLAG1=1 THEN GOTO 2720
2610 PORT = &H3F8
2620 LPRINT "A126"
2630 FOR I= 1 TO 50 : I=I+1 : NEXT
2640 LPRINT "A?"
2650 X1-INP(PORT)-32
2660 FOR I= 1 to 50 : I=I+1 : NEXT
2670 LPRINT "A?"
2680 X2= INP(PORT)-32
2690 IF X1 <> X2 THEN GOTO 2640
2700 LPRINT "AX"
2705 FOR I= 1 TO 50 :I=I+1 : NEXT
2710 LPRINT "A-2"
2711 FLAG1=1
2712 JPOS(2)=1
2720 RETURN
2721 LPRINT "A-20"
2722 FLAG1=0
2723 JPOS(2)=2
2724 RETURN
2725 END
```

9.6.4 Program PLAY BACK A MOVE FILE

This program enables us to play back, in any desired manner, a move file that has already been taught to a robot in conjunction with a conveyer belt.

```
10 PROGRAM PMOVE(PLAY BACK A MOVE FILE)
20
```

```
30 PURPOSE-PLAYBACK A MOVE FILE
40
50 DEFINT A-Z
60 DIM IPOS(8),MOVES(100,9),DMAP(8),JPOS(8)
61 DIM MK$(8)
70
80 DATA 1,1,-1,-1,1,1,-1,-1
90
91 FLAG1=0 'SET GRIPPER IN OPEN STATUS
100 NCON = 8
110 NAXIS = 6
120 ISTEP = 40
130 PORT = &H3F8
131 MK$(1) = "CCW CW"
132 MK$(2) = "DOWNUP"
133 MK$(3) = "DOWNUP"
134 MK$(4) = "DOWNUP"
135 MK$(5) = "CCW CW"
136 MK$(6) = "LEFTRIGHT"
140
150 FOR I = 1 TO NCON : READ DMAP(I) : NEXT I
170
180 MR$="HAFEDCBG"
190
200 CLS
210 PRINT "PROGRAM PHU:"
220 PRINT
230 INPUT "ENTER FILENAME";MFILE$
240 OPEN "I",1,MFILE$
250
260 I = 1
270 IF EOF(1) THEN 310
280  FOR J = 1 TO NCON+1:INPUT#1, MOVES(I,J): NEXT J
290  I = I + 1
300 GOTO 270
310 IROW = I - 1
320
330 K$ = INKEY$ 'CLEAR KEYBOARD
340 CLS          'CLEAR SCREEN
350 PRINT
360 PRINT "PROGRAM PMOVE:"
370 PRINT
380 PRINT "      [1] = PLAY MOVE FILE FORWARD"
390 PRINT "      [2] = PLAY MOVE FILE REVERSE"
400 PRINT "      [3] = CYCLE THROUGH MOVE FILE"
```

```
410 PRINT "       [4] = RETURN TO THE TEST PROGRAM"
420 PRINT "       [5] = EXIT"
430 PRINT
440 PRINT "PLEASE SELECT ONE OF THE ABOVE"
450 PRINT
451 PRINT "                    BLOCK #"
452 PRINT "
453 PRINT "---------------------------------------------------------------------"
454 PRINT ":  BASE   : SHOULDER :  ELBOW   :  PITCH   :   ROLL   : CONVEYER :"
455 PRINT "---------------------------------------------------------------------"
456 PRINT ":  CW     :   UP     :   UP     :   UP     :   CW     :   RIGHT  :"
457 PRINT "---------------------------------------------------------------------"
458 PRINT ":  CCW    :   DOWN   :   DOWN   :   DOWN   :   CCW    :   LEFT   :"
459 PRINT "---------------------------------------------------------------------"
460 PRINT ":         :          :          :          :          :          :"
461 PRINT "---------------------------------------------------------------------"
462 PRINT ":         :          :          :          :          :          :"
463 PRINT "---------------------------------------------------------------------"
474 K$=INKEY$ : IF K$ = " " THEN 474
475 FLAGO=1 'SET FORWARD MOTION STATUS
476 IR1 = 1 : IR2 = IROW : IR3 = 1
480 IF K$ = "1" OR K$ = "3" THEN GOSUB 550
490 IR1 = IROW : IR2 = 1: IR3 = -1 :FLAGO = -1
500  IF K$ = "2" THEN GOSUB 550
510 IF K$ = "4" THEN CHAIN "TEST"
520 IF K$ = "5" THEN END
530
540 GOTO 474
550
560 PLAYBACK THE MOVE FILE
570
580 FOR I - IR1 TO IR2 STEP IR3
620 FOR J = 3 TO NCON
621   LOCATE 12,36 : PRINT MOVES(I,9)
630  M$ = MID$(MR$,J,1)
640   IMOVE = MOVES(I,J)*FLAGO
641   JPOS(J)=IMOVE : TSGN=SGN(IMOVE)
642   IF TSGN = O THEN GOTO 770
643   COLOR 0,7,5
644  H$=MID$(MK$(J-2), TSGN*2+3,4)
645   LOCATE -TSGN +17,11*J-28 :PRINT USING "&"; H$
650   SIZE = ABS(IMOVE)
660   SIGN = DMAP(J)*SGN(IMOVE)
670   LPRINT M$ + "?"
680     IF INP(PORT) - 32 > ISTEP THEN 670
690     IF SIZE < ISTEP THEN 730
700     LPRINT M$ + STR$(SIGN*ISTEP)
710  SIZE = SIZE - ISTEP
```

```
711  IPOS(J)=IPOS(J)+SIGN*ISTEP*DMAP(J)
712  JPOS(J)=JPOS(J)-SIGN*ISTEP*DMAP(J)
713  COLOR 7,0,5
714  LOCATE 20,11*J-28 : PRINT JPOS(J)
715  LOCATE 22,11*J-28 : PRINT IPOS(J)
720   GOTO 670
730   LPRINT M$ + STR$(SIGN*SIZE)
740   LPRINT M$ + "?"
750     IF INP(PORT) - 32 > 0 THEN 740
751   COLOR 7,0,5
752  LOCATE -TSGN + 17,11*J-28 : PRINT USING "&"; H$
753   IPOS(J)=IPOS(J)+SIGN*SIZE*DMAP(J)
754   JPOS(J)=JPOS(J)-SIGN*SIZE*DMAP(J)
755   LOCATE 20,11*J-28 : PRINT JPOS(J)
756   LOCATE 22,11*J-28 : PRINT IPOS(J)
770     Q$ = INKEY$
780     IF K$ = "3" AND Q$ <> " " THEN RETURN
790   NEXT J
791 IF MOVES(I,2) = 1 THEN GOSUB 990
792 IF MOVES(I,2) = 2 THEN GOSUB 1110
793 IF MOVES(I,2) = 3 THEN GOSUB 880
794 FOR K=1 TO 40 : K=K+1 : NEXT
800 NEXT I
810 IF K$ = "1" OR K$ = "2" THEN RETURN
820 IR3 = -IR3 : FLAGO = -FLAGO
830 IF IR3 > 0 THEN IR1 = 1  ELSE IR1 = IROW
840 IF IR3 > 0 THEN IR2 = IROW ELSE IR2 = 1
850 GOTO 580
860
870 'GRIP RESET ROUTINE
880 LPRINT "A126"
890 FOR K=1 TO 10 : K=K+1 : NEXT
900 LPRINT "A?"
910 X1= INP(PORT)-32
920 FOR K= 1 TO 10 : K=K+1 : NEXT
930 PRINT "A?"
940 X2= INP(PORT)-32
950 IF X1 <> X2 THEN GOTO 900
960 LPRINT "AX"
970 LPRINT "A-65"
971 FLAG1=0
980 RETURN
990 'GRIP ROUTINE
991 IF FLAG1=1 THEN GOTO 1100
1000 LPRINT "A126"
1010 FOR K-1 TO 10 : K=K+1 : NEXT
```

```
1020 LPRINT "A?"
1030 X1= INP(PORT)-32
1040 FOR K=1 TO 10: K=K+1 :NEXT
1050 LPRINT "A?"
1060 X2= INP(PORT)-32
1070 IF X1<> X2 THEN GOTO 1020
1080 LPRINT "AX"
1090 LPRINT "A-2"
1091 FLAG1=1
1100 RETURN
1110 'RELEASE ROUTINE
1120 LPRINT "A-20"
1121 FLAG1=0
1130 RETURN
1140 END
```

9.6.5 Program EDIT A MOVE FILE

This program enables the user to edit completely, move by move, a robot move file that has already been taught to a robot and been saved. It does so by giving the user the option of modifying any move by forward, backward, insertion, and replacement of steps.

```
100 PROGRAM EMOVE(Edit A Move File)
110
120 PURPOSE-EDIT A MOVE FILE
130
140 DIM IPOS(8),MOVES(100,8),NMOVES(100,8),DMAP(8)
150 DIM JPOS(8),MK$(4),STEPS(8)
160
170 DATA 1,1,-1,-1,1,1,1,1
180 DATA 0,0,0,0,0,0,0,0
190 DATA 10,1,-1,-10,100,10,-10,-100
200
210
220 NCON = 8
230 NAXIS = 6
240 2ISTEP = 30
250 PORT = &H3F8
260 FOR I = 1 TO NCON
270  MOVES(0,I) = IPOS(I)
280  NMOVES(0,I) = IPOS(I)
290 NEXT I
300
310 FOR I = 1 TO NCON : READ DMPA(I) : NEXT I
```

```
320 FOR I = 1 TO NCON : READ JPOS(I) : NEXT I
330 FOR I = 1 TO 8 : READ STEPS(I) : NEXT I
340
350 MR$ = "HGFEDCBA"
360 MK$(1) = "78123456"
370 MK$(2) = "UIQWERTY"
380 MK$(3) = "JKASDFGH"
390 MK$(4) = "M,ZXCVBN"
400
410 CLS
420 PRINT "PROGRAM EMOVE:"
430 PRINT
440 INPUT "ENTER FILENAME";MFILE$
450 OPEN "I",1,MFILE$
460
470 I = 1
480 IF EOF(1) THEN 550
490   FOR J = 1 TO NCON
500     INPUT#1, MOVES(I,J)
510     NMOVES(I,J) = MOVES(I,J)
520  NEXT J
530  I = I + 1
540 GOTO 480
550 IROW = I -1
560 NROW = IROW
570 IR = 0
580
590 K$ = INKEY $ 'CLEAR KEYBOARD
600 CLS            'CLEAR SCREEN
610 PRINT
620 PRINT "PROGRAM EMOVE:"
630 PRINT
640 PRINT "     [1] = SINGLE STEP FORWARD"
650 PRINT "     [2] = SINGLE STEP REVERSE"
660 PRINT "     [3] = INSERT STEPS"
670 PRINT "     [4] = REPLACE STEPS"
680 PRINT "     [5] = END THE FILE"
700 PRINT
710 PRINT "PLEASE SELECT ONE OF THE ABOVE"
720 PRINT
730 K$ = INKEY$ : IF K$ = " " THEN 730
740 IF K$ = "1" THEN INC = 1
750 IF K$ = "2" THEN INC = -1
760 IF K$ = "3" THEN INC = 0
770 IF K$ = "4" THEN INC = 1
780 IF K$ = "5" THEN GOTO 1940
```

```
790 IF K$ = "6" THEN GOTO 1940
800 IR = IR + INC
810 IF IR < 0 OR IR > IROW THEN 730
820 IF K$ = "1" OR K$ = "2" THEN GOSUB 850
830 IF K$ = "3" OR K$ = "4" THEN GOSUB 1040
840 GOTO 600
850
860 'PLAYBACK A MOVE FROM THE FILE
870
880  FOR J = 3 TO NCON
890   M$ = MID$(MR$,J,1)
900   IMOVE = MOVES(IR,J) - IPOS(J)
910   SIZE = ABS(IMOVE)
920   SIGN = DMAP(J)*SGN(IMOVE)
930   LPRINT M$ + "?"
940    X = ABS((INP(PORT) AND 127) - 32)
950    IF X > ISTEP THEN 930
960    IF SIZE < ISTEP THEN 1000
970    LPRINT M$ + STR$(SIGN*ISTEP)
980    SIZE = SIZE - ISTEP
990   GOTO 930
1000   LPRINT M$ + STR$(SIGN*SIZE)
1010   IPOS(J) = IPOS(J) + IMOVE
1020   NEXT J
1030  RETURN
1040
1050 'INSERT OR REPLACE MOVES
1060
1070 K$ = INKEY$ 'CLEAR KEYBOARD
1080 CLS          'CLEAR SCREEN
1090
1100 PRINT "---------------------------------------------------------------------"
1110 PRINT ":   BASE    : SHOULDER :   ELBOW    :   PITCH    :    ROLL    : CONVEYER :"
1120 PRINT "---------------------------------------------------------------------"
1130 PRINT ":    [1]    :    [2]    :    [3]    :    [4]    :    [5]    :    [6]    :"
1140 PRINT ":   CW 10   :   UP 10   :   UP 10   :   UP 10   :   CW 10   :   IN 10   :"
1150 PRINT "---------------------------------------------------------------------"
1160 PRINT ":    [Q]    :    [W]    :    [E]    :    [R]    :    [T]    :    [Y]    :"
1170 PRINT ":   CW 1    :   UP 1    :   UP 1    :   UP 1    :   CW 1    :   IN 1    :"
1180 PRINT "---------------------------------------------------------------------"
1190 PRINT ":    [A]    :    [S]    :    [D]    :    [F]    :    [G]    :    [H]    :"
1200 PRINT ":  CCW 1    : DOWN 1    : DOWN 1    : DOWN 1    :   CCW 1   :  OUT  1   :"
1210 PRINT "---------------------------------------------------------------------"
1220 PRINT ":    [Z]    :    [X]    :    [C]    :    [V]    :    [B]    :    [N]    :"
1230 PRINT ":  CCW 10   : DOWN 10   : DOWN 10   : DOWN 10   :  CCW 10   :  OUT 10   :"
1240 PRINT "---------------------------------------------------------------------"
1250 PRINT ":           :           :           :           :           :           :"
1260 PRINT "---------------------------------------------------------------------"
1270 PRINT ":           :           :           :           :           :           :"
1280 PRINT "---------------------------------------------------------------------"
```

```
1300 PRINT "[L] TOGGLES THE SENSITIVITY"
1310 PRINT "[P] STORES EACH MOVE"
1320 PRINT "[O] RETURNS TO PLAYBACK MODE"
1330
1340 FOR I = 3 TO NCON
1350  LOCATE 18,11*I-25 : PRINT " "
1360  LOCATE 18,11*I-28 : PRINT IPOS(I)
1370 NEXT I
1380
1390 FOR I = 3 TO NCON
1400  LOCATE 16,11*I-25 : PRINT " "
1410  LOCATE 16, 11*I-28 : PRINT JPOS(I)
1420 NEXT I
1430
1440 K$ = INKEY$ : IF K$ = " " THEN 1440
1450 IF K$ = "L" THEN GOSUB 2210 : GOTO 1440
1460 IF K$ = "P" THEN GOSUB 1750 : RETURN
1470 IF K$ = "O" THEN RETURN
1480 '
1490 FOR I = 3 TO NCON
1500  FOR J = 1 TO 4
1510   IF K$ = MID$(MK$(J),I,1) THEN 1560
1520 NEXT J
1530 NEXT I
1540 GOTO 1440
1550 '
1560 GOSUB 1580 'MOVE MOTOR
1570 GOTO 1390
1580 '
1590 'MOVE MOTOR STEPS(J) HOLES
1600 '
1610 SIZE = ABS(STEPS(J))
1630 M$ = MID$(MR$,I,1)
1640 LPRINT M$ + "?"
1650  X = ABS((INP(PORT) AND 127) - 32)
1660  IF X > ISTEP THEN 1640
1670  IF SIZE < ISTEP THEN 1710
1680  LPRINT M$ + STR$(SIGN*ISTEP)
1690  SIZE = SIZE - ISTEP
1700 GOTO 1640
1710 LPRINT M$ + STR$(SIGN*SIZE)
1720 IPOS(I) = IPOS(I) + STEPS(J)
1730 JPOS(I) = JPOS(I) + STEPS(J)
1740 RETURN
1750 '
1760 'SAVE NEW MOVE
```

```
1770 '
1780 FOR L = 1 TO NCON
1790   NMOVES(IR,L=) = IPOS(L)
1800 NEXT L
1810 IF INC = 1 THEN 1880
1820 FOR L = IR TO IROW
1830  FOR M = 1 TO NCON
1840    NMOVES(L-1,M) = MOVES(L,M)
1850  NEXT M
1860 NEXT L
1870 NROW = NROW + 1
1880 FOR L = 3 TO NCON
1890  JPOS(L) = 0
1900  LOCATE 18,11*L-25 : PRINT " "
1910  LOCATE 18,11*L-28 : PRINT IPOS(L)
1920 NEXT L
1930 RETURN
1940 '
1950 CLOSE FILE
1960 '
1970 CLOSE#1
1980 OPEN "O",1,MFILE$
1990 FOR I = 1 TO NROW
2000  FOR J = 1 TO NCON
2010    WRITE#1,NMOVES(I,J)
2020  NEXT J
2030 NEXT I
2040 CLOSE#1
2050 CLS
2060 K$ = INKEY$
2070 PRINT "THE FILE";MFILE$;"HAS BEEN UPDATED"
2080 PRINT
2090 PRINT "     [1] = EDIT ANOTHER MOVE FILE"
2100 PRINT "     [2] = RETURN TO THE TEACH PROGRAM"
2110 PRINT "     [3] = EXIT"
2120 PRINT
2130 PRINT "PLEASE SELECT ONE OF THE ABOVE"
2140 PRINT
2150 '$ = INKEY$ : IF K$ = " " THEN 2150
2160
2170 'F K$ = "1" THEN 410
2180 'F K$ = "2" THEN CHAIN "TEST"
2190 'F K$ = "3" THEN END
2200 CHAIN "TEACH"
2210 '
```

```
2220 CHANGE THE SENSITIVITY OF THE KEYS :   [I]FOR L = 1
     TO 4
2230 '
2240 FOR L = 1 TO 4
2250  X = STEPS(L)
2260  STEPS(L) = STEPS(L+4)
2270  STEPS(L+4) = X
2280  FOR K = 1 TO NAXIS
2290   LOCATE 2+3*L,11*K-4
2300   PRINT ABS(STEPS(L))
2310  NEXT K
2320 NEXT L
2330 RETURN
2340 '
2350 END
```

9.6.6 Program INVERSE KINEMATICS

This program asks the user to specify the position and orientation (i.e., the kinematic attitude) of the robotic hand. The program then solves for the vector of the joint angles necessary to achieve such kinematic attitudes and executes the joint motions necessary to move the robotic hand to the desired position and orientation.

```
10 PROGRAM INVERSE KINEMATICS
15
20 MOHSEN SHAHINPOOR
25
30 PURPOSE-IMPLEMENT THE INVERSE KINEMATICS EQUATIONS
40
50 DIM T(4,4),CMAP(8),RES(8),W(5),PHI(8),D(5),A(5)
60 DIM HOLES(5),SIGN(5),MOT$(5)
70
80 DATA 45,&H3F8,"HGFEDCBA"
90 DATA 1.570797,57.29578
100 DATA 9.75,0,0,0,7.125
110 DATA 0,9,9,9.375,0.0
120 DATA 1,1,1,1,-1,-1,-1,1
130 DATA 1.0,1.0,0.1374,0.1145,0.1145,0.0793,0.2406,1.0
140 DATA 0.0,0.0,0.0,-90.0,90.0,-90.0,0.0,0.0
150 '
160 READ ISTEP, PORT,MR$
170 READ P2,RAD
180 FOR I = 1 TO 5: READ D(I) : NEXT I
190 FOR I = 1 TO 5: READ A(I) : NEXT I
```

```
200 FOR I = 1 TO 8: READ CMAP(I) : NEXT I
210 FOR I = 1 TO 8: READ RES(I) : NEXT I
220 FOR I = 1 TO 8: READ PHI(I) : NEXT I
230 '
240 CLS
250 LOCATE 1,1 : PRINT "PROGRAM INVERSE KINEMATICS:"
260 PRINT
270 PRINT "Select the values of the transformation
    matrix"
280 PRINT
290 PRINT "                    TRANSFORMATION MATRIX"
300 PRINT "------------------------------------------------------------"
310 PRINT ":            :             :             :             :"
320 PRINT ":     Nx     :      Ox     :      Ax     :      Px     :"
330 PRINT ":            :             :             :             :"
340 PRINT "------------------------------------------------------------"
350 PRINT ":            :             :             :             :"
360 PRINT ":     Ny     :      Oy     :      Ay     :      Py     :"
370 PRINT ":            :             :             :             :"
380 PRINT "------------------------------------------------------------"
390 PRINT ":            :             :             :             :"
400 PRINT ":     Nz     :      Oz     :      Az     :      Pz     :"
410 PRINT ":            :             :             :             :"
420 PRINT "------------------------------------------------------------"
430 PRINT ":            :             :             :             :"
440 PRINT ":            :             :             :             :"
450 PRINT ":     0.0    :      0.0    :      0.0    :      1.0    :"
460 PRINT "------------------------------------------------------------"
470 LOCATE 9,17,1 : INPUT;" ",NX
480 LOCATE 13,17,1:INPUT;" ",NY
490 LOCATE 17,17,1:INPUT;" ",NZ
500 LOCATE 9,33,1 :INPUT;" ",OX
510 LOCATE 13,33,1:INPUT;" ",OY
520 LOCATE 17,33,1:INPUT;" ",OZ
530 LOCATE 9,48,1 :INPUT;" ",AX
540 LOCATE 13,48,1:INPUT;" ",AY
550 LOCATE 17,48,1:INPUT;" ",AZ
560 LOCATE 9,64,1 :INPUT;" ",X
570 LOCATE 13,64,1:INPUT;" ",Y
580 LOCATE 17,64,1:INPUT;" ",Z
590 LOCATE 23,1,1
600 IF (AX > 1 OR AX < -1) THEN 700
605 IF (AY > 1 OR AY < -1) THEN 700
610 IF (AZ > 1 OR AZ < -1) THEN 700
615 IF (OX > 1 OR OX < -1) THEN 700
620 IF (OY > 1 OR OY < -1) THEN 700
625 IF (OZ > 1 OR OZ < -1) THEN 700
630 IF (NX > 1 OR NX < -1) THEN 700
635 IF (NY > 1 OR NY < -1) THEN 700
```

```
640 IF (NZ > 1 OR NZ < -1) THEN 700
645 GOTO 725
650 '
652 Q-1.570797 : IF (O<0) THEN Q=Q-3.141597
654 IF (A=O) THEN 685
655 Q=ATN(O/A)
660 S=O/A : IF (S<O) THEN 675
665 IF (O<0) THEN Q=Q-3.141597 : GOTO 685
675 IF (O>0) THEN Q=Q+3.141597 : GOTO 685
685 RETURN
700 CLS
702 PRINT " "
705 PRINT "              INVALID ROTATION MATRIX"
710 PRINT " "
715 PRINT "     SELECT PROPER ROTATION MATRIX VALUES
    (-1 < X < 1)"
720 GOTO 260
725 A=-1*AZ : O=SQR(1-A^2) : GOSUB 650
730 W(4)=Q-(2*P2) : A234=W(4)
735 A=X : O=Y : GOSUB 650
740 W(1)=Q
742 IF (Q>3.141597) THEN Q=Q-3.141597
743 A1=Q
744 W(1)=Q
750 C234=COS(A234) : S234=SIN(A234)
760 IF (S234=0) THEN 780
770 A=NZ : O=-OZ : GOSUB 650 : A5=Q : GOTO 790
780 A=NX : O=OX : GOSUB 650 : A5=A1-Q
790 W(5)=A5
800 ALPHA=(D(1)-D(5)*C234-A(4)*S234-Z)/A(2)
810 BETA=(COS(A1)*X+SIN(A1)*Y+D(5)*S234-A(4)*C234)/A(2)
820 DEL=ALPHA^2+BETA^2 : IF (DEL>4) THEN DEL=4
830 A=(DEL/2-1) : O=SQR(1-A^2) : GOSUB 650
840 A3=Q
850 C3=COS(A3) : S3=SIN(A3)
860 O=ALPHA/2-BETA*S3/DEL
865 D=ABS(1-O^2)
870 A=SQR(D) : GOSUB 650
880 A2=Q
890 W(2)=A2
895W(3)=A2+A3+P2
900 CLS
1110 PRINT "
1120 PRINT "                    THE SOLUTION"
1130 PRINT "
```

```
1140 PRINT " "
1150 PRINT "            THE JOINT ANGLES:"
1160 PRINT
1170 PRINT "     BASE          = " ; W(1)*RAD , "degrees"
1180 PRINT "     SHOULDER      = " ; W(2)*RAD , "degrees"
1190 PRINT "     ELBOW         = " ; W(3)*RAD , "degrees"
1200 PRINT "     WRIST PITCH = " ; W(4)*RAD , "degrees"
1210 PRINT "     WRIST ROLL  = " ; W(5)*RAD , "degrees"
1220 '
1230 '
1240 'EXECUTE MOVE BY USE OF THESE EQUATIONS
1250 '
1260 FOR I = 1 TO 5
1270  PSI(I) = RAD*W(I) - PHI(I+2)
1280  HOLES(I) = ABS(FIX(PSI(I)/RES(I+2)))
1290  SIGN(I) = SGN(CMAP(I+2)*PSI(I))
1300  MOT$(I) = MID$(MR$,I+2,1)
1310 NEXT I
1320 FOR I = 1 TO 5
1330  IF HOLES(I) = 0 THEN 1380
1340  IF HOLES(I) < ISTEP THEN QSTEP = HOLES(I) ELSE
      QSTEP = ISTEP
1350  LPRINT MOT$(I) + "?" : IF INP(PORT) - 32 > ISTEP
      THEN 1350
1360  LPRINT MOT$(I) + STR$(SIGN(I)*QSTEP)
1370  HOLES(I) = HOLES(I) - QSTEP
1380 NEXT I
1390 FOR I = 1 TO 5
1400  IF HOLES(I) > 0 THEN 1320
1410 NEXT I
1420 FOR I = 1 TO 5
1430   PHI(I+2) = PHI(I+2) + PSI(I)
1440 NEXT I
1450 '
1460 LOCATE 20,1 : PRINT "Press ~ to RESET"
1465 PRINT "PRESS : Y TO MAKE ANOTHER MOVE"
1470 K$ = INKEY$ : IF K$ = " " THEN 1470
1475 IF K$ = "~" THEN 1485
1480 IF K$ = "Y" THEN 240
1485 W(1) = 0: W(2) = -90/RAD : W(3) = 90/RAD : W(4) =
     -90/RAD: W(5) = 0
1490 CLS
1500 GOTO 1110
1510
1520 END
```

BIBLIOGRAPHY

Craig, J. J. (1980). JARS: JPL Autonomous Robot System, Robotics and the Operators Group, JPL, Pasadena, CA.

Darringer, J. A. and Blasgen, M. W. (1975). "MAPLE: A High Level Language for Research in Mechanical Assembly," IBM Thomas J. Watson Research Center, Tech. Rept. RC 5606 (Sept.)

Ernst, H. A. (1981). "A Computer-Controlled Mechanical Hand," Sc.D. thesis, M.I.T., Cambridge, MA.

Evans, R. C., Garnett, D. C., and Grossman, D. D. (1976). Software System for a Computer Controlled Manipulator, IBM Thomas J. Watson Research Center, Tech. Rept. RC 6210 (May).

Falek, D. and Parent, M. (1980). An Evolutive Language for an Intelligent Robot, *Indus. Robots*, **7**, 168–171.

Finkel, R., Taylor, R., Bolles, R., Paul, R. P., and Feldman, A. (1974). "AL: Programming System for Automation," Stanford Artificial Intelligence Laboratory, Rept. AIM-177 (Nov.).

Franklin, J. W. and Vanderberg, G. J. (1982). Programming Vision and Robotics Systems with RAIL, *SME Robots*, **VI**, 392–406.

General Electric (1981). *A12 Assembly Robot Maintenance and Installation Manual*, #P50VE005 (Dec.), General Electric Co., Orlando, FL.

General Electric (1982). *Allegro Operator's Manual (A12 Assembly Robot)*, #P50VE025 (Feb.) General Electric Co., Orlando, FL.

General Electric (1982). "GE Allegro Documentation," Tech. Rept., 1982.

Gilbert, A., et al. (1984). "AR-BASIC: An Advanced and User Friendly Programming System for Robots," American Robot Corp., Clinton, PA (June).

Gini, G., Gini, M., Gini, R., and Gruse, D. (1979). Introducing Software Systems in Industrial Robots, in *Proceedings of the 9th International Symposium on Industrial Robots*, Washington, DC, Mar., pp. 309–321.

Grossman, D. D. (1982). Robotic Software, *Mech. Eng.*, **104**(8), 46–47.

Gruver, W. A., Soroka, B. I., Craig, J. J., and Turner, T. (1984). Industrial Robot Programming Languages: A Comparative Evaluation, *IEEE Trans. Syst. Man Cybernet.*, **SMC-14**(4), 565–570.

IBM Corporation (1981/1982). *IBM Robot System/1: AML, Reference Manual*, SC34-0410-1 (Sept. 1981); *IBM Robot System/1: AML, Concepts and User's Guide*, 1st ed., SC34-0411-0 (Sept. 1981); *IBM Robot System/1: General Information Manual and User's Guide*, 3rd ed., 0180-2 (Mar. 1982); IBM Corporation, Boca Raton, FL.

Intelledex (1983). *Model 605 Operator's Manual*, Section 3, Intelledex Inc., Corvallis, OR.

Latombe, J. C. and Mazer, E. (1981). LM: A High-Level Language for Controlling Assembly Robots, in *Proceedings of the 11th International Symposium on Industrial Robots*, Tokyo, Japan, Oct.

Lozano-Perez, T. (1983). Robot Programming, *Proc. IEEE*, **71**(7), 821–841.

McDonnel–Douglas (1981). Robotic System for Aerospace Batch Manufacturing, Task B—High Level Language User Manual, Wright Patterson Air Force Base, OH, Tech. Rept. AFML-JR-79-4202 (Oct.)

Mujtaba, S. and Goldman, R. (1979). "AL User's Manual," Stanford Artificial Intelligence Laboratory, Rept. AIM-323 (Jan.).

Park, W. T. (1979). *How to Generate a New Version of the SRI Robot Programming System (RPS)*, SRI International, Menlo Park, CA, Aug. 6.

Park, W. T. (1981). *The SRI Robot Programming System (RPS)—An Executive Summary*, SRI International, Menlo Park, CA, Mar. 29.

Paul, R. P. (1977). WAVE: A Model-Based Language for Manipulator Control, *Indust. Robot*, **4**(1), 10–17

Ruoff, C. F. (1979). TEACH—A Concurrent Robot Control Language, in *Proceedings of the IEEE COMPSAC*, Chicago, IL, Nov., pp. 442–445.

Ruoff, C. F. (1980). An Advanced Multitasking Robot System, *Indus. Robot*, **7**(1), 26–32, (June).

Salmon, M. (1981). SIGLA: The Olivetti SIGMA Robot Programming Language, in *Proceedings of the 8th International Symposium on Industrial Robots*, Stuttgart, West Germany, June.

Shimano, B. (1979). VAL: An Industrial Robot Programming and Control System, in *Progress IRIA Seminar on Languages and Methods of Programming Industrial Robots*, Roequencourt, France, pp. 47–59.

Silver, D. (1973). "The Littler Robot System," M.I.T. Artificial Intelligence Lab, Rept. AIM273 (Jan.).

Takase, K., Paul, R. P., and Berg, E. J. (1979). A Structured Approach to Robot Programming and Teaching, in *Proceedings of the IEEE COMPSAC*, Chicago, IL, Nov.

Taylor, R. H., Summers, P. D., and Meyer, J. M. (1982). AML: A Manufacturing Language, *Robotics Res.*, **1**(3), 31–38.

Unimation Inc. (1980). *User's Guide to VAL: A Robot Programming and Control System*, Version 12, Unimation Inc., Danbury, CT (June).

Unimation Inc. (1983). *User's Guide to VAL-II, Part 1: Control from the System Terminal*, Version X2; *Part 2: Communications with a Supervisory System;* Version X2; *Part 3: Real-Time Path Control*, Version X2. Unimation Inc., Danbury, CT (Apr.).

Will, P. M. and Grossman, D. D. (1975). An Experimental System for Computer Controlled Mechanical Assembly, *IEEE Trans. Comput.* **C-24**(9), 879–888.

HOMEWORK PROBLEMS

1. Write a DANCE program in BASIC for a robot whose move command is LPRINT and who has six motors designated A, B, C, D, E, and F. Note that LPRINT "A50" would imply that motor A should move by 50 degrees.
2. Write a program for the Rhino XR-2 to dance to music such as a waltz.
3. Describe tying a tie by simple commands and use the constructed set of commands as a basis for robotic software.
4. Write a program in BASIC for the Rhino XR-2 on an XY table to write in block letters any desired sentence on a horizontal surface.
5. Write a program in BASIC for the Rhino XR-2 to spell any word by arranging cubic letter blocks in front of itself and by picking up the blocks from two gravity-fed chutes.

6. Design the necessary software and hardware for two Rhino XR-2 robots to perform an assembly task by cooperating with each other.
7. Write a program for the Rhino XR-2 to stack blocks with four different colors on their sides such that no color is duplicated on each vertical wall (Instant Insanity problem!)
8. Write an inverse kinematics program in C language for the Rhino XR-2.
9. Design the necessary software and hardware for a Rhino XR-2 gripper to sense the rigidity of an object and to communicate the information to a host computer.
10. Write a program in AML for an IBM 7565 robotic manipulator to pick up block letters from two gravity chutes and spell out a sentence by arranging the block letters in a straight line.

Author Index

Subject Index